Data Modeling for the Sciences

With the increasing prevalence of big data and sparse data, and rapidly growing data-centric approaches to scientific research, students must develop effective data analysis skills at an early stage of their academic careers. This detailed guide to data modeling in the sciences is ideal for students and researchers keen to develop their understanding of probabilistic data modeling beyond the basics of p-values and fitting residuals. The textbook begins with basic probabilistic concepts. Models of dynamical systems and likelihoods are then presented to build the foundation for Bayesian inference, Monte Carlo samplers, and filtering. Modeling paradigms are then seamlessly developed, including mixture models, regression models, hidden Markov models, state-space models and Kalman filtering, continuous time processes, and uniformization. The text is self-contained and includes practical examples and numerous exercises. This would be an excellent resource for courses on data analysis within the natural sciences, or as a reference text for self-study.

Steve Pressé is Professor of Physics and Chemistry at Arizona State University, Tempe. His research lies at the interface of biophysics and chemical physics with an emphasis on inverse methods. He has extensive experience in teaching data analysis and modeling at both undergraduate and graduate level with funding from the National Institutes of Health (NIH) and NSF in data modeling applied to the interpretation of single molecule dynamics and image analysis.

Ioannis Sgouralis is Assistant Professor of Mathematics at the University of Tennessee, Knoxville. His research is focused on computational modeling and applied mathematics, particularly the integration of data acquisition with data analysis across biology, chemistry, and physics.

Data Modeling for the Sciences

Applications, Basics, Computations

STEVE PRESSÉ
Arizona State University

IOANNIS SGOURALIS
University of Tennessee

CAMBRIDGE
UNIVERSITY PRESS

CAMBRIDGE
UNIVERSITY PRESS

Shaftesbury Road, Cambridge CB2 8EA, United Kingdom

One Liberty Plaza, 20th Floor, New York, NY 10006, USA

477 Williamstown Road, Port Melbourne, VIC 3207, Australia

314–321, 3rd Floor, Plot 3, Splendor Forum, Jasola District Centre, New Delhi – 110025, India

103 Penang Road, #05-06/07, Visioncrest Commercial, Singapore 238467

Cambridge University Press is part of Cambridge University Press & Assessment, a department of the University of Cambridge.

We share the University's mission to contribute to society through the pursuit of education, learning and research at the highest international levels of excellence.

www.cambridge.org
Information on this title: www.cambridge.org/9781009098502

DOI: 10.1017/9781009089555

First published 2023

A catalogue record for this publication is available from the British Library.

Library of Congress Cataloging-in-Publication Data
Names: Pressé, Steve, author. | Sgouralis, Ioannis, author.
Title: Data modeling for the sciences : applications, basics, computations / Steve Pressé, Ioannis Sgouralis.
Description: Cambridge ; New York, NY : Cambridge University Press, 2023. | Includes bibliographical references and index.
Identifiers: LCCN 2023002178 (print) | LCCN 2023002179 (ebook) | ISBN 9781009098502 (hardback) | ISBN 9781009089555 (epub)
Subjects: LCSH: Research–Statistical methods. | Science–Statistical methods. | Probabilities. | Mathematical statistics.
Classification: LCC Q180.55.S7 P74 2023 (print) | LCC Q180.55.S7 (ebook) | DDC 001.4/22–dc23/eng20230506
LC record available at https://lccn.loc.gov/2023002178
LC ebook record available at https://lccn.loc.gov/2023002179

ISBN 978-1-009-09850-2 Hardback

In memoriam

SP: Adelina "Zia Adi" D'Orta (1912–2000);
IS: Pinelopi Sgourali (1906–2001).

Si l'on considère les méthodes analytiques auxquelles la théorie des probabilités a déjà donné naissance, et celles qu'elle peut faire naître encore, [...], si l'on observe ensuite que dans les choses même qui ne peuvent être soumises au calcul, cette théorie donne les aperçus les plus sûrs qui puissent nous guider dans nos jugemen[t]s, et qu'elle apprend à se garantir des illusions qui souvent nous égarent; on verra qu'il n'est point de science plus digne de nos méditations, et dont les résultats soient plus utiles.

[Considering analytical methods already engendered by the theory of probability, and those that could still arise, [...], and then considering that in those matters that do not lend themselves to [exact] calculation, this theory yields the surest of insights guiding us in our judgments, and teaching us to warrant against those illusions driving us astray; we will see that there exists no science worthier of our inquiry, and whose results are as useful.]

Comte Pierre-Simon de Laplace, *Théorie analytique des probabilités*, 1812

Contents

Preface page xi
Acknowledgments xiii
Expanded Note for Instructors xiv

Part I Concepts from Modeling, Inference, and Computing

1 Probabilistic Modeling and Inference 3
1.1 Modeling with Data 3
1.2 Working with Random Variables 8
1.3 Data-Driven Modeling and Inference 28
1.4 Exercise Problems 33
Additional Reading 39

2 Dynamical Systems and Markov Processes 40
2.1 Why Do We Care about Stochastic Dynamical Models? 40
2.2 Forward Models of Dynamical Systems 41
2.3 Systems with Discrete State-Spaces in Continuous Time 44
2.4 Systems with Discrete State-Spaces in Discrete Time 74
2.5 Systems with Continuous State-Spaces in Discrete Time 80
2.6 Systems with Continuous State-Spaces in Continuous Time 93
2.7 Exercise Problems 101
Additional Reading 107

3 Likelihoods and Latent Variables 108
3.1 Quantifying Measurements with Likelihoods 108
3.2 Observations and Associated Measurement Noise 117
3.3 Exercise Problems 125
Additional Reading 130

4 Bayesian Inference 131
4.1 Modeling in Bayesian Terms 131
4.2 The Logistics of Bayesian Formulations: Priors 139
4.3 EM for Posterior Maximization 147
4.4 Hierarchical Bayesian Formulations and Graphical Representations 148
4.5 Bayesian Model Selection 152
4.6 Information Theory 157

4.7 Exercise Problems 160
Additional Reading 162

5 Computational Inference 163
5.1 The Fundamentals of Statistical Computation 163
5.2 Basic MCMC Samplers 170
5.3 Processing and Interpretation of MCMC 189
5.4 Advanced MCMC Samplers 195
5.5 Exercise Problems 208
Additional Reading 212

Part II Statistical Models

6 Regression Models 215
6.1 The Regression Problem 215
6.2 Nonparametric Regression in Continuous Space: Gaussian Process 219
6.3 Nonparametric Regression in Discrete Space: Beta Process Bernoulli Process 230
6.4 Exercise Problems 234
Additional Reading 243

7 Mixture Models 245
7.1 Mixture Model Formulations with Observations 246
7.2 MM in the Bayesian Paradigm 250
7.3 The Infinite MM and the Dirichlet Process 258
7.4 Exercise Problems 260
Additional Reading 262

8 Hidden Markov Models 264
8.1 Introduction 264
8.2 The Hidden Markov Model 265
8.3 The Hidden Markov Model in the Frequentist Paradigm 268
8.4 The Hidden Markov Model in the Bayesian Paradigm 282
8.5 Dynamical Variants of the Bayesian HMM 291
8.6 The Infinite Hidden Markov Model 296
8.7 A Case Study in Fluorescence Spectroscopy 299
8.8 Exercise Problems 312
Additional Reading 316

9 State-Space Models 318
9.1 State-Space Models 318
9.2 Gaussian State-Space Models 320
9.3 Linear Gaussian State-Space Models 322
9.4 Bayesian State-Space Models and Estimation 328
9.5 Exercise Problems 330
Additional Reading 332

10 Continuous Time Models 333
10.1 Modeling in Continuous Time 333
10.2 MJP Uniformization and Virtual Jumps 335
10.3 Hidden MJP Sampling with Uniformization and Filtering 337
10.4 Sampling Trajectories and Model Parameters 339
10.5 Exercise Problems 342
Additional Reading 343

Part III Appendices

Appendix A Notation and Other Conventions 347
A.1 Time and Other Physical Quantities 347
A.2 Random Variables and Other Mathematical Notions 347
A.3 Collections 348

Appendix B Numerical Random Variables 349
B.1 Continuous Random Variables 349
B.2 Discrete Random Variables 357

Appendix C The Kronecker and Dirac Deltas 361
C.1 Kronecker Δ 361
C.2 Dirac δ 361

Appendix D Memoryless Distributions 364

Appendix E Foundational Aspects of Probabilistic Modeling 366
E.1 Outcomes and Events 366
E.2 The Measure of Probability 371
E.3 Random Variables 376
E.4 The Measurables 378
E.5 A Comprehensive Modeling Overview 381
Additional Reading 382

Appendix F Derivation of Key Relations 383
F.1 Relations in Chapter 2 383
F.2 Relations in Chapter 3 384
F.3 Relations in Chapter 5 387
F.4 Relations in Chapter 6 389
F.5 Relations in Chapter 7 392
F.6 Relations in Chapter 8 395
F.7 Relations in Chapter 9 401
F.8 Relations in Chapter 10 406

Index 408
Back Cover 416

Preface

Data analysis courses that go beyond teaching elementary topics, such as fitting residuals, are rarely offered to students of the natural sciences. As a result, data analysis, much like programming, remains improvised. Yet, with an explosion of experimental methods generating large quantities of diverse data, we believe that students and researchers alike would benefit from a clear presentation of methods of analysis, many of which have only become feasible due to the practical needs and computational advances of the last decade or two.

The framework for data analysis that we provide here is inspired by new developments in data science, machine learning and statistics in a language accessible to the broader community of natural scientists. As such, this text is ambitiously aimed at making topics such as statistical inference, computational modeling, and simulation both approachable and enjoyable to natural scientists.

It is our goal, if nothing else, to help develop an appreciation for data-driven modeling and what data analysis choices are available alongside what approximations are inherent to the choices explicitly or implicitly made. We do so because theoretical modeling in the natural sciences has traditionally provided limited emphasis on data-driven approaches. Indeed, the prevailing philosophy is to first propose models and then verify or otherwise disprove these by experiments or simulations. But this approach is not data-centric. Nor is it rigorous except for the cleanest of data sets as one's perceived choice in how to compare, say, models and experiments may have dramatic consequences in whether the model is ultimately shown improbable. As we move toward monitoring events on smaller or faster timescales or complex events otherwise sparsely sampled, examples of clean data are increasingly few and far between.

Organization of the Text

We designed the text as a self-contained single semester course in data analysis, statistical modeling, and inference. Earlier versions have been used as class notes in a course at Arizona State University since 2017 for first-year chemistry and physics graduate students as well as upper-level undergraduates across the sciences and engineering. Since 2020, they have also been used in mathematics at the University of Tennessee. While the text is appropriate for upper-level undergraduates in the sciences, its intended audience is at the master's level. The concepts presented herein are self-contained, though a basic course in computer programming and prior knowledge of undergraduate-level calculus is assumed.

Our text places equal emphasis on explaining the foundations of existing methods and their implementation. It correspondingly places little emphasis on formal proofs and research topics yet to be settled. Along core sections, we have interspersed sections and topics designated by an asterisk. These contain more advanced materials that may be included at the instructor's discretion and are otherwise not necessary upon a first reading. Similarly, we avoid long derivations in the text by marking designated equations with asterisks; these lengthy derivations are relegated to Appendix F.

Part I begins with a survey of modeling concepts to motivate the problem of parameter estimation from data. This leads to a discussion of frequentist and Bayesian inference tools. Along the way, we introduce computational techniques including Monte Carlo methods necessary for a comprehensive exposition of the most recent advances. Part II is devoted to specific models starting from basic mixture models followed by Gaussian processes, hidden Markov models and their adaptations, as well as models appropriate to continuous space and time.

In writing some end-of-chapter exercises, we are reminded of a quote from J. S. Bach (1723) as a prefatory note to his own keyboard exercises (two and three part inventions). That is, we not only wish to inspire a clear way by which to tackle data analysis problems but also create a strong foretaste for the proper independent development of the reader's own analysis tools. Indeed, some end-of-chapter exercises bring together notions intended to broaden the reader's scope of what is possible and spark their interest in developing further inference schemes as complex and realistic as warranted by the application at hand.

Finally, we made clear choices on what topics to include in the book. These were sometimes based on personal interest, though, most often, these choices were based on what we believe is most relevant. To keep our presentation streamlined, however, we have excluded many topics. Some of these we perceive as easier for students to understand after reading this book, such as specialized cases of topics covered herein.

Acknowledgments

First and foremost, we thank our families for their support and for patiently staying by our side as we undertook this effort.

We also thank Weiqing Xu for helping generate some of the figures and Alex Rojewski for her critical help with the solution manual; Weiqing Xu, Sina Jazani, and Zeliha Kilic for suggesting scientific arguments used in select chapters; Banu Ozkan, Julian Lee, and Corey Weistuch for reading over the entirety of earlier drafts of the text; John Fricks for his thoughts on the chapter on foundations; Bob Zigon for pointing us to analysis topics that eventually grew into our interest in nonparametrics; Oliver Beckstein for his thoughts on likelihood maximization; Carlos Bustamante for providing his time to share his contagious passion for (stochastic) single molecule data; Ken Dill for his inspiration in pursuing inverse methods; anonymous referees who provided feedback on the text; and funding agencies for their continued support (Pressé acknowledges the NSF, NIH, and Army Research Office (ARO) and Sgouralis acknowledges his University of Tennessee startup and Eastman Chemical Company).

We also thank the many members of the Pressé lab and students at Arizona State University taking CHM/PHY 598 (“Unraveling the noise”) who have proposed and inspired homework problems and identified typos and confusions across earlier drafts of the text. Finally, we thank our many experimental colleagues, often first encountered over the course of Telluride (TSRC) workshops, who have suggested multiple topics of interest and inspired us to learn many of the methods presented herein.

Any remaining typos and omissions are ours alone.

Expanded Note for Instructors

Various iterations of this course have been taught since 2017 at ASU where the course was cross-listed in both physics and chemistry. The course was taken by senior undergraduates and first year graduate students from both of these disciplines as well as by students from various engineering disciplines. The course has also been offered for two years at UTK in mathematics.

In a one-semester course, we cover Chapters 1–7 and the beginning of Chapter 8 (as time allows) in the order in which the material is presented. We exclude all topics labeled advanced and skip all extensive derivations, which are relegated to the appendices. We do otherwise summarize the simpler derivations appearing in the text. Taught from start to finish, without skipping advanced material, the text stands as a complete two-semester course in data modeling.

In presenting the material, we created boxed environments specifically with students and instructors in mind. For instance, the Note environments highlight special issues we hope will not elude the reader. The Example environments are meant to tie the mathematical material back to applications. Most important are the Algorithm environments, which focus on implementation of the ideas presented. The algorithms are intentionally presented in a general manner independent of any coding language. For this reason, students in class have typically used the coding language of their choice in the problem sets.

The end-of-chapter problems appear in two forms: exercises and projects. The exercises are simpler and we provide detailed solutions of sample exercises to verified instructors that we have used in our own offering of the course. The solutions we provide exceed what would be needed to cover weekly or biweekly problem sets for a one-semester course.

The projects, however, are more demanding and are meant to inspire the formulation of broader and deeper questions that may be addressed using the tools presented in each chapter. Indeed, some projects may be worthy of independent publication when completed.

The index is ordered by topic for ease of reference. For instance, "friction coefficient" will be found as a subitem under "physics" Similarly, specific algorithms (that may otherwise introduce confusion by being indexed by long or abbreviated names) simply appear as subitems under "algorithm."

Finally, we appreciate that the material we present is normally not available to students of the natural sciences as it would require advanced prerequisites in probability and stochastic processes that cannot otherwise fit into their schedules. For this reason, we pay special attention to notation and concepts that would not be familiar to students of the sciences in order to allow them to work from the very basics (Chapter 1) to state-of-the-art modeling (Chapters 6–10). We relegate theoretical topics in probability to an appendix entitled "Foundational Aspects of Probabilistic Modeling."

PART I

CONCEPTS FROM MODELING, INFERENCE, AND COMPUTING

1 Probabilistic Modeling and Inference

By the end of this chapter, we will have presented

- *Data oriented modeling*
- *Random variables and their properties*
- *An overview of inverse problem solving*

1.1 Modeling with Data

If experimental observations or, put concretely, binaries on a screen were all we ever cared about, then no experiment would require modeling or interpretation and the remainder of this book would be unnecessary. But binaries on a screen do not constitute knowledge. They constitute *data*. Put differently, quantum mechanics, like any scientific knowledge, is not self-evident from the pixelated outcome on a camera chip of a modern incarnation of a Young's two-slit interference experiment.

In the natural sciences, *models* of physical systems provide mathematical frameworks from which we unify disparate pieces of information. These include conceptual notions such as symmetries, fundamental constituents, and other postulates, as well as scientific *measurements* and, even more generally, empirical observations of any form. If we think of direct observations as data in particular, at least for now, we can think of mathematical models as a way of compressing or summarizing these data.

Data summaries may be used to make predictions about physical conditions we may encounter in the future, such as in new experiments, or to interpret and describe an underlying physical system already probed in past experiments. For example, with time-ordered data we may be interested in learning equations of motion or kinetic schemes. Or, already knowing a kinetic scheme sufficiently well from past experiments or fundamental postulates, we may only be interested in learning the noise characteristics of a new piece of equipment on which future experiments will be run. Thus, models may be aimed at discovering new science as well as at devising careful controls to get a better handle on error bars and, more broadly, even at designing new experiments altogether.

1.1.1 Why Do We Obtain Models from Raw Data?

Experimental data rarely provide direct insight into the physical conditions and systems of interest. At the very least, measurements are *corrupted* by unavoidable noise and, as a result, models obtained from experimental data are unavoidably probabilistic. So, we ask: *How should we, the scientific community, go about obtaining models from imperfect data?*

> **Note 1.1** Obtaining models from data
>
> Data can be time and labor intensive to acquire. Perhaps more importantly, every datum in a dataset encodes information. In light of this, we re-pitch our question and ask: *How should we go about obtaining models* efficiently *and* without compromising *the information encoded in the data?*

The key is to start from data acquired in experiments and arrive at models with a minimal amount of data preprocessing, if at all. This is because obtaining a model from quantities derived from the data, as opposed to directly from the data, is necessarily *equal to* or *worse than* obtaining the model from the data directly since derived quantities contain as much as or less information than the data themselves. For instance, fitting histogrammed data is an information-inefficient and unreliable approach to obtaining models as it demands downsampling via binning and an arbitrary choice of bin sizes.

Besides information efficiency, obtaining models from unprocessed data also has another critical advantage that gets to the heart of scientific practice. While error bars around individual data points may be imperfectly known, they are, by construction, *better characterized* than error bars around derived quantities. Thus error bars around models determined from derived quantities are necessarily only as good as, but often less reliable, than error bars around models determined from the raw data. Unfortunately, as error bars around derived quantities can become too difficult to compute in practice, they are often ignored altogether. Nevertheless, error bars are a cornerstone of modern scientific research. They not only help quantify reproducibility but also directly inform error bars around the models obtained and, as such, inspire the formulation of new competing models.

Putting it all together, it becomes clear that a model is *best informed*, and has the *most reliable error bars*, when learned from the data available in as raw a form as accessible from the experiments. This is true so long as it is computationally feasible to obtain models from such raw data and, as we will see in subsequent chapters, we are far from reaching computational bottlenecks in most problems of interest across the natural sciences.

1.1.2 Why Do We Formulate Models with Random Variables?

If there is no uncertainty involved, a physical system is adequately described using deterministic variables. For example, Newtonian mechanics are expressed in terms of momenta, positions, and forces. However, when a system involves any degree of uncertainty, either due to noise, poor characterization of some or all of its constituents, or features as of yet unresolved or otherwise fundamentally stochastic, then it is better described using *random variables*. This is true of the probabilistic nature of quantum mechanics as well as statistical physics and, as we illustrate herewith, also of data analysis.

Random variables are used to represent observations generated by stochastic systems. Stochasticity in data analysis arises due to inherent randomness in the physical phenomena of interest or due to measurement noise or both. Random variables are useful constructs because, as we will see, they are mathematical notions that reproduce naturally stochastic relationships between uncertain effects and observations, while their deterministic counterparts cannot.

Note 1.2 Measurement noise

It is sometimes thought that models with probabilistic formulations are only required when the quantities of interest are inherently probabilistic. Nevertheless, measurement noise corrupts experimental observations irrespective of whether the quantities themselves are probabilistic or not. Consequently, probabilistic models are *always required* whenever models are informed by experimental output.

Random variables are abstract notions that most often represent numbers or collections of numbers. However, more generally, random variables can be generic notions that may include nonnumeric quantities such as: labels for grouping data, *e.g.*, group A, group B; logical indicators, *e.g.*, true, false; functions, *e.g.*, trajectories or energy potentials. In all cases, numeric or not, random variables may be *discrete*, *e.g.*, dice rolls, coin flips, photon counts, bound energy states, or *continuous*, such as temperatures, pressures, or distances. Further, random variables may be finite collections of individual quantities, *e.g.*, measurements acquired during an experiment or infinite quantities, *e.g.*, successive positions on a *Brownian* particle's trajectory. At any rate, random variables have unique properties, which we will shortly explore, that allow us to use them in the construction and evaluation of meaningful probabilistic models.

Commonly, we imagine a random variable, which we denote with W, as being instantiated or assigned a specific value realized at w as a result of performing a measurement that amounts to a *stochastic event*. That is, we think of a measurement output w as a *stochastic realization* of W. Our stochastic events entail randomness inherited through W and influencing the assigned values w. We therefore distinguish between a random variable W and its realizations, w, *i.e.*, the particular values that W attains or may attain.

Stochastic events may encompass *physical* events, like the occurrence of chemical reactions or events in a cell's life cycle. Stochastic events may also encompass *conceptual* events, like an idealized version of a real-life system expressed in terms of fair coin tosses or even like instantaneously learning the spin orientation of a faraway particle given a local measurement of another spin to which the first is entangled.

Example 1.1 The photoelectric effect

When a photon falls onto certain materials photoelectrons are sometimes emitted. Such a phenomenon provides the basis for a stochastic event.

In the photoelectric setting, it is often convenient to formulate a random variable W that counts the number of photoelectrons emitted. This random variable may take values $w = 0, 1, 2, \ldots$.

To develop a model, we imagine a *prototype experiment* as a sequence of stochastic events that produce N distinct numeric measurements or, more generally, observations of any kind. We typically use w_n to denote the nth observation and use $n = 1, \ldots, N$ to index them. As we highlighted earlier, individual observations in our experiment may be scalars, for example $w_n = 20.1°\text{C}$ or $w_n = 0.74\ \mu\text{m}^3$ for typical measurements of room temperature or an *E. coli*'s volume, respectively, or even nonnumeric, such as w_n = p.R83SfsX15 for descriptions of gene mutations. In general, we do not require that each observation in our experiment be of the same type; that is, w_1 may be a temperature while w_2 may be a volume.

As we will often do, we gather every observation conveniently together in a list,

$$w_{1:N} = \{w_1, w_2, \ldots, w_N\},$$

and use subscripts $1:N$ to indicate that the list $w_{1:N}$ gathers every single w_n with an index n ranging between 1 and N. Unless explicitly needed to help draw attention to the subscript, for clarity we may sometimes suppress this subscript and write simply w for the entire list.

As we have already mentioned, the observations $w_{1:N}$ are better understood as realizations of appropriate random variables $W_{1:N} = \{W_1, W_2, \ldots, W_N\}$ that we use to formulate our model.

1.1.3 Why Do Our Models Have Parameters?

Models are mathematical formulations to which we associate parameters. Both models and their associated parameters are specialized to particular systems, experiment types, and experimental setups. Assuming a model structure encoded in $W_{1:N}$ and provided observed values $w_{1:N}$, our main objective in data analysis becomes the estimation of the model's associated parameters.

Example 1.2 Normal random variables

The mean of a sequence of identical random variables W_n is only probabilistically related to each measured value w_n. For the simple example of a normally distributed sequence W_n, what we call the model is the normal distribution, often termed the Gaussian distribution. The associated parameters are the mean μ and variance $v = \sigma^2$, with standard deviation σ, that indicate the center and spread, respectively, of the values $w_{1:N}$. These are collectively described by the list of model parameters $\theta = \{\mu, v\}$. As illustrated in Fig. 1.1, and as we will see in detail in later chapters, θ can be estimated from $w_{1:N}$.

In the Example 1.2, the Gaussian forms a simple model that contains two parameters, namely the mean μ and the variance v, that we gather in θ. More generally, our models may contain K individual parameters that we may also gather in a list $\theta_{1:K} = \{\theta_1, \theta_2, \ldots, \theta_K\}$.

Typically, the parameters $\theta_{1:K}$ represent quantities we care to *estimate*, for example μ and v. A model is deemed *specified* when *numerical values* are assigned to $\theta_{1:K}$. Thus, specifying a model is understood as being equivalent to assigning values to $\theta_{1:K}$. Similarly, deriving error bars around the assigned values of $\theta_{1:K}$ is equivalent to deriving error bars around the model.

As we invariably face some degree of measurement noise, we formulate an experiment's results $w_{1:N}$ as probabilistically related to the parameters $\theta_{1:K}$. In the context of our *prototype experiment*, we incorporate such relations through the random variables $W_{1:N}$ and in the next section we lay down some necessary concepts.

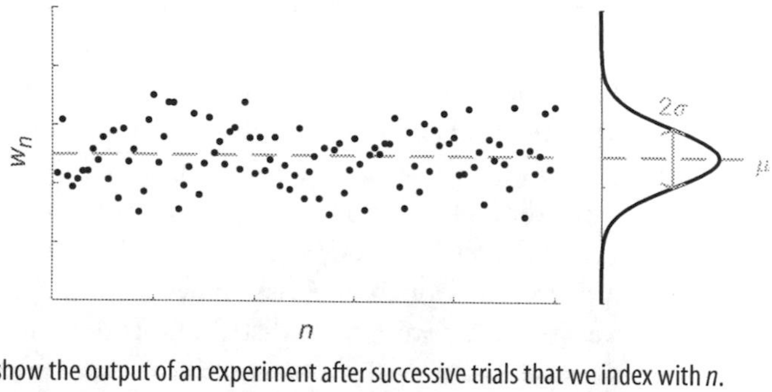

Fig. 1.1 (Left) We show the output of an experiment after successive trials that we index with n. (Right) We find a histogram of the data with very fine bin sizes that assumes the shape of a Gaussian distribution. We denote the mean of this distribution by μ and the standard deviation by σ.

Note 1.3 Modeling terminology

In this chapter, when we use the term model, we mean the mathematical formulation itself alongside numerical values for its associated parameters. When we speak of measurements, observations, assessments, or data points, we refer to the random variables $W_{1:N}$ and their realizations $w_{1:N}$. Similarly, by calibrating a model we imply selecting the correct values for its associated parameters (and sometime also characterizing their uncertainty). Determining both model parameters and their uncertainty is collectively referred to as *model estimation* or model training.

1.2 Working with Random Variables

Before we embark on specific modeling and estimation strategies, we begin by exploring some important notions that we need in order to work with random variables and the distributions from which they are sampled. That is, just as we can easily deduce derivatives and integrals of complicated functions by remembering a few simple rules of calculus, we can similarly deduce probability distributions of complicated models by remembering a few simple rules of probability that we put forth in this section.

As we will soon start using random variables not only to represent measurements W, but also other relevant quantities of our model, we begin using R to label generic random variables.

1.2.1 How to Assign Probability Distributions

In any model, a random variable R is *drawn* or *sampled from* some *probability distribution*. We label such a distribution with $\mathbb{P}$ and we write

$$R \sim \mathbb{P}.$$

In the language of statistics this reads "the random variable R is sampled from the probability distribution $\mathbb{P}$" or "R follows the statistics of $\mathbb{P}$."

In statistical notation, in writing $R \sim \mathbb{P}$, we use $\mathbb{P}$ as a notational shorthand that summarizes the most important properties of the variable R. These include a description of the values r that R may take and a recipe to compute probabilities associated with them. As we will see, most often we work with probability distributions that are associated with probability density functions. In such cases, it is more convenient to think of $R \sim \mathbb{P}$ as a compact way of communicating: the allowed values r of R obey the probability density $p(r)$ associated with $\mathbb{P}$.

Example 1.3 The normal distribution

We previously encountered the normal distribution, Normal(μ, v). A shorthand like

$$R \sim \text{Normal}\left(\mu, v\right)$$

captures the following pieces of information:

- The particular values r that R attains are real numbers ranging from $-\infty$ to $+\infty$.
- The probability density $p(r)$ of R depends on two parameters, μ and v, and has the form

$$p(r) = \frac{1}{\sqrt{2\pi v}} \exp\left(-\frac{1}{2}\frac{(r-\mu)^2}{v}\right).$$

Furthermore, the two parameters μ and v can be interpreted as the mean and the variance of R, respectively, since integration of the density leads to

$$(\text{Mean of } R) = \int_{-\infty}^{+\infty} dr\, r p(r) = \mu,$$

$$(\text{Variance of } R) = \int_{-\infty}^{+\infty} dr\, (r-\mu)^2 p(r) = v.$$

Using the density $p(r)$, we can also compute the probability of measuring any value r between some specified r_{min} and r_{max}. In particular, this is

$$\int_{r_{\text{min}}}^{r_{\text{max}}} dr\, p(r) = \frac{1}{2}\left[\text{erf}\left(\frac{r_{\text{max}}-\mu}{\sqrt{2v}}\right) - \text{erf}\left(\frac{r_{\text{min}}-\mu}{\sqrt{2v}}\right)\right], \qquad (1.1)$$

where erf$(\cdot)$ is the error function defined by an integral

$$\text{erf}\,(r) = \frac{2}{\sqrt{\pi}} \int_0^r dr'\, e^{-\frac{1}{2}(r')^2}.$$

Example 1.4 The exponential distribution

The exponential distribution arises in many applications. A shorthand like

$$R \sim \text{Exponential}\,(\lambda)$$

captures the following pieces of information:

- The particular values r that R attains are real numbers ranging from 0 to ∞.
- The probability density $p(r)$ of R depends on one positive parameter, λ, and has the form

$$p(r) = \lambda e^{-\lambda r}.$$

The parameter λ can be interpreted as the reciprocal of the mean of R, since integration of the density leads to

$$(\text{Mean of } R) = \int_0^{\infty} dr\, r p(r) = \frac{1}{\lambda}.$$

Through the density $p(r)$, we can also compute the probability of measuring any value r between some specified r_{min} and r_{max}. In particular, this is

$$\int_{r_{\text{min}}}^{r_{\text{max}}} dr\, p(r) = e^{-\lambda r_{\text{min}}} - e^{-\lambda r_{\text{max}}}. \tag{1.2}$$

Example 1.5 The multivariate Normal$_M$ distribution

The multivariate Normal$_M$ distribution is a generalization of the univariate Normal of Example 1.3. In fact, the two definitions coincide for $M = 1$. The shorthand

$$\boldsymbol{R} \sim \text{Normal}_M\left(\boldsymbol{\mu}, \boldsymbol{V}\right)$$

captures the following pieces of information:

- The particular values $\boldsymbol{r}$ that $\boldsymbol{R}$ attains are real vectors of size M.
- The probability density $p(\boldsymbol{r})$ of $\boldsymbol{R}$ depends on two parameters, $\boldsymbol{\mu}$ and $\boldsymbol{V}$, and has the form

$$p(\boldsymbol{r}) = \frac{1}{\sqrt{(2\pi)^M|\boldsymbol{V}|}} \exp\left(-\frac{1}{2}(\boldsymbol{r}-\boldsymbol{\mu})\boldsymbol{V}^{-1}(\boldsymbol{r}-\boldsymbol{\mu})^T\right).$$

The parameter $\boldsymbol{\mu}$ is also a vector of size M and the parameter $\boldsymbol{V}$ is a positive definite square matrix of size M. Here, $|\cdot|$ is the matrix determinant. Similar to the univariate case, the two parameters $\boldsymbol{\mu}$ and $\boldsymbol{V}$ can be interpreted as the mean and the covariance of $\boldsymbol{R}$, respectively.

In the simplest case, a normally distributed bivariate random variable $\boldsymbol{R} = (R_1, R_2)$ may be written as

$$(R_1, R_2) \sim \text{Normal}_2\left((\mu_1, \mu_2), \begin{pmatrix} v_1 & \rho\sqrt{v_1 v_2} \\ \rho\sqrt{v_1 v_2} & v_2 \end{pmatrix}\right).$$

In this parametrization $\mu_1, \mu_2, v_1, v_2, \rho$ are scalars, v_1, v_2 are positive, and ρ is bounded between -1 and $+1$. In this case, the density takes the equivalent form

$$p(r_1, r_2) = \frac{1}{2\pi\sqrt{v_1 v_2(1-\rho^2)}} \times \exp\left(-\frac{1}{2(1-\rho^2)}\left(\frac{(r_1-\mu_1)^2}{v_1} + \frac{(r_2-\mu_2)^2}{v_2} - 2\rho\frac{(r_1-\mu_1)(r_2-\mu_2)}{\sqrt{v_1 v_2}}\right)\right).$$

Throughout this book, we extensively use several common distributions. In Examples 1.3 and 1.4 we introduced two of them, though many more are to come. As these will appear frequently, to refer back to them we adopt a convention that we summarize in Appendix B. Briefly, we use $R \sim \text{Normal}(\mu, v)$ and $\text{Normal}(\mu, v)$ to denote a normal random variable and the normal distribution, respectively. Furthermore, we use $\text{Normal}(r; \mu, v)$ to help distinguish this associated density with its distribution. According to our convention, the values r of the random variable R do *not* appear in the distribution Normal(μ, v); while, they *do* appear in the density Normal$(r; \mu, v)$. In the latter, we separate with ";" the variable values r from the parameters μ and v. We apply the same convention to the other distributions and densities.

As, for clarity, we distinguish between a random variable R and its values r, in this chapter we also distinguish between a probability distribution $\mathbb{P}$ and its associated density $p(r)$. However, in subsequent chapters, we relax this convention whenever there is no ambiguity.

Distributions on Random Variables with Probability Density Functions

For a random variable R whose distribution has a probability density, we can compute the probability of attaining any of the values gathered in η, where η is a collection of r values, by the integral

$$P_\eta = \int_\eta dr\, p(r). \tag{1.3}$$

In this integral, $p(r)$ is the *probability density function* of $R \sim \mathbb{P}$ and its precise form is characteristic of the distribution $\mathbb{P}$. For instance, as we saw in Examples 1.3 and 1.4, a normal distribution Normal(μ, v) has a normal density $p(r) = \exp\left(-\frac{(r-\mu)^2}{2v}\right)/\sqrt{2\pi v}$ and an exponential distribution Exponential(λ) has an exponential density $p(r) = \lambda e^{-\lambda r}$. For these two, Eq. (1.3) reduces to Eqs. (1.1) and (1.2), respectively.

> **Note 1.4** Probability density over inadmissible values
>
> By convention, the density $p(r)$ is assumed equal to zero over inadmissible values r. For example, if the random variable models a distance, then $p(r) = 0$ for $r < 0$; similarly, if the random variable models a temperature reported in Kelvin units, $p(r) = 0$ for $r < 0$.

By definition, the area, or more generally the volume, underneath the entire probability density $p(r)$ must be equal to 1. This is called the *normalization condition* and implies that an η including *every* admissible value r has probability 1. For instance, from Eqs. (1.1) and (1.2) we can see that the probability of sampling any real scalar value is equal to 1 for normal and exponential random variables.

As can be seen from Eq. (1.3), a density $p(r)$ is *unitful* and its units are determined by normalization. Since $\int dr\, p(r) = 1$, where the region of integration includes every admissible value, the density $p(r)$ has the *reciprocal units* of r. So, if r is a length (in cm), the density $p(r)$ has units of reciprocal length (1/cm); similarly, if r is a time (in s), the density $p(r)$ has units of frequency (Hz).

Note 1.5 Resampling the same value of a random variable

Equation (1.3) already signals that the probability of sampling a continuous scalar random variable between some values $r_{\min}$ and $r_{\max}$ is

$$P_{r_{\min},r_{\max}} = \int_{r_{\min}}^{r_{\max}} dr\, p(r). \tag{1.4}$$

This brings up an interesting point: there is a vanishingly small probability for the same value of a continuous scalar random variable to be sampled twice with finite samplings. In fact, the probability of sampling *any particular value* is 0, as can be seen by having coinciding $r_{\min}$ and $r_{\max}$ in the integral Eq. (1.4). This indicates that when thinking about probabilities over continuous variables, we need to consider *intervals* of values rather than isolated values.

This feature generalizes to any continuous random variable that need not necessarily be scalar, but, as we will see shortly, it does not carry over the values of discrete random variables that can reoccur even in finite samplings. For example, a roll of 4 will almost certainly reoccur multiple times in a total of 1,000 rolls of a fair dice.

For a random variable R, it is also possible, and often useful, to *transform* its density $p(r)$ into a density $q(r')$ over another random variable R' with values related by a given function $r' = f(r)$. For example, such a transformation occurs when we want to apply a change to our coordinate system or simply otherwise re-parametrize our model. As we require the transformation $R \mapsto R'$ to leave the probabilities we compute either using the initial or transformed variables unaffected, the two densities must satisfy

$$\int_{\eta} dr\, p(r) = \int_{f(\eta)} dr'\, q(r').$$

In this equality, $f(\eta)$ is a collection that contains the transformed values $r' = f(r)$ of all r in the initial collection η. In the most general setting, it is hard to relate mathematically the densities $p(r)$ and $q(r')$ any further. However, provided $f(r)$ is a *differentiable* function that can be *inverted uniquely*, as is often the case, we may apply a change of variables to the right-hand side integral to reach $\int_{\eta} dr\, p(r) = \int_{\eta} dr\, |J_{r\mapsto r'}| q(r')$. Here, $|J_{r\mapsto r'}|$ is the absolute value of the determinant of the transformation's Jacobian. In turn, since such an equality holds for any η, we may drop the integrals to reach a simpler form

$$q(r') = \frac{1}{|J_{r\mapsto r'}|} p(r). \tag{1.5}$$

Example 1.6 **Rescaling of random variables**

Any physical quantity measured in real-life experiments almost always carries units. For practical reasons, we often need to convert from quantities reported in one system of units to another. Unit conversion itself is a simple example of variable transformation.

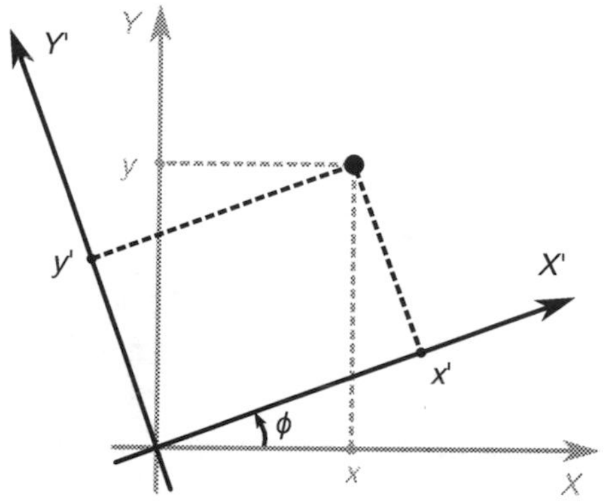

Fig. 1.2 A random Cartesian position in the initial (X, Y) and the transformed (X', Y') frames of reference.

For concreteness, we consider a random variable R reported in some units and suppose $r' = \xi r$ where r' is expressed in different units from r. Here, ξ is the conversion factor, for example ξ could be 100 cm/m for r' expressed in terms of centimeters and r in terms of meters. In this example, both random variables R and R' are scalar, and so the Jacobian reduces to a simple derivative. More specifically, $|J_{r\mapsto r'}| = |df(r)/dr| = \xi$, and so the densities are

$$q(r') = \frac{p(r)}{\xi}.$$

Example 1.7 Coordinate transformation of spatial random variables

Measurements of position are reported with respect to certain frames of reference. Changing the frame of reference is another example of a variable transformation.

For concreteness, we consider a bivariate random variable (X, Y) that models a location in the Cartesian plane, and suppose that (X', Y') is the same location in another Cartesian frame of reference rotated through an angle ϕ about the origin; see Fig. 1.2. In this case, the original and transformed positions are related through

$$x' = x\cos\phi + y\sin\phi, \qquad y' = -x\sin\phi + y\cos\phi,$$

and the Jacobian of the transformation has the form

$$J_{(x,y)\mapsto(x',y')} = \begin{pmatrix} \frac{\partial x'}{\partial x} & \frac{\partial x'}{\partial y} \\ \frac{\partial y'}{\partial x} & \frac{\partial y'}{\partial y} \end{pmatrix} = \begin{pmatrix} \cos\phi & \sin\phi \\ -\sin\phi & \cos\phi \end{pmatrix}.$$

Since $|J_{(x,y)\mapsto(x',y')}| = \cos^2\phi + \sin^2\phi = 1$, the densities in the two coordinate systems are

$$q(x', y') = p(x, y).$$

Distributions on Random Variables with Discrete Values

If $\rho_{1:M} = \{\rho_1, \rho_2, \ldots, \rho_M\}$ gathers every admissible value of a *discrete* random variable R, then its probability density has the generic form

$$p(r) = \pi_{\rho_1}\delta_{\rho_1}(r) + \cdots + \pi_{\rho_M}\delta_{\rho_M}(r) = \sum_{m=1}^{M} \pi_{\rho_m}\delta_{\rho_m}(r), \qquad (1.6)$$

where π_{ρ_m} are the probabilities of the individual values ρ_m contained in $\rho_{1:M}$. The *Dirac* terms $\delta_\rho(r)$ are specified by the properties

$$\delta_\rho(r) = 0, \qquad r \neq \rho,$$
$$\int dr\, \delta_\rho(r) = 1,$$

where the integral is taken over every allowed value of r; see Appendix C.

Note 1.6 What is a discrete random variable?

One way to gain some intuition about discrete random variables is to consider *limiting* cases of continuous random variables. For instance, we may consider acquiring measurements where we wish to distinguish between M distinct scalar values $\rho_{1:M}$. In a real-life experiment, our acquisitions are contaminated with noise and, for this reason, our measurements are generally scattered *around* the values $\rho_{1:M}$. As such, we may model our measurements with a random variable $R \sim \mathbb{P}$ that, due to the noise, attains continuous values; Fig. 1.3.

In a noisy scenario, the scattering of r around $\rho_{1:M}$ is wide, and our measurements are found generally anywhere around and between $\rho_{1:M}$. In this case, a fine separation between the outcomes $\rho_{1:M}$ might be impossible. However, in increasingly clearer scenarios, our measurement distribution $\mathbb{P}$ concentrates around the outcomes giving rise to cleanly isolated peaks; Fig. 1.3.

In the extreme limit of an idealized noiseless scenario, the distribution $\mathbb{P}$ places all of its probability around $\rho_{1:M}$ and its density, $p(r)$, becomes a train of Dirac terms as in Eq. (1.6).

Normalization, in the case of a discrete random variable, reads $1 = \sum_{m=1}^{M} \pi_{\rho_m} \int dr\, \delta_{\rho_m}(r)$, where the integral over r spans any admissible and inadmissible value. Since probabilities π_{ρ_m} are dimensionless, this implies that each $\delta_{\rho_m}(r)$ on the right-hand side of Eq. (1.6) has dimensions of reciprocal r, similar to the density $p(r)$. As such, normalization of a discrete random variable's density can also take the equivalent form $1 = \sum_{m=1}^{M} \pi_{\rho_m}$.

One way to represent the distribution of a random variable with a density as in Eq. (1.6) is

$$R \sim \text{Categorical}_{\rho_{1:M}}(\pi_{\rho_{1:M}}),$$

where $\text{Categorical}_{\rho_{1:M}}(\pi_{\rho_{1:M}})$ denotes the categorical *distribution* with outcomes $\rho_{1:M}$ and associated probabilities $\pi_{\rho_{1:M}}$. A random variable drawn from this distribution samples an outcome, ρ_m, in proportion to that outcome's probability, π_{ρ_m}; see Fig. 1.4.

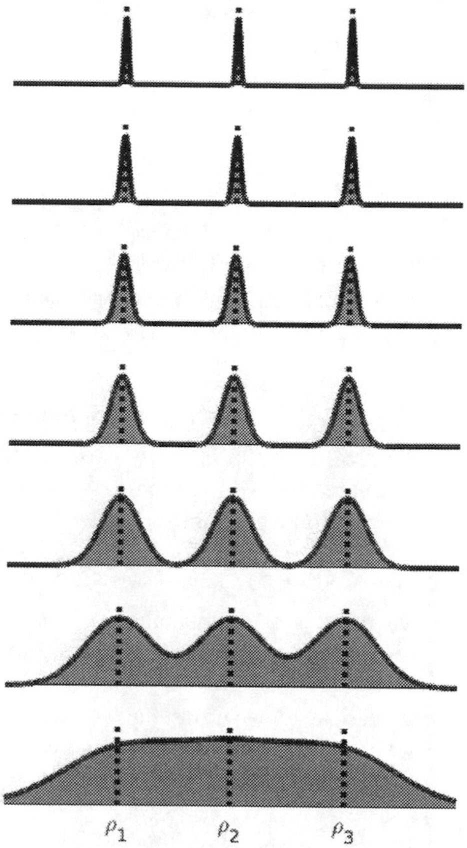

Fig. 1.3 A discrete random variable arises as an idealization of increasingly less noisy measurements. Here, the bottom panel shows the probability distribution of highly noisy measurements. Upper panels show the probability distribution of successively clearer scenarios.

Example 1.8 Dice rolls modeled as categorical random variables

Rolling a common dice leads to one of six outcomes that we idealize as the faces marked with the numbers "1" through "6." Provided we identify these outcomes with the categories ρ_m, $m = 1, \ldots, 6$, we can model a dice roll as a categorical random variable

$$R \sim \text{Categorical}_{\rho_{1:6}}(\pi_{\rho_{1:6}}),$$

where the probability a face marked with "m," or category ρ_m, is π_{ρ_m}. As we know, fair dice have equiprobable faces, $\pi_{\rho_1} = \cdots = \pi_{\rho_6} = 1/6$; but loaded dice do not follow these probabilities.

The simplest example of a categorical distribution is the Bernoulli distribution, which is the special case having just two outcomes that, conventionally, we identify with the numbers 1 and 0, and respective probabilities $\pi_1 = \pi$ and $\pi_2 = 1 - \pi$; see Fig. 1.4. We often write Bernoulli(π) instead of the more elaborate $\text{Categorical}_{1,0}(\pi, 1 - \pi)$.

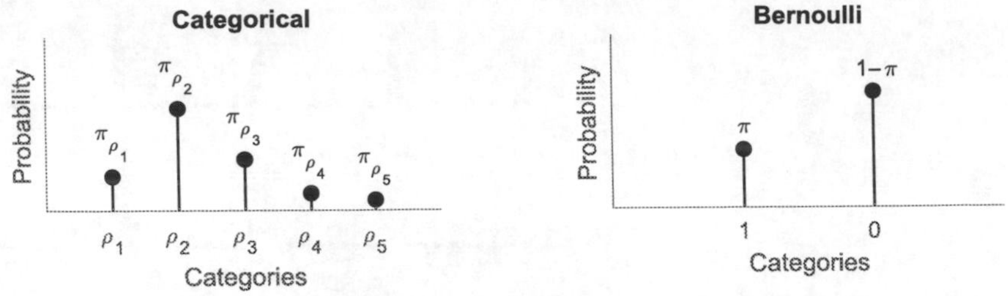

Fig. 1.4 (Left) We plot the associated probabilities $\pi_{\rho_{1:5}}$ with five outcomes, $\rho_{1:5}$, of the categorical distribution. (Right) We show a Bernoulli distribution, a special case of the categorical distribution, with associated probabilities π and $1 - \pi$ at its two possible outcomes.

Example 1.9 Coin flips modeled as Bernoulli random variables

An ideal coin flip has only two outcomes: "heads" or "tails." Provided we identify these with the numbers 1 and 0, respectively, we can model a coin flip as a Bernoulli random variable,

$$R \sim \text{Bernoulli}(\pi), \tag{1.7}$$

where π is the probability of "heads." Here, specifying the probability of tails, $1 - \pi$, is redundant since, by normalization, it is uniquely determined by π.

If, instead, we want to avoid identifying "heads" and "tails" with 1 and 0, we can also model a coin flip as a categorical random variable

$$R \sim \text{Categorical}_{\text{heads,tails}}(\pi, 1 - \pi). \tag{1.8}$$

Essentially, the only difference between Eq. (1.7) and Eq. (1.8) is in the meaning we assign to the values r, with the latter representation here having an *interpretational advantage* over the former.

Note 1.7 The categorical and Bernoulli densities

An alternative to Eq. (1.6) for expressing a categorical density is via the product

$$p(r) \propto \pi_{\rho_1}^{\Delta_{\rho_1}(r)} \cdots \pi_{\rho_M}^{\Delta_{\rho_M}(r)} = \prod_{m=1}^{M} \pi_{\rho_m}^{\Delta_{\rho_m}(r)}.$$

Here, the Kronecker terms $\Delta_\rho(r)$ are specified such that $\Delta_\rho(r) = 0$ when $r \neq \rho$ and $\Delta_\rho(r) = 1$ when $r = \rho$; see Appendix C. In the special case of Bernoulli random variables, this reduces to

$$p(r) \propto \pi^r (1 - \pi)^{1-r}.$$

These expressions involve products as opposed to the sums of Eq. (1.6). As we will see in subsequent chapters, these product forms are more convenient when we need to derive analytic formulas associated with discrete variables.

Distributions on Random Variables without Probability Density Functions*

Since in subsequent chapters we will formulate models with random variables to which we *cannot* assign a probability density function, for example random variables that are functions or trajectories or even random variables that are probability distributions themselves, we also need to account for appropriate distributions over these. In such cases, recipes to compute probabilities are case specific and, in general, the description of the associated distributions is necessarily more demanding. In Examples 1.10 and 1.11 we provide only a sneak preview.

Example 1.10 The standard Brownian motion

We will examine Brownian motion in more detail in Chapter 2. As we will see, standard Brownian motions in one dimension are random variables that represent functions from a time interval spanning 0 to some positive T to the real line. To denote them we write

$$X(\cdot) \sim \text{BMotion}_T^{1\text{D}}(D),$$

where the parameter D in the Brownian motion is a positive real scalar and, as we will see, can be interpreted as the diffusion coefficient of a particle diffusing in one dimension.

A shorthand like this captures the following pieces of information:

- The realizations of X are functions $x(\cdot)$ that, to any time t between 0 and T, assign $x(t)$ that is a position on the real line.
- Any realization of X is initialized at the origin, *i.e.*, $x(0) = 0$.
- For any choice of times t and t' between 0 and T, the difference $x(t) - x(t')$ between the values $x(t)$ and $x(t')$ of any realization $x(\cdot)$ is itself a random variable.
- The random variable $x(t) - x(t')$ has a probability density given by

$$p\left(x(t) - x(t')\right) = \frac{1}{\sqrt{4\pi D|t - t'|}} \exp\left(-\frac{(x(t) - x(t'))^2}{4D|t - t'|}\right).$$

Example 1.11 The Gaussian process

We will examine *Gaussian processes* in more detail in Chapter 6. As we will see, Gaussian processes are random variables that represent functions from a space S to the real numbers. To denote them we write

$$F(\cdot) \sim \text{GaussianP}_S\left(\mu(\cdot), C(\cdot,\cdot)\right).$$

A shorthand like this captures the following pieces of information:

- The realizations of F are functions $f(\cdot)$ that, to any point x in S, assign $f(x)$, which is a real number.

* This is an advanced topic and could be skipped on a first reading.

- The parameter $\mu(\cdot)$ is a function that, to every point x in S, assigns $\mu(x)$, which is also a real number.
- The parameter $C(\cdot,\cdot)$ is a function that, to all points x and x' in S, assigns $C(x, x')$, which is a nonnegative real number.
- For any choice $x_1, \ldots, x_M$ of any finite number M of points in S, the values $\boldsymbol{f} = (f(x_1), \ldots, f(x_M))$ form a random array.
- The random array $\boldsymbol{f} = (f(x_1), \ldots, f(x_M))$ has a multivariate normal probability density given by

$$p\left(\boldsymbol{f}\right) = \text{Normal}_M\left(\boldsymbol{f}; \boldsymbol{\mu}, \boldsymbol{C}\right).$$

The parameters of this density depend upon the points $x_1, \ldots, x_M$ and are given by

$$\boldsymbol{\mu} = \left(\mu(x_1), \ldots, \mu(x_M)\right), \qquad \boldsymbol{C} = \begin{pmatrix} C(x_1, x_1) & \cdots & C(x_1, x_M) \\ \vdots & \ddots & \vdots \\ C(x_M, x_1) & \cdots & C(x_M, x_M) \end{pmatrix}.$$

1.2.2 How to Sample from Probability Distributions

So far we have discussed random variables and probability distributions from which random variables are drawn. What we discuss next is how to run *simulations*. That is, how to generate values or sample random variables in a computer from specified probability distributions. Random simulations are useful when we seek to recreate in silico repetitions of our prototype experiment. In subsequent chapters, we will see that we can use random sampling not only to recreate an experiment's results but also to draw inferences *from* an experiment's results.

Continuous Random Variables

For a random variable $R \sim \mathbb{P}$ that takes scalar real values r, its *probability cumulative function* (commonly termed the cumulative distribution function or, simply, CDF) is a function $C(r)$ given by

$$C(r) = \int_{-\infty}^{r} dr'\, p(r'), \tag{1.9}$$

where $p(r)$ is the probability density associated with $\mathbb{P}$. From this definition, we see that a CDF is dimensionless and increases monotonically between 0 and 1. This is a characteristic that we can use to develop a method from which to generate random values r of R on a computer.

For instance, starting with the density $p(r)$, we first calculate its CDF, $C(r)$. We then generate a random value, call it u, uniformly between 0 and 1, and ask: *For what value r is $C(r)$ equal to u?* In other words, we find $r = C^{-1}(u)$, where $C^{-1}(u)$ is the inverse function of $C(r)$. This method, often termed the *fundamental theorem of simulation*, is summarized in Algorithm 1.1 and is visually illustrated in Fig. 1.5.

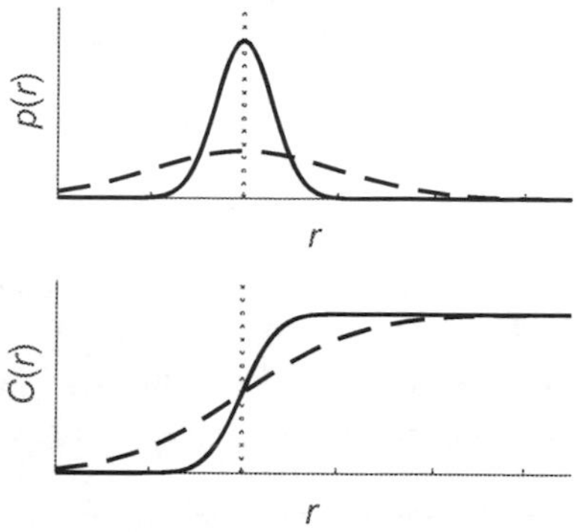

Fig. 1.5 (Top) We have PDFs broadly (dashed line) and tightly (solid line) centered around the same mean indicated by a vertical line. (Bottom) The corresponding CDFs. As seen, the CDF of the tighter PDF has a sharper slope near the mean, so in applying Algorithm 1.1, most values of u would coincide with values of r near the mean. By contrast, the CDF of the wider PDF has a broader slope near its mean. In this case, the same values of u coincide with a larger range of values of r.

Algorithm 1.1 Fundamental theorem of simulation for continuous variables

To simulate a continuous random variable $R \sim \mathbb{P}$:

- First, find the cumulative function $C(r)$ and its inverse $C^{-1}(u)$.
- Then, repeat the following steps:
  - Generate $U \sim \text{Uniform}_{[0,1]}$.
  - Set $r = C^{-1}(u)$.

Upon completion, this algorithm generates values r according to $\mathbb{P}$.

Example 1.12 shows a concrete example for an exponential distribution.

Example 1.12 Simulating from an exponential distribution

We consider a random variable $R \sim \text{Exponential}(\lambda)$. As we saw in Example 1.4, this random variable takes real scalar values and its density is

$$p(r) = \begin{cases} \lambda e^{-\lambda r}, & r \geq 0 \\ 0, & r < 0 \end{cases}.$$

To apply the fundamental theorem of simulation, we first compute the CDF and its inverse:

$$C(r) = 1 - e^{-\lambda r}, \qquad C^{-1}(u) = -\frac{1}{\lambda}\log(1-u).$$

By means of Algorithm 1.1, we then sample R as follows:

- First, sample

$$U \sim \text{Uniform}_{[0,1]}.$$

- Then, we compute r from

$$r = -\frac{1}{\lambda}\log(1-u). \quad (1.10)$$

Since $\tilde{U} = 1 - U$ is also uniformly distributed between 0 and 1, for computational efficiency, when sampling exponential random variables, we generate $\tilde{U} \sim \text{Uniform}_{[0,1]}$ in the first place and then use $r = -\frac{1}{\lambda}\log\tilde{u}$ instead of using Eq. (1.10). In this way, we speed up the algorithm's execution by avoiding computation of the difference $1 - u$.

Note 1.8 Distribution functions

So far, we have encountered three important functions $p(r)$, $C(r)$, and $C^{-1}(u)$ associated with a distribution $\mathbb{P}$. These are very common in the literature and now we summarize some terms used to designate them:

- $p(r)$ is termed the *probability density function* or *PDF*.
- $C(r)$ is termed the *probability cumulative function* or *cumulative distribution function* (*CDF*).
- $C^{-1}(u)$ is termed the *probability quantile function, inverse cumulative distribution function*, or *ICDF*.

These functions may characterize the associated distribution $\mathbb{P}$. For this reason, they are often termed *probability distribution functions.*

Why the Fundamental Theorem of Simulation for Continuous Variables Works?*

In Algorithm 1.1, we have a uniform random variable u whose density is $g(u) = 1$ for any u between 0 and 1. When we set $r = C^{-1}(u)$, effectively we perform a transformation of random variables. As we saw in Section 1.2.1, the density $h(r)$ of the transformed variable is given by Eq. (1.5) which, in this setting, takes the form

$$h(r) = \frac{1}{|J_{u\mapsto r}|}g(u).$$

Here, because both of our variables are scalar, the Jacobian of the transformation is given by the derivative

$$J_{u\mapsto r} = \frac{d}{du}C^{-1}(u) = \frac{1}{\frac{d}{dr}C(r)} = \frac{1}{p(r)}.$$

Considered together, the last two equations lead to $h(r) = p(r)$. In other words, the values r generated in Algorithm 1.1 do indeed, follow the desired density $p(r)$.

Discrete Random Variables

We can use a similar procedure to sample discrete random variables too. In particular, for $R \sim \text{Categorical}_{\rho_{1:M}}(\pi_{\rho_{1:M}})$, the cumulative function is

* This is an advanced topic and could be skipped on a first reading.

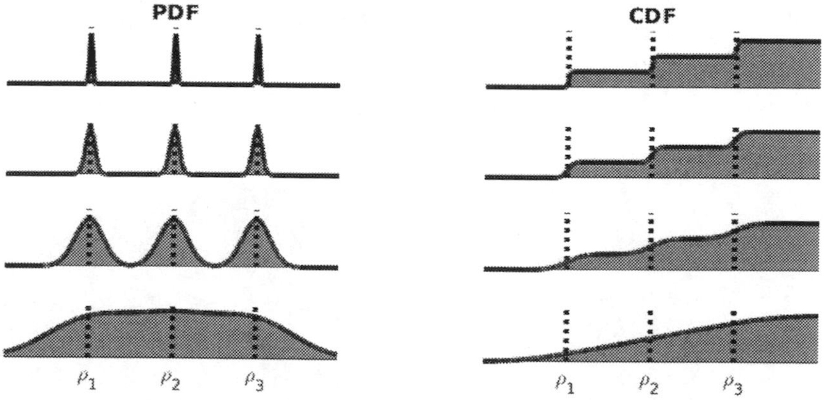

Fig. 1.6 The CDF of a discrete random variable arises as an idealization of increasingly less noisy measurements. Here, the left panels show the probability distributions of noisy measurements and the right panels illustrate the corresponding CDFs.

$$C(\rho_m) = \sum_{m'=1}^{m} \pi_{\rho_{m'}}$$

or, more concretely, it adopts the form

$$C(\rho_1) = \pi_{\rho_1}, \quad C(\rho_2) = \pi_{\rho_1} + \pi_{\rho_2}, \quad \cdots \quad C(\rho_M) = \pi_{\rho_1} + \pi_{\rho_2} + \cdots + \pi_{\rho_M}.$$

To sample an outcome r, as with continuous random variables, we also need to generate $U \sim \text{Uniform}_{[0,1]}$. However, now a problem concerning the inversion of $C(r)$ arises. Namely, there may be no r such that $C(r) = u$. For this reason, instead of searching for outcomes such that $C(r) = u$, we search for the *lowest* value r such that $u \leq C(r)$. This version of the fundamental theorem of simulation is summarized in Algorithm 1.2 and is visually illustrated in Fig. 1.6.

Algorithm 1.2 Fundamental theorem of simulation for discrete variables

To simulate a random variable $R \sim \text{Categorical}_{\rho_{1:M}}(\pi_{\rho_{1:M}})$:

- Generate $U \sim \text{Uniform}_{[0,1]}$.
- Find the lowest m such that $u \leq \pi_{\rho_1} + \pi_{\rho_2} + \cdots + \pi_{\rho_m}$.
- Set $r = \rho_m$.

At first, it might appear that this algorithm depends on the particular labeling of $\rho_{1:M}$ and that it would lead to different realizations r if the labels m had been assigned differently over the categories ρ_m. However, since relabeling of $\rho_{1:M}$ also involves a similar relabeling of $\pi_{\rho_{1:M}}$, this is *not* the case. In other words, this algorithm realizes each outcome $r = \rho_m$ with probability π_{ρ_m}, even when the labels m are reassigned over ρ_m.

Example 1.13 shows a concrete example for a Bernoulli distribution.

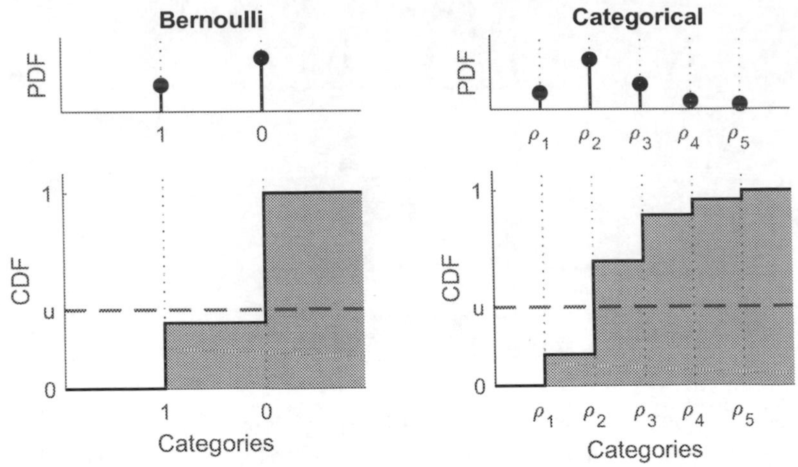

Fig. 1.7 (Left) We consider a probability ranging from zero to one segmented into two portions of weight π and $1-\pi$ separated by a break point. We imagine these to be the probability of sampling outcome 1 or outcome 0 in a Bernoulli trial. To determine which outcome we select, we draw a uniform random number u. In this figure, the u sampled falls above the break point. As such, we select outcome 0 for this Bernoulli trial. This argument carries over to the right panel with more than two portions.

Example 1.13 Simulation of Bernoulli random variables

Consider $R \sim \text{Bernoulli}(\pi)$. In this case, the cumulative function has a very simple form

$$C(1) = \pi, \qquad C(0) = 1.$$

To sample r, according to the fundamental theorem of simulation:

- First, we generate $U \sim \text{Uniform}_{[0,1]}$.
- Then, if $u \leq \pi$ we set $r = 1$, else we set $r = 0$.

The two steps are illustrated in Fig. 1.7.

Note 1.9 Intuition as to how the fundamental theorem of simulation works

There exists an intuitive explanation for the fundamental theorem of simulation that we illustrate in Fig. 1.7. That is, we imagine an interval from 0 to 1 with a break point just as shown. The portion of the interval before the break point has length π and the remainder has length $1-\pi$. We now sample a uniform random variable, u. If u falls before the break point, outcome 0 is realized. Otherwise outcome 1 is realized.

A similar logic holds for the categorical distribution. Figure 1.7 shows the steps in the CDF of a discrete distribution. A draw from the uniform distribution can be visualized as the dotted horizontal line. The value of the abscissa that coincides with the location where the dotted line intersects with the CDF dictates the value realized by the discrete random variable.

Why the Fundamental Theorem of Simulation for Discrete Variables Works?*

We consider the limiting scenario of Note 1.6. Namely, we have a continuous random variable $R \sim \mathbb{P}$ in an experiment aiming to distinguish between the outcomes $\rho_{1:M}$. Generally, in a noisy experiment, $\mathbb{P}$ has a wide PDF and a CDF that increases smoothly from 0 to 1 as seen in Fig. 1.6. In a less noisy experiment, however, the PDF has peaks and this means that the CDF retains smooth albeit clearly visible steps. As the noise level reduces, the PDF's peaks become more prominent, resulting in the CDF's steps becoming sharper. In the extreme case of a noiseless scenario, the CDF forms perfectly sharp steps mathematically represented by discontinuities. Seen as a limiting case of sampling continuous random variables, Algorithm 1.2 is the direct analog of Algorithm 1.1.

1.2.3 Manipulating Probability Distributions

In a prototype experiment, we most often have random variables with different properties. When handling such a complex model, we need to work simultaneously with more than one distribution. Here, we present book-keeping rules that help us combine and manipulate multiple distributions.

Joint and Marginal Distributions

Provided the random variables in our model are independent from each other, for example because they may encode physical processes or observations that exert no influence upon each other, we may write

$$R_1 \sim \mathbb{P}_1, \qquad R_2 \sim \mathbb{P}_2, \qquad \dots \qquad R_N \sim \mathbb{P}_N.$$

Each distribution $\mathbb{P}_1, \mathbb{P}_2, \dots, \mathbb{P}_N$ is, in turn, associated with its own density $p_1(r_1), p_2(r_2), \dots, p_N(r_N)$. As there is little chance of confusion, usually we simply write $p(r_n)$ instead of $p_n(r_n)$.

Occasionally, we also encounter models with multiple random variables,

$$R_1 \sim \mathbb{P}, \qquad R_2 \sim \mathbb{P}, \qquad \dots \qquad R_N \sim \mathbb{P}, \tag{1.11}$$

that are independent and that also follow identical distributions $\mathbb{P}$, for example random variables that may model independent observations obtained from a time invariant system. On such occasions, we might abbreviate Eq. (1.11) to $R_1, R_2, \dots, R_N \overset{iid}{\sim} \mathbb{P}$ and speak of *independent and identically distributed*, or simply *iid*, random variables. Essentially, we mean that all densities $p(r_1), p(r_2), \dots, p(r_N)$ happen to have the same form. In the iid setting, and only when there is no chance of confusion, we might refer to each one of the densities simply as $p(r)$. However, as we will see shortly, even with iid variables, it is often necessary to clarify what is meant by $p(r)$.

* This is an advanced topic and could be skipped on a first reading.

Following our convention, we denote the density of a single variable R_n as $p(r_n)$ and call it a *marginal density*. When multiple random variables $R_{1:N}$ arise in the same setting and a density gathers all of them, we write $p(r_{1:N}) = p(r_1, r_2, \ldots, r_N)$ and we refer to $p(r_{1:N})$ as a *joint density*.

A marginal density $p(r_n)$ is related to a joint density $p(r_{1:N})$ through an integration over the entire range spanned by $r_{1:n-1}$ and $r_{n+1:N}$. That is,

$$p(r_n) = \underbrace{\int dr_1 \cdots \int dr_{n-1} \int dr_{n+1} \cdots \int dr_N}_{\text{everything but } r_n} p(r_{1:N}). \tag{1.12}$$

Colloquially, we refer to the integration over variables, *i.e.*, from right-to-left in Eq. (1.12), as a *marginalization*. We refer to the reverse process, *i.e.*, from left-to-right in Eq. (1.12), as a *de-marginalization* or a *completion*.

Example 1.14 Marginalization

Similarly to how we obtain distributions over one random variable, we may obtain distributions over any subset of the variables in $R_{1:N}$. For concreteness, we consider a total of $N = 5$ variables and suppose that we wish to obtain a distribution over R_2 and R_4 only. In this case, marginal and joint densities are linked by

$$p(r_2, r_4) = \underbrace{\int dr_1 \int dr_3 \int dr_5}_{\text{everything but } r_2 \text{ and } r_4} p(r_{1:5}).$$

Now that we have discussed marginal and joint distributions, we are ready to discuss sampling of random variables from a normal distribution.

Note 1.10 Box–Muller simulation of normal random variables

We consider a random variable $X \sim \text{Normal}(\mu, v)$. As the CDF of X does not have a closed analytic form, we cannot use the fundamental theorem of simulation. For this reason, we follow a different approach. Namely, we start by considering two iid random variables,

$$X, Y \overset{iid}{\sim} \text{Normal}(\mu, v).$$

The associated joint density reads

$$p(x, y) = p(x)p(y) = \left(\frac{1}{\sqrt{2\pi v}} \exp^{-\frac{(x-\mu)^2}{2v}}\right)\left(\frac{1}{\sqrt{2\pi v}} e^{-\frac{(y-\mu)^2}{2v}}\right).$$

On X and Y we perform three successive transformations:

- A linear transformation from x and y to

$$x' = \frac{x - \mu}{\sqrt{v}}, \qquad y' = \frac{y - \mu}{\sqrt{v}}.$$

- A nonlinear transformation from Cartesian (x', y') to polar coordinates (ρ, ϕ) with

$$x' = \rho \cos\phi, \qquad y' = \rho \sin\phi.$$

- A nonlinear transformation from ρ to λ with

$$\lambda = \rho^2.$$

The advantage of applying these transformations is that the resulting density over λ and ϕ is separable,

$$p(\lambda, \phi) = \frac{1}{2} \exp\left(-\frac{\lambda}{2}\right) \frac{1}{2\pi} = \text{Exponential}\left(\lambda; \frac{1}{2}\right) \text{Uniform}_{[0,2\pi]}(\phi),$$

where we have made explicit that the exponential and uniform distributions are on the variables λ and ϕ, respectively.

The cumulative function over λ and ϕ can now be computed analytically. Thus, by generating two uniform random samples $U_1, U_2 \overset{iid}{\sim} \text{Uniform}_{[0,1]}$, we can readily obtain random samples from the radial and polar angle distribution

$$\rho = \sqrt{\lambda} = \sqrt{-2 \log u_1}, \qquad \phi = 2\pi u_2.$$

Transforming back to our original variables, we obtain

$$x = \mu + \sigma x' = \mu + \sigma \rho \cos\phi = \mu + \sigma\sqrt{-2 \log u_1} \cos(2\pi u_2).$$

This algorithm for sampling normal random variables is termed the *Box–Muller algorithm*. As can be seen, with little additional computational cost, this method also provides another normal sample

$$y = \mu + \sigma\sqrt{-2 \log u_1} \sin(2\pi u_2)$$

that is independent of x.

Conditional Distributions

The order in which random variables arise in a model may be irrelevant, for example random variables modeling an experiment's observations that exert no influence upon each other, such as individual test scores, biometric measurements collected from a group of unrelated participants, or a temporal sequence of observations generated by a system at equilibrium. On the other hand, the order in which random variables arise *may be important*, for example random variables modeling observations of time-dependent phenomena, such as successive measurements of the number of cells in a growing cell culture, or the number of molecules available to react in a chain of chemical reactions.

To express *dependencies* among two random variables, R_1 and R_2, we write

$$R_2 | r_1 \sim \mathbb{P}(r_1). \tag{1.13}$$

This reads as "the random variable R_2, given the realization r_1 of the random variable R_1, is sampled from the probability distribution $\mathbb{P}(r_1)$" and means that the values of r_2 are associated with a density $p(r_2|r_1)$ depending upon r_1. We designate a distribution that depends upon the value of another random variable like $\mathbb{P}(r_1)$ and the associated density $p(r_2|r_1)$ as *conditionals*.

Note 1.11 How to avoid inaccuracies in specifying variable dependencies

In the setting of Eq. (1.13), the random variable R_1 is sampled from its own (marginal) distribution that needs to be specified *separately*. In a complete model, both random variables $R_1 \sim \mathbb{P}_1$ and $R_2|r_1 \sim \mathbb{P}_2(r_1)$ need to be specified.

In a properly formulated model, the distribution of R_1 *must not* depend upon r_2 and, for this reason, the description of R_1 should precede that of $R_2|r_1$. If this is not possible, then we need to describe the random variables together through a single (joint) distribution $(R_1, R_2) \sim \mathbb{P}$.

Ideally, proper descriptions of models involving multiple random variables, which depend upon each other, should be provided in a nested fashion. For example,

$$R_1 \sim \mathbb{P}_1,$$

$$R_2|r_1 \sim \mathbb{P}_2(r_1),$$

$$R_3|r_2, r_1 \sim \mathbb{P}_3(r_2, r_1),$$

$$\ldots$$

A necessary condition, although not always sufficient, for a reliable description of a probabilistic model, no matter how convincing the involved arguments may be, is that *every single distribution* $\mathbb{P}_1, \mathbb{P}_2(r_1), \mathbb{P}_3(r_2, r_1), \ldots$ be specified *clearly and explicitly*.

Whenever a model cannot be put in a nested form as above, even when random variables are grouped and joint distributions are applied, the model most likely contains flaws such as tautologies or contradictions. Consequently, such a model is, even qualitatively, inappropriate.

In Note 1.11, we consider nested variable dependencies. In particular, R_3 depends on the realizations r_2 and r_1, while R_2 depends on the realization r_1, and, finally, R_1 depends on no other realization. In the most general case, the probability distribution over our last random variable, say R_N, may depend on the realization of all previous random variables, $r_{1:N-1}$, and the same happens for all other variables up to the very first one, r_1. On account of this hierarchy, resembling successive generations of variables, simulating a nested model requires a sampling algorithm termed *ancestral sampling* detailed in Algorithm 1.3.

Algorithm 1.3 Ancestral sampling

To draw values for a group of random variables $R_{1:N}$, we proceed as follows:

- Find the density $p(r_1)$ associated with $R_1 \sim \mathbb{P}_1$.
- Sample r_1 using $p(r_1)$.
- For n from 2 to N, repeat:
  - Find the density $p(r_n|r_{1:n-1})$ associated with $R_n|r_{1:n-1} \sim \mathbb{P}_n(r_{1:n-1})$.
  - Sample r_n using $p(r_n|r_{1:n-1})$.

Since we use ancestral sampling and hierarchical models extensively in Chapter 2 and subsequent chapters, we describe methods here to obtain the necessary conditional densities. Our starting point is the full joint density $p(r_{1:N})$, whose arguments we conveniently reorder and write as $p(r_{N:1})$.

Conditional and joint densities are related to each other through the *chain rule* which, in the most general setting, reads as

$$p(r_{N:1}) = p(r_N|r_{N-1:1}) \cdots p(r_2|r_1)p(r_1).$$

In the simplest case, consisting of only two random variables, the chain rule is

$$p(r_2, r_1) = p(r_2|r_1)p(r_1).$$

From this, we immediately see that a conditional density over r_2 is normalized irrespective of r_1, *i.e.*,

$$\int dr_2\, p(r_2|r_1) = \int dr_2 \frac{p(r_2, r_1)}{p(r_1)} = \frac{\int dr_2\, p(r_2, r_1)}{p(r_1)} = \frac{p(r_1)}{p(r_1)} = 1.$$

Additionally, from the chain rule, we obtain two equalities, $p(r_2, r_1) = p(r_2|r_1)p(r_1)$ and $p(r_1, r_2) = p(r_1|r_2)p(r_2)$, that we can combine to obtain another important rule, namely *Bayes' rule*, which most often is written in the form

$$p(r_2|r_1) = \frac{p(r_1|r_2)p(r_2)}{p(r_1)}, \qquad p(r_1) \neq 0. \tag{1.14}$$

As we will see in subsequent chapters, Eq. (1.14) is an indispensable tool in data analysis.

Example 1.15 Modeling dynamical systems

Dependency among variables is especially important when the physical system of interest evolves over time. In this dynamical setting, explored in depth in Chapter 2 and also in Chapters 8–10, our prototype experiment is temporally structured: causality indicates that the last measurement W_N may be influenced by all preceding measured values, $w_{1:N-1}$; the penultimate measurement, W_{N-1}, may be influenced by all of its preceding ones $w_{1:N-2}$; and so forth.

With the rules of joint and conditional distributions, we can work out the densities of such models in the most general setting. For instance,

$$p(w_{1:N}) = p(w_N|w_{1:N-1})p(w_{N-1}|w_{1:N-2}) \cdots p(w_2|w_1)p(w_1).$$

It follows that if we need to sample realizations of $W_{1:N}$, we need 1 marginal and $N-1$ *different* conditional distributions. As a result, this sampling may become infeasible unless we make some assumptions:

- One drastic assumption, often too crude for realistic dynamical systems, is to assume that all variables are independent, which in this particular case is equivalent to assuming $p(w_n|w_{1:n-1}) = p(w_n)$. Under this assumption, the joint density factorizes into a product of densities,

$$p(w_{1:N}) = p(w_1)p(w_2) \cdots p(w_N) = \prod_{n=1}^{N} p(w_n). \quad (1.15)$$

- Another, less drastic and often quite realistic, assumption is to consider $p(w_n|w_{1:n-1}) = p(w_n|w_{n-1})$. Under this assumption, the joint density also factorizes into a product of densities,

$$p(w_{1:N}) = p(w_1)p(w_2|w_1) \cdots p(w_N|w_{N-1}) = p(w_1) \prod_{n=2}^{N} p(w_n|w_{n-1}). \quad (1.16)$$

Under these two assumptions, the total number of different probability distributions needed to sample $W_{1:N}$ reduces from N in general to a single marginal, for Eq. (1.15) or a marginal and a conditional, for Eq. (1.16). This is under the assumption that the marginals, over each variable, and conditionals, over each pair of variables, are all the same.

Somewhat pedantically, in deriving Eq. (1.15), we invoked a so-called zeroth-order Markov assumption

$$p(w_n|w_{1:n-1}) = p(w_n),$$

while, in deriving Eq. (1.16) we invoked a so-called first-order Markov assumption often abbreviated simple as a *Markov assumption*

$$p(w_n|w_{1:n-1}) = p(w_n|w_{n-1}).$$

In principle, we can also invoke higher order assumptions where a measurement w_n is influenced by more than 1 past measurement; however, as we will see in the subsequent chapters, such assumptions are rarely used in practice, either because a first-order assumption is already sufficient or because they lead to models with prohibitive computational cost.

1.3 Data-Driven Modeling and Inference

Having introduced the necessary formalism, our emphasis from now on is not as much on mathematical rigor as it is focused on problem formulation and problem solving. But, *of what problem exactly?* With our basic notions laid down, we are now ready to define and address our problem more concretely.

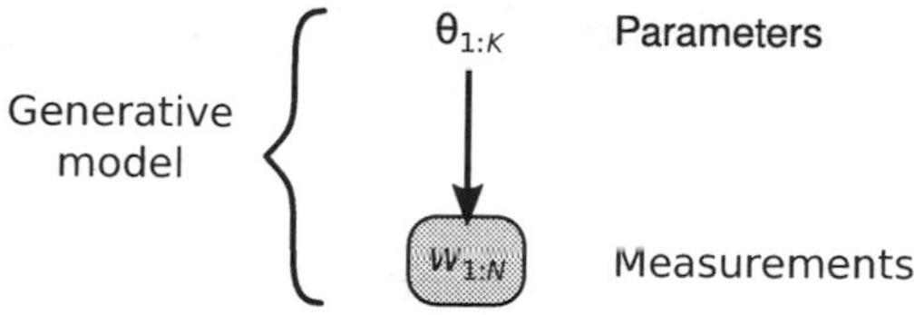

Fig. 1.8 A generative model describes how measurements are generated. Implicitly, it encodes any influence the parameters $\theta_{1:K}$ exert upon the measurements $w_{1:N}$.

In the data-centric context that is most appropriate for the physical and natural sciences, we envision being provided information on a physical system such as:

- *How this system behaves* under relevant, well or poorly characterized, conditions.
- *How observations are acquired* on this system.
- *Specific values* of acquired observations.

These are the *data* and they serve as our input or starting point. Our primary task is to analyze the data and we tackle data analysis with the framework introduced in Section 1.1.2. More specifically, within the framework set by the prototype experiment, which we adapt to real-life scenarios, our goal is to use the data to infer a model. However, before we can infer a model, we first need to go through a *synthesis stage* in order to develop the necessary mathematical formulation.

During the synthesis stage, we utilize the available information on our system to formulate the probability distribution $p(w_{1:N}|\theta_{1:K})$ that best describes our experiment. For example, in this stage we consider physical laws, dynamics, and noise properties, which, although nonnumeric, in a very concrete sense are part of our given data. At this stage, we also decide on parameters $\theta_{1:K}$ and ascribe physical meaning to all or some of them.

The synthesis stage concludes with a concrete *generative model*; that is, a quantitive description of *how our experiment's measurements are generated*; see Fig. 1.8. Our generative model mathematically links our unknowns, $\theta_{1:K}$, with our knowns, $w_{1:N}$, and, in principle, could be simulated on a computer.

Note 1.12 A model's likelihood

The probability distribution $p(w_{1:N}|\theta_{1:K})$, mathematically established in a generative model, is a key quantity. This distribution is termed the *likelihood* or, colloquially, the likelihood function. The term follows from the notion that $p(w_{1:N}|\theta_{1:K})$ quantifies the likelihood of observing (sampling) the sequence of observations $w_{1:N}$ in our prototype experiment given the parameters $\theta_{1:K}$ that influence their realizations.

During the analysis stage, once we have formulated $p(w_{1:N}|\theta_{1:K})$ we apply the measured values of $w_{1:N}$ to compute parameter estimates, $\theta_{1:K}$.

Traditionally, we call these values *estimators* and denote them as $\hat{\theta}_{1:K}$. We will see in Chapter 3 that a likelihood provides us with a *universal* strategy to obtain $\hat{\theta}_{1:K}$ needed to specify uniquely a model we wish to learn. The challenge, however, is that we are also often interested in error bars around $\hat{\theta}_{1:K}$ or, put differently, probability distributions over all possible values of the random variable. For this reason, in Chapter 4, we will consider an extended strategy that uses more than an experiment's likelihood.

The first stage in our workflow, namely setting up the generative model, constitutes a *modeling task*; while, the second stage, namely obtaining parameter estimates, constitutes a *computational task*. As we discuss in Example 1.16, both stages in the solution of our problem are important and both stages pose unique challenges. As we will see in subsequent chapters, often we have to devise comprehensive approaches that deal with the challenges arising in both stages simultaneously.

Example 1.16 Likelihood based modeling and inference

As a concrete example, we imagine an experiment idealized as having one of two measurement outcomes, for example the emission or not of a photoelectron as described in Example 1.1. For simplicity, we may encode these outcomes with $\rho_1 = 1$ and $\rho_2 = 0$, respectively.

If we idealize individual assessments as iid, meaning that each measurement is independent of the others as in Eq. (1.15), then the mathematical form of the likelihood is readily derived. In particular, the model responsible for generating the data takes the form

$$W_n|\pi \sim \text{Bernoulli}(\pi), \qquad n = 1 : N,$$

and, as of yet, has one unspecified parameter, namely π, to which we ascribe the meaning of probability that a single assessment results in a photoelectron emission. From now on, our goal is to estimate π.

To achieve our goal, we ask: *Given this generative model, what is the likelihood of our measurements?* This likelihood is the probability of observing the sequence $w_{1:N}$ and we may compute it as

$$p(w_{1:N}|\pi) = \prod_{n=1}^{N} p(w_n|\pi) = \prod_{n=1}^{N} \text{Bernoulli}(w_n; \pi) \propto \prod_{n=1}^{N} \pi^{w_n}(1-\pi)^{1-w_n}$$
$$= \pi^M (1-\pi)^{N-M},$$

where we assumed that, within $w_{1:N}$, the first outcome, ρ_1, has been observed in total M times and the second outcome, ρ_2, has been observed the remainder of the times, namely $N - M$.

Finally, we estimate a value for the parameter π by asking: *Which value of π makes our observations most likely?* This is equivalent to asking which value of π makes $p(w_{1:N}|\pi)$ highest. Essentially, we seek the maximizer of $p(w_{1:N}|\pi)$. For instance, solving $\frac{d}{d\pi}p(w_{1:N}|\pi) = 0$, we recover $\hat{\pi} = M/N$, as intuitively expected.

For this example, we assumed that both outcomes, ρ_1 and ρ_2, are observed at least once and so $0 < \hat{\pi} < 1$ because $0 < M < N$. Yet had $M = 0$ or $M = N$, we may have erroneously concluded, due to limited data, that $\hat{\pi} = 0$ or $\hat{\pi} = 1$. Thus, even this toy example presages our need to go beyond approaches that rely exclusively on likelihoods.

In the natural sciences, data-driven approaches are sometimes termed *inverse methods*, *inverse problems*, or *inverse modeling*. Yet, as Example 1.16 illustrates, there is nothing backward about obtaining models starting from the data and these somewhat unfortunate terms arose only because traditional approaches, namely obtaining models from the ground up with a combination of first principles and data-fitting, came first historically and are now termed forward (or direct) methods.

Note 1.13 Inverse modeling

Data-driven model inference is essentially an inverse problem. Solving an inverse problem is the opposite of solving a direct problem. Briefly, in a *direct problem*, also termed forward problem, we seek to determine an effect knowing its cause; while, in an *inverse problem* we seek to recover the cause knowing only the effect.

Inverse problems arise mainly when we need to interpret indirect physical measurements of unknown or partially known origin. For instance, when we are interested in elucidating the dynamics of complex biomolecules observed indirectly through fluorescence microscopy. In an experiment, we acquire images (measurements) with all sorts of artifacts that subsequently need to be removed in order to reveal the positions or dynamics of the biomolecules of interest. By contrast, simulating possible measurements (by, say, molecular dynamics simulations) and invoking a physical model, established or tentative, that *predicts* system behavior and subsequently checking whether predictions are in agreement or disagreement with the observed measurements constitutes direct modeling; see Fig. 1.9.

A problem, whether direct or inverse, is *well-posed* when it meets the following conditions:

- The problem has a solution.
- The solution is unique.
- The solution does not differ substantially unless the supplied data also differs substantially.

These conditions are mathematically known as *existence, uniqueness,* and *stability*, respectively. If a problem fails to satisfy one or more of them, it is *ill-posed*.

Direct problems are well-posed when the effects (data) are well defined, single-valued, and depend continuously on their causes. Often this is the case when we seek to reproduce observations mathematically or computationally. On the other hand, solutions to inverse problems do *not always exist*, or when they exist they are *almost never unique* or may *change dramatically*

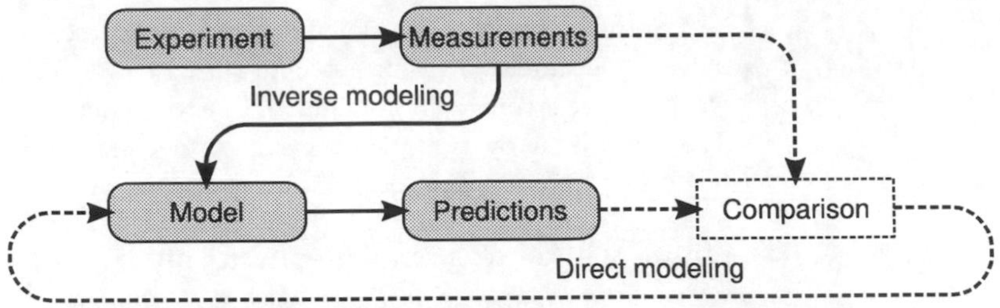

Fig. 1.9 Here, we show an illustration of the direct and inverse modeling paradigms. In the direct paradigm, a model is adjusted until its predictions agree with an experiment's measurements. By contrast, in the inverse paradigm, a model is inferred from the experiment's measurements without adjustments.

even when the supplied data (effects) differ only insignificantly. As a result, inverse problems are commonly ill-posed and solving them can be far more challenging.

Throughout the following chapters, with the use of the appropriate random variables and probability distributions, we will see how inverse problems can be formulated statistically and how solutions to these problems can be computed robustly and efficiently.

Forward modeling, also termed direct modeling, has had its role to play and is heavily showcased throughout physics where disparate observations were unified into predictive frameworks inspired by logic, symmetries, and fundamental postulates. Undoubtedly, the forward approach has been tremendously successful. To wit, among others, it predicted the magnetic moment of the electron to a spectacular number of significant digits. But there are limitations to this historically successful approach.

While forward modeling historically came first, inverse methods, spurred in equal parts by advances in probability theory and motivating data-centric questions in the natural sciences, transiently gained prominence in mainstream physics thanks to Laplace. Today, large swathes of complicated enough physical and chemical systems, in addition to life and social sciences, are not naturally modeled from the ground up, *i.e.*, starting from first principles. Instead, in these cases, observations often only suggest loose couplings between variables of interest and probabilistic relations between various quantities implied by the data.

The forward approach is different from the philosophy we adopt in this textbook. Instead, we use first principles only to motivate forms for our generative models. Beyond this, we are motivated by the practice of statistics that instead attempts, from the onset, to be as agnostic about the model parameters (or the model itself) as possible and learn parameters and models self-consistently from the available data as efficiently as computationally possible.

1.4 Exercise Problems

Exercise 1.1 Product of normal densities

Show that the product of normal probability densities remains proportional to a normal probability density.

Exercise 1.2 Calculus warm-up

Evaluate the following.

1. Gaussian integral: $\int_{-\infty}^{+\infty} dx\, e^{-(x-\mu)^2/(2v)}$, assume μ and v are real scalars and $v > 0$.
2. Gaussian moments: $\int_{-\infty}^{+\infty} dx\, x^2 e^{-(x-\mu)^2/(2v)}$, assume μ and v are real scalars and $v > 0$.
3. Gaussian convolution: $\int_{-\infty}^{+\infty} dx\, e^{-(y-x)^2/(2v)} e^{-(x-\mu)^2/(2v)}$, assume μ, y, and v are real scalars and $v > 0$.
4. Gamma function integral: $\int_0^{\infty} dx\, x^n e^{-x/a}$, assume n is a positive integer and $a > 0$.
5. Poisson moments: $\sum_{n=0}^{\infty} n^2 \lambda^n \exp(-\lambda)/n!$, assume $\lambda > 0$.

Exercise 1.3 Matrix algebra warm-up

Complete the following matrix algebra operations and, in doing so, state the conditions on the individual matrices (A, B, C, and D) required for the operation to be well defined. Further, assume that all matrices and vectors are of the appropriate dimension to support the operations required.

1. Derive the matrix transpose relation: $(AB)^T = B^T A^T$, where T denotes the transpose.
2. Using a Taylor expansion, demonstrate that $e^A e^A = e^{2A}$.
3. Complete the squares in f: $fAf^T - fB^T - Bf^T$, where f is a row vector.
4. Perform the following matrix normal convolution: $\int e^{-(f-\mu)A^{-1}(f-\mu)^T} e^{-(h-f)B^{-1}(h-f)^T} df$ where f, μ, and h are vectors with real coordinates and the integral is over every coordinate of f taken over the whole real line.
5. Verify the "inverse of sum" identity: $(A+B)^{-1} = A^{-1} - A^{-1}\left(B^{-1} + A^{-1}\right)^{-1} A^{-1}$.
6. Verify the "push through" identity: $(\mathbb{1} + AB)^{-1} = \mathbb{1} - A(\mathbb{1} + BA)^{-1}B$.
7. Verify the Woodbury identity: $(A+BCD)^{-1} = A^{-1} - A^{-1}B\left(C^{-1} + DA^{-1}B\right)^{-1} DA^{-1}$.

Exercise 1.4 Permutations and combinations

Consider integers $N = 1, 2, \ldots$ and $M = 0, 1, \ldots, N$.

1. Show that the total number of distinct arrangements (permutations) of M objects selected out of N distinct objects is $\frac{N!}{(N-M)!}$.

2. Show that if we ignore the arrangement of the objects (combinations) the total number drops to $\frac{N!}{M!(N-M)!}$.
3. Show that the total number of different combinations of N distinct objects is 2^N.

Exercise 1.5 Cumulative probability function

Explain why the cumulative probability function in Eq. (1.9) takes only values between 0 and 1.

Exercise 1.6 Cumulative and quantile functions of exponential random variables

Verify the formulas of $C(r)$ and $C^{-1}(r)$ in Example 1.12.

Exercise 1.7 Joint distribution

Show that a joint distribution encodes all information required to reconstruct all relevant marginals and conditionals. For concreteness, consider a model with three random variables $R_{1:3}$.

Exercise 1.8 Sum of random variables

Consider two independent random variables R_1 and R_2 with densities $p_1(r_1)$ and $p_2(r_2)$, respectively. Use a transformation to show that the density $p_3(r_3)$ of a random variable R_3, with values $r_3 = r_1 + r_2$, is equal to the convolution $p_3(r_3) = (p_1 * p_2)(r_3)$.

Exercise 1.9 Minimum of exponential random variables

Computing the minimum of two exponential random variables is often relevant when considering the time of arrival of the first event given two competing events. For this reason, here we consider two exponential random variables $R_1 \sim \text{Exponential}(\lambda_1)$ and $R_2 \sim \text{Exponential}(\lambda_2)$. Show that the random variable R_3, with values $r_3 = \min(r_1, r_2)$, follows an $\text{Exponential}(\lambda_1 + \lambda_2)$ distribution.

Exercise 1.10 A sanity check on random variable rescaling

Verify that the density $q(r')$ of the rescaled random variable R' in Example 1.6 has the correct units and that it is properly normalized.

Exercise 1.11 Linear transformations

In Examples 1.6 and 1.7 we have seen how to obtain the probability density of random variables under rescaling and rotation. These two separate operations can be combined into a single one. For instance, consider a bivariate random variable (X, Y) and suppose that (X', Y') is the random variable under a linear transformation

$$x' = Ax + By + C, \qquad y' = Dx + Ey + F,$$

where $A, B, C, D, E,$ and F are scalar constants. Find the probability density of (X', Y') in terms of $p(x, y)$ and to avoid degeneracies consider only the case with $AE - BD \neq 0$.

Exercise 1.12 Division of random variables

Consider a random variable R with values $r = 1/x$, where X is a scalar random variable with density $p(x)$. Compute the density $q(r)$ in terms of $p(x)$.

Exercise 1.13 Spherical coordinate transformations

Consider a trivariate random variable (X, Y, Z) that models a position in Cartesian space. Use a transformation of random variables to relate the probability density $p(x, y, z)$ with the probability density $q(r, \phi, \theta)$ of the same position in spherical coordinates (R, Φ, Θ).

Exercise 1.14 Gamma random variables and derivatives

Suppose $R_1 \sim \text{Gamma}(\alpha_1, \beta)$ and $R_2 \sim \text{Gamma}(\alpha_2, \beta)$ are independent gamma random variables; see Appendix B for the definition of the gamma distribution. Find the probability densities of the random variables V_1, V_2, V_3 with values

$$v_1 = r_1 + r_2, \qquad v_2 = \frac{r_1}{r_1 + r_2}, \qquad v_3 = \frac{r_1}{r_2}.$$

Exercise 1.15 Bernoulli random variables

Show that the sum of Bernoulli random variables is distributed according to a binomial distribution; see Appendix B for the description of a binomial distribution.

Exercise 1.16 The Gibbs inequality

Consider two categorical probability distributions over the same categories $\rho_{1:M}$, one with parameters $\pi_{\rho_{1:M}}$ and the other with parameters $\pi'_{\rho_{1:M}}$. Prove the Gibbs inequality

$$-\sum_{m=1}^{M} \pi_{\rho_m} \log \frac{\pi'_{\rho_m}}{\pi_{\rho_m}} \geq 0.$$

Hint: Use the fact that $\log x \leq x - 1$ for $x > 0$.

Exercise 1.17 Poisson convolution

Show that the sum of Poisson random variables remains a Poisson random variable. Hint: If necessary, see Appendix B for the description of a Poisson distribution.

Exercise 1.18 Manipulating transformed densities

Consider iid random variables R_1, R_2, R_3 with a common density $p(r)$. Assume R_1, R_2, R_3 are positive real scalar random variables. Find, in terms of $p(r)$, the probability that the polynomial $r_1 x^2 + r_2 x + r_3$ has real roots.

Exercise 1.19 A fair dice

Use the fundamental theorem of simulation to simulate a roll of a fair dice. Generate several rolls and verify that indeed the dice simulated is fair.

Exercise 1.20 The Weibull distribution

Consider a Weibull random variable X. This variable takes real scalar values and its probability density reads as

$$p(x) = \frac{\alpha}{\beta} \left(\frac{x}{\beta} \right)^{\alpha - 1} e^{-\left(\frac{x}{\beta}\right)^{\alpha}}$$

for appropriate values of α and β.

1. Describe and implement an algorithm that uses the fundamental theorem of simulation to simulate the random variable X.
2. Use your algorithm to generate a large number of random realizations x of X.
3. Construct a histogram of your output and compare it to the analytic form of the Weibull density.

Exercise 1.21 Label invariance of the fundamental theorem of simulation

1. Apply the fundamental theorem of simulation to simulate draws from $\text{Categorical}_{\rho_1,\rho_2,\rho_3}(\pi_{\rho_1}, \pi_{\rho_2}, \pi_{\rho_3})$.
2. Verify that ρ_1, ρ_2, ρ_3 are realized with probabilities $\pi_{\rho_1}, \pi_{\rho_2}, \pi_{\rho_3}$, respectively.
3. Apply a relabeling of ρ_1, ρ_2, ρ_3 and verify that the fundamental theorem of simulation keeps yielding realizations with the correct probabilities.

Exercise 1.22 Normal random variables

1. Implement the Box–Muller algorithm of Note 1.10 and generate a large number of $\text{Normal}(\mu, v)$ random values.
2. Use your generated values to construct histograms and verify that your implementation yields the correct statistics.

Exercise 1.23 Poisson random variables

1. Show that if the time between successive events is exponentially distributed, with rate λ, then the number of events expected within any time interval, T, is distributed according to $\text{Poisson}(\mu)$ where $\mu = T\lambda$.
2. Develop a method to draw samples from a Poisson distribution by leveraging the fact that the time between events is exponentially distributed. Repeat the exercise for various values of λ, construct a histogram of your samples and compare your histogram to the coinciding Poisson density.

Exercise 1.24 A loaded dice

A dice is rolled 120 times yielding the results:

face	"1"	"2"	"3"	"4"	"5"	"6"
number of appearances	15	34	18	19	19	15.

Reason, based on likelihoods, that the dice is loaded.

Exercise 1.25 Random variable convolutions: the instrument response function

Consider a single photon detector recording exponentially distributed photon inter-arrival times, with rate λ. Detector electronics add a stochastic delay to the photon detection time distributed according to a normal distribution with mean μ and variance v. Find the resulting probability distribution over photon detection times.

Project 1.1 The point spread function in fluorescence microscopy

In fluorescence microscopy, photons are detected at positions that differ probabilistically from the point at which they are emitted. In particular, under ideal imaging conditions each photon emitted from a position $(x_\star, y_\star)$ is detected independently of the other photons at a position (X, Y) randomly distributed according to the *Airy probability density*

$$p(x,y) = \frac{4\pi n_\alpha^2}{\lambda^2}\left(\frac{J_1\left(\frac{2\pi N_A}{\lambda}\sqrt{(x-x_\star)^2+(y-y_\star)^2}\right)}{\frac{2\pi N_A}{\lambda}\sqrt{(x-x_\star)^2+(y-y_\star)^2}}\right)^2,$$

where λ is the photon's wavelength, N_A is the microscope's numerical aperture, and $J_1(\cdot)$ is the 1st Bessel function of the first kind. Typical values for the parameters are $\lambda = 510$ nm and $N_A = 1.40$.

1. Verify that the probability density $p(x, y)$ is properly normalized.
2. Apply a transformation from Cartesian to polar coordinates and change the photon detection position from (X, Y) to radius and azimuth (R, Φ) relative to $(x_\star, y_\star)$.
3. Use the Airy density of (X, Y) to derive the density of (R, Φ).
4. Verify that the Airy density is radially symmetric.
5. Evaluate the Airy density at a set of fixed grid point at radii $r_{1:M}$.
6. Use your tabulation of the Airy density at these radii and numerical integration to approximate the CDF $C(r_m)$ at the grid's radii $r_{1:M}$.
7. Use interpolation to approximate the CDF $C(r)$ at radii between $r_{1:M}$.
8. Use the fundamental theorem of simulation and your interpolated $C(r)$ to simulate the detection position of a large number of photons.
9. Summarize your simulated positions in a histogram and verify that your implementation produces photon detections from the correct distribution.

Project 1.2 EMCCD signal amplification

Light detectors based on *electron multiplication charge coupled devices* (EMCCD) are widely used in both telescope and microscope cameras. EMCCDs perform well under low light conditions as they amplify the detected signal. Signal amplification by an EMCCD is modeled as follows:

- A light source emits a Poisson distributed number of photons n_ϕ that strike the detector at a rate λ for a period τ_{exp}.
- Each of these photons may induce the transport of an electron into the electron multiplication (EM) register with probability q independently of the other photons resulting in a total of n_e electrons transported into the EM register.
- The electrons in the EM register are subsequently multiplied through an electron cascade process, which is well approximated by a gamma distribution with shape n_e and scale G, outputting n_o electrons that are transferred to the analog-to-digital (A/D) converter.

- Due to thermal noise, the final readout w, resulting from A/D conversion, is normally distributed around n_o.

The full generative model is provided below:

$$N_\phi|\lambda \sim \text{Poisson}\left(\lambda\tau_{\text{exp}}\right),$$
$$N_e|n_\phi, q \sim \text{Binomial}\left(qn_\phi\right),$$
$$N_o|n_e, G \sim \text{Gamma}\left(n_e, G\right),$$
$$W|n_o, v \sim \text{Normal}\left(n_o, v\right).$$

1. Derive the joint distribution over N_e and N_ϕ.
2. Derive $p(w, n_\phi, n_e, n_o|\lambda, q, G, v)$ and, from this density, compute $p(w|\lambda, q, G, v)$ by marginalization.
3. Explain your reasoning in words, justifying which of the two likelihoods above you would hypothetically maximize in order to determine λ.

Additional Reading

C. Bishop. *Pattern recognition and machine learning*. Springer, 2006.

D. S. Sivia, J. Skilling. *Data analysis: a Bayesian tutorial*. Oxford University Press, 2006.

J. A. Rice. *Mathematical statistics and data analysis*. 3rd ed. Duxbury Press, 2007.

L. Wasserman. *All of statistics: a concise course in statistical inference*. Reprint. Springer, 2005.

M. Hirsch, R. J. Wareham, M. L. Martin-Fernandez, M. P. Hobson, D. J. Rolfe. A stochastic model for electron multiplication charge-coupled devices – from theory to practice. *PLoS One*, 8:e53671, 2013.

2 Dynamical Systems and Markov Processes

By the end of this chapter, we will have presented

- *Systems with stochastic dynamics*
- *The Markov property*
- *Common models of dynamical systems*

In this chapter we focus on models of dynamical systems that we will be using in subsequent chapters. For clarity, we present specific, tractable examples that give rise to the development of Markov processes.

Note 2.1 Processes

Before getting into details, we want to clarify a subtlety in the terminology adopted in the literature as it pertains to dynamics. In mathematics and statistics, the term *process*, most often an abbreviation for a *random process*, has a *very particular and technical* meaning. It is used to designate a *collection* of random variables that is, most commonly, *infinite*.

Although intuitive in its abstraction, the term process might be misleading in the context of dynamical systems as it conflicts with the way the word process is used in everyday language in the natural sciences. For example, in everyday language a process preassumes some degree of temporal arrangement and, as such, entails some sense of causality. However, a temporal arrangement, or any structure at all, is most often absent in the technical sense.

In this chapter we come across processes in the second, less technical, capacity that might not necessarily coincide with the mathematical one. Following this convention, a process here is nothing more than a physical effect that extends over time. Of course, as we will see shortly, such effects are naturally formulated mathematically with collections of random variables and, for this reason, our formulations become random processes in the technical capacity as well.

2.1 Why Do We Care about Stochastic Dynamical Models?

The systems we consider are *stochastic*. That is, our systems are influenced by various random events, and their time-courses are random as well. As with all other random quantities, our systems' time-courses are sampled from appropriate probability distributions. This is in contrast to deterministic systems, for example dictated by Newtonian dynamics, where their courses are certain and sampling is unnecessary. Our main objective, for now, is to introduce such distributions and to do so we develop appropriate generative models. However, before we jump into the fine details, we ask: *Why do we care about stochastic models?*

One answer to this question is that stochastic models are genuinely interesting and important as they often exhibit behavior *different* from their deterministic counterparts. For example, genes in a living cell may be either actively transcribing or suppressed and stochastic models may capture the random toggling between both gene states. A deterministic model, such as one developed on mean-field or mass-action principles, can instead capture only an "average" behavior over time of a gene. This fails to provide insight into the fact that the gene is either active or not at any given time instant. While receiving much attention, arguments such as these on the inadequacy of averages are only a small part of why we need stochastic models.

Another, more general, reason for studying stochastic systems is because measurement noise adds uncertainty, which invariably introduces stochasticity. Thus, stochastic models are required to quantify the uncertainty introduced not only by the random events affecting the system dynamics, but also by the unavoidable measurement noise. As we will see shortly, stochastic models can be translated into probabilities of sequences (termed likelihoods), or otherwise interrelated events, that are the starting point for quantitative data-driven analyses.

Note 2.2 Signal processing

The analysis of a dynamical system's time-course is sometimes termed *signal processing*. From our perspective, a signal is a trajectory of a stochastic system. In practice, we encounter four types of signal:

discrete courses studied in discrete time	discrete courses studied in continuous time
continuous courses studied in discrete time	continuous courses studied in continuous time

2.2 Forward Models of Dynamical Systems

Most often, we use *state variables* to describe a dynamical system and its trajectory. The term *state* designates the properties or features of the system that evolve in time. The precise physical meaning of these features, of course, depends on the specifics of the system at hand. For example,

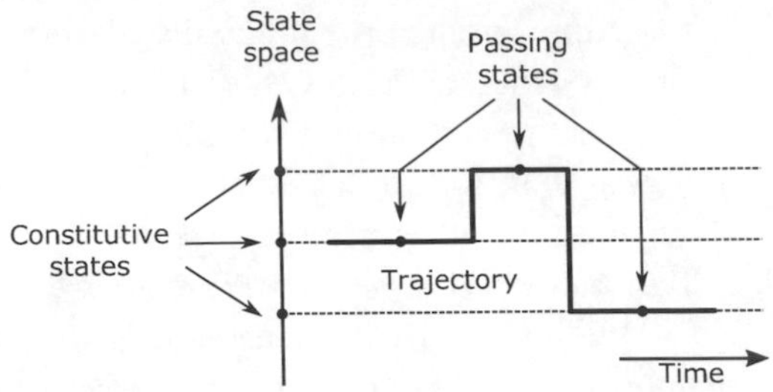

Fig. 2.1 A generic description of a dynamical system. For clarity, on the vertical axis the state-space is shown as having one dimension; however, a state-space may more generally have any number of dimensions.

when studying a cell culture, our state may simply be the total population of cells in the culture. Alternatively, when studying particle motion, such as particle diffusion in solution, our state may be the particle's position with respect to some frame of reference.

For any given system, all possible values that may be attained by its state are termed the *state-space*. For example, for a cell culture, in which we keep track of cell population, the state-space consists of all nonnegative integers. Similarly, for a diffusing particle, the state-space consists of every point in the volume available to this particle.

As our system evolves in time, the value of its state changes. Put differently, a system's state moves across the state-space. Monitored over a period of time, successive state values form the system's *trajectory*; see Fig. 2.1. State positions within the state-space may be revisited from time to time and so, in general, a trajectory may attain and then reattain the same values multiple times. As a system may visit only a portion of its entire state-space, however, some state positions within the state-space may be absent from or remain unexplored within a single trajectory.

Note 2.3 Distinction between constitutive and passing states

When we describe a dynamical system, it is essential to distinguish between a state that is an element of the system's state-space and a state that is an element of its trajectory. For such subtle cases, we will refer to the former as *constitutive* and the latter as *passing* states. See Fig. 2.1.

The distinction between these two is made clearer if we consider that two different trajectories that the system may follow consist of the *same* constitutive states; however, they contain *different* passing states. As we will see shortly, mathematically we represent the passing states as random variables and the constitutive states as the specific values that these random variables may attain.

The distinction between a constitutive state and a passing state is made clearer through Example 2.1.

Example 2.1 A cell culture as a dynamical system

Suppose that we are interested in studying how a population of cells in a culture changes over time and that we assess the culture at regular time intervals. In this setting, we consider the cell number measured at each time point as our system's state. Therefore, the state-space consists of the nonnegative integers $0, 1, 2, \ldots$. In other words, the constitutive states are $0, 1, 2, \ldots$.

Further, suppose that s_n denotes the population measured at the nth assessment. In this case, our system's trajectory is $(s_1, s_2, \ldots)$ and, following our convention, each of $s_1, s_2, \ldots$ is a passing state.

Different cell cultures, for example grown in separate petri dishes, have the *same* constitutive states; however, because cell division happens at random times, different cell cultures generally follow *different* trajectories. Accordingly, although the constitutive states are the same for all cultures, the passing states at any given time may differ from culture to culture.

In subsequent chapters we will see how to study systems where the state is only *indirectly* observed. For example, the population s_n of a cell culture is often assessed by measuring the surface area of the petri dish covered by the cells. However, because cells may grow over one another or because of the variability in the surface area covered by a single cell, the assessment of the culture's population s_n carried out this way is generally inaccurate. In such cases, we may speak of s_n as being a *hidden* or *latent* state and we will learn how to formulate and analyze models with hidden passing states.

Depending on the context and nature of the state-space, a system's trajectory may consist of *discrete* or *continuous* values. For example, the population of cells in a growing culture is discrete while the position of a diffusing particle is continuous. However, despite the values of their trajectories, all natural systems, including most physical and chemical ones, evolve *continuously in time*. This is certainly true of a classical mechanical system whose state (*i.e.*, positions and momenta) evolves according to Newton's equations and is equally true of a quantum system whose state (*i.e.*, wavefunction) evolves according to the Schrödinger equation. Nevertheless, as an idealization to continuous time evolution, it is sometimes advantageous to model systems and their states as evolving *discretely in time* as well.

Example 2.2 Discrete and continuous time evolution

The position $\boldsymbol{r}(t) = (x(t), y(t), z(t))$ of a diffusing particle in solution is well defined for every time instance t. Thus, we say that the system's state, *i.e.*, the particle's position, evolves continuously; see Fig. 2.2. However, positions may be assessed only at discrete time points $t_1, t_2, \ldots$. It may therefore be sufficient to consider a discrete time evolution where we only keep track of $\boldsymbol{r}(t_1), \boldsymbol{r}(t_2), \ldots$ which, for simplicity, we may denote $\boldsymbol{r}_1, \boldsymbol{r}_2, \ldots$. In the latter setting, the position at times *between* the time points $t_1, t_2, \ldots$ may remain undefined. Idealizations leading to time discretization are sometimes preferable as they facilitate the development of mathematically compact formulations.

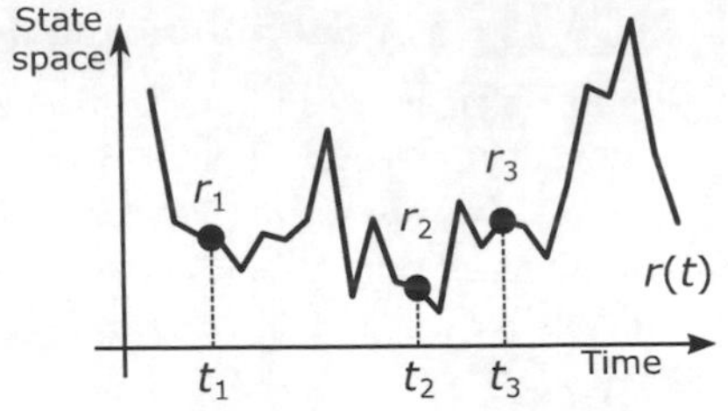

Fig. 2.2 Although the state $r(t)$ of a system may change continuously over time, we may model the system only at discrete times $t_1, t_2, t_3, \ldots$

A system's dynamical properties depend on whether the system has a discrete or continuous state-space as well as whether it evolves in discrete or continuous time. All four possibilities, now concretely reformulated from Note 2.2 using state-space language, include: systems with discrete state-spaces evolving in continuous time; systems with discrete state-spaces evolving in discrete time; systems with continuous state-spaces evolving in discrete time; and systems with continuous state-spaces evolving in continuous time. All cases are of practical interest and, due to their unique characteristics, in the following sections we present each one separately.

Note 2.4 Description of a dynamical system

Irrespective of the details, a description of a dynamical system that is appropriate for quantitative analysis must specify:

- the state-space; *i.e., what are the system's states?*
- the initialization rule; *i.e., where does the system start?*
- the transition rules; *i.e., how does the system evolve?*

In subsequent chapters, where we will encounter systems with hidden states, we will have to include an additional feature:

- the assessment rules; *i.e., how is the system's state related to the measurements?*

2.3 Systems with Discrete State-Spaces in Continuous Time

By discrete system we mean a system whose constitutive states are separated from each other and do *not* form a continuum. Since the system's state-space is discrete, we may universally denote the constitutive states with σ_m labeled $m = 1{:}M$, where M is the size of the state-space.

Example 2.3 Examples of discrete systems

The analysis of imaging experiments often requires models of fluorescing molecules or *fluorophores* as they are commonly termed. In the simplest of cases, fluorophores can be modeled as attaining $M = 2$ states, namely:

- σ_1 = light emitting (bright) state.
- σ_2 = light non-emitting (dark) state.

More detailed models may demand a greater number of constitutive states. For example, the state-space could be as follows:

- σ_1 = bright state.
- σ_2 = short-lasting dark state.
- σ_3 = long-lasting dark state.
- σ_4 = permanent dark state.

Here, $M = 4$. The latter case may more faithfully model real fluorophores as it accounts for multiple dark states, capturing the phenomenon of photo-switching, and a permanent dark state that captures a phenomenon termed photo-bleaching.

As another example, large biological molecules, such as proteins, can attain multiple conformational states approximated as discrete on measurement timescales. For a protein undergoing transitions between folded and unfolded states, the state-space may be modeled as:

- σ_1 = folded.
- σ_2 = unfolded.
- σ_3 = partially folded.

Biology may require additional states. In such cases, extra states may be recruited to more faithfully model the protein.

Note 2.5 A labeling convention

The labels we use to distinguish the constitutive states, σ_m, of a discrete system are a mere convention and, generally, carry no particular meaning. In fact, instead of $\sigma_1, \sigma_2, \ldots$ we could very well chose a convention that does not rely on numerical labels. For example, we could denote the constitutive states with $\alpha, \beta, \ldots$ or even $\spadesuit, \clubsuit, \ldots$ and still develop a framework identical to that described.

We emphasize that we denote the states with $\sigma_1, \sigma_2, \ldots$ and adopt *numerical* indices $1, 2, \ldots$ for typographical reasons only. Unfortunately, in subtle situations, such numerical labels may be misleading as: they may suggest that the underlying state-space has an arrangement that, in general, is not needed; and shift the attention to the labels m rather than the states σ_m, which are the primary objects of interest.

2.3.1 Modeling a System with Discrete Events

The trajectory of a system with a discrete state-space in continuous time, as shown in Fig. 2.3, consists of *phases* or *epochs* during which the system

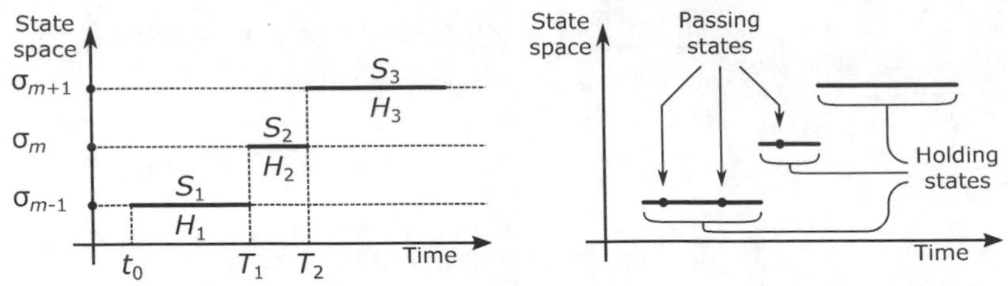

Fig. 2.3 A system with a discrete state-space following continuous time dynamics. This system is described by holding states S_n and either jump times T_n or holding periods H_n.

occupies the same constitutive states. The phases start and end precisely at the times at which the system switches from one constitutive state to another. For this reason, the system is fully described by the sequence of states attained in each phase $S_1, S_2, S_3, \ldots$ and the sequence of times $T_1, T_2, \ldots$ at which the system jumps from state to state. Specifically, we use T_n to denote the time at which the system jumps from S_n to S_{n+1} and, for this reason, call it the *jump time*. As $S_1, S_2, S_3, \ldots$ and $T_1, T_2, \ldots$ are random variables, we need to describe the probability distributions from which they are sampled.

Our description is simplified if we adopt variables S_n, H_n rather than S_n, T_n as shown in Fig. 2.3. In this description, S_n is the random variable representing the system's state *before* the nth jump and H_n is the random variable representing the period of time for which our system stays in S_n. From now on, we will refer to H_n as a *holding period* and S_n as its coinciding *holding state*. As we can see from Fig. 2.3, jump times and holding periods are related by

$$T_n = t_0 + H_1 + \cdots + H_n = t_0 + \sum_{n'=1}^{n} H_{n'}, \tag{2.1}$$

where t_0 is a reference time that is, typically, not random. From Eq. (2.1), we can easily deduce holding periods from jump times and vice versa.

Note 2.6 What is the trajectory?

A system's trajectory, with discrete state-space evolving in continuous time, is a function of time. We may denote this function with $\mathcal{S}(\cdot)$ and use functional notation to avoid any confusion that might be caused with the passing state $\mathcal{S}(t)$ evaluated at a particular time instant t.

According to Fig. 2.4, the trajectory or its passing states are given piecewise by

$$\mathcal{S}(t) = \begin{cases} S_1, & t_0 \le t < T_1 \\ S_2, & T_1 \le t < T_2 \\ \ldots \end{cases} \quad \text{or} \quad \mathcal{S}(t) = \begin{cases} S_1, & t_0 < t \le T_1 \\ S_2, & T_1 < t \le T_2 \\ \ldots \end{cases}.$$

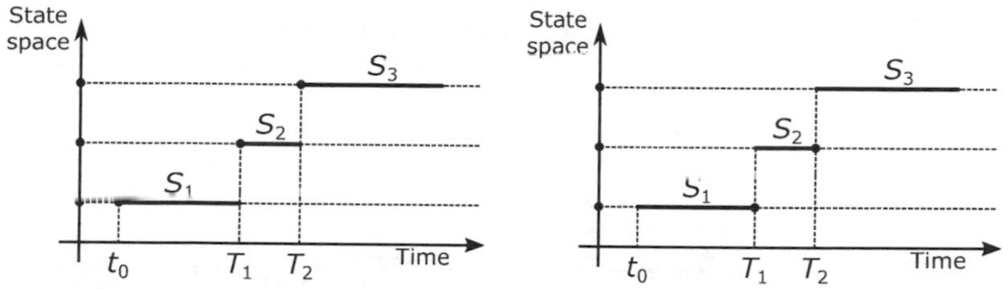

Fig. 2.4 Examples of càllàl trajectories: the trajectory on the left is a càdlàg; while the trajectory on the right is a càglàd.

Both of these are special cases of functions termed càllàl from a French acronym for *"continuous from one side with limit from the other."* In particular, functions like $\mathcal{S}(\cdot)$ on the left are termed càdlàg for *"continuous on the right with limits on the left,"* and functions like $\mathcal{S}(\cdot)$ on the right are termed *càglàd for "continuous on the left with limits on the right."*

For the examples we will see in this chapter, which side (left or right) we place the equality on the jump times will be unimportant. However, in subtle modeling cases, where multiple dynamical processes evolve simultaneously, such a choice may have important consequences. When it is necessary to emphasize the distinction in schematic diagrams, as we demonstrate in Fig. 2.4, we place filled dots at the jump times on the appropriate state.

Our statistical formulation of the system's trajectory begins with the sequence of holding states $S_1, S_2, S_3, \ldots$. In this sequence, the initial state S_1, which is unaffected by the system's transitions at later times, is sampled separately from the future holding states. As there are only discretely many choices for S_1, in general, our initialization of the system is encoded in

$$S_1 \sim \text{Categorical}_{\sigma_{1:M}}(\boldsymbol{\rho}), \tag{2.2}$$

where the probability array $\boldsymbol{\rho} = [\rho_{\sigma_1}, \ldots, \rho_{\sigma_M}]$ provides initial weights for particular constitutive states.

Causality suggests that sampling of a subsequent holding state may be affected only by preceding states. Under the *Markov assumption*, we make the more drastic assumption that holding states may only be affected by their immediate predecessor. As such, the system's transitions described by

$$S_{n+1}|s_n \sim \text{Categorical}_{\sigma_{1:M}}(\boldsymbol{\pi}_{s_n}) \tag{2.3}$$

capture the evolution rules of what is sometimes termed a *Markov process*. Here, the probability array $\boldsymbol{\pi}_{s_n} = [\pi_{s_n \to \sigma_1}, \ldots, \pi_{s_n \to \sigma_M}]$ provides weights for particular transitions out of s_n. Since there are M constitutive states in the system's state-space, the transition rules entail M different probability arrays $\boldsymbol{\pi}_{\sigma_m} = [\pi_{\sigma_m \to \sigma_1}, \ldots, \pi_{\sigma_m \to \sigma_M}]$. As the holding states change at every jump time, these arrays must exclude self-transitions. Accordingly, our $\boldsymbol{\pi}_{\sigma_m}$ need to have $\pi_{\sigma_m \to \sigma_m} = 0$.

Together, Eqs. (2.2) and (2.3) entail $M+1$ probability arrays each consisting of M scalar values. Gathered in these arrays, we have M initial weights

ρ_{σ_m} coupled to a normalization constraint $\sum_m \rho_{\sigma_m} = 1$. Similarly, we have M^2 transition weights $\pi_{\sigma_m \to \sigma_{m'}}$ coupled to M normalization constraints $\sum_{m'} \pi_{\sigma_m \to \sigma_{m'}} = 1$ and M weights fixed to zero. In total, these leave $M^2 - M - 1$ free weights ρ_{σ_m} and $\pi_{\sigma_m \to \sigma_{m'}}$ that we may choose in order to adjust our model's dynamics.

Note 2.7 Transition probability matrix

It is often convenient to tabulate the state-space and the transition probability arrays as

$$\begin{array}{c} \\ \sigma_1 \\ \vdots \\ \sigma_M \end{array} \begin{array}{c} \begin{array}{ccc} \sigma_1 & \cdots & \sigma_M \end{array} \\ \begin{bmatrix} \pi_{\sigma_1 \to \sigma_1} & \cdots & \pi_{\sigma_1 \to \sigma_M} \\ \vdots & \ddots & \vdots \\ \pi_{\sigma_M \to \sigma_1} & \cdots & \pi_{\sigma_M \to \sigma_M} \end{bmatrix} \end{array} = \mathbf{\Pi}.$$

In this tabulation, all individual probability arrays out of one state are encoded in rows of $\mathbf{\Pi}$ commonly termed the *transition probability matrix*.

To avoid confusion with similar notions introduced later, it is useful to summarize some important characteristics of a transition probability matrix:

- It is a square matrix of size equal to the size of the state-space.
- Its row and column elements are arranged with the ordering of the state-space.
- It gathers unitless parameters.
- Each row must sum to 1, *i.e.*, it is normalized to the unit row sum.
- Its diagonal elements are 0.

Sampling of the holding states $S_1, S_2, \ldots$ under the Markov assumption offers little modeling flexibility besides choosing the probability arrays ρ and π_{σ_m}; however, sampling of the holding periods $H_1, H_2, \ldots$ is open to a variety of modeling choices that we will explore below. As each H_n attains positive scalar values, the only universal requirement for the sampling distributions we adopt in our model is to allow exclusively positive values.

Example 2.4 The birth process

Here, we begin by defining a creation process, akin to particle creation in physics, in which a source is responsible for the generation of new members of a population. This process is often termed a *birth process.* In particular, with a birth process we model the creation events of elements of some species $\mathcal{A}$ and identify the state-space with the integers $0, 1, 2, \ldots$. The state s_n defines the population of species $\mathcal{A}$ just before the nth birth event. In other words, the size of the state-space is infinite, $M = \infty$, and the constitutive states are $\sigma_m = m - 1$, for $m = 1, 2, \ldots$.

Typically, in a birth process we initially have k elements of $\mathcal{A}$. Subsequently, one element is added following each birth event. Under these conditions, the holding states read

$$s_n = k + n - 1, \qquad n = 1, 2, \ldots.$$

Although the holding states in a birth process are specified deterministically, we can also express them within our probabilistic framework. In particular, Eqs. (2.2) and (2.3) for birth processes become

$$S_1 \sim \text{Categorical}_{0,1,2,\ldots}(\rho),$$
$$S_{n+1}|s_n \sim \text{Categorical}_{0,1,2,\ldots}(\pi_{s_n}),$$

where the transition probability arrays are given by the matrix

$$\begin{array}{c} \\ 0 \\ 1 \\ 2 \\ \vdots \end{array} \begin{array}{c} \begin{array}{ccccc} 0 & 1 & 2 & 3 & \cdots \end{array} \\ \begin{bmatrix} 0 & 1 & 0 & 0 & \cdots \\ 0 & 0 & 1 & 0 & \cdots \\ 0 & 0 & 0 & 1 & \cdots \\ \vdots & \vdots & \vdots & \vdots & \ddots \end{bmatrix} \end{array} = \mathbf{\Pi},$$

and, initiating with $k = 0$ elements, the initial probability array is given by

$$\begin{array}{c} \begin{array}{ccccc} 0 & 1 & 2 & 3 & \cdots \end{array} \\ \begin{bmatrix} 1 & 0 & 0 & 0 & \cdots \end{bmatrix} \end{array} = \rho.$$

In a birth process, the holding periods are iid and modeled by

$$H_n \sim \text{Exponential}(\lambda_b),$$

with a common birth rate λ_b.

Dynamical systems such as in Example 2.4 are known as *renewal processes*. In these systems each holding period H_n is sampled from the same distribution independently of the others. In other words, the variables $H_1, H_2, \ldots$ are iid. The exponential distribution of Example 2.4 is a very special case. In general, the particular form for the sampling distribution differs from problem to problem. For instance, in Example 2.5, sampling of each holding period H_n depends on its holding state s_n; unlike a renewal process where such dependence is disallowed. Such more general dynamical systems are special cases of stochastic processes known as *Markov renewal processes*. In this case, the holding periods $H_1, H_2, \ldots$ remain independent, similar to a renewal process; however, only holding periods associated with holding states in the same constitutive state are identically distributed.

Example 2.5 The death process

With a death process, we model the annihilation events of elements of species $\mathcal{A}$. In a death process, similar to a birth process, the state-space is identified with the integers $0, 1, 2, \ldots$. The state s_n designates the population of species $\mathcal{A}$ before the nth death event.

As with birth processes, in a death process we initially have k elements of $\mathcal{A}$ with the understanding that one element is removed following each death event. Under these conditions, the holding states are specified

$$s_n = k - n + 1, \qquad n = 1, 2, \ldots, k.$$

As a population cannot be negative, the death process terminates at the kth event. Just as with birth processes, there is no need to sample the state as the subsequent state is deterministic.

In a death process, the holding periods are independently sampled from an exponential distribution

$$H_n | s_n \sim \text{Exponential}\left(\lambda_{s_n}\right),$$

the rate of which depends on the holding states $\lambda_s = s\lambda_d$ where λ_d is the death rate.

Sampling each holding period using a rate proportional to the current population ($\lambda_s = s\lambda_d$) indicates that each element of $\mathcal{A}$ is annihilated due to a cause that is independent of the causes of the other elements. This result is derived from the understanding that the rate of each holding period for s elements is obtained from the minimum holding period for each of these elements treated separately; see Appendix D.

Birth and death processes can be combined to model a realistic scenario involving simultaneous creation and annihilation events. We study this scenario within the unified framework of Markov jump processes. These, as we see next, are used to model systems with exclusively exponentially distributed holding periods.

2.3.2 Markov Jump Processes

Dynamical systems known as *Markov jump processes* are a specific kind of Markov renewal process most commonly encountered when modeling physical, chemical, and biological systems. Due to their prominence, we discuss their properties in detail.

Modeling Systems without Memory

As Markov jump processes are examples of Markov renewal processes, each holding period H_n depends on its holding state s_n. In Markov jump processes, however, holding periods are sampled from *memoryless* distributions. That is, the time period until the system jumps out of a passing state is independent of the time already spent in this state. As shown in Appendix D, this means that the holding periods are sampled according to

$$H_n | s_n \sim \text{Exponential}\left(\lambda_{s_n}\right). \tag{2.4}$$

In general, with M constitutive states a Markov jump process entails M rates $\lambda_{\sigma_1}, \ldots, \lambda_{\sigma_M}$. Since these rates determine when a holding period ends, we call them *escape rates*.

Combining the birth and death processes of Examples 2.4 and 2.5 yields *birth-death processes*, which are important examples of Markov jump processes.

Example 2.6 The birth-death process

With a birth-death process, we model the creation and annihilation events of elements of some species $\mathcal{A}$. In a birth-death process, the state-space is identified by integers $0, 1, 2, \ldots$ and each state S_n with the population of species $\mathcal{A}$ before the nth event, which now combines both births and deaths. In other words, the size of the state-space is infinite, $M = \infty$, and the constitutive states are $\sigma_m = m - 1$, for $m = 1, 2, \ldots$.

In a birth-death process, we initially have k elements of species $\mathcal{A}$. Subsequently, one element is added or removed depending on whether a birth or death event was sampled, respectively. Unlike with pure birth or pure death processes, the holding states S_n cannot be expressed deterministically. Nevertheless, Eqs. (2.2) and (2.3) remain valid

$$S_1 \sim \text{Categorical}_{0,1,2,\ldots}(\rho),$$
$$S_{n+1}|s_n \sim \text{Categorical}_{0,1,2,\ldots}(\pi_{s_n}),$$

where the transition probability arrays are given by the matrix

$$\begin{array}{c} \\ 0 \\ 1 \\ 2 \\ 3 \\ \vdots \end{array} \begin{array}{c} \begin{array}{ccccc} 0 & 1 & 2 & 3 & \cdots \end{array} \\ \begin{bmatrix} 0 & p_0 & 0 & 0 & \cdots \\ q_1 & 0 & p_1 & 0 & \cdots \\ 0 & q_2 & 0 & p_2 & \cdots \\ 0 & 0 & q_3 & 0 & \cdots \\ \vdots & \vdots & \vdots & \vdots & \ddots \end{bmatrix} \end{array} = \mathbf{\Pi},$$

with individual probabilities $p_s = \lambda_b/(s\lambda_d + \lambda_b)$ and $q_s = s\lambda_d/(s\lambda_d + \lambda_b)$. Here, λ_b and λ_d are the birth and death rates, respectively.

In a birth-death process, the holding periods are independently sampled from an exponential distribution

$$H_n|s_n \sim \text{Exponential}(\lambda_{s_n}),$$

the rate of which depends on the holding states $\lambda_s = s\lambda_d + \lambda_b$; see Appendix D.

Reparametrizing the Markov Jump Process*

As we have seen, to fully describe a Markov jump process on a state-space of size M, we need $M^2 - 2M$ scalar values $\pi_{\sigma_m \to \sigma_{m'}}$ to specify the transition probabilities that characterize the sequence of holding states $S_1, S_2, \ldots$ and M rates λ_{σ_m} that characterize the sequence of holding periods $H_1, H_2, \ldots$. The $M^2 - 2M$ originates from the total number of matrix elements $\pi_{\sigma_m \to \sigma_{m'}}$ minus the normalization conditions and self-transitions on each row of $\mathbf{\Pi}$.

* This is an advanced topic and could be skipped on a first reading.

Despite its mathematical convenience, however, a representation containing parameters $\pi_{\sigma_m \to \sigma_{m'}}$ and λ_{σ_m} is not always the most physically interpretable. This is because, in this scheme, the sampling of S_n and H_n is decoupled, suggesting that our system selects separately how long it stays in a state and the state to which it jumps. However, from a physical point of view, these two events are better understood as a single event.

Markov jump processes have an *equivalent* parametrization that more accurately reflects the selection event. In particular, since our system has M constitutive states, at any given time there are $M - 1$ constitutive states to which the system may jump. Accordingly, starting from a holding state s_n we can envision $M - 1$ concurrent *reactions* that compete to attract the system each to a different σ_m. Provided these reactions are memoryless, following Appendix D, each one occurs after a period that is exponentially distributed with some rate that we denote $\lambda_{s_n \to \sigma_m}$. From this viewpoint, the reaction that occurs *first* determines the next holding state s_{n+1} as well as the holding period h_n. Of course, consistency requires $\lambda_{s_n \to s_n} = 0$, but otherwise the remaining $\lambda_{s_n \to \sigma_m}$ are free parameters.

In this representation, the transitions of a Markov jump process are fully specified by $M^2 - M$ reaction rates $\lambda_{\sigma_m \to \sigma_{m'}}$ that can be formed between any possible pair of constitutive states. At first, this agrees with the total number of transition weights $\pi_{\sigma_m \to \sigma_{m'}}$ *and* escape rates λ_{σ_m} needed in the earlier representation; however, as we spell out next, the connection is deeper.

As s_{n+1} is the state that yields the minimum of the individual reaction periods, following Note D.1, it is selected out of the constitutive states with a probability proportional to the corresponding rates $\lambda_{s_n \to \sigma_m}$. Formally, this implies

$$S_{n+1}|s_n \sim \text{Categorical}_{\sigma_{1:M}}\left(\frac{\lambda_{s_n \to \sigma_1}}{\sum_m \lambda_{s_n \to \sigma_m}}, \ldots, \frac{\lambda_{s_n \to \sigma_M}}{\sum_m \lambda_{s_n \to \sigma_m}}\right). \tag{2.5}$$

Further, because the holding period h_n is the minimum of exponential periods, according to Note D.1, it follows that it is exponentially distributed itself with a rate equal to the sum of the individual reaction rates

$$H_n|s_n \sim \text{Exponential}\left(\sum_m \lambda_{s_n \to \sigma_m}\right). \tag{2.6}$$

Comparing Eq. (2.5) with Eq. (2.3) and Eq. (2.6) with Eq. (2.4), we immediately see that S_n and H_n are sampled from similar distributions and the parameters of these distributions are interrelated:

$$\pi_{\sigma_m \to \sigma_{m'}} = \frac{\lambda_{\sigma_m \to \sigma_{m'}}}{\sum_{m'} \lambda_{\sigma_m \to \sigma_{m'}}}, \qquad \lambda_{\sigma_m} = \sum_{m'} \lambda_{\sigma_m \to \sigma_{m'}}.$$

As such, we can readily convert from one representation of a Markov jump process to the other.

Note 2.8 Transition rate matrix

It is often convenient to tabulate the state-space and the reaction rates as

$$\begin{array}{c} \sigma_1 \\ \vdots \\ \sigma_M \end{array} \overset{\begin{array}{ccc} \sigma_1 & \cdots & \sigma_M \end{array}}{\begin{bmatrix} \lambda_{\sigma_1 \to \sigma_1} & \cdots & \lambda_{\sigma_1 \to \sigma_M} \\ \vdots & \ddots & \vdots \\ \lambda_{\sigma_M \to \sigma_1} & \cdots & \lambda_{\sigma_M \to \sigma_M} \end{bmatrix}} = \mathbf{\Lambda}.$$

In this tabulation, the reaction rates $\lambda_{\sigma_m \to \sigma_{m'}}$, which are nonnegative scalars, are placed in the off diagonals. Further, self-reaction rates $\lambda_{\sigma_m \to \sigma_m}$, which are by definition zero, are placed along the diagonal. Commonly, $\mathbf{\Lambda}$ is termed the *transition rate matrix*.

To avoid confusion with similar notions that we will introduce later on, we summarize important characteristics of the transition rate matrix:

- It is a square matrix of size equal to the size of the state-space.
- Its row and column elements are arranged with the ordering of the state-space.
- It gathers parameters with units of reciprocal time.
- Rows need *not* be normalized.
- Its diagonal elements are zero.

Example 2.7 **The switching process**

With a *switching process*, we model a system that cycles between a fixed number of constitutive states $\sigma_{1:M}$. For concreteness, here we use $M = 3$. In this setting, the transition rate matrix is

$$\mathbf{\Lambda} = \begin{bmatrix} 0 & \lambda_{\sigma_1 \to \sigma_2} & \lambda_{\sigma_1 \to \sigma_3} \\ \lambda_{\sigma_2 \to \sigma_1} & 0 & \lambda_{\sigma_2 \to \sigma_3} \\ \lambda_{\sigma_3 \to \sigma_1} & \lambda_{\sigma_3 \to \sigma_2} & 0 \end{bmatrix}.$$

Given $\mathbf{\Lambda}$, we can readily determine the escape rates

$$\lambda_{\sigma_1} = \lambda_{\sigma_1 \to \sigma_2} + \lambda_{\sigma_1 \to \sigma_3}, \quad \lambda_{\sigma_2} = \lambda_{\sigma_2 \to \sigma_1} + \lambda_{\sigma_2 \to \sigma_3}, \quad \lambda_{\sigma_3} = \lambda_{\sigma_3 \to \sigma_1} + \lambda_{\sigma_3 \to \sigma_2},$$

as well as the transition probability matrix

$$\mathbf{\Pi} = \begin{bmatrix} \pi_{\sigma_1 \to \sigma_1} & \pi_{\sigma_1 \to \sigma_2} & \pi_{\sigma_1 \to \sigma_3} \\ \pi_{\sigma_2 \to \sigma_1} & \pi_{\sigma_2 \to \sigma_2} & \pi_{\sigma_2 \to \sigma_3} \\ \pi_{\sigma_3 \to \sigma_1} & \pi_{\sigma_3 \to \sigma_2} & \pi_{\sigma_3 \to \sigma_3} \end{bmatrix} = \begin{bmatrix} 0 & \lambda_{\sigma_1 \to \sigma_2}/\lambda_{\sigma_1} & \lambda_{\sigma_1 \to \sigma_3}/\lambda_{\sigma_1} \\ \lambda_{\sigma_2 \to \sigma_1}/\lambda_{\sigma_2} & 0 & \lambda_{\sigma_2 \to \sigma_3}/\lambda_{\sigma_2} \\ \lambda_{\sigma_3 \to \sigma_1}/\lambda_{\sigma_3} & \lambda_{\sigma_3 \to \sigma_2}/\lambda_{\sigma_3} & 0 \end{bmatrix}.$$

Note 2.9 Notation

In statistical notation, we may abbreviate a Markov jump process as

$$\mathcal{S}(\cdot) \sim \mathrm{MJP}_{\sigma_{1:M}}(\rho, \mathbf{\Lambda}),$$

while we need to ensure that all necessary information, such as state-space and parametrization, has already been provided.

2.3.3 Structured Markov Jump Processes*

As we have seen already, Markov jump processes are used to model systems without memory and we have already presented the basic theory needed to formulate a simple memoryless system. Now, we consider the formulation of *composite* systems. Specifically, we consider systems formed by multiple subsystems, which we call *elements*. Each one of our elements is memoryless and so follows its own Markov jump process. As we will see, the composite system itself also forms a Markov jump process that, due to the additional structure inherited by the elements, exhibits properties that simple systems do not.

Composite Markov Jump Processes

To begin, we consider a case where our system is formed by J similar elements. In other words, each element follows a Markov jump process with common constitutive states and transition rates. That is, we model a *congruent* system defined as a system whose elements have the same state-space. To avoid confusion, we term the state-space of the single element *elementary state-space*.

When J elements are modeled together, they form a system whose holding states are best described by a J-tuple

$$\bar{s}_n = \left(s_n^1, \ldots, s_n^j, \ldots, s_n^J\right),$$

where s_n^j is the passing state of the jth element before the nth jump event; see Fig. 2.5. Of course, since in this description we consider the full system, n now counts jump events from *all* elements.

For clarity, we denote the composite state-space with $\bar{\sigma}_{1:\bar{M}}$ and the elementary one with $\sigma_{1:M}$. Since each element has M constitutive states, the state-space of the composite system has $\bar{M} = M^J$ states.

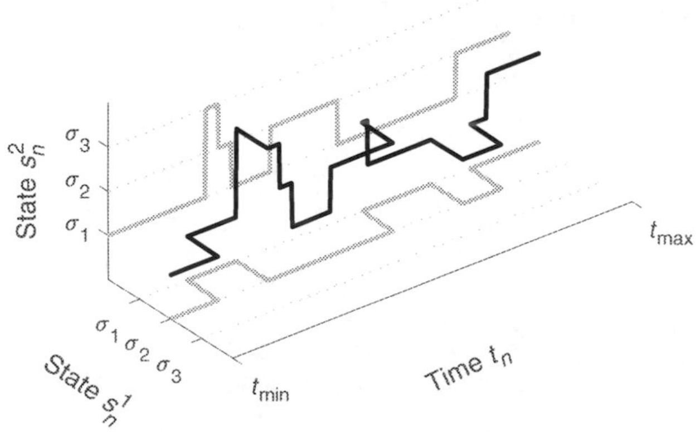

Fig. 2.5 A composite system that moves over a state-space with two components.

* This is an advanced topic and could be skipped on a first reading.

Example 2.8 The composite state-space and rate matrix

For instance, if $M = 3$, the constitutive states of each element are $\sigma_1, \sigma_2, \sigma_3$ and the transition rate matrix of each element is

$$\mathbf{\Lambda} = \begin{bmatrix} 0 & \lambda_{\sigma_1\to\sigma_2} & \lambda_{\sigma_1\to\sigma_3} \\ \lambda_{\sigma_2\to\sigma_1} & 0 & \lambda_{\sigma_2\to\sigma_3} \\ \lambda_{\sigma_3\to\sigma_1} & \lambda_{\sigma_3\to\sigma_2} & 0 \end{bmatrix}.$$

For the special case of $J = 2$ elements, the state-space of the composite system has size $\bar{M} = 9$ and consists of

$$\begin{aligned} &\bar{\sigma}_1 = (\sigma_1, \sigma_1), && \bar{\sigma}_2 = (\sigma_1, \sigma_2), && \bar{\sigma}_3 = (\sigma_1, \sigma_3), \\ &\bar{\sigma}_4 = (\sigma_2, \sigma_1), && \bar{\sigma}_5 = (\sigma_2, \sigma_2), && \bar{\sigma}_6 = (\sigma_2, \sigma_3), \\ &\bar{\sigma}_7 = (\sigma_3, \sigma_1), && \bar{\sigma}_8 = (\sigma_3, \sigma_2), && \bar{\sigma}_9 = (\sigma_3, \sigma_3). \end{aligned}$$

This is the *Cartesian product* $\bar{\sigma}_{1:9} = \sigma_{1:3} \times \sigma_{1:3}$ of the elementary state-spaces. In a composite state-space such as this one, it is customary to list the composite states $\bar{\sigma}_m$ in *lexicographical* order.

With a careful examination of all possible transitions $\bar{\sigma}_m \to \bar{\sigma}_{m'}$, we can see that the composite system is described by the rate matrix $\mathbf{\Lambda}$ that has the form

$$\begin{array}{c|ccc|ccc|ccc|} & \bar{\sigma}_1 & \bar{\sigma}_2 & \bar{\sigma}_3 & \bar{\sigma}_4 & \bar{\sigma}_5 & \bar{\sigma}_6 & \bar{\sigma}_7 & \bar{\sigma}_8 & \bar{\sigma}_9 \\ \bar{\sigma}_1 & 0 & \lambda_{\sigma_1\to\sigma_2} & \lambda_{\sigma_1\to\sigma_3} & \lambda_{\sigma_1\to\sigma_2} & 0 & 0 & \lambda_{\sigma_1\to\sigma_3} & 0 & 0 \\ \bar{\sigma}_2 & \lambda_{\sigma_2\to\sigma_1} & 0 & \lambda_{\sigma_2\to\sigma_3} & 0 & \lambda_{\sigma_1\to\sigma_2} & 0 & 0 & \lambda_{\sigma_1\to\sigma_3} & 0 \\ \bar{\sigma}_3 & \lambda_{\sigma_3\to\sigma_1} & \lambda_{\sigma_3\to\sigma_2} & 0 & 0 & 0 & \lambda_{\sigma_1\to\sigma_2} & 0 & 0 & \lambda_{\sigma_1\to\sigma_3} \\ \hline \bar{\sigma}_4 & \lambda_{\sigma_2\to\sigma_1} & 0 & 0 & 0 & \lambda_{\sigma_1\to\sigma_2} & \lambda_{\sigma_1\to\sigma_3} & \lambda_{\sigma_2\to\sigma_3} & 0 & 0 \\ \bar{\sigma}_5 & 0 & \lambda_{\sigma_2\to\sigma_1} & 0 & \lambda_{\sigma_2\to\sigma_1} & 0 & \lambda_{\sigma_2\to\sigma_3} & 0 & \lambda_{\sigma_2\to\sigma_3} & 0 \\ \bar{\sigma}_6 & 0 & 0 & \lambda_{\sigma_2\to\sigma_1} & \lambda_{\sigma_3\to\sigma_1} & \lambda_{\sigma_3\to\sigma_2} & 0 & 0 & 0 & \lambda_{\sigma_2\to\sigma_3} \\ \hline \bar{\sigma}_7 & \lambda_{\sigma_3\to\sigma_1} & 0 & 0 & \lambda_{\sigma_3\to\sigma_2} & 0 & 0 & 0 & \lambda_{\sigma_1\to\sigma_2} & \lambda_{\sigma_1\to\sigma_3} \\ \bar{\sigma}_8 & 0 & \lambda_{\sigma_3\to\sigma_1} & 0 & 0 & \lambda_{\sigma_3\to\sigma_2} & 0 & \lambda_{\sigma_2\to\sigma_1} & 0 & \lambda_{\sigma_2\to\sigma_3} \\ \bar{\sigma}_9 & 0 & 0 & \lambda_{\sigma_3\to\sigma_1} & 0 & 0 & \lambda_{\sigma_3\to\sigma_2} & \lambda_{\sigma_3\to\sigma_1} & \lambda_{\sigma_3\to\sigma_2} & 0 \end{array}.$$

In this matrix, besides self-transitions, all rates that correspond to double elementary jumps are zero, indicating that such transitions are excluded. In other words, a composite system *cannot* undergo transitions for which more than one of its elements moves across constitutive states.

The composite rate matrix has a particular block form that, informally, is expressed as

$$\begin{array}{c} \\ \sigma_1 \times \sigma_{1:3} \\ \sigma_2 \times \sigma_{1:3} \\ \sigma_3 \times \sigma_{1:3} \end{array} \begin{array}{c} \begin{array}{ccc} \sigma_{1:3} \times \sigma_1 & \sigma_{1:3} \times \sigma_2 & \sigma_{1:3} \times \sigma_3 \end{array} \\ \begin{bmatrix} \mathbf{\Lambda} & \lambda_{\sigma_1\to\sigma_2}\mathbb{1} & \lambda_{\sigma_1\to\sigma_3}\mathbb{1} \\ \lambda_{\sigma_2\to\sigma_1}\mathbb{1} & \mathbf{\Lambda} & \lambda_{\sigma_2\to\sigma_3}\mathbb{1} \\ \lambda_{\sigma_3\to\sigma_1}\mathbb{1} & \lambda_{\sigma_3\to\sigma_3}\mathbb{1} & \mathbf{\Lambda} \end{bmatrix} \end{array} = \bar{\mathbf{\Lambda}}.$$

Formally, such a matrix is obtained by

$$\bar{\mathbf{\Lambda}} = \mathbf{\Lambda} \otimes \mathbb{1}_M + \mathbb{1}_M \otimes \mathbf{\Lambda},$$

where $\mathbb{1}_M$ is the identity matrix of size M and $\otimes$ is the matrix Kronecker product.

In general, every time the composite system leaves a holding state $\bar{s}_n$, we may expect there to be $\bar{M}-1$ choices from which to select the subsequent holding state $\bar{S}_{n+1}|\bar{s}_n$. However, the majority of these choices involve more than one simultaneous elementary jump. Such choices are *nonelementary* and our system does *not* support them. Consequently, when leaving $\bar{s}_n$ the composite system may choose one out of at most $J(M-1)$ composite constitutive states $\bar{\sigma}_m$. These are precisely those $\bar{\sigma}_m$ that involve only a single element's jump.

Example 2.9 Composite jump events

Continuing in the same setting as in Example 2.8, if for some holding state $\bar{s}_n = \bar{\sigma}_1 = (\sigma_1, \sigma_1)$, then $\bar{S}_{n+1}|\bar{s}_n$ has four choices. Namely, these are

$$\bar{\sigma}_2 = (\sigma_1, \sigma_2), \qquad \bar{\sigma}_3 = (\sigma_1, \sigma_3), \qquad \bar{\sigma}_4 = (\sigma_2, \sigma_1), \qquad \bar{\sigma}_7 = (\sigma_3, \sigma_1).$$

The first two, $\bar{\sigma}_2$ and $\bar{\sigma}_3$, are triggered by jumps of the first element, $j = 1$; while, the remaining two, $\bar{\sigma}_4$ and $\bar{\sigma}_7$, are triggered by jumps of the second element, $j = 2$.

Similarly, if $\bar{s}_n = \bar{\sigma}_2 = (\sigma_1, \sigma_2)$, then $\bar{S}_{n+1}|\bar{s}_n$ has four different choices. Namely, these now read

$$\bar{\sigma}_1 = (\sigma_1, \sigma_1), \qquad \bar{\sigma}_3 = (\sigma_1, \sigma_3), \qquad \bar{\sigma}_5 = (\sigma_2, \sigma_2), \qquad \bar{\sigma}_8 = (\sigma_3, \sigma_2).$$

The first two, $\bar{\sigma}_1$ and $\bar{\sigma}_3$, are now triggered by jumps of the second element, $j = 2$; while, the remaining two, $\bar{\sigma}_5$ and $\bar{\sigma}_8$, are triggered by jumps of the first element, $j = 1$.

Each one of the J elements may trigger the transition $\bar{s}_n \to \bar{s}_{n+1}$. Therefore, a composite Markov jump process can be advanced by first sampling the identity J_n of the element triggering the nth event and subsequently sampling the reaction causing this element's jump. In view of Note D.1, a sampling scheme for composite systems proceeds as follows:

- First, sample J_n through

$$J_n|\bar{s}_n \sim \text{Categorical}_{1:J}\left(\frac{\lambda_{s_n^1}}{\sum_j \lambda_{s_n^j}}, \ldots, \frac{\lambda_{s_n^J}}{\sum_j \lambda_{s_n^j}}\right). \tag{2.7}$$

- Then, sample $S_{n+1}^{j_n}$ through

$$S_{n+1}^{j_n}|j_n, \bar{s}_n \sim \text{Categorical}_{\sigma_{1:M}}\left(\frac{\lambda_{s_n^{j_n} \to \sigma_1}}{\lambda_{s_n^{j_n}}}, \ldots, \frac{\lambda_{s_n^{j_n} \to \sigma_M}}{\lambda_{s_n^{j_n}}}\right).$$

- Finally, the remaining elements maintain their states:

$$s_{n+1}^j = s_n^j, \qquad\qquad j \neq j_n.$$

This scheme is satisfactory for sampling systems consisting of few elements and for keeping track of the identity of each element. However, for large systems we can derive a far more efficient sampling scheme. Before describing this scheme, we first introduce some formalism.

With an elementary state-space of size M, there are at most $K = M^2 - M$ elementary reactions supported. Specifically, these are the reactions $\sigma_m \to \sigma_{m'}$ between any possible pair of different constitutive states. For clarity, we will index these reactions with $\gamma^{1:K}$. In view of Note D.1, each one of these reactions may cause the nth jump with a rate equal to $c_{\bar{s}_n}^{\sigma_m} \lambda_{\sigma_m \to \sigma_{m'}}$, where $c_{\bar{s}_n}^{\sigma_m}$ is the total number of elements in $\bar{s}_n$ occupying σ_m. We will denote the rate of a reaction γ^k with $\mu_{\bar{s}_n}^k$ and, to distinguish it from the rates λ_{γ^k}, we will call it the *propensity*. As with $c_{\bar{s}_n}^{\sigma_m}$, we use subscripts to emphasize that propensities depend upon the holding state, $\bar{s}_n$.

Example 2.10 Elementary reactions and propensities

In the setting of the previous examples, there are $K = 6$ elementary reactions that may be supported. Explicitly, these are

$$\begin{array}{cccccc} \gamma^1 & \gamma^2 & \gamma^3 & \gamma^4 & \gamma^5 & \gamma^6 \\ \sigma_1 \to \sigma_2 & \sigma_1 \to \sigma_3 & \sigma_2 \to \sigma_1 & \sigma_2 \to \sigma_3 & \sigma_3 \to \sigma_1 & \sigma_3 \to \sigma_2. \end{array}$$

The reactions triggered by $\bar{s}_n = \bar{\sigma}_1 = (\sigma_1, \sigma_1)$ and their respective propensities are

$$\begin{array}{cccccc} \mu_{\bar{\sigma}_1}^1 & \mu_{\bar{\sigma}_1}^2 & \mu_{\bar{\sigma}_1}^3 & \mu_{\bar{\sigma}_1}^4 & \mu_{\bar{\sigma}_1}^5 & \mu_{\bar{\sigma}_1}^6 \\ 2 \cdot \lambda_{\sigma_1 \to \sigma_2} & 2 \cdot \lambda_{\sigma_1 \to \sigma_3} & 0 \cdot \lambda_{\sigma_2 \to \sigma_1} & 0 \cdot \lambda_{\sigma_2 \to \sigma_3} & 0 \cdot \lambda_{\sigma_3 \to \sigma_1} & 0 \cdot \lambda_{\sigma_3 \to \sigma_2}, \end{array}$$

while the reactions triggered by $\bar{s}_n = \bar{\sigma}_2 = (\sigma_1, \sigma_2)$ and their respective propensities are

$$\begin{array}{cccccc} \mu_{\bar{\sigma}_2}^1 & \mu_{\bar{\sigma}_2}^2 & \mu_{\bar{\sigma}_2}^3 & \mu_{\bar{\sigma}_2}^4 & \mu_{\bar{\sigma}_2}^5 & \mu_{\bar{\sigma}_2}^6 \\ 1 \cdot \lambda_{\sigma_1 \to \sigma_2} & 1 \cdot \lambda_{\sigma_1 \to \sigma_3} & 1 \cdot \lambda_{\sigma_2 \to \sigma_1} & 1 \cdot \lambda_{\sigma_2 \to \sigma_3} & 0 \cdot \lambda_{\sigma_3 \to \sigma_1} & 0 \cdot \lambda_{\sigma_3 \to \sigma_2}. \end{array}$$

Under these definitions, an alternative scheme for advancing a composite Markov jump process proceeds by first sampling the reaction G_n causing the nth event and subsequently sampling the element that triggers this reaction. In detail, the scheme is as follows:

- First, sample the reaction G_n inducing the transition through

$$G_n | \bar{s}_n \sim \text{Categorical}_{\gamma^{1:K}} \left(\frac{\mu_{\bar{s}_n}^1}{\sum_k \mu_{\bar{s}_n}^k}, \ldots, \frac{\mu_{\bar{s}_n}^K}{\sum_k \mu_{\bar{s}_n}^k} \right). \tag{2.8}$$

- Then, sample the identity J_n of the element causing this transition. The identity of the triggering element $J_n | g_n, \bar{s}_n$ is now sampled among the elements in $\bar{s}_n$ consistent with g_n. Since all of these are associated with the same escape rate λ_{g_n}, the sampling of $J_n | g_n, \bar{s}_n$ is uniform among the elements consistent with g_n.
- Finally, the remaining elements maintain their states:

$$s_{n+1}^j = s_n^j, \qquad j \neq j_n.$$

In general, the total number, K, of supported reactions γ^k in a composite system cannot exceed $M^2 - M$; however, provided some elemental rates $\lambda_{\sigma_m \to \sigma_{m'}}$ are zero, K is lower. Although the second scheme is more complicated than the first, it can be implemented and executed faster. This is because, most often, the number of elementary reactions in a composite system is considerably lower than the total number of its elements, $K \ll J$.

Note 2.10 Holding periods in a composite Markov jump process

In the *first scheme* the transition $\bar{s}_n \to \bar{s}_{n+1}$ is triggered by the element sampled in Eq. (2.7), indicating that the holding period is determined by

$$H_n|\bar{s}_n \sim \text{Exponential}\left(\sum_j \lambda_{s^j_n}\right).$$

Similarly, in the *second scheme*, the transition $\bar{s}_n \to \bar{S}_{n+1}$ is caused by the reaction sampled in Eq. (2.8), indicating that the holding period is determined by

$$H_n|\bar{s}_n \sim \text{Exponential}\left(\sum_k \mu^k_{\bar{s}_n}\right).$$

Both exponential rates are equal as we can see from

$$\sum_k \mu^k_{\bar{s}_n} = \sum_m \sum_{m'} c^{\sigma_m}_{\bar{s}_n} \lambda_{\sigma_m \to \sigma_{m'}} = \sum_m c^{\sigma_m}_{\bar{s}_n} \sum_{m'} \lambda_{\sigma_m \to \sigma_{m'}} = \sum_m c^{\sigma_m}_{\bar{s}_n} \lambda_{\sigma_m} = \sum_j \lambda_{s^j_n}.$$

Therefore, the holding periods are sampled similarly in both schemes.

Up to this point, we have focused on composite Markov jump processes formed by similar elements. However, this is not a necessary requirement. In fact, we may encounter composite systems whose elementary state-spaces differ. For example, the jth element can have a state-space $\sigma^j_{1:M^j}$ of different size M^j than the others and may also have different elementary reaction rates $\lambda_{\sigma^j_m \to \sigma^j_{m'}}$. As the possibilities of combining elementary systems are endless, we give a specific example open to generalizations in Example 2.11.

Example 2.11 Incongruent composite systems

In this example, we consider an incongruent system formed by $J = 2$ elements where the first element has $M^1 = 2$ constitutive states that we denote σ^1_1, σ^1_2 and the second element has $M^2 = 3$ constitutive states that we denote $\sigma^2_1, \sigma^2_2, \sigma^2_3$.

As the two elements differ, their reactions are described by different rate matrices

$$\mathbf{\Lambda}^1 = \begin{bmatrix} 0 & \lambda_{\sigma^1_1 \to \sigma^1_2} \\ \lambda_{\sigma^1_2 \to \sigma^1_1} & 0 \end{bmatrix}, \qquad \mathbf{\Lambda}^2 = \begin{bmatrix} 0 & \lambda_{\sigma^2_1 \to \sigma^2_2} & \lambda_{\sigma^2_1 \to \sigma^2_3} \\ \lambda_{\sigma^2_2 \to \sigma^2_1} & 0 & \lambda_{\sigma^2_2 \to \sigma^2_3} \\ \lambda_{\sigma^2_3 \to \sigma^2_1} & \lambda_{\sigma^2_3 \to \sigma^2_2} & 0 \end{bmatrix}.$$

The composite state-space contains $M^1 M^2 = 6$ constitutive states resulting from the Cartesian product of the elementary state-spaces $\bar{\sigma}_{1:6} = \sigma^1_{1:2} \times \sigma^2_{1:3}$. In particular,

$$\bar{\sigma}_1 = (\sigma^1_1, \sigma^2_1), \qquad \bar{\sigma}_2 = (\sigma^1_1, \sigma^2_2), \qquad \bar{\sigma}_3 = (\sigma^1_1, \sigma^2_3),$$
$$\bar{\sigma}_4 = (\sigma^1_2, \sigma^2_1), \qquad \bar{\sigma}_5 = (\sigma^1_2, \sigma^2_2), \qquad \bar{\sigma}_6 = (\sigma^1_2, \sigma^2_3).$$

The transition rate matrix of the composite system is

$$\begin{array}{c} \\ \bar{\sigma}_1 \\ \bar{\sigma}_2 \\ \bar{\sigma}_3 \\ \bar{\sigma}_4 \\ \bar{\sigma}_5 \\ \bar{\sigma}_6 \end{array} \begin{array}{c} \begin{array}{cccccc} \bar{\sigma}_1 & \bar{\sigma}_2 & \bar{\sigma}_3 & \bar{\sigma}_4 & \bar{\sigma}_5 & \bar{\sigma}_6 \end{array} \\ \left[\begin{array}{ccc|ccc} 0 & \lambda_{\sigma^2_1 \to \sigma^2_2} & \lambda_{\sigma^2_1 \to \sigma^2_3} & \lambda_{\sigma^1_1 \to \sigma^1_2} & 0 & 0 \\ \lambda_{\sigma^2_2 \to \sigma^2_1} & 0 & \lambda_{\sigma^2_2 \to \sigma^2_3} & 0 & \lambda_{\sigma^1_1 \to \sigma^1_2} & 0 \\ \lambda_{\sigma^2_3 \to \sigma^2_1} & \lambda_{\sigma^2_3 \to \sigma^2_2} & 0 & 0 & 0 & \lambda_{\sigma^1_1 \to \sigma^1_2} \\ \hline \lambda_{\sigma^1_2 \to \sigma^1_1} & 0 & 0 & 0 & \lambda_{\sigma^2_1 \to \sigma^2_2} & \lambda_{\sigma^2_1 \to \sigma^2_3} \\ 0 & \lambda_{\sigma^1_2 \to \sigma^1_1} & 0 & \lambda_{\sigma^2_2 \to \sigma^2_1} & 0 & \lambda_{\sigma^2_2 \to \sigma^2_3} \\ 0 & 0 & \lambda_{\sigma^1_2 \to \sigma^1_1} & \lambda_{\sigma^2_3 \to \sigma^2_1} & \lambda_{\sigma^2_3 \to \sigma^2_2} & 0 \end{array}\right] \end{array} = \bar{\mathbf{\Lambda}}.$$

As can be seen, the composite matrix maintains its block form that is still obtained through Kronecker products

$$\bar{\mathbf{\Lambda}} = \mathbf{\Lambda}^1 \otimes \mathbb{1}_{M^2} + \mathbb{1}_{M^1} \otimes \mathbf{\Lambda}^2.$$

The composite transition rate matrix can also written as

$$\bar{\mathbf{\Lambda}} = \mathbf{\Lambda}^1 \oplus \mathbf{\Lambda}^2,$$

where $\oplus$ denotes the matrix Kronecker sum that, by definition, is equal to $\mathbf{\Lambda}^1 \oplus \mathbf{\Lambda}^2 = \mathbf{\Lambda}^1 \otimes \mathbb{1}_{M^2} + \mathbb{1}_{M^1} \otimes \mathbf{\Lambda}^2$.

As illustrated by this example, using Cartesian products and Kronecker sums, we can readily obtain the state-space and rate matrix resulting from the combination of any number of elementary systems. For instance, in the most general case, combining J different state-spaces

$$\sigma^1_{1:M^1}, \ldots, \sigma^j_{1:M^j}, \ldots, \sigma^J_{1:M^J},$$

with rate matrices

$$\mathbf{\Lambda}^1, \ldots, \mathbf{\Lambda}^j, \ldots, \mathbf{\Lambda}^J,$$

respectively, results in a composite state-space of size

$$\bar{M} = M^1 \cdots M^j \cdots M^J.$$

The composite state-space is given by

$$\bar{\sigma}_{1:\bar{M}} = \sigma^1_{1:M^1} \times \cdots \times \sigma^j_{1:M^j} \times \cdots \times \sigma^J_{1:M^J}$$

and the composite rate matrix is given by

$$\bar{\mathbf{\Lambda}} = \mathbf{\Lambda}^1 \oplus \cdots \oplus \mathbf{\Lambda}^j \oplus \cdots \oplus \mathbf{\Lambda}^K.$$

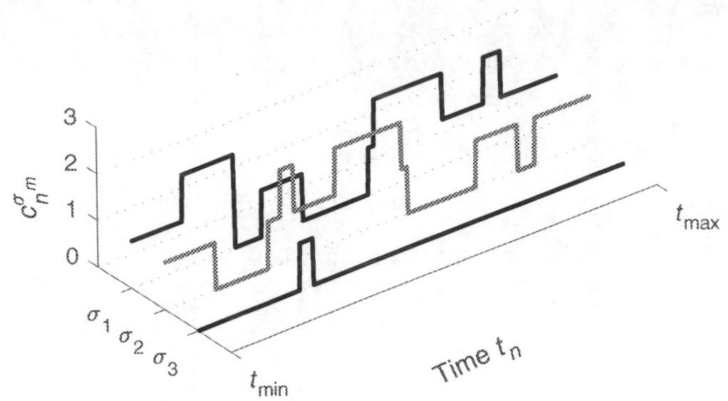

Fig. 2.6 A collapsed description of a composite system.

Collapsed Markov Jump Processes

Up to this point, we have considered composite systems under the assumption that their elements do not interact. However, this scenario is quite restrictive. For example, chemical systems involving multimolecular reactions or ecosystems involving multiple species contain interacting elements and cannot be accommodated within our existing framework.

In this section, we relax this assumption. To do so, we adopt a new formulation of composite systems that ignores the identities of the individual elements forming the system but rather focuses on the *total population* of elements occupying each elementary state. This is a bulk formulation and, whenever we can afford it, *i.e.*, whenever we can relax keeping track of the precise identity of the elements, as we will see, allows for a particularly convenient way to study our system. Indeed, we have already seen a special case of this collapsed formulation in birth and death processes.

As before, we consider J elements, each having M constitutive states $\sigma_{1:M}$. Further, we consider K elemental reactions $\sigma_m \to \sigma_{m'}$ labeled with $\gamma^{1:K}$. Now, instead of seeking a detailed description of the composite system through $\bar{s}_n$, we keep track of a *collapsed* state. This is an M-tuple

$$\tilde{c}_n = [c_n^{\sigma_1}, \ldots, c_n^{\sigma_m}, \ldots, c_n^{\sigma_M}],$$

where $c_n^{\sigma_m}$ is the population of elements occupying state σ_m before the nth event; see Fig. 2.6.

Note 2.11 The collapsed state-space

The state-space of a collapsed Markov jump process consists of all M-tuples $[c^{\sigma_1}, \ldots, c^{\sigma_m}, \ldots, c^{\sigma_M}]$, formed by the nonnegative integers c^{σ_m} with $\sum_m c^{\sigma_m} = J$. For example, for $M = 3$ and $J = 2$, the collapsed state-space consists of

$$[2, 0, 0], \quad [0, 2, 0], \quad [0, 0, 2], \quad [1, 1, 0], \quad [1, 0, 1], \quad [0, 1, 1].$$

As it only keeps track of the total population in each state and not the states of all individual elements, a collapsed state-space is smaller than a composite state-space.

As we saw in the preceding section, the propensities of the elementary reactions $\gamma^{1:K}$ depend only on the populations of elements in each σ_m, *i.e.*, the collapsed state $\tilde{c}_n$, and reaction rates. For clarity, from now on, we will denote the propensity of a reaction γ^k with $\mu^k_{\tilde{c}_n} = c_n^{\sigma_m} \lambda_{\gamma^k}$, where λ_{γ^k} is the reaction rate of γ^k and, as earlier, we will use subscripts to emphasize that propensities depend upon $\tilde{c}_n$.

In view of Eq. (2.8), the reaction G_n triggering the transition $\tilde{c}_n \to \tilde{c}_{n+1}$ is sampled according to

$$G_n | \tilde{c}_n \sim \text{Categorical}_{\gamma^{1:K}} \left(\frac{\mu^1_{\tilde{c}_n}}{\sum_k \mu^k_{\tilde{c}_n}}, \ldots, \frac{\mu^K_{\tilde{c}_n}}{\sum_k \mu^k_{\tilde{c}_n}} \right).$$

Once g_n is sampled, $\tilde{c}_{n+1}$ is determined by updating the populations of the elements involved in g_n without sampling any additional random variables. According to Note 2.10, the holding period H_n is then sampled from

$$H_n | \tilde{c}_n \sim \text{Exponential} \left(\sum_k \mu^k_{\tilde{c}_n} \right).$$

Note 2.12 Encoding elementary reactions

At this point we summarize some nomenclature regarding the formulation of elementary reactions.

A composite system, irrespective of whether we model it through detailed $\bar{S}_1, \bar{S}_2, \ldots$ or collapsed $\tilde{C}_1, \tilde{C}_2, \ldots$ holding states, entails an elementary state-space $\sigma_{1:M}$. This state-space may support at most $M^2 - M$ elementary reactions. We encode the elementary reactions with $\sigma_m \to \sigma_{m'}$, where σ_m and $\sigma_{m'}$ are two different elementary constitutive states and associate them with an elementary reaction rate $\lambda_{\sigma_m \to \sigma_{m'}}$. In $\sigma_m \to \sigma_{m'}$, we designate σ_m as the *departing* state and $\sigma_{m'}$ as the *arriving* state.

A system may have some elementary reaction rates equal to zero. Accordingly, we may say that the system cannot support these reactions. In such cases, which are very common in applications involving composite systems, the number of supported elementary reactions K is lower than $M^2 - M$. Occasionally, we denote the supported elementary reactions with $\gamma^{1:K}$, *i.e.*, each γ^k stands for some $\sigma_m \to \sigma_{m'}$. Following this convention, we may also denote the elementary reaction rates using λ_{γ^k} instead of the more elaborate $\lambda_{\sigma_m \to \sigma_{m'}}$, and we may further refer to the associated σ_m and $\sigma_{m'}$ as the reaction's departing and arriving states, respectively.

Besides a rate λ_{γ^k}, each γ^k is also associated with a propensity. Depending on the modeling description chosen, we denote the propensity of γ^k with either $\mu^k_{\bar{s}_n}$ or $\mu^k_{\tilde{c}_n}$. In either case, a reaction's propensity is equal to the product of elements occupying the departing state and the reaction's rate.

In a collapsed description of a composite system, it is convenient to associate each reaction γ^k with a *stoichiometric array* $\tilde{\zeta}^k$ that indicates the population changes induced by a reaction. Similar to a collapsed state $\tilde{c}$, each stoichiometric array is also an M-tuple,

$$\tilde{\zeta}^k = \left[\zeta^k_{\sigma_1}, \ldots, \zeta^k_{\sigma_m}, \ldots, \zeta^k_{\sigma_M}\right],$$

with each ζ^k_σ encoding how the population of an elementary constitutive state σ is affected by the reaction γ^k. For reactions where an element simply switches its state, we have $\zeta^k_\sigma = -1$ for the departing and $\zeta^k_\sigma = +1$ for the arriving state of γ^k. In subsequent sections, we will encounter reactions with more complicated stoichiometry. For this reason, it is most convenient to tabulate a system's reactions $\gamma^{1:K}$ and their stoichiometric arrays $\tilde{\zeta}^{1:K}$ as

$$\begin{array}{c} \gamma^1 \\ \vdots \\ \gamma^K \end{array} \overset{\begin{array}{ccc} \sigma_1 & \cdots & \sigma_M \end{array}}{\begin{pmatrix} \zeta^1_{\sigma_1} & \cdots & \zeta^1_{\sigma_M} \\ \vdots & \ddots & \vdots \\ \zeta^K_{\sigma_1} & \cdots & \zeta^K_{\sigma_M} \end{pmatrix}} = \begin{pmatrix} \tilde{\zeta}^1 \\ \vdots \\ \tilde{\zeta}^K \end{pmatrix}.$$

For instance, a system with $M = 3$ elementary states, can support at most $K = 6$ elementary reactions. These are tabulated in

$$\begin{array}{c} \gamma^1 \\ \gamma^2 \\ \gamma^3 \\ \gamma^4 \\ \gamma^5 \\ \gamma^6 \end{array} \overset{\begin{array}{ccc} \sigma_1 & \sigma_2 & \sigma_3 \end{array}}{\begin{pmatrix} -1 & +1 & 0 \\ -1 & 0 & +1 \\ +1 & -1 & 0 \\ 0 & -1 & +1 \\ +1 & 0 & -1 \\ 0 & +1 & -1 \end{pmatrix}} = \begin{pmatrix} \tilde{\zeta}^1 \\ \tilde{\zeta}^2 \\ \tilde{\zeta}^3 \\ \tilde{\zeta}^4 \\ \tilde{\zeta}^5 \\ \tilde{\zeta}^6 \end{pmatrix}.$$

The advantage of tabulating $\tilde{\zeta}^{1:K}$ is that, following the sampling of a reaction G_n, the new occupying populations are readily obtained from $\tilde{c}_{n+1} = \tilde{c}_n + \tilde{\zeta}^{g_n}$.

A Case Study in Chemical Systems

Modeling a Chemical System

Composite systems are useful when modeling the joint states of different molecules in a chemical system and wishing to keep track of their individual states as a function of time. That is, retaining molecular identity. Retaining molecular identity is important when we deal with molecules, such as DNA plasmids or mRNAs, that are present in small numbers and whose behavior from molecule to molecule may vary. For example, one DNA plasmid, of which there may be only a few replicate copies, may be transcriptionally inhibited for some time while another may be actively transcribing its DNA.

However, retaining molecular identity can quickly become computationally cumbersome if we are dealing with hundreds, thousands, or more molecules. In this case, we may only be interested in modeling how a population of *chemical species* evolves without keeping track of the state of each individual molecule within each species. As such, the collapsed state formalism may be preferred simply on account of its reduced computational cost. Keeping track of and propagating in time the collapsed state, *i.e.*, the population, of all species is strikingly less expensive than the state of each molecule. For example, when dealing with three chemical species, whose state-space is denoted with σ_1, σ_2, and σ_3, the collapsed state before the nth jump is $\tilde{\boldsymbol{c}}_n = [c_n^{\sigma_1}, c_n^{\sigma_2}, c_n^{\sigma_3}]$, where each element coincides with the population of the superscripted species.

Note 2.13 Chemical species

The term "chemical species" is not uniquely defined and depends on the problem at hand. Typically, when we speak of different chemical species, we may be designating molecules with a different number of atoms, atom types, or bond arrangement. It is possible, indeed common depending on the level of modeling fidelity, to refer to all RNA, irrespective of sequence and thus composition, as one species in order to distinguish it from protein, another species. At the other extreme, *i.e.*, at a finer scale, it is even possible to speak of different species as subtly different electronic states of molecules.

The term chemical species also captures abstract differences that are difficult to describe in terms of atom numbers, bond arrangements, or quantum states. For example, different chemical species may include actively transcribing DNA loci versus transcriptionally inhibited loci irrespective of how transcription or inhibition is achieved.

Besides the reduction of computational cost, working with populations is especially critical when considering interacting molecules. For example, in constructing the composite transition rate matrix of two systems, such as two molecules, we previously wrote $\bar{\boldsymbol{\Lambda}} = \boldsymbol{\Lambda}^1 \oplus \boldsymbol{\Lambda}^2$. As all rates of double transitions in $\bar{\boldsymbol{\Lambda}}$ are zero, this pre-assumes that the molecules are noninteracting. Yet bimolecular reactions, where two molecules react to create a new molecule or a molecule dissociates releasing two molecules, cannot be feasibly treated within a composite formulation. Yet, as we will see, bimolecular and higher order molecular reactions are easily treated with a collapsed state-space formalism.

Example 2.12 Bimolecular reaction

As an illustrative example, we consider a simple *bimolecular reaction*

$$\mathcal{A} + \mathcal{B} \xrightarrow{\lambda_{+1}} \mathcal{AB},$$

where 1 molecule of species $\mathcal{A}$ and 1 molecule of species $\mathcal{B}$ collide, with a rate λ_{+1}, to form 1 molecule of species $\mathcal{AB}$. For clarity, we assume that initially our system contains $J = 10$ molecules with 5 $\mathcal{A}$'s and 5 $\mathcal{B}$'s. That is, we take $\tilde{c}_1 = [c_1^{\sigma_1}, c_1^{\sigma_2}, c_1^{\sigma_3}] = [5, 5, 0]$, where the elementary states are σ_1 for $\mathcal{A}$, σ_2 for $\mathcal{B}$, and σ_3 for $\mathcal{AB}$. Once a reaction occurs, we are left with $J = 9$ molecules, and $\tilde{c}_2 = [4, 4, 1]$.

Similarly, if we consider the reverse reaction,

$$\mathcal{AB} \xrightarrow{\lambda_{-1}} \mathcal{A} + \mathcal{B},$$

and initiate from $\tilde{c}_1 = [0, 0, 5]$, say, after one reaction, we reach $\tilde{c}_2 = [1, 1, 4]$.

Note 2.14 Propensities

Here, we list the rules for computing the propensities of some common reactions:

$$\begin{array}{lccl} \gamma^1 & \emptyset \xrightarrow{\lambda_{\gamma^1}} \mathcal{A} & & \mu^1_{\tilde{c}_n} = \lambda_{\gamma^1}, \\ \gamma^2 & \mathcal{A} \xrightarrow{\lambda_{\gamma^2}} \emptyset & & \mu^2_{\tilde{c}_n} = c_n^{\sigma_1} \lambda_{\gamma^2}, \\ \gamma^3 & \mathcal{A} \xrightarrow{\lambda_{\gamma^3}} \mathcal{B} & & \mu^3_{\tilde{c}_n} = c_n^{\sigma_1} \lambda_{\gamma^3}, \\ \gamma^4 & \mathcal{A} + \mathcal{B} \xrightarrow{\lambda_{\gamma^4}} \mathcal{AB} & & \mu^4_{\tilde{c}_n} = c_n^{\sigma_1} c_n^{\sigma_2} \lambda_{\gamma^4}, \\ \gamma^5 & \mathcal{AB} \xrightarrow{\lambda_{\gamma^5}} \mathcal{A} + \mathcal{B} & & \mu^5_{\tilde{c}_n} = c_n^{\sigma_3} \lambda_{\gamma^5}, \end{array}$$

where the populations of species $\mathcal{A}$, $\mathcal{B}$, and $\mathcal{AB}$ are tabulated in $\tilde{c}_n = [c_n^{\sigma_1}, c_n^{\sigma_2}, c_n^{\sigma_3}]$, as in Example 2.12. We can also encode the net change in each population induced by the reactions in $\tilde{\zeta}^{1:5}$ through the following stoichiometric array:

$$\begin{array}{c} \\ \gamma^1 \\ \gamma^2 \\ \gamma^3 \\ \gamma^4 \\ \gamma^5 \end{array} \begin{array}{c} \begin{array}{ccc} \sigma_1 & \sigma_2 & \sigma_3 \end{array} \\ \begin{pmatrix} +1 & 0 & 0 \\ -1 & 0 & 0 \\ -1 & +1 & 0 \\ -1 & -1 & +1 \\ +1 & +1 & -1 \end{pmatrix} \end{array} = \begin{pmatrix} \tilde{\zeta}^1 \\ \tilde{\zeta}^2 \\ \tilde{\zeta}^3 \\ \tilde{\zeta}^4 \\ \tilde{\zeta}^5 \end{pmatrix}.$$

An additional remark on propensity is in order here: preceding the rate of the reaction is the number of molecules in all chemical species involved in the chemical reaction. It is for this reason that reaction γ^4, which involves two reactant molecules, contains the product $c_n^{\sigma_1} c_n^{\sigma_2}$ in the propensity.

We are now ready to consider the propensity of reactions involving a reactant reacting with its same species. For example, we consider adding to the collection of reactions above the additional reaction $\mathcal{A} + \mathcal{A} \xrightarrow{\lambda_{\gamma^6}} \mathcal{A}_2$. As the propensity depends on the number of molecules involved in the reaction, $\mu^6_{\tilde{c}_n} = \left(c_n^{\sigma_1}(c_n^{\sigma_1} - 1)/2!\right) \lambda_{\gamma^6}$. Following a similar logic, the propensity of a reaction such as $\mathcal{A} + \mathcal{A} + \mathcal{A} \xrightarrow{\lambda_{\gamma^7}} \mathcal{A}_3$ reads $\mu^7_{\tilde{c}_n} = \left(c_n^{\sigma_1}(c_n^{\sigma_1} - 1)(c_n^{\sigma_1} - 2)/3!\right)\lambda_{\gamma^7}$. Similarly, the propensity for $a\mathcal{A} + b\mathcal{B} \xrightarrow{\lambda_{\gamma^8}} \mathcal{A}_a\mathcal{B}_b$ is $\mu^8_{\tilde{c}_n} = \left(c_n^{\sigma_1}!/[a!\,(c_n^{\sigma_1} - a)!]\right)\left(c_n^{\sigma_2}!/[b!\,(c_n^{\sigma_2} - b)!]\right)\lambda_{\gamma^8}$.

As interactions are common in chemical applications, we adjust our terminology from the collapsed state Markov jump processes to accommodate them. In particular, we consider a system of interacting molecules distributed among M species $\sigma_{1:M}$. Further, we consider K possible reactions $\gamma^{1:K}$ that may lead from one species to another and may involve single *or* multiple molecules. Of course, unlike before, in $\gamma^{1:K}$ we now allow reactions that may alter the total population of molecules; that is, reactions that may not necessarily be elemental.

Bimolecular or higher order reactions are still described by Markov jump processes. The reason for this is that the probability of forming a species $\mathcal{AB}$ at the $(n+1)$th jump only depends on the number of its $\mathcal{A}$ and $\mathcal{B}$ components available after the nth jump. However, the rationale for why higher order reactions, *e.g.*, $\mathcal{A} + \mathcal{A} + \mathcal{A} \rightarrow \mathcal{A}_3$, are considered Markovian is somewhat more opaque. That is because reactions beyond second order are normally considered *effective* reactions. As such, these can, in principle, be broken down in terms of simpler bimolecular reactions except that the rate of such reactions may be so fast, as compared to any measurement timescale, that no bimolecular event would ever be observed. Thus $\mathcal{A} + \mathcal{A} \rightarrow \mathcal{A}_2$ followed by $\mathcal{A}_2 + \mathcal{A} \rightarrow \mathcal{A}_3$ would be abbreviated as $\mathcal{A} + \mathcal{A} + \mathcal{A} \rightarrow \mathcal{A}_3$.

Note 2.15 Well-stirred approximation

By invoking propensities dependent on the number of molecules available of each relevant species, we inherently make a *well-stirred* approximation. That is, we assume that molecules randomize in physical space on timescales far exceeding the largest supported propensity. The error of this approximation is difficult to bound if the number of molecular species is allowed to grow to high numbers or indefinitely (such as in a simple birth process) and the time between reactions can, in practice, shrink to zero.

Simulating a Chemical System

To avoid unnecessarily complex formulations, in the chemical setting we typically avoid forming composite states $\bar{s}_n$. Instead, we describe our system directly with collapsed states $\tilde{c}_n = [c_n^{\sigma_1}, \ldots, c_n^{\sigma_m}, \ldots, c_n^{\sigma_M}]$ that track down only the populations of the molecules of each species. Provided that the propensities $\mu^k_{\tilde{c}_n}$ can be computed, we can readily simulate the transitions $\tilde{c}_n \rightarrow \tilde{C}_{n+1}$. The scheme is known as the *Gillespie simulation* and follows the recipe provided in Section 2.3.3 naturally extended to include bimolecular and higher order reactions. Algorithm 2.1 summarizes the steps involved in a typical implementation.

Algorithm 2.1 Gillespie simulation

Initialize time $t = t_0$ and the population of the species $\bar{c} = [c^{\sigma_1}, \dots, c^{\sigma_M}]$. Use $\bar{c}$ to initialize the propensities of the reactions $(\mu^1, \dots, \mu^K)$ and compute $\mu^* = \sum_k \mu^k$.

As long as $t < t_{\max}$, iterate the following:

- Sample the period until the next event

$$H|\bar{c} \sim \text{Exponential}\left(\mu^*\right).$$

- Sample the reaction triggering the next event

$$G|\bar{c} \sim \text{Categorical}_{\gamma^{1:K}}\left(\frac{\mu^1}{\mu^*}, \dots, \frac{\mu^K}{\mu^*}\right).$$

- Update t according to h.
- Update $\bar{c}$, $(\mu^1, \dots, \mu^K)$, and μ^* according to the reaction g that was sampled using the associated $\bar{\zeta}$ array.

2.3.4 Global Descriptions of Markov Jump Processes*

All descriptions of Markov jump processes encountered so far remain at the trajectory level, *i.e.*, they rely on parameterizations and sampling schemes for the system's trajectory. In this sense, they are a *local description*. Although they are accurate and mathematically complete, they may not always be the most convenient descriptions for further analyses. Here we turn to an alternative. Unlike earlier, we now try to conceal by integrating over or marginalizing the trajectory from our formulation. In this way, we provide a *global description* of the system's dynamics.

The Master Equation

We already studied Markov jump processes and we have seen how to sample their holding states $S_1, S_2, \dots$ and holding periods $H_1, H_2, \dots$. In view of Eq. (2.1) and Note 2.6, these are effectively complete descriptions of the system's trajectory and could be useful in simulating trajectories or computing probabilities associated with given trajectories.

Note 2.16 Probability of a Markov jump process trajectory

According to Note 2.6, a trajectory $\mathcal{S}(\cdot)$ of a Markov jump process is a random function of time. Here, we assume that this function has been observed between t_0 and some later time that we denote $t_{\max}$. Observing

* This is an advanced topic and could be skipped on a first reading.

$\mathcal{S}(\cdot)$ means that we have three pieces of information available: how many holding states the system occupied during the observation window; the precise sequence of these states; and the precise sequence of jump times. For clarity, we denote the number of states by N, the sequence of states undergone by $s_1, \ldots, s_N$, and the sequence of jump times by $t_1, \ldots, t_{N-1}$. Through Eq. (2.1) we can recover the values of the first $N-1$ holding periods,

$$h_1 = t_1 - t_0, \qquad h_2 = t_2 - t_1, \qquad \ldots \qquad h_{N-1} = t_{N-1} - t_{N-2},$$

while, for the last holding period, we can recover only a lower threshold $h_N > t_{\max} - t_{N-1}$. Therefore, according to the sampling scheme of Eqs. (2.2)–(2.4), the probability of sampling this observed trajectory $\mathcal{S}(\cdot)$ for a fixed number of jumps with that specific sequence of holding states and holding periods is proportional to the product

$$\overbrace{\rho_{s_1} \lambda_{s_1} e^{-\lambda_{s_1}(t_1-t_0)}}^{\text{sample } s_1 \text{ and } h_1} \cdot \overbrace{\pi_{s_1 \to s_2} \lambda_{s_2} e^{-\lambda_{s_2}(t_2-t_1)}}^{\text{sample } s_2 \text{ and } h_2} \cdots$$

$$\cdots \underbrace{\pi_{s_{N-2} \to s_{N-1}} \lambda_{s_{N-1}} e^{-\lambda_{s_{N-1}}(t_{N-1}-t_{N-2})}}_{\text{sample } s_{N-1} \text{ and } h_{N-1}} \cdot \underbrace{\pi_{s_{N-1} \to s_N} \left(1 - e^{-\lambda_{s_N}(t_{\max}-t_{N-1})}\right)}_{\text{sample } s_N \text{ and } h_N}.$$

The last term in parentheses is obtained from the probability of not transitioning between $t_{\max}$ and t_{N-1}.

In practice, however, we may not always be interested in quantifying the probability of a system's entire trajectory. Instead, we may be interested in the probability of the system occupying a particular constitutive state at only a given sequence of times. This naturally involves integrating over all jump events that may occur between the given times. To compute such probabilities, it is convenient to derive an appropriate set of differential equations that, as we will see, give rise to the *master equation*.

To be more precise, we consider a Markov jump process with initial probabilities $\boldsymbol{\rho}$ and rate matrix $\boldsymbol{\Lambda}$ on a state-space $\sigma_{1:M}$. As usual, we denote the trajectory by $\mathcal{S}(\cdot)$ to emphasize that the trajectory is a function of time. Since $\mathcal{S}(\cdot)$ is random, the passing state evaluated at a time t, namely $\mathcal{S}(t)$, is also random. Following our convention, the values $\mathcal{S}(t)$ may attain are precisely the constitutive states σ_m. To arrive at the probabilities of the system passing from each σ_m, we first write down the equation satisfied by $P_{\sigma_m}(t)$, which is informally defined by

$$P_{\sigma_m}(t) = \begin{pmatrix} \text{probability of} \\ \mathcal{S}(t) = \sigma_m \end{pmatrix}.$$

To begin, we consider the outcomes that contribute to the probability difference $P_{\sigma_m}(t+dt) - P_{\sigma_m}(t)$ provided dt is small enough: the system starts at σ_m and jumps to some other $\sigma_{m'}$, this negatively contributes to $P_{\sigma_m}(t+dt) - P_{\sigma_m}(t)$; or, the system starts at any other $\sigma_{m'}$ and jumps to σ_m, this positively contributes to $P_{\sigma_m}(t+dt) - P_{\sigma_m}(t)$.

Since the holding periods of Markov jump process are memoryless, we can readily compute the probabilities of these two outcomes. In particular, by considering holding periods that start at t, following Eq. (2.4), the probabilities are

$$P_{\sigma_m}(t)\int_0^{dt} dh\,\text{Exponential}\left(h;\lambda_{\sigma_m}\right) = P_{\sigma_m}(t)\left(1-e^{-\lambda_{\sigma_m}dt}\right),$$

$$\sum_{m'\neq m} P_{\sigma_{m'}}(t)\int_0^{dt} dh\,\text{Exponential}\left(h;\lambda_{\sigma_{m'}\to\sigma_m}\right) = \sum_{m'\neq m} P_{\sigma_{m'}}(t)\left(1-e^{-\lambda_{\sigma_{m'}\to\sigma_m}dt}\right).$$

Combining them, we see that

$$P_{\sigma_m}(t+dt)-P_{\sigma_m}(t)\approx P_{\sigma_m}(t)\left(e^{-\lambda_{\sigma_m}dt}-1\right)+\sum_{m'\neq m}P_{\sigma_{m'}}(t)\left(1-e^{-\lambda_{\sigma_{m'}\to\sigma_m}dt}\right), \tag{2.9}$$

with the approximation valid provided dt is sufficiently small such that no more than one jump can occur between t and $t+dt$. Dividing both sides by dt and taking the limit $dt\to 0^+$ yields an exact equality. This equality is the differential equation

$$\frac{d}{dt}P_{\sigma_m}(t) = -\lambda_{\sigma_m}P_{\sigma_m}(t)+\sum_{m'\neq m}\lambda_{\sigma_{m'}\to\sigma_m}P_{\sigma_{m'}}(t). \tag{2.10}$$

Gathering the probabilities of all constitutive states into a probability array

$$\boldsymbol{P}(t) = \left[P_{\sigma_1}(t),\ldots,P_{\sigma_M}(t)\right], \tag{2.11}$$

the differential equation in Eq. (2.10) takes a particularly simple form,

$$\frac{d}{dt}\boldsymbol{P}(t) = \boldsymbol{P}(t)\boldsymbol{G}, \tag{2.12}$$

known as the *master equation*.

Note 2.17 Generator matrix

It is often convenient to tabulate the state-space and all rates as

$$\begin{array}{c} \\ \sigma_1 \\ \vdots \\ \sigma_M \end{array}\begin{array}{c} \begin{array}{ccc} \sigma_1 & \cdots & \sigma_M \end{array} \\ \begin{bmatrix} -\lambda_{\sigma_1} & \cdots & \lambda_{\sigma_1\to\sigma_M} \\ \vdots & \ddots & \vdots \\ \lambda_{\sigma_M\to\sigma_1} & \cdots & -\lambda_{\sigma_M} \end{bmatrix} \end{array} = \boldsymbol{G}.$$

In this tabulation, the reaction rates $\lambda_{\sigma_m\to\sigma_{m'}}$, which are nonnegative scalars, are placed in the off-diagonals. By contrast, *negative* escape rates λ_{σ_m}, which are nonpositive scalars, are placed along the diagonal. Commonly, $\boldsymbol{G}$ is termed the *generator matrix*. The generator $\boldsymbol{G}$ and the transition rate $\boldsymbol{\Lambda}$ matrices, which we have already seen, are tightly related and we can interconvert from one to the other.

To avoid confusion with similar notions, we summarize some important characteristics of the generator matrix:

- It is a square matrix of size equal to the size of the state-space.
- Its row and column elements are arranged with the ordering of the state-space.
- It gathers parameters with units of reciprocal time.
- Each row must sum to 0.
- Its diagonal elements may not be zero, except for the uninteresting case where all row elements are also zero.

At the initial time t_0, our trajectory passes from the first holding state, *i.e.*, $\mathcal{S}(t_0) = S_1$, which is sampled according to Eq. (2.2). As such, we immediately get $P_{\sigma_m}(t_0) = \rho_{\sigma_m}$, which provides the initial condition

$$\boldsymbol{P}(t_0) = \boldsymbol{\rho}. \tag{2.13}$$

Under this initial condition, we can integrate the master equation in time. Using the matrix exponential, the solution attains a general form

$$\boldsymbol{P}(t) = \boldsymbol{\rho} \exp\left((t - t_0)\boldsymbol{G}\right), \tag{2.14}$$

where the exponential of the matrix $\exp(\cdot)$ is defined in Note 2.18.

Note 2.18 The matrix exponential

The matrix exponential $\exp(\ell \boldsymbol{L})$, where ℓ is a scalar and $\boldsymbol{L}$ is a square matrix, is defined by a sum

$$\exp(\ell \boldsymbol{L}) = \mathbb{1} + \ell \boldsymbol{L} + \frac{\ell^2}{2!}\boldsymbol{L}\boldsymbol{L} + \frac{\ell^3}{3!}\boldsymbol{L}\boldsymbol{L}\boldsymbol{L} + \cdots = \sum_{j=0}^{\infty} \frac{\ell^j}{j!}\boldsymbol{L}^j,$$

where $\mathbb{1}$ is the identity matrix of size equal to the size of $\boldsymbol{L}$.

By its definition, the mth element of $\boldsymbol{P}(t)$ is the probability of starting in a state dictated by $\boldsymbol{\rho}$ and, after some elapsed time t, landing in state σ_m *irrespective of what path* the system may take from t_0 to t. From Eq. (2.14), we can see that the state probabilities $\boldsymbol{P}(t')$ at some time t' are related to the state probabilities $\boldsymbol{P}(t)$ as some time $t < t'$ through

$$\begin{aligned}\boldsymbol{P}(t') &= \boldsymbol{\rho} \exp\left((t' - t_0)\boldsymbol{G}\right) = \boldsymbol{\rho} \exp\left((t - t_0)\boldsymbol{G}\right) \exp\left((t' - t)\boldsymbol{G}\right) \\ &= \boldsymbol{P}(t) \exp\left((t' - t)\boldsymbol{G}\right).\end{aligned}$$

For this reason, the matrix $\exp\left((t' - t)\boldsymbol{G}\right)$ is termed a *propagator*. From now on, we will denote the propagator from an earlier time t to a subsequent time t' by

$$\boldsymbol{Q}^{t \to t'} = \exp\left((t' - t)\boldsymbol{G}\right). \tag{2.15}$$

Note 2.19 A sanity check on the master equation

The array of probabilities $\boldsymbol{P}(t)$ is, by definition, unitless. By contrast, $\boldsymbol{G}$ has units of reciprocal time, similar to d/dt, as can be verified from Eq. (2.12).

As our system must occupy some constitutive state at all times, the vector elements of $\boldsymbol{P}(t)$ must be normalized to unity, *i.e*, $\sum_m P_{\sigma_m}(t) = 1$, for all t. We can verify that the total probability does not change over time using Eq. (2.10). As such, the rate of change of the total probability is

$$\frac{d}{dt}\sum_m P_{\sigma_m}(t) = \sum_m \frac{d}{dt} P_{\sigma_m}(t) = 0. \qquad (*2.16*)$$

Therefore, we can verify that the total probability is equal to unity. In particular, we immediately see that

$$\sum_m P_{\sigma_m}(t) = \sum_m P_{\sigma_m}(t_0) = \sum_m \rho_{\sigma_m} = 1,$$

where we invoked the initial condition of Eq. (2.13).

The master equation in matrix form, Eq. (2.12), and the propagator $\boldsymbol{Q}^{t\to t'}$ are valid provided $\boldsymbol{P}(t)$ is a *row* array as in Eq. (2.11) *and* the generator matrix $\boldsymbol{G}$ is a formed as in Note 2.17. In a computational implementation, these give rise to row-matrix operations. If, instead of a row array, we consider $\boldsymbol{P}(t)$ as a column array, as is customary in programming environments better suited for column-matrix operations, then the master equation in Eq. (2.12), the propagator $\boldsymbol{Q}^{t\to t'}$, the form of the generator $\boldsymbol{G}$, and the general solution in Eq. (2.11) must be adapted appropriately.

Master Equations for Structured Markov Jump Processes

As composite Markov jump processes composed of independent elements are Markov jump processes in their own right, appropriate master equations can also be used for their study. Here, we derive these equations. As usual, we consider a congruent composite system formed by J elements, each on a common elementary state-space $\sigma_{1:M}$ with elementary reaction rates gathered in $\boldsymbol{\Lambda}$.

Composite Markov Jump Process

In a detailed description of a composite Markov jump process, the probabilities we seek are

$$\bar{P}_{\bar{\sigma}_m}(t) = \begin{pmatrix} \text{probability of} \\ \bar{\mathcal{S}}(t) = \bar{\sigma}_m, \end{pmatrix},$$

where now $\bar{\mathcal{S}}(\cdot)$ denotes the composite trajectory, which is a function of time and attains values in the composite state-space. The composite state-space

* Reminder: The asterisks by some equations indicates that the detailed derivation can be found in Appendix F.

$\bar{\sigma}_{1:\bar{M}} = \sigma_{1:M} \times \cdots \times \sigma_{1:M}$ is formed by Cartesian products and the composite rate matrix $\bar{\mathbf{\Lambda}} = \mathbf{\Lambda} \oplus \cdots \oplus \mathbf{\Lambda}$ is formed by Kronecker sums. It is easy to verify that the composite generator $\bar{\boldsymbol{G}} = \boldsymbol{G} \oplus \cdots \oplus \boldsymbol{G}$ is also formed by Kronecker sums. Therefore, the master equation, Eq. (2.14), for composite systems is

$$\frac{d}{dt}\bar{\boldsymbol{P}}(t) = \bar{\boldsymbol{P}}(t)\bar{\boldsymbol{G}} = \bar{\boldsymbol{P}}(t)\left(\boldsymbol{G} \oplus \cdots \oplus \boldsymbol{G}\right),$$

and, because $\oplus$ commutes with $\exp(\cdot)$, its general solution is

$$\bar{\boldsymbol{P}}(t) = \bar{\boldsymbol{\rho}} \exp\left((t - t_0)\bar{\boldsymbol{G}}\right) = \bar{\boldsymbol{\rho}} \exp\left((t - t_0)\mathbf{\Lambda}^1\right) \otimes \cdots \otimes \exp\left((t - t_0)\mathbf{\Lambda}^J\right),$$

where $\bar{\boldsymbol{\rho}}$ gathers the initial probabilities of the composite system.

Collapsed Markov Jump Process

Unlike composite Markov jump processes whose master equation we can derive by exclusively relying on properties of the matrix exponential, for collapsed Markov jump processes we need to repeat the derivation from the onset. For this, we consider K reactions $\gamma^{1:K}$ with associated stoichiometric array $\tilde{\zeta}^{1:K}$ as in Note 2.12. The derivation parallels the earlier one and, as such, we only highlight key steps.

In the collapsed formulation, the probabilities we seek are

$$\tilde{P}_{\tilde{c}}(t) = \begin{pmatrix} \text{probability of} \\ \tilde{\mathcal{S}}(t) = \tilde{c} \end{pmatrix},$$

where now $\tilde{\mathcal{S}}(\cdot)$ denotes the collapsed trajectory. This is a function of time that takes values $\tilde{c} = [c^{\sigma_1}, \ldots, c^{\sigma_M}]$. To derive the master equation for $\tilde{P}_{\tilde{c}}(t)$ we consider changes occurring between time intervals of duration dt. Provided dt is sufficiently small such that at most one of the reactions $\gamma^{1:K}$ occurs, we need to consider only the changes induced by individual reactions. Formally, these are captured in

$$\tilde{P}_{\tilde{c}}(t + dt) - \tilde{P}_{\tilde{c}}(t) \approx \tilde{P}_{\tilde{c}}(t)\left(\exp\left(-dt \sum_k \mu^k_{\tilde{c}}\right) - 1\right) + \sum_k \tilde{P}_{\tilde{c}-\tilde{\zeta}^k}(t)\left(1 - \exp\left(-dt\mu^k_{\tilde{c}-\tilde{\zeta}^k}\right)\right).$$

Rearranging this equation and taking the limit $dt \to 0^+$, we obtain the differential form of the master equation

$$\frac{d}{dt}\tilde{P}_{\tilde{c}}(t) = -\left(\sum_k \mu^k_{\tilde{c}}\right)\tilde{P}_{\tilde{c}}(t) + \sum_k \mu^k_{\tilde{c}-\tilde{\zeta}^k}\tilde{P}_{\tilde{c}-\tilde{\zeta}^k}(t). \tag{2.17}$$

As this form is commonly found in chemical applications, it is most commonly called the *chemical master equation*.

A Case Study in the Laws of Mass Action

Master equations describe probabilities of having $\tilde{c}$ elements at some time t given some initial conditions. From these equations, we can derive *mass action laws* familiar across the natural sciences. Mass action laws describe how the average population $\langle c^{\sigma_m}(t)\rangle$ of a species σ_m in a composite Markov jump process changes over time. This expectation is customarily designated by

$$\langle c^{\sigma_m}(t)\rangle = \sum_{\tilde{c}} c^{\sigma_m} \tilde{P}_{\tilde{c}}(t),$$

where the sum is taken over every feasible $\tilde{c}$, *i.e.*, all M-tuples of non-negative integers. Here, $\tilde{P}_{\tilde{c}}(t)$ satisfies the chemical master equation of Eq. (2.17). We can use the master equation to study the evolution of $\langle c^{\sigma_m}(t)\rangle$ for each σ_m. For this, we first multiply Eq. (2.17) by $c^{\sigma_m}(t)$ and sum over all $\tilde{c}$. We illustrate the steps involved in Example 2.13.

Example 2.13 Mass action laws for the birth-death process

We consider the birth-death events of Example 2.6. In this setting, our system consists of only one species whose population $c^{\mathcal{A}}$ has a birth rate λ_b and death rate λ_d. To derive the mass action law, we start with the master equation

$$\frac{d}{dt}\tilde{P}_{c^{\mathcal{A}}}(t) = -\lambda_b \tilde{P}_{c^{\mathcal{A}}}(t) - \lambda_d c^{\mathcal{A}} \tilde{P}_{c^{\mathcal{A}}}(t) + \lambda_b \tilde{P}_{c^{\mathcal{A}}-1}(t) + \lambda_d \left(c^{\mathcal{A}}+1\right) \tilde{P}_{c^{\mathcal{A}}+1}(t).$$

Multiplying both sides of the equation by $c^{\mathcal{A}}$ and summing, we obtain

$$\frac{d}{dt}\sum_{c^{\mathcal{A}}=0}^{\infty} c^{\mathcal{A}}\tilde{P}_{c^{\mathcal{A}}}(t) = -\lambda_b \sum_{c^{\mathcal{A}}=0}^{\infty} c^{\mathcal{A}}\tilde{P}_{c^{\mathcal{A}}}(t) - \lambda_d \sum_{c^{\mathcal{A}}=0}^{\infty} \left(c^{\mathcal{A}}\right)^2 \tilde{P}_{c^{\mathcal{A}}}(t)$$

$$+ \lambda_b \sum_{c^{\mathcal{A}}=0}^{\infty} c^{\mathcal{A}}\tilde{P}_{c^{\mathcal{A}}-1}(t) + \lambda_d \sum_{c^{\mathcal{A}}=0}^{\infty} c^{\mathcal{A}}\left(c^{\mathcal{A}}+1\right)\tilde{P}_{c^{\mathcal{A}}+1}(t).$$

We can immediately simplify this to obtain

$$\frac{d}{dt}\left\langle c^{\mathcal{A}}(t)\right\rangle = -\lambda_b \left\langle c^{\mathcal{A}}(t)\right\rangle - \lambda_d \left\langle \left(c^{\mathcal{A}}(t)\right)^2\right\rangle$$

$$+ \lambda_b \sum_{c^{\mathcal{A}}=0}^{\infty} c^{\mathcal{A}}\tilde{P}_{c^{\mathcal{A}}-1}(t) + \lambda_d \sum_{c^{\mathcal{A}}=0}^{\infty} c^{\mathcal{A}}(c^{\mathcal{A}}+1)\tilde{P}_{c^{\mathcal{A}}+1}(t). \quad (2.18)$$

The third and fourth terms on the right-hand side here, which have not yet been simplified, are averages with respect to a probability taken at $c^{\mathcal{A}}-1$ or $c^{\mathcal{A}}+1$. In order to rewrite these as averages over a probability in $c^{\mathcal{A}}$, we define a new dummy index $\ell = c^{\mathcal{A}}-1$ or $\ell = c^{\mathcal{A}}+1$. For example, for the third term on the right-hand side of Eq. (2.18), we have

$$\lambda_b \sum_{c^{\mathcal{A}}=0}^{\infty} c^{\mathcal{A}}\tilde{P}_{c^{\mathcal{A}}-1}(t) = \lambda_b + \lambda_b \left\langle c^{\mathcal{A}}(t)\right\rangle. \quad (*2.19*)$$

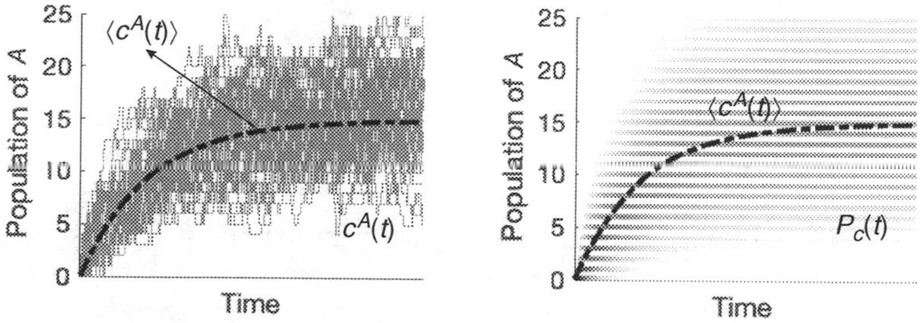

Fig. 2.7 Here, we show cartoon of a birth-death process where the birth rate exceeds the death rate such that the population increases until it reaches steady-state. The left panel compares Gillespie trajectories with mass action, while the right panel compares master equation with mass action.

Substituting the expression above into Eq. (2.18) and repeating the exercise for the fourth term of Eq. (2.18), we obtain

$$\frac{d}{dt}\left\langle c^A(t)\right\rangle = \lambda_b - \lambda_d\left\langle c^A(t)\right\rangle,$$

which is the mass action law for the birth-death process.

Figure 2.7 summarizes the difference between Gillespie simulations, master equations, and mass action laws in the study of birth-death processes. Starting from a zero initial population, we can see that:

- The Gillespie simulation captures individual birth and death events. The population $c^A(t)$ is an integer-valued variable that varies stochastically in time. Different stochastic realizations give rise to different trajectories. By contrast, the mass action law describes only the average population over every possible realization of $c^A(t)$. The average population $\left\langle c^A(t)\right\rangle$ is a continuous variable that changes deterministically across time.
- The master equation itself describes the probability of having a population $c^A(t)$ at every time. Put differently, had we run an infinite number of Gillespie simulations and assessed the probability over each $c^A(t)$, from 0 to ∞, at each time, this probability would be the solution to the coinciding master equation. In particular, the panel on the right-hand side of Fig. 2.7 shows the probability over each $c^A(t)$ for any given time. For example, at small times probabilities for lower values of $c^A(t)$ are higher. We can assess this by considering how the shade of the horizontal lines changes as we move from horizontal line to horizontal line for a fixed time. By contrast, at longer times we have highest probability of being at the average population coinciding with $\left\langle c^A(t)\right\rangle$.

Note 2.20 Mass action laws for more complex reactions

For the birth-death process, as we saw, the mass action law becomes a differential equation of the form $\frac{d}{dt}\left\langle c^A(t)\right\rangle = f\left(\left\langle c^A(t)\right\rangle\right)$ where $f(\cdot)$ is a function of $\left\langle c^A(t)\right\rangle$. In a more general system, however, such closed-form expressions may not exist. That is, depending on how the involved propensities, $\mu^{1:K}_{\tilde{c}}$, depend on $\tilde{c}$, we may very well obtain differential equations where the evolution of the average, $\frac{d}{dt}\langle c^{\sigma_m}(t)\rangle$, is related to the evolution of higher moments, such as $\langle (c^{\sigma_m}(t))^n\rangle$, or in cases with more than one species on mixed moments, such as $\left\langle (c^{\sigma_m}(t))^{n-\ell}\,(c^{\sigma_{m'}}(t))^{\ell}\right\rangle_{m\neq m'}$.

Inevitably, when the evolution of $\langle c^{\sigma_m}(t)\rangle$ depends on higher moments, we have non-closed-form expressions and, to attain closed-form ones, we impose additional assumptions. For instance, suppose that $\frac{d}{dt}\langle c^{\sigma_m}(t)\rangle = f\left(\langle c^{\sigma_m}(t)\rangle, \left\langle (c^{\sigma_m}(t))^2\right\rangle\right)$ and that our goal is to obtain an expression where $f(\cdot,\cdot)$ has no dependency on $\left\langle (c^{\sigma_m}(t))^2\right\rangle$. In this case, it is common to assume that second moments can be reduced to first moments squared, $\left\langle (c^{\sigma_m}(t))^2\right\rangle = \langle c^{\sigma_m}(t)\rangle^2$, and thus that variances, $\left\langle (c^{\sigma_m}(t))^2\right\rangle - \langle c^{\sigma_m}(t)\rangle^2$, are negligible as compared to means $\langle c^{\sigma_m}(t)\rangle$.

2.4 Systems with Discrete State-Spaces in Discrete Time

Now we consider modeling a system's trajectory at *discrete* times only. This may be needed either because our system inherently progresses in steps, for example as an idealization of reoccurring peaks of tidal waves, or as a simplification of a system inherently evolving in continuous time but assessed only at discrete steps. As we will see, formulations of discrete time systems are often artificial and, for their interpretation, we frequently invoke *analogous* continuous time models.

2.4.1 Modeling a System at Discrete Times

To model a system in discrete time, Fig. 2.8, we need to consider a *fixed* grid of time points $t_0, t_1, t_2, \ldots$. We require only that these time points be ordered such that they form a strictly increasing sequence,

$$t_0 < t_1 < t_2 < \cdots < t_n < t_{n+1} < \cdots .$$

Beyond this temporal arrangement, additional assumptions are unnecessary. In most applications, of course, the grid of time points t_n is *regular*. That is, time points repeat at constant periods $t_{n+1} = t_n + \tau$, where τ is either

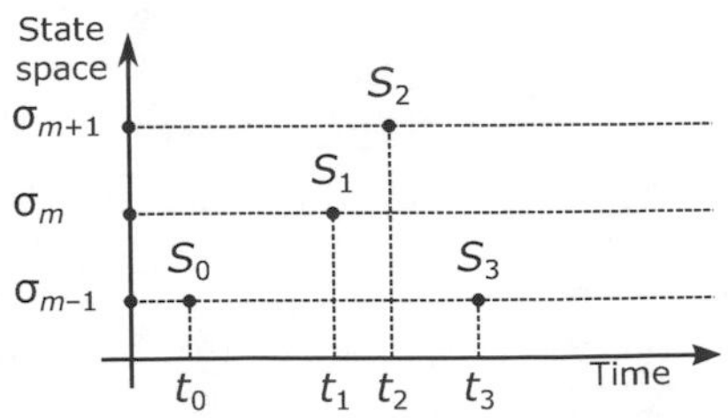

Fig. 2.8 A system with a discrete state-space following discrete time dynamics. This system is described by passing states S_n at given times t_n.

set by a periodically occurring physical phenomenon or the acquisition protocol of individual measurements.

A system that evolves in discrete time is best described by the sequence of successive states

$$S_0 \to S_1 \to S_2 \to \cdots \to S_n \to S_{n+1} \to \cdots,$$

through which the system passes at *precisely* the grid time points $t_0, t_1, t_2, \ldots$. Since our system may be affected by random events, all $S_0, S_1, S_2, \ldots$ are random variables.

Unlike continuous time systems where we focused mostly on holding states, here we emphasize that S_n are *passing* states and that they may maintain the same value over successive steps. This may happen either because the underlying system does not change between successive time points t_n and t_{n+1} or because it changes and changes back again.

Note 2.21 What is a discrete-time trajectory?

The trajectory of a system with discrete state-space evolving in discrete time is a sequence $S_0, S_1, S_2, \ldots$. Whenever our system results from the discretization of an analogous system that evolves in continuous time, in view of Note 2.6, the passing states are related to the underlying continuous time trajectory $\mathcal{S}(\cdot)$ by $S_n = \mathcal{S}(t_n)$. As the trajectory $\mathcal{S}(\cdot)$ is random, the states S_n remain random too, despite being evaluated at fixed times.

As we focus on systems with discrete state-spaces, we will continue denoting constitutive states with $\sigma_{1:M}$. To complete our description, we need to specify an initialization rule for the selection of S_0 and the transition rules for the selection of all subsequent S_n.

Similar to systems evolving in continuous time, the initialization rule takes the form

$$S_0 \sim \text{Categorical}_{\sigma_{1:M}}(\boldsymbol{\rho}), \tag{2.20}$$

where the probability array $\boldsymbol{\rho} = [\rho_{\sigma_1}, \ldots, \rho_{\sigma_M}]$ can model a preference for particular initializations. Under the Markov assumption, the transition rules also take a similar form,

$$S_{n+1}|s_n \sim \text{Categorical}_{\sigma_{1:M}}(\boldsymbol{\pi}_{n,s_n}), \qquad n = 1, 2, \ldots, \tag{2.21}$$

which, in the most general case, allow for transition probability arrays $\boldsymbol{\pi}_{n,\sigma_m} = [\pi_{n,\sigma_m\to\sigma_1}, \ldots, \pi_{n,\sigma_m\to\sigma_M}]$ that may be time dependent and hence we have the added subscript n on the transition probability arrays.

Since our transition rules may allow for self-transitions, $\pi_{n,\sigma_m\to\sigma_m}$ need not be zero. As such, transition probability matrices

$$\begin{array}{c} \\ \sigma_1 \\ \vdots \\ \sigma_M \end{array} \begin{array}{c} \begin{array}{ccc} \sigma_1 & \cdots & \sigma_M \end{array} \\ \begin{bmatrix} \pi_{n,\sigma_1\to\sigma_1} & \cdots & \pi_{n,\sigma_1\to\sigma_M} \\ \vdots & \ddots & \vdots \\ \pi_{n,\sigma_M\to\sigma_1} & \cdots & \pi_{n,\sigma_M\to\sigma_M} \end{bmatrix} \end{array} = \mathbf{\Pi}_n$$

now, in general, assume *nonzero* diagonal elements. As our grid of time points is fixed, the notion of jump times or holding periods that we encountered earlier do not carry over. Accordingly, advancing our system is fully specified by Eqs. (2.20) and (2.21) without additional random variables.

Note 2.22 Markov chains

A sequence of states $S_0, S_1, S_2, \ldots$ sampled as described in Eqs. (2.20) and (2.21) is termed a *Markov chain.* We refer to a Markov chain as *inhomogeneous* when the transition probability matrices $\mathbf{\Pi}_n$ differ from step to step. By contrast, we refer to a Markov chain as *homogeneous* when the transition probability matrices remain the same for all steps. In the former case, we may denote them simply with $\mathbf{\Pi}$ and the corresponding arrays with $\pi_{\sigma_m} = [\pi_{\sigma_m \to \sigma_1}, \ldots, \pi_{\sigma_m \to \sigma_M}]$.

2.4.2 Modeling Kinetic Schemes

To specify a kinetic scheme for our system, there are two major modeling approaches that we may adopt. We may choose to ascribe values directly to the transition probabilities $\pi_{n,\sigma_m \to \sigma_{m'}}$ or we may invoke an analogous system evolving in continuous time and derive these probabilities based on the kinetics of the underlying system. Below, we describe both approaches.

Ascribing Transition Probabilities*

By specifying the values of each $\pi_{n,\sigma_m \to \sigma_{m'}}$, we have direct control of the allowed or disallowed transitions that the system may take. For example, by setting certain weights $\pi_{n,\sigma_m \to \sigma_{m'}}$ to zero, we can explicitly cancel particular transitions. Of course, each array π_{n,σ_m} must remain a probability array, so it must be normalized. This means that every π_{n,σ_m} contains at least one nonzero weight such that there will always be at least one constitutive state available for sampling $S_{n+1}|s_n$.

Example 2.14 Random walks

In this example, we consider a system with $M = 5$ constitutive states that evolves according to a transition matrix

$$\mathbf{\Pi} = \begin{bmatrix} 1/5 & 4/5 & 0 & 0 & 0 \\ 2/5 & 1/5 & 2/5 & 0 & 0 \\ 0 & 2/5 & 1/5 & 2/5 & 0 \\ 0 & 0 & 2/5 & 1/5 & 2/5 \\ 0 & 0 & 0 & 4/5 & 1/5 \end{bmatrix}$$

that remains fixed in time. In other words, our system is a homogenous Markov chain where each state may transition only to its nearby states. Such Markov chains are termed *random walks*. As can be seen in Fig. 2.9, once the system reaches the state-space's edges at σ_1 or σ_5, it can only transition back.

* This is an advanced topic and could be skipped on a first reading.

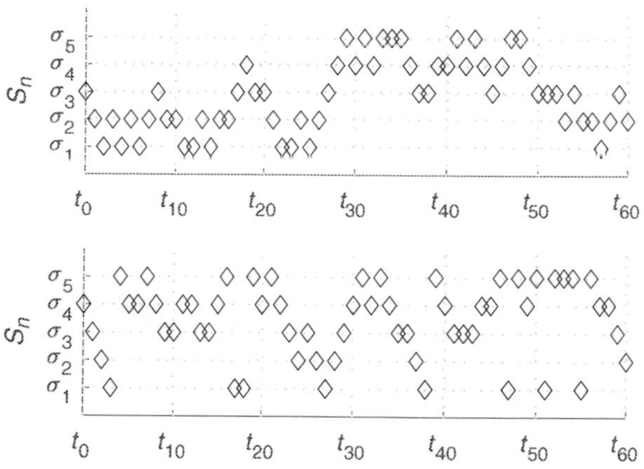

Fig. 2.9 Two Markov chains representing random walks. In the upper panel the system transitions back at the boundaries, while in the lower panel the system exits and reenters from the other end.

If instead, we allow the system to exit and reenter from the other side of the state-space, the transition matrix becomes

$$\mathbf{\Pi} = \begin{bmatrix} 1/5 & 2/5 & 0 & 0 & 2/5 \\ 2/5 & 1/5 & 2/5 & 0 & 0 \\ 0 & 2/5 & 1/5 & 2/5 & 0 \\ 0 & 0 & 2/5 & 1/5 & 2/5 \\ 2/5 & 0 & 0 & 2/5 & 1/5 \end{bmatrix}.$$

Such Markov chains are termed *circular random walks*.

Ascribing Transition Rates

Directly ascribing the values of each $\pi_{n,\sigma_m \to \sigma_{m'}}$ might be cumbersome or, most importantly, might be highly demanding for real-life systems, especially when our time points $t_0, t_1, t_2, \ldots$ are irregular. An alternative approach that avoids these pitfalls proceeds by starting our modeling with an analogous system that evolves in continuous time and uses its propagator to arrive at transition matrices in discrete time. Example 2.15 demonstrates this approach.

Example 2.15 Discretizing a Markov jump process

In this example, we consider an analogous system in continuous time that is modeled by a Markov jump process. For this system, we consider a state-space $\sigma_{1:M}$ and a transition rate matrix $\mathbf{\Lambda}$. As we saw in Note 2.17, from $\mathbf{\Lambda}$ we can readily obtain the generator $\boldsymbol{G}$ and, through Eq. (2.15), we can obtain its propagator $\boldsymbol{Q}^{t \to t'}$ relating the state probabilities at times t and t'.

According to Eq. (2.21), each transition $s_n \to S_{n+1}$ entails a probability matrix $\mathbf{\Pi}_n$. Given that this matrix relates system states at times t_n and t_{n+1}, it is given by

$$\mathbf{\Pi}_n = \boldsymbol{Q}^{t_n \to t_{n+1}} = \exp\left((t_{n+1} - t_n)\boldsymbol{G}\right). \tag{2.22}$$

Since, in general, the matrix exponential does not have a simpler form for systems with large state-spaces, in most cases the matrices $\mathbf{\Pi}_n$ need to be evaluated numerically, even when all rates contained within $\mathbf{\Lambda}$ are known.

As a concrete example, we consider an underlying Markov jump process with $M = 2$ constitutive states. Such a system is characterized by two rates $\lambda_{\sigma_1 \to \sigma_2}, \lambda_{\sigma_2 \to \sigma_1}$ and the rate matrix

$$\mathbf{\Lambda} = \begin{bmatrix} 0 & \lambda_{\sigma_1 \to \sigma_2} \\ \lambda_{\sigma_2 \to \sigma_1} & 0 \end{bmatrix}.$$

Provided that such a Markov jump process is assessed at regular times $t_n = t_0 + n\tau$, the resulting transition probability matrices are related to the rates $\lambda_{\sigma_1 \to \sigma_2}$ and $\lambda_{\sigma_2 \to \sigma_1}$ by

$$\mathbf{\Pi}_n = \exp\left(\tau \begin{bmatrix} -\lambda_{\sigma_1 \to \sigma_2} & \lambda_{\sigma_1 \to \sigma_2} \\ \lambda_{\sigma_2 \to \sigma_1} & -\lambda_{\sigma_2 \to \sigma_1} \end{bmatrix}\right).$$

Owing to the lack of memory of the underlying Markov jump process *and* the regularity of the time grid, the resulting transition matrices are homogenous. Namely, the resulting $\mathbf{\Pi}_n$ are the same for all steps, so we may denote them with $\mathbf{\Pi}$. Due to the small size of the state-space, in this case we can compute the transition matrix analytically,

$$\mathbf{\Pi} = \frac{1}{\lambda} \begin{bmatrix} \lambda_{\sigma_2 \to \sigma_1} + \lambda_{\sigma_1 \to \sigma_2} e^{-\lambda\tau} & \lambda_{\sigma_1 \to \sigma_2} - \lambda_{\sigma_1 \to \sigma_2} e^{-\lambda\tau} \\ \lambda_{\sigma_2 \to \sigma_1} - \lambda_{\sigma_2 \to \sigma_1} e^{-\lambda\tau} & \lambda_{\sigma_1 \to \sigma_2} + \lambda_{\sigma_2 \to \sigma_1} e^{-\lambda\tau} \end{bmatrix},$$

where $\lambda = \lambda_{\sigma_1 \to \sigma_2} + \lambda_{\sigma_2 \to \sigma_1}$.

Note 2.23 Interpreting transition probabilities

Starting with some transition probability matrix $\mathbf{\Pi}_n$, it is tempting to use Eq. (2.22) to obtain the generator $\boldsymbol{G}$ from which, in turn, we may extract rates $\lambda_{\sigma_m \to \sigma_{m'}}$. For example, according to Note 2.18, we may approximate $\exp(\ell \boldsymbol{L}) \approx \mathbb{1} + \ell \boldsymbol{L}$ and arrive at

$$\boldsymbol{G} \approx \frac{\mathbf{\Pi}_n - \mathbb{1}}{t_{n+1} - t_n}.$$

Although this approximation is valid provided $(t_{n+1} - t_n)\lambda_{\sigma_m \to \sigma_{m'}} \ll 1$ holds for all involved reactions, generally such an approach is unsafe. Despite the errors it introduces in the analysis, it pre-assumes that the underlying dynamical system behaves as a Markov jump process.

2.4.3 Quantifying State Persistence*

A system with pronounced state persistence often exhibits self-transitions. In these systems, $\pi_{n,\sigma_m \to \sigma_m}$ are the dominant weights in each π_{n,σ_m} forming $\mathbf{\Pi}_n$. Figure 2.10 illustrates two representative cases. The trajectories shown in the upper and lower panels are sampled with respective matrices

$$\mathbf{\Pi} = \begin{bmatrix} 1/3 & 1/3 & 1/3 \\ 1/3 & 1/3 & 1/3 \\ 1/3 & 1/3 & 1/3 \end{bmatrix}, \qquad \mathbf{\Pi} = \begin{bmatrix} 7/9 & 1/9 & 1/9 \\ 1/9 & 7/9 & 1/9 \\ 1/9 & 1/9 & 7/9 \end{bmatrix}.$$

* This is an advanced topic and could be skipped on a first reading.

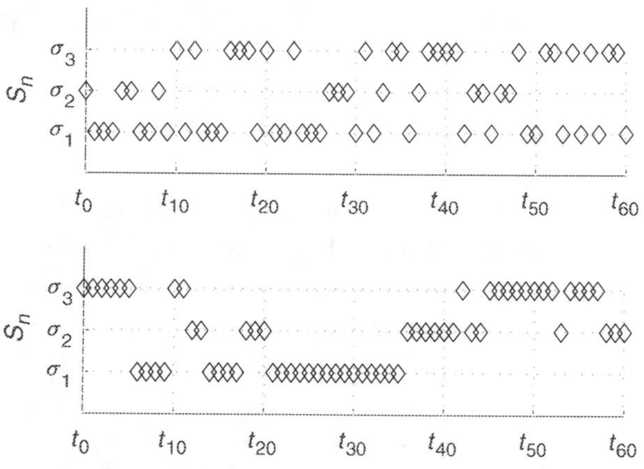

Fig. 2.10 Two systems with different state persistences.

As can be seen, the first system jumps to a different constitutive state at almost every step. By contrast, the second system is more "sticky" and typically stays in the same constitutive state for several steps before escaping.

Below, we show how to quantify state persistence. To begin, we suppose that our system reaches a constitutive state σ_m at some step n. In other words, $s_n = \sigma_m$. We also assume that $s_{n-1} \neq \sigma_m$, so the transition $s_{n-1} \to S_n$ initiates a dwell phase. This phase may be terminated immediately, in which case it is followed by a transition $s_n \to S_{n+1} \neq \sigma_m$; or may be terminated after one step, in which case it is followed by $s_n \to S_{n+1} = \sigma_m \to S_{n+2} \neq \sigma_m$; or after two steps and so on.

To be more quantitative, we denote by D_{n,σ_m} the number of steps that our system needs in order to escape σ_m given that it enters σ_m at the nth step. This is a random variable with values $0, 1, 2, \ldots$ that correspond to the following sequences:

$$
\begin{aligned}
D_{n,\sigma_m} &= 0 \quad \cdots \to s_{n-1} \neq \sigma_m \to s_n = \sigma_m \to s_{n+1} \neq \sigma_m \to \cdots, \\
D_{n,\sigma_m} &= 1 \quad \cdots \to s_{n-1} \neq \sigma_m \to s_n = \sigma_m \to s_{n+1} = \sigma_m \to s_{n+2} \neq \sigma_m \to \cdots, \\
D_{n,\sigma_m} &= 2 \quad \cdots \to s_{n-1} \neq \sigma_m \to s_n = \sigma_m \to s_{n+1} = \sigma_m \to s_{n+3} = \sigma_m \to s_{n+3} \\
&\neq \sigma_m \to \cdots,
\end{aligned}
$$

which we can use to derive precise dwell probabilities. For instance, according to Eq. (2.21), these probabilities are proportional to

$$
\begin{aligned}
D_{n,\sigma_m} &= 0 \qquad (1 - \pi_{n,\sigma_m \to \sigma_m}), \\
D_{n,\sigma_m} &= 1 \qquad \pi_{n,\sigma_m \to \sigma_m}(1 - \pi_{n+1,\sigma_m \to \sigma_m}), \\
D_{n,\sigma_m} &= 2 \qquad \pi_{n,\sigma_m \to \sigma_m}\pi_{n+1,\sigma_m \to \sigma_m}(1 - \pi_{n+2,\sigma_m \to \sigma_m}),
\end{aligned}
$$

from which we readily deduce, more generally, that a dwell lasting for d steps in σ_m starting at the nth step has probability

$$
P_{n,\sigma_m}(d) = \left(\prod_{n'=n}^{n+d-1} \pi_{n',\sigma_m \to \sigma_m} \right) (1 - \pi_{n+d,\sigma_m \to \sigma_m}).
$$

For homogenous Markov chains, the transition probabilities are constant in time and $P_{n,\sigma_m}(d)$ as well as D_{n,σ_m} become independent of n. In particular, $P_{n,\sigma_m}(d)$ takes the form

$$P_{\sigma_m}(d) = (\pi_{\sigma_m \to \sigma_m})^d (1 - \pi_{\sigma_m \to \sigma_m}),$$

which indicates that $D_{\sigma_m} \sim \text{Geometric}(1 - \pi_{\sigma_m \to \sigma_m})$; see Appendix B for the definition of the geometric distribution. We can conclude that once our system enters a constitutive state σ_m, on average it remains for an additional $\pi_{\sigma_m \to \sigma_m}/(1 - \pi_{\sigma_m \to \sigma_m})$ steps.

Note 2.24 Interpreting dwell times

The notion of dwell or holding *time* for a system in discrete time lacks physical meaning. As seen from $P_{n,\sigma_m}(d)$ and $P_{\sigma_m}(d)$, we can *only* quantify the statistics of steps d needed to escape a constitutive state. Of course, we may invoke the reference grid of time points t_n and always interpret these statistics in temporal terms. For example, with a homogenous Markov chain at regular time points $t_n = t_0 + n\tau$, we may transform the mean number of steps needed to escape a state to a mean dwell time $\tau \pi_{\sigma_m \to \sigma_m}/(1 - \pi_{\sigma_m \to \sigma_m})$. However, we need to be careful with our definitions, as what is meant by "time" or "period" in the context of a discrete system is ambiguous.

2.5 Systems with Continuous State-Spaces in Discrete Time

We now turn to systems whose state-spaces form a *continuum.* In this section, we consider modeling such systems using discrete time steps. That is, only at a predefined grid of time points. To be more specific, we exclusively consider fixed time points that we denote $t_0, t_1, t_2, \ldots$. As before, we assume this grid is temporally arranged $t_n < t_{n+1}$, although not necessarily regularly. For clarity, we denote by r a generic constitutive state and with $R_0, R_1, R_2, \ldots$ the states through which the system *passes precisely* at the grid times. As our systems are stochastic, each R_n is a random variable that attains specific values that we denote r_n. Such r_n may be equal to any r in the state-space's continuum.

Our passing states R_n are now continuous random variables. For this reason, the categorical distributions that we have used so far are inadequate to formulate initialization and transition rules for our system. Instead, these rules must now be formulated using probability distributions that allow for continuous variables. To describe the initialization rule and transition rules in a continuous state-space, we need the distributions of the random variables

$$\begin{aligned} R_0 &\sim \mathbb{R}, \\ R_{n+1}|r_n &\sim \mathbb{Q}_n(r_n), \qquad n = 1, 2, \ldots, \end{aligned} \tag{2.23}$$

associated with probability densities $\rho(r_0)$ and $q_n(r_{n+1}|r_n)$. These are the direct analogs of Eqs. (2.20) and (2.21) for the discrete state-space systems discussed earlier.

Note 2.25 Initialization and transition densities

Informally, the densities specifying our system's dynamics can be seen as

$$\begin{pmatrix} \text{probability of} \\ \text{starting } dr \text{ near } r \end{pmatrix} = \rho(r)dr, \qquad \begin{pmatrix} \text{probability of jumping} \\ \text{from } r' \text{ to } dr \text{ near } r \end{pmatrix} = q_n(r|r')dr.$$

In this setup, we have *one* initial density $\rho(\cdot)$ and *a family* of transition densities $q_n(\cdot|r')$. From the mathematical point of view, these are functions mapping the system's state-space to the nonnegative real numbers. These functions are also probability densities for which the normalization conditions read

$$\int dr\, \rho(r) = 1, \qquad \int dr\, q_n(r|r') = 1.$$

Each member in the family of transition densities $q_n(\cdot|r')$ corresponds to each constitutive state r'. Occasionally, the entire family $q_n(\cdot|\cdot)$ is collectively termed a *transition kernel*. For systems with a kernel $q_n(\cdot|\cdot)$ that changes over time, the dynamics described are termed *inhomogeneous;* while, for systems with a kernel that does not change over time, which we may denote simply by $q(\cdot|\cdot)$, the dynamics described are termed *homogeneous*.

The exact form of the densities $\rho(r)$ and $q_n(r|r')$ depends on the system at hand. Over the next sections we explore some common cases. Initially, we show how such distributions can be obtained to describe a physical system for which equations of motion are available. Subsequently, we show how such distributions may be modeled for artificial systems for which equations of motion are not available.

2.5.1 Modeling under Equations of Motion

We consider a classical mechanical system for which the constitutive states are given by $\boldsymbol{r} = (q, p)$. Here, q is a position and p a momentum variable in a continuous state-space. For simplicity, we initially consider a system whose q and p are both scalars. In our setting, the passing states of the system are $\boldsymbol{R}_n = (Q_n, P_n)$ and are evaluated at our grid of times t_n.

A Point Mass under State Independent Forces

We begin by first discussing systems with no stochasticity. For example, a classical point mass subject to a force $F(t)$ that may vary temporally but *not* spatially evolves according to

$$dq = \frac{p}{m}dt, \tag{2.24}$$

$$dp = F(t)dt. \tag{2.25}$$

These are the equations of motion as given by Newton's second law and mathematically represent the evolution of our point mass m within the *phase-space* continuum of p and q.

Note 2.26 Differential equations

Equations (2.24) and (2.25) are equivalent to the more familiar

$$\dot{q} = \frac{p}{m}, \tag{2.26}$$

$$\dot{p} = F(t), \tag{2.27}$$

where $\dot{q}$ and $\dot{p}$ stand for the temporal derivatives dq/dt and dp/dt, respectively.

Similar to all descriptions mediated by differential equations, both Eqs. (2.24) and (2.25) and Eqs. (2.26) and (2.27) need to be integrated in order to have stand-alone meaning. That is,

$$\int_{t'}^{t''} dq = \int_{t'}^{t''} dt \frac{p}{m}, \tag{2.28}$$

$$\int_{t'}^{t''} dp = \int_{t'}^{t''} dt\, F(t). \tag{2.29}$$

As we will see, integral forms of the equations of motion are central to the derivation of transition rules. What we want to emphasize here is that mathematical models expressed in terms of differential equations *involve a notion of integration*, even when integrals are only implicit. Simple integrals such as those on the left-hand side are easy to interpret:

$$\int_{t'}^{t''} dq = q\left(t''\right) - q\left(t'\right),$$

$$\int_{t'}^{t''} dp = p\left(t''\right) - p\left(t'\right).$$

However, the interpretation of the integrals on the right-hand side is more challenging and very often depends critically on the properties of the integrands.

We can use the equations of motion to relate the passing states of this system across time. In particular, integrated between times t' and t'', Eqs. (2.24) and (2.25) lead to

$$q(t'') = q(t') + \frac{t'' - t'}{m} p(t') + \frac{1}{m} \int_{t'}^{t''} d\tau \int_{t'}^{\tau} d\tau' \, F(\tau'), \tag{2.30}$$

$$p(t'') = p(t') + \int_{t'}^{t''} d\tau \, F(\tau), \tag{2.31}$$

where we obtained Eq. (2.30) by first solving for $p(t)$ from Eq. (2.29) to obtain Eq. (2.31) and inserting this expression into Eq. (2.28).

Applied at two successive times in our grid, $t' = t_n$ and $t'' = t_{n+1}$, these equalities indicate that successive passing states are linked via

$$Q_{n+1} = Q_n + (t_{n+1} - t_n)\frac{P_n}{m} + \frac{1}{m}\int_{t_n}^{t_{n+1}} d\tau \int_{t_n}^{\tau} d\tau' \, F(\tau'),$$

$$P_{n+1} = P_n + \int_{t_n}^{t_{n+1}} d\tau \, F(\tau).$$

Provided that t_{n+1} and t_n are fairly close to each other, then we may approximate $\int_{t_n}^{t_{n+1}} d\tau \, F(\tau) \approx (t_{n+1} - t_n)F(t_n)$ and $\int_{t_n}^{t_{n+1}} d\tau \int_{t_n}^{\tau} d\tau' \, F(\tau') \approx 0$. If we ignore the errors introduced by the approximations, we see that successive passing states are linked via different expressions,

$$Q_{n+1} = Q_n + (t_{n+1} - t_n)\frac{P_n}{m}, \tag{2.32}$$

$$P_{n+1} = P_n + (t_{n+1} - t_n)F(t_n). \tag{2.33}$$

Note 2.27 What are the transition distributions?

Since our toy system evolves deterministically, in statistical language we may summarize the transition rules by Dirac deltas. In particular, Eqs. (2.32) and (2.33) read as

$$Q_{n+1}|(q_n, p_n) \sim \delta_{q_n + (t_{n+1} - t_n)\frac{p_n}{m}},$$

$$P_{n+1}|(q_n, p_n) \sim \delta_{p_n + (t_{n+1} - t_n)F(t_n)}.$$

Somewhat pedantically, the corresponding transition kernel is

$$q_n\,(q, p|q', p') = \delta_{q' + (t_{n+1} - t_n)\frac{p'}{m}}(q)\delta_{p' + (t_{n+1} - t_n)F(t_n)}(p).$$

A Point Mass under State Dependent Forces

According to Newton's second law, a point mass subject to a general force $F(t, \mathbf{r})$ that may be state-dependent evolves according to

$$dq = \frac{p}{m}dt,$$

$$dp = F(t, \mathbf{r})dt.$$

Just as with state-independent forces, we now show how to derive appropriate transition rules relating our system's passing states across time. As before, by integrating the equations of motion between times t' and t'' we obtain

$$q(t'') = q(t') + (t'' - t')\frac{p(t')}{m} + \frac{1}{m}\int_{t'}^{t''} d\tau \int_{t'}^{\tau} d\tau' \, F\,(\tau', \mathbf{r}(\tau')),$$

$$p(t'') = p(t') + \int_{t'}^{t''} d\tau \, F\,(\tau, \mathbf{r}(\tau)).$$

Provided $F\,(\tau, \mathbf{r}(\tau))$ does not change considerably between t' and t'', we may invoke an approximation $\int_{t'}^{\tau} d\tau' \, F\,(\tau', \mathbf{r}(\tau')) \approx (\tau - t')F\,(t', \mathbf{r}(t'))$ and readily arrive at

$$q(t'') \approx q(t') + (t'' - t')\frac{p(t')}{m} + \frac{(t'' - t')^2}{2m}F\left(t', \boldsymbol{r}(t')\right),$$
$$p(t'') \approx p(t') + (t'' - t')F\left(t', \boldsymbol{r}(t')\right).$$

In turn, applied at successive times, $t' = t_n$ and $t'' = t_{n+1}$, and ignoring the errors introduced by our approximation, these equalities indicate that successive passing states are linked via

$$Q_{n+1} = Q_n + (t_{n+1} - t_n)\frac{P_n}{m} + \frac{(t_{n+1} - t_n)^2}{2m}F\left(t_n, \boldsymbol{R}_n\right),$$
$$P_{n+1} = P_n + (t_{n+1} - t_n)F\left(t_n, \boldsymbol{R}_n\right).$$

Provided that t_{n+1} and t_n are fairly close to each other, then we may further approximate $(t_{n+1} - t_n)^2 \approx 0$. In this case, we see that successive passing states are linked via simpler expressions,

$$Q_{n+1} = Q_n + (t_{n+1} - t_n)\frac{P_n}{m},$$
$$P_{n+1} = P_n + (t_{n+1} - t_n)F\left(t_n, \boldsymbol{R}_n\right).$$

Note 2.28 Temporal discretization

To derive the transition rules of this system we had to invoke certain *approximations.* Initially, we assumed that $F(t, \boldsymbol{r}(t)) \approx F\left(t_n, \boldsymbol{r}(t_n)\right)$ holds for any τ between t_n and t_{n+1} and, additionally, that $(t_{n+1} - t_n)^2 \approx 0$ and derived a different rule. Clearly, approximations and numerical schemes like these determine when the resulting rule is valid or invalid as well as its functional form. While different approximations, for instance $F(t, \boldsymbol{r}(t)) \approx F\left(t_{n+1}, \boldsymbol{r}(t_{n+1})\right)$ or even $F(t, \boldsymbol{r}(t_{n+1})) \approx \frac{1}{2}\left(F\left(t_n, \boldsymbol{r}(t_n)\right) + F\left(t_{n+1}, \boldsymbol{r}(t_{n+1})\right)\right)$, may lead to more accurate transition rules, they may result in a more complex relationship between passing states.

A Point Mass under State-Dependent Forces and Noise

While the evolution of Newtonian systems like point masses in isolation is inherently deterministic, uncertainty that necessitates a stochastic approach may stem from the initial conditions. More generally, however, interactions that may be developed between the system of interest and its environment motivate the incorporation of uncertainty impacting the system's transitions as well. In this case, the equation of motion satisfied by our system needs to be supplemented by a randomly fluctuating component, often colloquially termed *noise*. Next, we describe systems with explicit noise representation.

One of the simplest models according to which a point mass may evolve under the influence of noise is

$$dq = \frac{p}{m}dt,$$
$$dp = F(t, \boldsymbol{r})dt + c\, dW_t. \tag{2.34}$$

This is a phenomenological modification of Newton's second law that accounts for a stochastic noise term $c\,dW_t$. The factor c is a positive *constant* that dictates the strength of the force fluctuations and the factor dW_t, termed the *Wiener process*, dictates its temporal effect. The latter is defined only indirectly by the characteristic property

$$\frac{\int_{t'}^{t''} dW_t}{\sqrt{|t''-t'|}} = \xi, \qquad \xi \sim \text{Normal}(0,1). \tag{2.35}$$

This means that the fluctuations accumulated over a time interval form a *random variable* with normal statistics of zero mean and *variance* growing linearly with the duration of the given interval

$$\int_{t'}^{t''} dW_t \sim \text{Normal}(0, |t''-t'|). \tag{2.36}$$

Note 2.29 Wiener process

Here, we briefly highlight some key properties of the *Wiener process*, dW_t, and begin by noting that dW_t is really only indirectly defined through its integral Eq. (2.35) or, equivalently, Eq. (2.36). From these definitions, it also follows that if $[t_1, t_2]$ and $[t_3, t_4]$ coincide with nonoverlapping intervals, then the dW_t integrated over these nonoverlapping regions are statistically independent. For this reason, the noise modeled by dW_t is sometimes termed uncorrelated or white when assessed over nonoverlapping regions. Finally, we end with a quick note on units. From Eq. (2.35), we see that ξ is unitless and thus dW_t has units of square root of time.

To derive transition densities under this equation of motion, just as before we integrate between times t' and t'' to obtain

$$q(t'') = q(t') + (t''-t')\frac{p(t')}{m} + \frac{1}{m}\int_{t'}^{t''} d\tau \int_{t'}^{\tau} d\tau' \left(F(\tau', \mathbf{r}(\tau')) + cW_\tau\right),$$

$$p(t'') = p(t') + \int_{t'}^{t''} d\tau\, F(\tau, \mathbf{r}(\tau)) + c\int_{t'}^{t''} dW_t.$$

Invoking the same approximations as before and Eq. (2.35), we arrive at

$$q(t'') \approx q(t') + (t''-t')\frac{p(t')}{m},$$

$$p(t'') \approx p(t') + (t''-t')F(t', \mathbf{r}(t')) + c\sqrt{|t''-t'|}\,\xi, \quad \xi \sim \text{Normal}(0,1).$$

Finally, applied at $t' = t_n$ and $t'' = t_{n+1}$ and ignoring the approximation, successive passing states are linked via

$$Q_{n+1} = Q_n + (t_{n+1}-t_n)\frac{P_n}{m}, \tag{2.37}$$

$$P_{n+1} = P_n + (t_{n+1}-t_n)F(t_n, \boldsymbol{R}_n) + c\sqrt{t_{n+1}-t_n}\,\xi_n, \quad \xi_n \sim \text{Normal}(0,1). \tag{2.38}$$

Note 2.30 What are the transition distributions?

In statistical language we may summarize the transition rules given by Eqs. (2.37) and (2.38) by

$$Q_{n+1}|(q_n, p_n) \sim \delta_{q_n+(t_{n+1}-t_n)\frac{p_n}{m}},$$
$$P_{n+1}|\boldsymbol{r}_n \sim \text{Normal}\left(p_n + (t_{n+1} - t_n)F(t_n, \boldsymbol{r}_n), (t_{n+1} - t_n)c^2\right).$$

The corresponding transition kernel is

$$q_n(\boldsymbol{r}|\boldsymbol{r}') = \delta_{q'+(t_{n+1}-t_n)\frac{p'}{m}}(q)\,\text{Normal}\left(p; p' + (t_{n+1} - t_n)F(t_n, \boldsymbol{r}'), (t_{n+1} - t_n)c^2\right).$$

2.5.2 Modeling under Increments

In the toy systems we saw, successive passing states ended up being linked in an additive manner,

$$R_{n+1} = R_n + \Delta R_n, \tag{2.39}$$

for appropriate increments $\Delta R_0, \Delta R_1, \ldots$. Such increments can be equivalently used as the primitive variables to model the successive transitions of our system. Namely, to describe the initialization rule and transition rules in a continuous state-space, it is sufficient if we have the distributions of the random variables

$$R_0 \sim \mathbb{R},$$
$$\Delta R_n | r_n \sim \mathbb{G}_n(r_n), \qquad n = 0, 1, 2, \ldots,$$

which are associated with probability densities $\rho(r_0)$ and $g_n(\Delta r_n|r_n)$.

Note 2.31 Incremental descriptions

In our setting, an increment ΔR_n is a random variable that takes specific values Δr_n. Such values, in general, are equal to some difference $\Delta r = r - r'$ that can be formed between any pair r, r' of constituent states in our system's state-space continuum. Normalization of the increment's density reads as

$$\int d\Delta r\, g_n(\Delta r|r') = 1.$$

Example 2.16 Transitions of a point mass system

In the point mass system moving under the influence of a general force and noise, Eqs. (2.37) and (2.38) adopt the form

$$Q_{n+1} = Q_n + \Delta Q_n,$$
$$P_{n+1} = P_n + \Delta P_n,$$

with increments

$$\Delta Q_n = (t_{n+1} - t_n)\frac{P_n}{m},$$

$$\Delta P_n = (t_{n+1} - t_n)F(t_n, \boldsymbol{R}_n) + c\sqrt{t_{n+1} - t_n}\xi_n, \qquad \xi_n \sim \text{Normal}(0, 1).$$

The increment's density then reads

$$g_n(\Delta q, \Delta p|q', p')$$
$$= \delta_{(t_{n+1}-t_n)\frac{p'}{m}}(\Delta q)\, \text{Normal}\left(\Delta p; (t_{n+1} - t_n)F(t_n, (q', p')), (t_{n+1} - t_n)c^2\right).$$

In an incremental description of a dynamical system, we have random variables R_0 and $\Delta R_0, \Delta R_1, \ldots$. Through the increments, we may describe transitions even in the absence of explicit equations of motion. Nevertheless, as increments and passing states are staggered,

$$R_0 \xrightarrow{\Delta R_0} R_1 \xrightarrow{\Delta R_1} R_2 \xrightarrow{\Delta R_2} \cdots,$$

we cannot avoid the initialization rule, which must be specified separately.

Although incremental descriptions may not always be available, generally they are preferred whenever possible as they are simpler. In Section 2.6 we will see that incremental descriptions generalize to continuous time as well. This is not generally the case for nonincremental descriptions derived from equations of motion.

2.5.3 A Case Study in Langevin Dynamics and Brownian Motion*

For microscopic systems idealized as point masses, the fluctuating force in Eq. (2.34) models the effect imparted by collisions from the environment and thus does not strictly satisfy energy conservation. More precisely, the fluctuating force captures electrostatic and other interactions of unobserved particles contained in the medium in which the point mass of interest is embedded. In practice, particles of the medium are often understood as being solvent molecules or other point masses not explicitly modeled interacting with the point mass of interest.

Thermal physics dictates that if the fluctuating force provides an energy source, then there must also exist an energy sink such as friction. Friction was not made explicit previously.

The Langevin Equation

One of the simplest dynamical models according to which a realistic point mass may evolve is the *Langevin model*. Put differently, we say the point mass satisfies Langevin dynamics. This model consists of the following phenomenological equations of motion:

$$dq = \frac{p}{m}dt,$$

$$dp = \underbrace{-\nabla_q U(q)dt - \zeta\frac{p}{m}dt}_{F(t,r)dt} + \sqrt{2}B\, dW_t, \qquad (2.40)$$

* This is an advanced topic and could be skipped on a first reading.

where ζ is the particle's *friction coefficient* and $\nabla_q U(\cdot)$ is the gradient of a conservative potential $U(\cdot)$ with respect to q. We note immediately that the force, $F(t, \boldsymbol{r})$, was partitioned into a portion depending on position and momentum. For thermal systems, it is customary to redesignate the strength of the noise, previously termed c, as $\sqrt{2}B$.

To derive transition densities under the Langevin model, just as before, we integrate position once, and following similar approximations as we did in Section 2.5.1, obtain

$$q_{n+1} = q_n + (t_{n+1} - t_n)\frac{p_n}{m}.$$

However, we now need to pay special attention to the integration of the momentum, which reads

$$p_{n+1} = p_n - \int_{t_n}^{t_{n+1}} dt' \, \nabla_q U\left(q(t')\right) - \frac{\zeta}{m}\int_{t_n}^{t_{n+1}} dt' \, p(t') + \sqrt{2}B \int_{t_n}^{t_{n+1}} dW_t.$$

The second and third terms on the right-hand side of the above may now be approximated as before, assuming $\nabla_q U\left(q(t)\right)$ and $p(t)$ vary negligibly over the time interval over which we integrate. However, the third term of the right-hand side is integrated exactly and, according to Eq. (2.35), yields $B\sqrt{2\left(t_{n+1} - t_n\right)}\xi_n$ where $\xi_n \sim \text{Normal}(0, 1)$. Putting it all together, we have

$$\begin{aligned} p_{n+1} &= p_n \left(1 - \frac{\zeta}{m}(t_{n+1} - t_n)\right) \\ &\quad - (t_{n+1} - t_n)\nabla_q U\left(q(t_n)\right) + B\sqrt{2\left(t_{n+1} - t_n\right)}\xi_n, \quad \xi_n \sim \text{Normal}(0, 1). \end{aligned}$$

In terms of Eq. (2.23), the Langevin transition densities can be summarized as

$$\begin{aligned} Q_{n+1}|(q_n, p_n) &\sim \delta_{q_n + (t_{n+1} - t_n)\frac{p_n}{m}}, \\ P_{n+1}|(q_n, p_n) &\sim \text{Normal}\left(p_n\left(1 - \frac{\zeta}{m}(t_{n+1} - t_n)\right)\right. \\ &\qquad \left. - (t_{n+1} - t_n)\nabla_q U\left(q_n\right), 2B^2(t_{n+1} - t_n)\right). \end{aligned}$$

The Physics behind the Langevin Equation

In the natural sciences, the Langevin equation, explicitly extended to multiple spatial dimensions here, is often written as

$$\dot{q} = \frac{p}{m}, \tag{2.41}$$

$$\dot{p} = -\nabla_q U(q) - \zeta\frac{p}{m} + F_p(t). \tag{2.42}$$

Here, $F_p(t)$, standing in $\sqrt{2}B\dot{W}_t$, is simply an instantaneous random force in time. The random force is understood to be a vector with as many components as we have momenta. For example, in three dimensions while working in Cartesian coordinates, $F_p(t)$ is $\left(F_p^x(t), F_p^y(t), F_p^z(t)\right)$. Similarly,

$\dot{W}_t$ is $\left(\dot{W}_t^x, \dot{W}_t^y, \dot{W}_t^z\right)$ where, for simplicity, we can assume isotropic noise. That is, we assume the strength of noise is the same in all directions.

While intuitive, Eqs. (2.41) and (2.42) are mathematically problematic. On the one hand, the left-hand side of the equation, $\dot{p}$, is to be interpreted in continuous time while it is unclear how to interpret $F_p(t)$ without a surrounding integral. For this reason, care must be provided in specifying the statistics satisfied by the force in order for Eqs. (2.41) and (2.42) to be consistent with Eq. (2.40). In this spirit, if each component of dW_t is to independently satisfy the statistics provided in Eq. (2.35), then $F_p(t)$ must satisfy

$$\left\langle F_p(t)\right\rangle = 0, \qquad \left\langle F_p(t)\cdot F_p(t')\right\rangle = 3\times 2B^2\delta_{t'}(t), \tag{2.43}$$

where, for clarity, we have emphasized the contraction of the three-component random vector using a dot product and will continue doing so, when necessary, for the remainder of this chapter. This contraction produces a factor of three also made explicit. The Dirac delta appearing on the right-hand side of Eq. (2.43) implies that the noise here is *time-decorrelated*.

The Langevin equation of Eqs. (2.41) and (2.42) can be made to look like the more familiar Newton's second law by taking a time derivative of $\dot{q}$ in Eq. (2.42) and inserting this expression into the expression for $\dot{p}$, resulting in

$$m\ddot{q} = -\nabla_q U(q) - \zeta\dot{q} + F_p(t). \tag{2.44}$$

Equation (2.44) is sometimes termed the *underdamped Langevin equation* as it contains the inertial mass of the point particle.

Note 2.32 Some assumptions of the Langevin model

We now highlight some assumptions inherent in Eq. (2.44) or, equivalently, Eqs. (2.41) and (2.42). First, as written in Eq. (2.44), this Langevin equation assumes a classical dynamical paradigm and models the particle's diffusion in a scalar potential $U(q)$ and otherwise homogeneous and isotropic medium. The diffusion coefficient, implicit in assuming the same B for each dimension, becomes independent of space. Perhaps less intuitive is the assumption that the solvent in which the point particle is embedded, which drives both dissipative and fluctuating forces, instantaneously readjusts to the position of the particle, of effective inertial mass m, causing a constant drag on the particle, ζ, and inducing $F_p(t)$. Relaxing the latter assumption, the explanation of which is beyond the scope of this case study, leads to the *generalized Langevin equation*.

Physics further dictates a relationship between ζ and B. Put differently, a particle cannot indefinitely accumulate energy if dissipation is too weak, nor can a particle's average kinetic energy drop to zero if dissipation is too strong. This is because equilibrium statistical physics insists that the average translational kinetic energy in three dimensions be $3k_BT/2$ where T is the equilibrium temperature and k_B is Boltzmann's constant. Taken together, these conditions impose the fact that ζ and B are related to each other through a relation called the fluctuation–dissipation theorem and both are related to the temperature.

Example 2.17 Fluctuation–dissipation theorem

To illustrate the fluctuation–dissipation theorem, we use the integrating factor method to rewrite Eq. (2.42) as

$$p(t) = e^{-\frac{\zeta}{m}t}p(0) + \int_0^t ds\, e^{-\frac{\zeta}{m}(t-s)}\left(-\nabla_q U(q(s)) + F_p(s)\right), \tag{2.45}$$

or, simply,

$$\frac{p(t)}{m} = v(t) = \frac{dq}{dt} = e^{-\frac{\zeta}{m}t}\left(\frac{dq(0)}{dt} - \frac{1}{m}\int_0^t ds\, e^{\frac{\zeta}{m}s}\nabla_q U(q(s))\right) + F_q(t), \tag{2.46}$$

where $v(t)$ is the velocity, and

$$F_q(t) = \frac{1}{m}\int_0^t ds\, e^{-\frac{\zeta}{m}(t-s)}F_p(s), \tag{2.47}$$

where it is understood that $F_q(t)$ has units of velocity.

In order to derive the relationship between ζ and B, we first note a result of statistical physics. That is, at equilibrium, the kinetic energy, written as $\frac{m}{2}v_{\text{eq}}^2$ and computed from $\lim_{t\to\infty}\frac{m}{2}\left\langle v^2(t)\right\rangle$, takes on the value $3k_BT/2$ where the expectation is over all realizations of the noise.

In order to leverage this result, we now take the square of Eq. (2.46) and average over noise, using Eq. (2.43), which yields

$$\begin{aligned}\left\langle v^2(t)\right\rangle &= \langle v(t)\cdot v(t)\rangle \\ &= \left(e^{-\frac{\zeta}{m}t}\left(\frac{dq(0)}{dt} - \frac{1}{m}\int_0^t ds\, e^{\frac{\zeta}{m}s}\nabla_q U(q(s))\right)\right)^2 + \left\langle F_q(t)\cdot F_q(t)\right\rangle.\end{aligned} \tag{2.48}$$

The final term can be written as

$$\left\langle F_q(t)\cdot F_q(t)\right\rangle = \frac{6B^2}{m^2}\int_0^t ds'\int_0^t ds\, e^{-\frac{\zeta}{m}(2t-s-s')}\delta_{s'}(s) = \frac{6B^2}{2m\zeta}(1-e^{-2t\zeta/m}). \tag{2.49}$$

Now, to use the result of statistical physics stated earlier, we take the $t\to\infty$ limit. When we do so, the first term on the right-hand side of Eq. (2.48) vanishes, as does the second term on the right-hand side of Eq. (2.49). From this it follows that

$$\lim_{t\to\infty}\frac{m}{2}\left\langle v^2(t)\right\rangle = \frac{m}{2}\frac{6B^2}{2m\zeta}.$$

As the right-hand side of the above must also be equal to $3k_BT/2$, then $B^2 = \zeta k_BT$, which is an expression of the fluctuation–dissipation theorem for our simplified case. Put differently,

$$\left\langle F_p(t)\cdot F_p(t')\right\rangle = 6B^2\delta_{t'}(t) = 6\zeta k_BT\delta_{t'}(t). \tag{2.50}$$

While the Langevin equation that we have written in Eq. (2.44) follows directly from Eq. (2.40), it is rarely written as is. In fact, it is the following Langevin equation termed the *overdamped Langevin equation*, that we most often encounter,

$$\zeta\frac{dq}{dt} = -\nabla_q U(q) + F_p(t). \tag{2.51}$$

Comparison of Eq. (2.44) with Eq. (2.51) reveals that Eq. (2.51) is recovered from Eq. (2.44) by setting $m\ddot{q} = 0$. The physical assumptions justifying the irrelevance of the "inertial term" ($m\ddot{q}$) in many scenarios are provided in Example 2.18.

Example 2.18 Dropping inertia

From Eq. (2.45) we see that the momentum at time t, $p(t)$, has exponentially decaying dependence on its initial condition, $p(0)$. In other words, momenta are randomized on m/ζ timescales. If we assume that this timescale is large compared to all other timescales relevant to this problem (such as any timescale intrinsic to the potential), then the integrals within Eqs. (2.46) and (2.47) (reproduced below),

$$\frac{dq(t)}{dt} = e^{-\frac{\zeta}{m}t}\frac{dq(0)}{dt} - \frac{1}{m}\int_0^t ds\, e^{-\frac{\zeta}{m}(t-s)}\nabla_q U(q(s)) + F_q(t),$$

$$F_q(t) = \frac{1}{m}\int_0^t ds\, e^{-\frac{\zeta}{m}(t-s)} F_p(s),$$

immediately simplify. That is, only that portion of the integrand where $s = t$ dominates and the gradient of the potential evaluated at this point can be pulled out of the integral. The remaining exponential integral can be performed yielding

$$\zeta\frac{dq}{dt} \approx -\nabla_q U(q) + F_p(t). \tag{2.52}$$

A comparison of Eq. (2.52) and Eq. (2.42) reveals that, in this limit,

$$\dot{p} \approx 0.$$

Or, equivalently, that the inertial term is irrelevant.

Now that we have related both B and ζ to temperature through the fluctuation–dissipation theorem, it is also worth relating these quantities to another core quantity appearing across disciplines: the *diffusion coefficient*, D. The diffusion coefficient is often computed from the *mean square displacement* in three dimensions,

$$6D = \frac{\left\langle (q(t_{n+1}) - q(t_n))^2 \right\rangle}{t_{n+1} - t_n}, \tag{2.53}$$

and this relation is sufficient, as we show in Example 2.19, to relate D to B and, thus, ζ. In general, for d dimensions the left-hand side of Eq. (2.53) reads $2dD$.

Example 2.19 Relating the diffusion coefficient to temperature

To relate the diffusion coefficient, D, to B we consider a D defined through Eq. (2.53) and a constant B, as before, independent of dimension.

We start by computing the right-hand side of Eq. (2.53) by considering Eq. (2.51) assuming no potential $U(q) = 0$. Integrating Eq. (2.51) once yields

$$q(t_{n+1}) - q(t_n) = \frac{1}{\zeta} \int_{t_n}^{t_{n+1}} ds\, F_p(s).$$

Taking the square of both sides and averaging over noise yields

$$\begin{aligned}\left\langle (q(t_{n+1}) - q(t_n))^2 \right\rangle &= \frac{1}{\zeta^2} \int_{t_n}^{t_{n+1}} ds \int_{t_n}^{t_{n+1}} ds' \left\langle F_p(s) \cdot F_p(s') \right\rangle \\ &= \frac{6B^2}{\zeta^2} \int_{t_n}^{t_{n+1}} ds \int_{t_n}^{t_{n+1}} ds' \delta_{s'}(s) = \frac{6B^2(t_{n+1} - t_n)}{\zeta^2}.\end{aligned}$$

Since $\left\langle (q(t_{n+1}) - q(t_n))^2 \right\rangle / (t_{n+1} - t_n) = 6D$, we have

$$6D = \frac{6B^2}{\zeta^2} \tag{2.54}$$

and, since from Eq. (2.50) we have $B^2 = \zeta k_B T$, then

$$D = \frac{k_B T}{\zeta}.$$

We end with a note on free Brownian motion in physics. The overdamped motion of a freely diffusing particle is termed *Brownian motion*. In the language of physics, considering the overdamped case as we did in Eq. (2.51), and setting $U(q) = 0$, we have

$$\dot{q} = \frac{1}{\zeta} F_p(t). \tag{2.55}$$

Integrating Eq. (2.55) over the time interval from t_n to t_{n+1}, we obtain

$$q_{n+1} - q_n = \frac{\sqrt{2}}{\zeta} \int_{t_n}^{t_{n+1}} dt\, B\dot{W}_t,$$

which, using Eq. (2.43), simplifies to

$$q_{n+1} = q_n + \frac{\sqrt{2}B}{\zeta} \sqrt{t_{n+1} - t_n}\xi_n.$$

Using Eq. (2.54), the equation satisfied by a particle's position is then $q_{n+1} = q_n + \sqrt{2D}\sqrt{t_{n+1} - t_n}\xi_n$.

Brownian Motion in Discrete Time

In the last section, we related the diffusion coefficient, D, to quantities derived earlier and arrived at $D = B^2/\zeta^2$ from fundamental considerations of statistical physics. We then arrived at an equation of motion satisfied by a freely diffusing particle's position, which reads $q_{n+1} = q_n + \sqrt{2D}\sqrt{t_{n+1} - t_n}\xi_n$.

Another way to write this is to say that the particle's position, $Q_n = (X_n, Y_n, Z_n)$, with respect to some Cartesian frame is sampled according to

$$X_{n+1}|x_n \sim \text{Normal}(x_n, 2D(t_{n+1} - t_n)),$$

$$Y_{n+1}|y_n \sim \text{Normal}(y_n, 2D(t_{n+1} - t_n)),$$

$$Z_{n+1}|z_n \sim \text{Normal}(z_n, 2D(t_{n+1} - t_n)).$$

In most applications, the initialization rule is unimportant, so we often omit it.

When we speak of a general state, $\boldsymbol{R}_n$ bolded for clarity, we say that it satisfies free Brownian motion when

$$\boldsymbol{R}_{n+1}|\boldsymbol{r}_n \sim \text{Normal}_3\left(\boldsymbol{r}_n, 2D(t_{n+1} - t_n)\mathbb{1}\right), \tag{2.56}$$

where we emphasize the dimensionality of the normal distribution through its subscript. Equivalently, Brownian motion can also be described as a random walk

$$d\boldsymbol{R}_n \sim \text{Normal}_3(0, 2D(t_{n+1} - t_n)\mathbb{1}).$$

We recover the states by $\boldsymbol{r}_{n+1} = \boldsymbol{r}_n + d\boldsymbol{r}_n$. This last representation leads, via ancestral sampling, to Algorithm 2.2.

Algorithm 2.2 Sampling Brownian motion

Given times $t_{1:N}$, a diffusion coefficient D, and an initial state $\boldsymbol{r}_0$, to sample a Brownian trajectory, proceed as follows:

For n from 0 to $N-1$, iterate the following

- Sample a $\xi_n \sim \text{Normal}_3(0, \mathbb{1})$.
- Set $\boldsymbol{r}_{n+1} = \boldsymbol{r}_n + \sqrt{2D(t_{n+1} - t_n)}\xi_n$.

2.6 Systems with Continuous State-Spaces in Continuous Time

In this section, we focus on modeling the evolution of systems with continuous state-spaces without the simplifying assumption of discrete time. This is a subtle point as, in Section 2.5, we had dealt with equations of motion, such as Eqs. (2.24) and (2.25), for which the system's evolution is described at every point in time. Nevertheless, we had to discretize these equations and, under suitable approximations, obtain transition rules for the system's passing states.

Our discussion draws an important point to the fore: time discretized transition rules of continuous dynamics are approximations and differ from their continuous time equations of motion. In other words, the passing states of a time discretized system need not, and generally do not, coincide with the exact trajectories of the underlying system. Indeed, this is explored in Exercise 2.4 where we use a simple harmonic oscillator to illustrate the artifacts introduced by the discretization on an important quantity: the energy of the system along its trajectories.

Now, we provide descriptions of continuous dynamical systems *without* discrete time approximations. Such descriptions allow us to formulate transition rules valid up to arbitrary accuracy. Specifically, we illustrate system descriptions that do not rely on the predefined grid of time points that had been central to our exposition thus far. Consequently, we reference time from now on with t and the system's passing states with $R(t)$. In this context, we use dt to denote an infinitesimally small period of time and $dR(t)$ to denote the corresponding change it induces to the system's state. As before, we use r to denote a generic constitutive state in the system's state-space continuum.

The question naturally arises as to why continuous time descriptions are useful at all if more complicated. This is for multiple reasons. First, such descriptions can be used to derive arbitrarily accurate discretization schemes. Second, and perhaps most importantly, they are the starting point for analyses of systems whose timescales may be unknown beforehand. For example, not knowing whether a fast timescale appears in the problem may incline us, out of precaution, to discretize models on a very fine temporal grid thereby introducing a high computational burden. As an alternative to fine temporal discretization, we may therefore opt for a continuous time description of the system.

2.6.1 Modeling with Stochastic Differential Equations

As we have already seen, system dynamics in discrete time may be described by increments. In such cases, successive passing states are linked by $R_{n+1} = R_n + \Delta R_n$. When the separation between successive times on our grid becomes arbitrarily small, we may generalize an incremental description to $R(t + dt) = R(t) + dR(t)$, which is the direct analog of Eq. (2.39).

Continuous time increments $dR(t)$ are most conveniently modeled by *stochastic differential equations* (SDEs) since they describe the evolution of a system with a stochastic component. A general SDE has the form

$$dr = \underbrace{\mu(t,r)dt}_{\substack{\text{deterministic}\\ \text{increment}}} + \underbrace{\sigma(t,r)dW_t}_{\substack{\text{stochastic}\\ \text{increment}}}, \tag{2.57}$$

where $\mu(t,r)$ is termed the drift velocity, $\sigma(t,r)$ is the noise standard deviation, and dW_t is the *Wiener process* satisfying Eq. (2.35).

Note 2.33 Stochastic integrals

An SDE like Eq. (2.57) can also be written as

$$\dot{r} = \mu(t, r) + \sigma(t, r)\dot{W}_t, \tag{2.58}$$

where $\dot{r}$ and $\dot{W}_t$ stand for the temporal derivatives dr/dt and dW_t/dt. As we pointed out in Note 2.26, either form of an SDE is nothing more than an abbreviation in place of

$$\int_{t'}^{t''} dr = \underbrace{\int_{t'}^{t''} dt\, \mu(t, r)}_{\text{deterministic integral}} + \underbrace{\int_{t'}^{t''} dW_t\, \sigma(t, r)}_{\text{stochastic integral}}.$$

In this equation, the left-hand side has a straightforward interpretation $\int_{t'}^{t''} dr = r(t'') - r(t')$, as does the deterministic integral on the right-hand side, $\int_{t'}^{t''} dt\, \mu(t, r)$. However, the stochastic integral $\int_{t'}^{t''} dW_t\, \sigma(t, r)$ is a random variable. For this reason, to fully make sense of Eq. (2.57) we need to be able to make explicit the probability distribution over this random variable.

In the simplest case, where $\sigma(t, r)$ is a constant, the stochastic integral is equal to the product $\sigma \int_{t'}^{t''} dW_t$ and we may compute its statistics using Eq. (2.35). However, for more general cases where $\sigma(t, r)$ truly depends on t and r, the stochastic integral may be assigned statistics in multiple physically meaningful ways giving rise to different interpretations of the SDE.

As before, to fully describe a system we need an initialization rule $R_0 \sim \mathbb{R}$ and an SDE that encodes transition rules. Essentially, an SDE is a proxy for the probability density of the increments dR. The precise probability density is given by the Fokker–Planck equation, which we discuss in the next section.

Note 2.34 Relating the Langevin equation to an SDE

The SDE is a general equation for which $\boldsymbol{r}(t)$ may coincide with position $q(t)$, momentum $p(t)$, both $(q(t), p(t))$, or other states relevant to nonmechanical systems. For sake of concreteness here, and under the assumption of constant $\sigma(t, \boldsymbol{r}(t))$, we start from a Langevin equation expressed as

$$\boldsymbol{r}(t+\tau) - \boldsymbol{r}(t) = \int_t^{t+\tau} ds\, \mu + \sigma \int_t^{t+\tau} dW_t. \tag{2.59}$$

More specifically, the various terms are given by

$$\begin{aligned}
\boldsymbol{r}(t) &= (q(t), p(t)),\\
\mu(t, \boldsymbol{r}(t)) &= \big(p(t)/m, -\nabla_q U(q(t)) - \zeta p(t)/m\big),\\
\sigma(t, \boldsymbol{r}(t)) &= (0, B),
\end{aligned}$$

from which we recover the underdamped Langevin equation, Eq. (2.40), that we saw earlier.

Similarly, reading off from Eq. (2.51), for the overdamped Langevin equation we would have

$$\boldsymbol{r}(t) = q(t), \qquad \mu(t, \boldsymbol{r}(t)) = -\nabla_q U(q(t))/\zeta, \qquad \sigma(t, \boldsymbol{r}(t)) = B/\zeta. \qquad (2.60)$$

2.6.2 The Fokker–Planck Equation

For systems with discrete state-spaces, Section 2.3 gave two parallel formulations. One is local and yields descriptions for the evolution of systems across constitutive states, thereby allowing us to sample individual trajectories. The other is global and allows us to compute probabilities over passing states without sampling particular trajectories through trajectory marginalization. The first formulation uses sampling equations that can readily implemented computationally via ancestral sampling or the Gillespie algorithm; while the second formulation uses master equations that typically need to be solved analytically or numerically.

For systems with continuous state-spaces in continuous time a similar picture also applies. SDEs describe the evolution of the system's state $\boldsymbol{R}(t)$ through time and are analogous to the transition rules. However, as the system evolves stochastically, $\boldsymbol{R}(t)$ is determined only probabilistically. As such, we can speak of a probability density of observing a realization $\boldsymbol{r}(t)$ of $\boldsymbol{R}(t)$ at a specific time t. That is, just as we did in deriving master equations, we may write an initialization rule for $\boldsymbol{R}_0$ and also write a probability of $\boldsymbol{r}(t)$ being inside some volume V as $\int_V d\boldsymbol{r}\, p(\boldsymbol{r}, t)$ where the volume element is $d\boldsymbol{r}$. This density, which is a function of both space and time and is written as $p(\boldsymbol{r}, t)$, satisfies a *Fokker–Planck equation*. The Fokker–Planck equation of a continuous state-space formulation is thus analogous to the master equation of a discrete state-space formulation.

In what follows, we derive the Fokker–Planck equation for the probability density $p(\boldsymbol{r}, t)$ for $\boldsymbol{r}$ given by the SDE in Eq. (2.57) to obtain

$$\frac{\partial}{\partial t} p(\boldsymbol{r}, t) + \nabla_q \cdot \left(\mu p(\boldsymbol{r}, t)\right) = \nabla_q \cdot \sigma \left(\nabla_q \cdot \sigma p(\boldsymbol{r}, t)\right). \qquad (2.61)$$

However, before we do so we explore perhaps the best known Fokker–Planck equation, namely the diffusion equation.

Note 2.35 The diffusion equation

One of the simplest forms of Eq. (2.61) is the case where $\mu = 0$ and σ is constant in space (spatially homogeneous) and time (sometimes termed *stationary*) and identical in all directions (isotropic). In this case, the resulting Fokker–Planck equation, called the *diffusion equation*, reads as

$$\frac{\partial}{\partial t} p(\boldsymbol{r}, t) = D\nabla_q^2 p(\boldsymbol{r}, t).$$

Here, from Eq. (2.60) we identify $\sigma = B/\zeta$ and from Eq. (2.54) we identify D as B^2/ζ^2. For B components constant but different along different directions, we would speak of an *anisotropic diffusion equation*. Similarly, for $\mu \neq 0$, we would speak of a *diffusion equation with drift*.

When starting from the (homogeneous, stationary, isotropic) diffusion equation and assuming *open boundary conditions* (density decays to zero at infinity) and initial conditions reflecting perfect certainty as to the location $\mathbf{r}(t_0)$ of the diffusing particle, $p(\mathbf{r}(t_0)) = \delta_{r_0}(\mathbf{r}(t_0))$, the three-dimensional diffusion equation is readily solved and its solution is

$$p(\mathbf{r}(t), t|\mathbf{r}(t_0)) = \frac{1}{\sqrt{4\pi D(t-t_0)}^3} e^{-\frac{(\mathbf{r}-\mathbf{r}(t_0))^2}{4D(t-t_0)}}. \tag{2.62}$$

This solution describes transition probability rules that we have already written in explicit form in Eq. (2.56).

For simplicity, in subsequent chapters the subscript on the Laplacian, ∇_q^2, of the diffusion equation will always be understood as q such that ∇^2 becomes synonymous with ∇_q^2.

2.6.3 Deriving the Fokker–Planck Equation*

As we did in deriving master equations in Eq. (2.9), to derive a Fokker–Planck equation, we begin with an argument on the evolution of the probability density. To do so, we focus on a volume V in the system's state-space and its boundary, which we denote by S. For mechanical systems in one dimension, $d\mathbf{r}$ takes the explicit form $dq\,dp$.

We begin by considering the amount of probability density flowing out of S. This is mathematically represented by

$$\int_S dS \cdot \dot{\mathbf{r}} p(\mathbf{r}, t),$$

where the normal to the surface is pointed outward, by convention, and the velocity $\dot{\mathbf{r}}$ is given by the SDE of Eq. (2.57). For clarity here, we also make dot products explicit. For positive $dS \cdot \dot{\mathbf{r}}$, probability flows out of the volume, V.

Now, the decrease in time of the probability enclosed in the volume, V, can also be written as

$$-\frac{\partial}{\partial t}\int_V d\mathbf{r}\, p(\mathbf{r}, t).$$

Equating the last two integrals and invoking the divergence theorem, we obtain

$$-\frac{\partial}{\partial t}\int_V d\mathbf{r}\, p(\mathbf{r}, t) = \int_V d\mathbf{r}\, \nabla_q \cdot (\dot{\mathbf{r}} p(\mathbf{r}, t)). \tag{2.63}$$

Since Eq. (2.63) holds for any volume V, the *continuity equation* follows,

$$\frac{\partial}{\partial t} p(\mathbf{r}, t) + \nabla_q \cdot (\dot{\mathbf{r}} p(\mathbf{r}, t)) = 0. \tag{2.64}$$

* This is an advanced topic and could be skipped on a first reading.

Now, inserting the SDE $\dot{r} = \mu + \sigma \dot{W}_t$ from Eq. (2.58) into Eq. (2.64) yields

$$\frac{\partial}{\partial t} p(\boldsymbol{r}, t) + \nabla_q \cdot \left(\left(\mu + \sigma \dot{W}_t \right) p(\boldsymbol{r}, t) \right) = 0. \tag{2.65}$$

From this point forward, our goal is to eliminate the fluctuating force from the expression above and arrive at the end result, Eq. (2.61). This derivation is provided in Note 2.36.

Note 2.36 Eliminating the fluctuating force from the Fokker–Planck equation

Here, we derive the Fokker–Planck equation for the probability, $p(\boldsymbol{r}, t)$, for $\boldsymbol{r}$ given by the SDE in Eq. (2.57). We start with Eq. (2.65) that we rewrite as

$$\frac{\partial}{\partial t} p(\boldsymbol{r}, t) + p(\boldsymbol{r}, t) \nabla_q \cdot \mu = -\mu \cdot \nabla_q p(\boldsymbol{r}, t) - \nabla_q \cdot \left(\sigma \dot{W}_t p(\boldsymbol{r}, t) \right). \tag{2.66}$$

Using the integrating factor method, in a way similar to the method of Example 2.17, we arrive at

$$p(\boldsymbol{r}, t) = p(\boldsymbol{r}_0) e^{-t \nabla_q \cdot \mu} - \int_0^t ds\, e^{-(t-s) \nabla_q \cdot \mu} \mu \cdot \nabla_q p(\boldsymbol{r}, s)$$
$$- \int_0^t ds\, e^{-(t-s) \nabla_q \cdot \mu} \nabla_q \cdot \left(\sigma \dot{W}_s p(\boldsymbol{r}, s) \right). \tag{2.67}$$

Now Eq. (2.67) can be inserted into the second term on the right-hand side of Eq. (2.66) to yield

$$\frac{\partial}{\partial t} p(\boldsymbol{r}, t) + \nabla_q \cdot \left(\mu p(\boldsymbol{r}, t) \right)$$
$$= -\nabla_q \cdot \left(\sigma \dot{W}_t \left(p(\boldsymbol{r}_0) e^{-t \nabla_q \cdot \mu} - \int_0^t ds\, e^{-(t-s) \nabla_q \cdot \mu} \mu \cdot \nabla_q p(\boldsymbol{r}, s) \right. \right.$$
$$\left. \left. - \int_0^t ds\, e^{-(t-s) \nabla_q \cdot \mu} \nabla_q \cdot \left(\sigma \dot{W}_s p(\boldsymbol{r}, s) \right) \right) \right), \tag{2.68}$$

which retains the fluctuating force on the right-hand side. Averaging both sides of Eq. (2.68) with respect to noise yields

$$\frac{\partial}{\partial t} p(\boldsymbol{r}, t) + \nabla_q \cdot \left(\mu p(\boldsymbol{r}, t) \right) = +\nabla_q \cdot \sigma \left(\nabla_q \cdot \sigma p(\boldsymbol{r}, t) \right). \tag{*2.69*}$$

The averaging resulting in the last term is provided in Appendix F.

A few remarks on noise averaging are warranted here. First, the noise average of the first term on the right-hand side of Eq. (2.68) is zero. This is because the initial condition, $p(\boldsymbol{r}_0)$, has no stochastic component and the average of W_t is, by definition, zero (otherwise the nonzero component would be lumped into the drift). Second, from Eq. (2.59) it is clear that $p(\boldsymbol{r}, t)$ itself has a stochastic component. As such, we could reinsert Eq. (2.59) into the second term on the right-hand side of Eq. (2.66). This would yield a series expansion termed the *Kramers–Moyall expansion.* By ignoring higher order terms in the Kramers–Moyall expansion, we are assuming that, up to second order in the fluctuating force, the probability appearing in the last term of Eq. (2.68) is independent of the fluctuating force. As such, we only compute the covariance of the fluctuating component as per Eq. (2.50), not, say, fourth moments.

2.6.4 A Case Study in Thermal Physics*

It is of interest in the sciences to ask under what circumstance we can derive the diffusion equation for systems with positions and momenta. As we now need to discriminate between momentum and the probability density ($p(\mathbf{r}, t)$), we use p for momentum and $p(\mathbf{r}, t)$ for probability density.

In this case, we consider the overdamped Langevin equation, Eq. (2.51), and following the recipe in Note 2.36, we arrive at the Fokker–Planck equation

$$\frac{\partial}{\partial t}p(\mathbf{r}, t) - \frac{1}{\zeta}\nabla_q \cdot \left(\nabla_q U(q)p(\mathbf{r}, t)\right) = D\nabla_q^2 p(\mathbf{r}, t),$$

which is sometimes called the *Smoluchowski equation.*

Another Fokker–Planck equation can be obtained by considering the underdamped Langevin equation, Eqs. (2.41) and (2.42). Again, following a recipe similar to Note 2.36, we arrive at a different Smoluchowski equation

$$\frac{\partial}{\partial t}p(\mathbf{r}, t) + \frac{q}{m} \cdot \nabla_q p(\mathbf{r}, t) + \nabla_p \cdot \left(\left(-\nabla_q U(q) - \frac{\zeta}{m}p\right) p(\mathbf{r}, t)\right) = D\zeta^2\nabla_p^2 p(\mathbf{r}, t), \tag{2.70}$$

where, to be clear, we have subscripted the gradient of the momentum as ∇_p.

It is now possible, for a mechanical system where $p = m\dot{q}$, to reparametrize the underdamped Smoluchowski equation from Eq. (2.70) into a partial differential equation that only depends on position and time. Indeed, the end result, derived in Note 2.37, reads

$$\frac{\partial p(\mathbf{r}, t)}{\partial t} = \frac{1}{\zeta}\nabla_q \cdot \left(\nabla_q U(q)p(\mathbf{r}, t)\right) + \frac{k_B T}{\zeta}\nabla_q^2 p(\mathbf{r}, t) - \frac{m}{\zeta}\frac{\partial^2}{\partial t^2}p(\mathbf{r}, t),$$

where the second time derivative of the probability suggests partial time-reversibility on timescales of m/ζ.

Note 2.37 The underdamped Fokker–Planck equation

To derive the Fokker–Planck equation coinciding with the underdamped Langevin equation for mechanical systems, Eq. (2.44), we start by inserting dynamics for $\dot{q}$ and $\dot{p}$, given by Eqs. (2.41) and (2.42), into the continuity equation, Eq. (2.64), yielding

$$\frac{\partial}{\partial t}p(\mathbf{r}, t) + \nabla_q \cdot (\dot{q}p(\mathbf{r}, t)) + \nabla_p \cdot (\dot{p}p(\mathbf{r}, t)) = 0,$$

* This is an advanced topic and could be skipped on a first reading.

which results in

$$\frac{\partial}{\partial t}p(\mathbf{r},t) + \frac{p}{m}\cdot\nabla_q p(\mathbf{r},t) + \nabla_p\cdot\left(\left(-\nabla_q U(q) - \frac{\zeta}{m}p + F_p(t)\right)p(\mathbf{r},t)\right) = 0. \tag{2.71}$$

As a first step toward eliminating the fluctuating force as we did earlier, we first average over all values of the momentum from Eq. (2.71), which yields

$$\frac{\partial}{\partial t}\left(\int dp\, p(\mathbf{r},t)\right) + \int dp\left(\frac{p}{m}\cdot\nabla_q p(\mathbf{r},t)\right) + 0 = 0,$$

where the third term on the left-hand side is zero as $p(\mathbf{r},t)$ vanishes when evaluated for infinite values of the momentum, p, as the density itself vanishes for such values. We can now rewrite the above as

$$\frac{\partial}{\partial t}\bar{p}(\mathbf{r},t) + \frac{1}{m}\nabla_q\cdot\langle p\rangle = 0, \tag{2.72}$$

where $\bar{p}(\mathbf{r},t)$ is the momentum averaged density. In other words, we say

$$\bar{p}(\mathbf{r},t) = \int dp\, p(\mathbf{r},t) = p(q,t).$$

Now, if we could express $\langle p\rangle$ of Eq. (2.72) in terms of position, we would be done as we would have functions of position on both sides of Eq. (2.72). This raises the question as to whether we can find an equation of motion for $\langle p\rangle$.

To find such an equation of motion, we multiply both sides of Eq. (2.71) by momentum and integrate over momentum,

$$\int dp\left(p\frac{\partial}{\partial t}p(\mathbf{r},t) + \frac{p}{m}p\cdot\nabla_q p(\mathbf{r},t) + p\nabla_p\cdot\left(\left(-\nabla_q U(q) - \frac{\zeta}{m}p + F_p(t)\right)p(\mathbf{r},t)\right)\right) = 0, \tag{2.73}$$

which we rewrite as

$$\frac{\partial}{\partial t}\langle p\rangle + \int dp\frac{p}{m}p\cdot\nabla_q p(\mathbf{r},t) - \int dp\left(\left(-\nabla_q U(q) - \frac{\zeta}{m}p + F_p(t)\right)p(\mathbf{r},t)\right) = 0.$$

The third term of Eq. (2.73) was simplified through integration by parts. Taking the divergence of both sides of the above with respect to q gives

$$\frac{\partial}{\partial t}\nabla_q\cdot\langle p\rangle + \frac{\mathbf{1}}{m}\int dp\left(p\cdot\nabla_q\right)^2 p(\mathbf{r},t)$$
$$-\nabla_q\cdot\int dp\left(\left(-\nabla_q U(q) - \frac{\zeta}{m}p + F_p(t)\right)p(\mathbf{r},t)\right) = 0.$$

Assuming momenta in all directions are decoupled and averaging over noise yields

$$\frac{\partial}{\partial t}\nabla_q\cdot\langle p\rangle + \frac{\mathbf{1}}{3m}\nabla_q^2\left\langle p^2\right\rangle - \nabla_q\cdot\left(-\nabla_q U(q)\bar{p}(\mathbf{r},t)\right) + \frac{\zeta}{m}\nabla_q\cdot\langle p\rangle = 0, \tag{2.74}$$

where in obtaining the second term on the right-hand side we wrote

$$\int dp\left(p\cdot\nabla_q\right)^2 p(\mathbf{r},t) = \partial_x^2\left\langle p_x^2\right\rangle + \partial_y^2\left\langle p_y^2\right\rangle + \partial_z^2\left\langle p_z^2\right\rangle = \nabla_q^2\frac{1}{3}\left\langle p^2\right\rangle$$

under the assumption that the average over momenta in any direction is the same, *e.g.*, $\partial_x^2 \langle p_y^2 \rangle = \partial_x^2 \langle p_z^2 \rangle$, where we invoked the usual shorthand notation for partial derivatives such as ∂_x^2 to be read as $\partial^2/\partial x^2$. Inserting Eq. (2.72) into Eq. (2.74), we recover

$$-m\frac{\partial^2}{\partial t^2}\bar{p}(\mathbf{r},t) + \frac{1}{3m}\nabla_q^2\langle p^2\rangle - \nabla_q\cdot\left(-\nabla_q U(q)\bar{p}(\mathbf{r},t)\right) - \zeta\frac{\partial}{\partial t}\bar{p}(\mathbf{r},t) = 0. \tag{2.75}$$

The above is still not a closed set of equations as it depends on $\langle p^2\rangle$. The next logical step would be to evaluate the evolution equation for $\langle p^2\rangle$ just as we had before for the first moment. Instead, we now invoke an approximation in the form of a moment closure relation. That is, we make the assumption that $p(\mathbf{r},t)$ is separable in position and momenta, $p(q,t)p(p,t)$, so that

$$\begin{aligned}\frac{1}{3m}\nabla_q^2\langle p^2\rangle &= \frac{1}{3m}\nabla_q^2\int dp\, p(\mathbf{r},t)p^2 \\ &= \frac{1}{3m}\nabla_q^2 p(q)\int dp\, p(p,t)p^2 = 3mk_BT\frac{1}{3m}\nabla_q^2\bar{p}(\mathbf{r},t).\end{aligned} \tag{2.76}$$

By inserting Eq. (2.76) into Eq. (2.75), we finally recover

$$\frac{\partial\bar{p}(\mathbf{r},t)}{\partial t} = \frac{1}{\zeta}\nabla_q\cdot\left(\nabla_q U(q)\bar{p}(\mathbf{r},t)\right) + \frac{k_BT}{\zeta}\nabla_q^2\bar{p}(\mathbf{r},t) - \frac{m}{\zeta}\frac{\partial^2}{\partial t^2}\bar{p}(\mathbf{r},t), \tag{2.77}$$

where k_BT/ζ is the diffusion coefficient. Equation (2.77) is a Smoluchowski equation describing underdamped dynamics where the effect of the fluctuating force has manifested itself in the magnitude of $\langle p^2\rangle$.

The Fokker–Planck equation given by Eq. (2.77) is more general than the overdamped form, Eq. (2.61), typically encountered when m/ζ is fast as compared to all timescales present in the problem.

2.7 Exercise Problems

Exercise 2.1 Biology's central dogma

Birth-death events are useful as they can be used to coarsely model complex chemical kinetic schemes. For example, we can employ birth-death processes to model the stochastic number of RNA transcribed from a specific gene locus (birth process) also subject to degradation (death process). Ignoring any gene activation/deactivation and degradation regulatory mechanisms, these reactions can be idealized as

$$\mathrm{DNA} \xrightarrow{\lambda_b} \mathrm{DNA} + \mathrm{RNA},$$

$$\mathrm{RNA} \xrightarrow{\lambda_d} \emptyset,$$

where λ_b is the transcription rate and λ_d is the coinciding degradation rate. Here:

1. Simulate this birth-death process for RNA (assuming a birth rate greater than the death rate).
2. Plot the amount of RNA as a function of time until the system reaches steady state.
3. Use mass action laws to relate this steady state to the production and degradation rates.

To develop a more realistic model that completes an idealization of biology's central dogma, we could also include the production (or translation) to proteins from RNA and the degradation of proteins.

Exercise 2.2 Stochastic binary decisions

Cell fate decisions may arise from small initial fluctuations amplified and reinforced (through feedback) over time. These events are called stochastic binary decisions (M. Artyomov J. Das, M. Kardar, A. K. Chakraborty. Purely stochastic binary decisions in cell signaling models without underlying deterministic bistabilities. *Proc. Natl. Acad. Sci. U.S.A.* 104:18958, 2007). We will now simulate this behavior. Consider the following set of chemical reactions:

$$\begin{aligned}
\mathcal{A}_1 &\overset{k_1}{\to} E + \mathcal{A}_1, \\
\mathcal{A}_2 &\overset{k_2}{\to} S + \mathcal{A}_2, \\
E + \mathcal{A}_1 &\overset{k_3}{\to} E + \mathcal{A}_1^*, \\
\mathcal{A}_1^* &\overset{k_4}{\to} E + \mathcal{A}_1^*, \\
S + \mathcal{A}_1 &\overset{k_5}{\to} S + \mathcal{A}_1^{\text{INACTIV}}, \\
S &\overset{k_d}{\to} \emptyset, \\
E &\overset{k_d}{\to} \emptyset,
\end{aligned}$$

where $\mathcal{A}_1$ is an agonist and $\mathcal{A}_2$ is its antagonist. E is an enzyme (*i.e.*, a catalyst) converting $\mathcal{A}_1$ into its protected form $\mathcal{A}_1^*$ and $\mathcal{A}_1^*$, in turn, stimulates the production of E (positive feedback). We are interested in the steady-state amount of $\mathcal{A}_1^*$. If S is present it can permanently deactivate $\mathcal{A}_1$.

1. Start with 10 agonists and 10 antagonists with $k_1 = k_2 = k_d = k_4 = 1$, $k_3 = 100, k_5 = 100$ (in units of inverse time). Simulate the process to completion using Gillespie's algorithm and produce a histogram for the final amount of $\mathcal{A}_1^*$. Explain, in words, the result you obtain.
2. Repeat the simulation and produce the required histogram but starting with 1,000 agonists and 1,000 antagonists. Explain, in words, how your histogram over of $\mathcal{A}_1^*$ now differs from what it was when starting with 10 agonists and 10 antagonists.
3. Had you solved the corresponding rate equations (mass action laws), explain in words the steady-state population of $\mathcal{A}_1^*$ you would expect.

Exercise 2.3 The genetic toggle switch and stochastic bistability

The genetic toggle switch (Gardner et al. Construction of a genetic toggle switch in *Escherichia coli*. *Nature*. 403:339, 2000) is a common feedback loop motif in systems biology and exhibits a behavior termed stochastic bistability. We will now simulate this behavior.

Consider the following chemical reactions involving two proteins, $\mathcal{A}$ and $\mathcal{B}$:

$$\mathcal{A} \xrightarrow{k_d} \emptyset, \qquad g_{\mathcal{A}} \xrightarrow{k_p} g_{\mathcal{A}} + \mathcal{A}, \qquad g_{\mathcal{A}} + \mathcal{B} \underset{k_r}{\overset{k_f}{\leftrightarrow}} g^*_{\mathcal{A}},$$

$$\mathcal{B} \xrightarrow{k_d} \emptyset, \qquad g_{\mathcal{B}} \xrightarrow{k_p} g_{\mathcal{B}} + \mathcal{B}, \qquad g_{\mathcal{B}} + \mathcal{A} \underset{k_r}{\overset{k_f}{\leftrightarrow}} g^*_{\mathcal{B}}$$

where k_d are degradation rates and k_p are production rates for both proteins. Here, $g_{\mathcal{A}}$ is understood as the gene responsible for the production of $\mathcal{A}$, which can be converted into an inactive form $g^*_{\mathcal{A}}$ by binding to $\mathcal{B}$ and vice versa for $g_{\mathcal{B}}$. Assume $k_d < k_p$ and $k_r < k_f$.

Also assume you only have one gene available such that $g_{\mathcal{A}} + g^*_{\mathcal{A}} = 1$ and $g_{\mathcal{B}} + g^*_{\mathcal{B}} = 1$. Also, assume throughout that $g^*_{\mathcal{A}} + g^*_{\mathcal{B}} = 1$. Hint: think of these constraints as reactions themselves.

1. Simulate the chemical reactions starting with $g_{\mathcal{A}} = 1$, $g_{\mathcal{B}} = 1$, $n_{\mathcal{A}} = 0$, and $n_{\mathcal{B}} = 0$. Adjust your rates until you see stochastic switching events between periods when $\mathcal{A}$ exceeds $\mathcal{B}$ in number and $\mathcal{B}$ exceeds $\mathcal{A}$ in number. That is, you should observe stochastic hopping between two solutions (termed "fixed points").
2. Would you expect to see this stochastic switching occur if you had started with a large amount of $n_{\mathcal{A}}$ and $n_{\mathcal{B}}$ initially? In technical language, qualitatively explain (in words) how the fixed point structure changes for the corresponding rate equations.
3. The condition that $g^*_{\mathcal{A}} + g^*_{\mathcal{B}} = 1$ is called the exclusive switch. Relax this condition and resimulate the toggle switch. What new fixed point appears?

Exercise 2.4 Numerical integrators

Consider a one-dimensional harmonic oscillator whose equations of motion are

$$dq = \frac{p}{m} dt, \qquad dp = -kq\, dt,$$

where k is the harmonic force constant. Assume the system is initialized deterministically at $Q_0 = q_0$ and $P_0 = p_0$.

1. Derive the general solution for $\boldsymbol{r}(t) = (q(t), p(t))$.
2. Derive the transition rules for the oscillator's trajectory $\boldsymbol{R}_n = (Q_n, P_n)$ at discrete times $t_n = t_0 + n\tau$ with the approximations mentioned in Note 2.28. Specifically, use
   - $F(t, q(t)) \approx F(t_n, q(t_n))$
   - $F(t, q(t)) \approx F(t_{n+1}, q(t_{n+1}))$
   - $F(t, q(t)) \approx \frac{1}{2} \left(F(t_n, q(t_n)) + F(t_{n+1}, q(t_{n+1}))\right)$.

3. Use $m = 1$, $k = 1$, $q_0 = 0$, $p_0 = 1$, and $\tau = 0.25$ (in the appropriate units) to compute specific trajectories and compare with the exact solution.
4. Compute the energy along the three trajectories determined above and compare with the energy along the exact trajectory.

Exercise 2.5 Fluctuation–dissipation theorem

Start with the following Langevin equation, where rather than a scalar friction coefficient, we now have what is termed a frequency-dependent memory kernel

$$m\ddot{q} = -\int_0^t dt' \zeta(t-t')\,\dot{q}(t') + F_p(t).$$

Show that the function $f(t)$, for which the fluctuation–dissipation theorem holds, *i.e.*, $\left\langle F_p(t) \cdot F_p(t') \right\rangle = f(t-t')$, is satisfied if $f(t) = 3k_B T\zeta(t)$.

Exercise 2.6 Transcription factor binding as an example of hop diffusion

Transcription factors are proteins that modulate gene expression. When tracking these in physical space, we often see transcription factors rapidly diffuse (within the cellular nucleus in the neighborhood of DNA) and then reversibly bind to DNA. This binding gives rise to a sudden decrease in their diffusion coefficient. As such, a minimal model describing transcription factor dynamics is to assume that transcription factors switch between two dynamical modes with both modes having no potential ($U = 0$). Simulate this spatially varying "hop diffusion" process by first considering a particle diffusing within a volume subdivided into two regions: one outer region with reflecting boundary conditions and an inner region meant to mimic DNA, embedded within the outer region, within which the particle may freely diffuse but within which it has a smaller diffusion coefficient.

Exercise 2.7 Ornstein–Uhlenbeck process

We now illustrate the versatility of models inspired by Brownian motion such as overdamped dynamics in a harmonic potential,

$$\zeta dq = -\nabla_q U(q)dt + BdW_t,$$

termed the *Ornstein–Uhlenbeck* (OU) process when the potential assumes the form $U(q) = k(q - q_c)^2/2$, where k is the harmonic force constant and q_c is the location of the potential's center. More explicitly, we write

$$\zeta dq = -k(q - q_c)dt + BdW_t.$$

The OU process is often used in describing diffusion within a confined environment or models of a microsphere, say, in an optical trap.

Derive the resulting Fokker–Planck equation for the OU process and solve the resulting Fokker–Planck assuming delta function distributed initial conditions and open boundaries. For simplicity, you can also define $q_c = 0$.

From this solution of your Fokker–Planck equation, write down the probability of observing a sequence $w_{1:N}$ of observations consisting of positions of a hypothetical particle in a harmonic trap. Finally, derive self-consistent equations for those values of k and the diffusion coefficient D maximizing your likelihood.

Exercise 2.8 Hop diffusion

The OU process discussed in Exercise 2.7 can be further generalized to model a particle hopping between M motion models indexed $\sigma_{1:M}$. Here, we consider hopping that can occur in continuous time according to a Markov jump process. One way to model this, following Eq. (2.5), is to select the holding states according to

$$S_n | s_{n-1}, \pi_{s_{n-1}} \sim \text{Categorical}_{\sigma_{1:M}}(\pi_{s_{n-1}}).$$

Next, following conventions from Eq. (2.6), we select a holding time for that state,

$$H_n | s_n \sim \text{Exponential}\,(\lambda_{s_n}).$$

The dynamics attained at each time point then depend on the state realized

$$\zeta dq = -\nabla_q U_{s_n}(q)dt + B_{s_n} dW_t,$$

where $U_{s_n}(q)$ and B_{s_n} set the potential and magnitude of the random force, respectively, in each passing state.

Reproduce the figure below, through simulation of the appropriate processes, illustrating position versus time for free diffusion (top), the OU process (middle) where the particle stays trapped close to the potential's mean, and a model describing the stochastic switching between two different harmonic potentials.

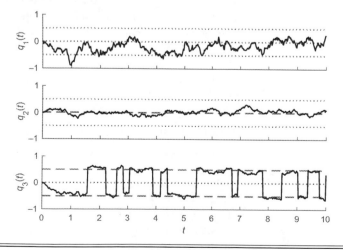

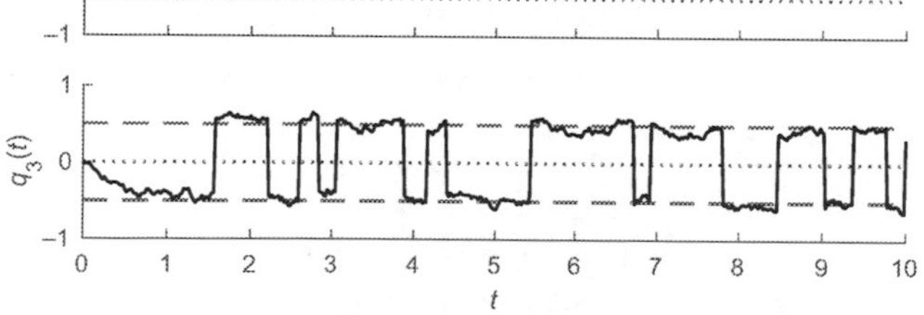

Exercise 2.9 Rotational diffusion

In this chapter we have introduced the diffusion equation and its solution with open boundaries is given in Note 2.35. While we focused on translation in space, a particle's orientation Φ can also diffuse assuming much of the same physics with rotational diffusion constant D_r, for which

$$p\left(\phi_n|\phi_{n-1}\right) = \frac{1}{\sqrt{4\pi D_r \tau}} e^{-\frac{(\phi_n-\phi_{n-1})^2}{4D_r\tau}}.$$

However, in reality, we typically cannot distinguish between ϕ and any integer multiple of 2π plus ϕ. Instead, we define another random variable Φ' whose range is limited to $(-\pi, \pi]$. Using $p\left(\phi_n|\phi_{n-1}\right)$, write down $p\left(\phi'_n|\phi'_{n-1}\right)$.

Project 2.1 Truncating infinite state-spaces

It is sometimes necessary to solve the master equation for systems with infinite state-spaces. In rare cases (such as the birth, death, and birth-death processes), the equation may be analytically solvable considering the full state-space. However, in general, the solution must be numerically approximated by truncating the state-space.

1. Simulate many trajectories of the birth-death process using the Gillespie algorithm, under the assumption that there is some maximum population at which the system is prohibited from sampling additional birth events (this is called a *Dirichlet boundary condition*). After some time, histogram the final population. Compare your histogram to an exact Gillespie simulation (with no imposed maximum population). How do your histograms compare? Repeat the procedure for when the maximum population allowed lies below, is on par with, or is well above the expected steady-state population.
2. In order to approximately solve the master equation, write down the (now finite) generator matrix for the birth-death system in a truncated state-space. Compute the probability distribution over populations at various times up to and beyond the steady state. Compare the result to the histograms obtained for the exact Gillespie when the maximum population allowed lies below, is on par with, or is well above the expected steady-state population.
3. Find the exact solution to the master equation for the birth-death process. Compare your results to the truncated master equation from part 2 of this Project when the maximum population allowed lies below, is on par with, or is well above the expected steady-state population.

Project 2.2 Algorithm: Finite State Projection

Very often we encounter systems with infinite state-spaces, $\sigma_{1:\infty}$. In this case, the finite state projection (FSP) algorithm (B. Munsky, M. Khammash. The finite state projection algorithm for the solution of the chemical master equation. *J. Chem. Phys.* 124:044104, 2006) and its variants are often invoked to approximate the solution of the master equation for a finite subset of the state-space, $\sigma_{1:M}$.

The algorithm consists of selecting an M and evolving the initial probability density, ρ_M, up to a preselected time, t, with a truncated approximation to the full generator matrix, which we call G_M. As the rows of the truncated G_M do not sum to unity, the total terminal probability will only be conserved up to some $1 - \epsilon_M$ where ϵ_M tends to zero in the infinite state-space limit.

Implement the FSP for the simplified kinetic scheme provided in Exercise 2.1 by first initializing M to a modest value. Also initialize, ρ_M, an acceptably small ϵ, and G_M. Then iterate the following:

1. Compute $\rho_M \exp(\mathrm{G}_M t)$.
2. From $\rho_M \exp(\mathrm{G}_M t)$, compute the total probability of remaining within this finite state-space at t.
3. If this probability is less than $1 - \epsilon$, increase M by 1 and repeat.

Project 2.3 Algorithm: the Finite State Projection for a composite system

Repeat Project 2.2 for the following kinetic scheme:

$$\mathrm{DNA} \xrightarrow{\lambda_b} \mathrm{DNA} + \mathrm{RNA},$$

$$\mathrm{RNA} \xrightarrow{\lambda_d} \emptyset,$$

$$\mathrm{RNA} \xrightarrow{\lambda_{+1}} \mathrm{RNA} + \text{protein},$$

$$\text{protein} \xrightarrow{\lambda_{-1}} \emptyset.$$

Consider the fact that both state-spaces for the RNA and protein will need to be truncated (presumably to a different value).

Additional Reading

S. Ross. *Introduction to probability models*, 11th ed. Academic Press, 2014.

N. G. Van Kampen. *Stochastic processes in physics and chemistry*, 3rd ed. North Holland, 2007.

R. Zwanzig. *Nonequilibrium statistical mechanics*. Oxford University Press, 2001.

H. Risken. *The Fokker–Planck equation: methods of solution and applications*. Springer-Verlag, 1984.

D. Gillespie. Exact stochastic simulation of coupled chemical reactions. *J. Phys. Chem.*, 81:2340, 1977.

E. Çinlar. *Probability and stochastics*. Springer, 2011.

G. A. Pavliotis. *Stochastic processes and applications: diffusion processes, the Fokker–Planck and Langevin equations*. Springer, 2014.

3 Likelihoods and Latent Variables

By the end of this chapter, we will have presented

- *How to manipulate likelihoods*
- *The principle of maximum likelihood*
- *Noise representations and latent variables*

3.1 Quantifying Measurements with Likelihoods

Models we formulate in analyzing data contain *variables* whose values are unknown and others whose values are known. Measurements, whose output we continue to denote by $w_{1:N}$, are examples of variables with known values.

In a model, variables of unknown values are of two kinds: those we wish to estimate, which we call *parameters* and often denote with $\theta_{1:K}$ or $\boldsymbol{\theta}$; and those that we may not care to estimate, which we term *latent* or *nuisance* variables. The distinction between parameters and latent variables is generally not mathematically clear cut. After all, both appear in the same equations and their properties are interrelated according to the rules developed in Section 1.2. Nevertheless, from a modeling perspective, it is helpful to be clear regarding which variables need to be estimated and which variables are needed only because they facilitate our mathematical formulation or our model's interpretation.

Example 3.1 Latent variables

Consider diffusing particles trapped in a harmonic potential, *i.e.*, whose trajectory is described by an Ornstein–Uhlenbeck process. In some situations, we may only care to learn the force constant trapping the particles and the particle's friction coefficients from grainy observations, not the actual particle trajectories themselves. In this case, the force constant and friction coefficient constitute parameters we care to learn and the particle's positions are latent variables.

We have previously seen cases with parameters inside generative models of dynamical systems. For example, in discrete space and time systems, such as those in Section 2.4, each constitutive state is sampled from a categorical distribution whose parameters include the probabilities of state-to-state transitions; while, in discrete space continuous time systems, see Section 2.3, each holding state is sampled from a categorical distribution whose parameters include the transition probabilities and each holding period is sampled from an exponential distribution whose parameters include the holding rates.

Using a model with measurements $w_{1:N}$, our goal is typically to estimate its parameters $\boldsymbol{\theta}$. Therefore, following the construction of a model, our main objective is to obtain *estimator* values that we denote with $\hat{\boldsymbol{\theta}}$. In this chapter, we do so within what is often termed the *frequentist paradigm*. In Chapter 4, we will discuss how to obtain full distributions over $\boldsymbol{\theta}$, not just "point" estimates $\hat{\boldsymbol{\theta}}$, and we will do so within the *Bayesian paradigm*.

Both paradigms, whether frequentist or Bayesian, employ *likelihoods*. Since formulating likelihoods given models and measurements is the starting point of frequentist inference, the focus of this chapter is in describing likelihoods.

Note 3.1 Conditional dependencies of likelihoods

The central object of this chapter is the *likelihood*, which, for a given model, is defined by $p(w_{1:N}|\boldsymbol{\theta})$. This is the probability of observing the measurements $w_{1:N}$ under our model, *provided* the parameters attain values $\boldsymbol{\theta}$. To emphasize that a likelihood can be evaluated for every possible parameter value for which the model remains meaningful, we occasionally write $\ell_{w_{1:N}}(\boldsymbol{\theta}) = p(w_{1:N}|\boldsymbol{\theta})$ and use subscripts in $\ell_{w_{1:N}}(\cdot)$ to denote that, as a function, the likelihood depends implicitly on our measurements.

Strictly, based on the definition of conditionals in Section 1.2.3, we only condition distributions on random variables. As such, we should not condition on the parameters $\boldsymbol{\theta}$, which, within the frequentist paradigm, are *not* random. Nevertheless, in this chapter we keep conditioning on parameters for two reasons, one good and one less so. We start with the latter. As we often maximize the likelihood, it is convenient to think of the likelihood as a function of $\boldsymbol{\theta}$ and make its dependency on $\boldsymbol{\theta}$ explicit. Now the former: in subsequent chapters, we will be working mostly within the Bayesian paradigm and start modeling $\boldsymbol{\theta}$ as a random variable itself. In doing so, writing likelihoods as $p(w_{1:N}|\boldsymbol{\theta})$ is demanded if only for the sake of consistency with the remainder of the book.

Finally, as our focus from earlier chapters has now shifted beyond random variables, we drop notational differences between capitalized random variables and their realizations denoted with lower case letters as this distinction should be clear from context.

3.1.1 Estimating Parameters with Maximum Likelihood

Within the frequentist paradigm, parameters are variables whose values are to be learned from the data $w_{1:N}$. To estimate parameters, we ask:

- What is the likelihood of observing $w_{1:N}$ under the assumptions of our model?
- Evaluated at $w_{1:N}$, what values of the parameters $\boldsymbol{\theta}$ maximize this likelihood?

Taken together, these two concepts constitute the *maximum likelihood principle* that we illustrate in Example 3.2.

Example 3.2 Estimating diffusion coefficients using maximum likelihood

Consider the position $r(t)$ of a freely diffusing particle in one dimension for now. We assume that the particle's position is assessed at discrete and equally spaced time levels $t_n = t_0 + n\tau$ such that, at the nth time level, our measurement is $r_n = r(t_n)$. From Eq. (2.62), we immediately obtain

$$p(r_n|r_{n-1}, D) = \frac{1}{\sqrt{4\pi D\tau}} e^{-\frac{(r_n - r_{n-1})^2}{4D\tau}}, \tag{3.1}$$

where the value of the diffusion coefficient D, our only unknown, remains to be determined.

One way by which to determine D is to consider the *likelihood* of observing the full trajectory $r_{1:N}$,

$$p(r_{1:N}|D) = \left(\prod_{n=2}^{N} p(r_n|r_{n-1}, D)\right) p(r_1|D). \tag{3.2}$$

As the initial position of the particle r_1 is independent of D, we have $p(r_1|D) = p(r_1)$ and, combining Eqs. (3.1) and (3.2), we arrive at

$$p(r_{1:N}|D) = \frac{1}{(4\pi D\tau)^{(N-1)/2}} e^{-\sum_{n=2}^{N} \frac{(r_n - r_{n-1})^2}{4D\tau}} p(r_1).$$

According to the maximum likelihood principle, our estimator $\hat{D}$ is the value of D that maximizes $p(r_{1:N}|D)$. To find this value, it is mathematically simpler to work with the *log likelihood* instead which immediately follows from the above,

$$\log p(r_{1:N}|D) = -\frac{(N-1)}{2} \log(4\pi D\tau) - \sum_{n=2}^{N} \frac{(r_n - r_{n-1})^2}{4D\tau} + \log p(r_1).$$

Now, maximizing the log likelihood by seeking the roots of its derivative yields

$$-\frac{(N-1)}{2\hat{D}} + \sum_{n=2}^{N} \frac{(r_n - r_{n-1})^2}{4\hat{D}^2\tau} = 0,$$

which we can easily rearrange to arrive at

$$\hat{D} = \frac{1}{2(N-1)\tau} \sum_{n=2}^{N} (r_n - r_{n-1})^2.$$

3.1.2 Likelihood Maximization as an Optimization Problem*

Mathematically, within the frequentist paradigm likelihoods are functions to be maximized with respect to their parameters. In Example 3.2 we were able to obtain a likelihood's maximizer analytically. Yet this example is somewhat atypical as likelihood maximization often gives rise to complicated equations for which a solution cannot be found analytically. Indeed, many important cases fall in this category that require numerical optimization techniques. This section is intended as an overview of such techniques. Specifically, we introduce a series of methods for likelihood maximization starting from naive approaches and progressing toward more sophisticated ones.

Note 3.2 Log likelihood

Throughout this section, while our intention is to maximize a likelihood $\ell_{w_{1:N}}(\boldsymbol{\theta}) = p(w_{1:N}|\boldsymbol{\theta})$, we often work with the corresponding log likelihood $L_{w_{1:N}}(\boldsymbol{\theta}) = \log p(w_{1:N}|\boldsymbol{\theta})$ instead, which is often abbreviated as $L(\boldsymbol{\theta})$. Indeed, we do so as likelihoods over long data sequences, $N \gg 1$, can numerically underflow. As $\log(\cdot)$ is a strictly increasing function, $\ell_{w_{1:N}}(\cdot)$ and $L_{w_{1:N}}(\cdot)$ lead to the same results.

While today, maximizing log likelihoods is a question of practical relevance, historically, maximizing logarithms may have arisen as the likelihood's logarithm assumes a mathematically simpler form. This logic explains why, in statistical physics, we often work with free energies and not their exponential, *i.e.*, partition functions.

Naive Optimization

In the simplest case, our model contains a single scalar unknown parameter θ. In this setting, our likelihood $p(w_{1:N}|\theta)$ is a univariate function. A simple strategy to find a maximizer $\hat{\theta}$ of such a function is by: first, establishing a range $\theta^{\min} \leq \theta \leq \theta^{\max}$ that may contain the maximizer $\hat{\theta}$; and then successively expanding or shrinking its end points $\theta^{\min}, \theta^{\max}$ until the maximizer we seek lies within it *and* the range's total size is sufficiently small. Upon termination, our estimator is approximated by the midpoint of the terminal range $\hat{\theta} \approx (\theta^{\min} + \theta^{\max})/2$.

This simple method of finding univariate maximizers can be generalized to multiple dimensions. Namely, to find the maximizer $\hat{\boldsymbol{\theta}}$ of a multivariate likelihood $p(w_{1:N}|\boldsymbol{\theta})$, the range of $\boldsymbol{\theta}$ values is replaced by a simplex, $\boldsymbol{\theta}^{1:K+1}$, determined by a total of $K + 1$ points in the parameter space where K is the dimension of $\boldsymbol{\theta}$. In particular, if $\boldsymbol{\theta}$ is a scalar with $K = 1$, then our simplex consists of two points, $\boldsymbol{\theta}^{1:2}$, defining the initial bracketed range for our scalar in one dimension.

* This is an advanced topic and could be skipped on a first reading.

Once an initial simplex is chosen, its end points are expanded or shrunk until they enclose the maximizer *and* are sufficiently close to each other. Upon termination, $\hat{\theta}$ is approximated by the center of the terminal simplex. This process is the basis for the heuristic *Nelder–Mead* method of Algorithm 3.1.

Algorithm 3.1 Nelder–Mead optimization

Given $L(\theta) = \log p(w_{1:N}|\theta)$ and initial points $\theta^{1:K+1}$ for all K dimensions, the Nelder–Mead method iterates until convergence the following steps:

- Arrange the points $\theta^{1:K+1}$ such that

$$L(\theta^1) \geq L(\theta^2) \geq \cdots \geq L(\theta^{K+1}).$$

- Get the center of the first K points

$$\theta^* = \frac{1}{K}\sum_{k=1}^{K} \theta^k.$$

- Get reflection, expansion, and contraction points by

$$\begin{aligned}\theta^R &= \theta^* + \lambda_R\left(\theta^* - \theta^{K+1}\right),\\ \theta^E &= \theta^* + \lambda_E\left(\theta^R - \theta^*\right),\\ \theta^{C1} &= \theta^* + \lambda_C\left(\theta^R - \theta^*\right),\\ \theta^{C2} &= \theta^* + \lambda_C\left(\theta^{K+1} - \theta^*\right).\end{aligned}$$

- Update the points $\theta^{1:K+1}$ according to:
  - If $L\left(\theta^1\right) \geq L\left(\theta^R\right)$ and $L\left(\theta^R\right) > L\left(\theta^K\right)$, replace θ^{K+1} with θ^R.
  - If $L\left(\theta^R\right) > L\left(\theta^1\right)$ and $L\left(\theta^E\right) > L\left(\theta^R\right)$, replace θ^{K+1} with θ^E.
  - If $L\left(\theta^R\right) > L\left(\theta^1\right)$ and $L\left(\theta^E\right) \leq L\left(\theta^R\right)$, replace θ^{K+1} with θ^R.
  - If $L\left(\theta^R\right) > L\left(\theta^{K+1}\right)$ and $L\left(\theta^{C1}\right) > L\left(\theta^R\right)$, replace θ^{K+1} with θ^{C1}.
  - If $L\left(\theta^{C2}\right) > L\left(\theta^{K+1}\right)$, replace θ^{K+1} with θ^{C2}.
  - Otherwise, maintain θ^1 and replace $\theta^{2:K+1}$ with $\theta^1 + \lambda_S(\theta^k - \theta^1)$.

In this algorithm, the gains λ_R, λ_E are preset such that $0 < \lambda_R, 1 < \lambda_E$ and the gains λ_C, λ_S are preset such that $0 < \lambda_C \leq \frac{1}{2}, 0 < \lambda_S \leq 1$.

Such a simple method can be applied to any tractable likelihood and properties like differentiability are not essential for its performance. Despite this generality, however, optimizers like Algorithm 3.1 rely on *heuristic* updating rules. Accordingly, they may fail to locate the maximizer of interest. In fact, they often become trapped in local maxima or cannot indicate the nonexistence of a maximum. Most importantly, however, they may be inefficient. This inefficiency can become especially pronounced when individual parameters in θ are not properly scaled or the evaluation of $\log p(w_{1:N}|\theta)$ is costly. For these reasons, more sophisticated approaches to likelihood maximization, that we see next, have been developed that take advantage of additional properties of the likelihood.

Newton–Raphson Optimization

Provided our log likelihood function is continuous and differentiable, elementary calculus indicates that the maximizer we seek is a root of its derivative. This is a condition that gives rise to Newton–Raphson optimizers. Optimizers of this type are iterative methods intended to approximate the zeroes of $\nabla \log p(w_{1:N}|\boldsymbol{\theta})$. Here, the gradient is evaluated with respect to $\boldsymbol{\theta}$.

Note 3.3 Gradient and Hessian

For any scalar function $L(\boldsymbol{\theta})$ of a multivariate variable $\boldsymbol{\theta} = (\theta_1, \theta_2, \ldots, \theta_K)$, the gradient and Hessian are given by the operators

$$\nabla L = \begin{pmatrix} \frac{\partial L}{\partial \theta_1} \\ \frac{\partial L}{\partial \theta_2} \\ \vdots \\ \frac{\partial L}{\partial \theta_K} \end{pmatrix}, \qquad HL = \begin{pmatrix} \frac{\partial^2 L}{\partial \theta_1^2} & \frac{\partial^2 L}{\partial \theta_1 \partial \theta_2} & \cdots & \frac{\partial^2 L}{\partial \theta_1 \partial \theta_K} \\ \frac{\partial^2 L}{\partial \theta_2 \partial \theta_1} & \frac{\partial^2 L}{\partial \theta_2^2} & \cdots & \frac{\partial^2 L}{\partial \theta_2 \partial \theta_K} \\ \vdots & \vdots & \ddots & \vdots \\ \frac{\partial^2 L}{\partial \theta_K \partial \theta_1} & \frac{\partial^2 L}{\partial \theta_K \partial \theta_2} & \cdots & \frac{\partial^2 L}{\partial \theta_K^2} \end{pmatrix}.$$

These are generalizations of a function's *first* and *second* derivatives. Indeed, for functions $L(\theta)$ of a univariate variable θ, the gradient and Hessian operators reduce to

$$\nabla L = \frac{dL}{d\theta}, \qquad HL = \frac{d^2 L}{d\theta^2}.$$

In particular, these methods start from an initial guess $\hat{\boldsymbol{\theta}}^{\text{old}}$ of the maximizer $\hat{\boldsymbol{\theta}}$ and improve upon this guess with $\hat{\boldsymbol{\theta}}^{\text{new}}$. Such iteration is established in an effort to solve $\nabla \log p\left(w_{1:N}|\hat{\boldsymbol{\theta}}\right) = 0$, which considers

$$\begin{aligned} 0 \approx \nabla \log p\left(w_{1:N}|\hat{\boldsymbol{\theta}}^{\text{new}}\right) &= \nabla \log p\left(w_{1:N}|\hat{\boldsymbol{\theta}}^{\text{old}} + \boldsymbol{h}\right) \\ &\approx \nabla \log p\left(w_{1:N}|\hat{\boldsymbol{\theta}}^{\text{old}}\right) + \left(H \log p\left(w_{1:N}|\hat{\boldsymbol{\theta}}^{\text{old}}\right)\right) \boldsymbol{h}^{\text{T}}, \end{aligned}$$

where $\boldsymbol{h} = \hat{\boldsymbol{\theta}}^{\text{new}} - \hat{\boldsymbol{\theta}}^{\text{old}}$, and where the superscript T denotes the transpose. A rearrangement of this yields

$$\hat{\boldsymbol{\theta}}^{\text{new}} \approx \hat{\boldsymbol{\theta}}^{\text{old}} - \left(\nabla \log p\left(w_{1:N}|\hat{\boldsymbol{\theta}}^{\text{old}}\right)\right)^{\text{T}} \left(H \log p\left(w_{1:N}|\hat{\boldsymbol{\theta}}^{\text{old}}\right)\right)^{-1}. \qquad (3.3)$$

As we now see, for models containing a single scalar parameter, Eq. (3.3) simplifies drastically.

Example 3.3 Newton–Raphson in one dimension

Since the log likelihood is a univariate function, the corresponding updates are given by

$$\hat{\theta}^{\text{new}} = \hat{\theta}^{\text{old}} - \frac{\frac{d}{d\theta}\log p\left(w_{1:N}|\hat{\theta}^{\text{old}}\right)}{\frac{d^2}{d\theta^2}\log p\left(w_{1:N}|\hat{\theta}^{\text{old}}\right)} = \hat{\theta}^{\text{old}} - \frac{L'_{w_{1:N}}\left(\hat{\theta}^{\text{old}}\right)}{L''_{w_{1:N}}\left(\hat{\theta}^{\text{old}}\right)},$$

where $L'_{w_{1:N}}(\cdot)$ and $L''_{w_{1:N}}(\cdot)$ denote the first and second derivative, respectively.

Equation (3.3) lends itself to Algorithm 3.2. The implementation of this algorithm requires both the Hessian and gradient of the log likelihood to be analytically computable and that the Hessian be, furthermore, invertible.

Algorithm 3.2 Newton–Raphson optimization

Given $L(\boldsymbol{\theta}) = \log p(w_{1:N}|\boldsymbol{\theta})$, $\nabla L(\boldsymbol{\theta}) = \nabla \log p(w_{1:N}|\boldsymbol{\theta})$, $HL(\boldsymbol{\theta}) = H \log p(w_{1:N}|\boldsymbol{\theta})$, and an initial point $\hat{\boldsymbol{\theta}}^{\text{old}}$, the Newton–Raphson method iterates until convergence the following steps:

- Evaluate $\boldsymbol{h}$ by solving the linear system

$$\boldsymbol{h}\left(HL\left(\hat{\boldsymbol{\theta}}^{\text{old}}\right)\right) = -\nabla L\left(\hat{\boldsymbol{\theta}}^{\text{old}}\right)^{\text{T}}.$$

- Evaluate $\hat{\boldsymbol{\theta}}^{\text{new}}$ by

$$\hat{\boldsymbol{\theta}}^{\text{new}} = \hat{\boldsymbol{\theta}}^{\text{old}} + \boldsymbol{h}.$$

The inversion of the Hessian at each iteration of Algorithm 3.2 is the main limitation of this algorithm. As this inversion is often costly, especially for high-dimensional problems, below we present alternative strategies that attempt to solve Eq. (3.3) in a less costly manner.

Gradient Optimization

The method of gradient descent is an iterative approach where we start from some $\hat{\boldsymbol{\theta}}^{\text{old}}$ and identify a direction and magnitude in which to move up toward the maximum of the likelihood or, equivalently, down toward the minimum of the negative log likelihood. We begin with the realization that the direction of steepest descent at a point $\boldsymbol{\theta}$ is always in the direction of the gradient $\boldsymbol{d} = -\left(\nabla \log p\left(w_{1:N}|\hat{\theta}\right)\right)^{\text{T}}$. As such, in an attempt to emulate Eq. (3.3), we may use

$$\hat{\boldsymbol{\theta}}^{\text{new}} = \hat{\boldsymbol{\theta}}^{\text{old}} + \alpha \boldsymbol{d}^{\text{old}},$$

where α is a positive scalar that determines the size of the step along the steepest direction $\boldsymbol{d}^{\text{old}} = -\left(\nabla \log p\left(w_{1:N}|\hat{\boldsymbol{\theta}}^{\text{old}}\right)\right)^{\text{T}}$. In order to avoid overshooting the minimum along the search line, we note that when we have reached an extremum along the search line, the gradient $-\nabla \log p\left(w_{1:N}|\hat{\boldsymbol{\theta}}^{\text{old}} + \alpha \boldsymbol{d}^{\text{old}}\right)$ will be exactly orthogonal to the search direction $\boldsymbol{d}^{\text{old}}$.

In general, it is possible to find an analytic expression for α only for log likelihoods assuming a quadratic form. As such, we must invoke Newton–Raphson iterations, or related methods, to locate the minimum of $-\log p\left(w_{1:N}|\hat{\boldsymbol{\theta}}^{\text{old}} + \alpha \boldsymbol{d}^{\text{old}}\right)$ with respect to α. This necessarily compounds both merits and demerits of Newton–Raphson used here in gradient descent. This is often achieved by finding the zeroes of the product $\boldsymbol{d}^{\text{old}}\nabla \log p\left(w_{1:N}|\hat{\boldsymbol{\theta}}^{\text{old}} + \alpha \boldsymbol{d}^{\text{old}}\right)$ by the minimization

$$\hat{\alpha} = -\underset{\alpha>0}{\operatorname{argmin}}\left(\boldsymbol{d}^{\text{old}}\nabla \log p\left(w_{1:N}|\hat{\boldsymbol{\theta}}^{\text{old}} + \alpha \boldsymbol{d}^{\text{old}}\right)\right). \tag{3.4}$$

Once we have identified the minimum, we set $\hat{\boldsymbol{\theta}}^{\text{new}} = \hat{\boldsymbol{\theta}}^{\text{old}} + \hat{\alpha}\boldsymbol{d}^{\text{old}}$. Algorithm 3.3 provides a method for the *gradient descent*.

Algorithm 3.3 Gradient descent

Given $L(\boldsymbol{\theta}) = \log p(w_{1:N}|\boldsymbol{\theta})$, $\nabla L(\boldsymbol{\theta}) = \nabla \log p(w_{1:N}|\boldsymbol{\theta})$, and an initial point $\hat{\boldsymbol{\theta}}^{\text{old}}$, the gradient descent method iterates until convergence the following steps:

- Evaluate a search direction

$$\boldsymbol{d}^{\text{old}} = -\left(\nabla L\left(\hat{\boldsymbol{\theta}}^{\text{old}}\right)\right)^{\text{T}}.$$

- Find $\hat{\alpha}$ according to the minimization of Eq. (3.4)
- Evaluate $\hat{\boldsymbol{\theta}}^{\text{new}}$ by

$$\hat{\boldsymbol{\theta}}^{\text{new}} = \hat{\boldsymbol{\theta}}^{\text{old}} + \hat{\alpha}\boldsymbol{d}^{\text{old}}.$$

Although gradient descent optimizers do not require analytic formulas for the likelihood's gradient and Hessian as Newton–Raphson does, they may suffer from the severe inefficiency of searching repeatedly along the same direction. To be clear, while alternate search directions are orthogonal by construction, it is not generally true that orthogonality of search directions is retained two steps ahead. Thus, successive directions $\boldsymbol{d}$ are not generally orthogonal. As the same directions may be revisited multiple times, the method may find itself inefficiently zigzagging its way to a (possibly local) optimum.

Conjugate Gradient Optimization

To improve upon the inefficiencies of gradient descent, we first consider computing an auxiliary point $\hat{\boldsymbol{\theta}}'$ in just the same way we had computed $\hat{\boldsymbol{\theta}}^{\text{new}}$ in gradient descent,

$$\hat{\boldsymbol{\theta}}' = \hat{\boldsymbol{\theta}}^{\text{old}} + \hat{\alpha}\boldsymbol{d}^{\text{old}}.$$

Next, we compute the gradient at $\boldsymbol{d}' = -\left(\nabla \log p\left(w_{1:N}|\hat{\boldsymbol{\theta}}'\right)\right)^{\text{T}}$. However, unlike in the case of gradient descent, we do not strictly take this direction next. This would be equivalent to insisting that $\boldsymbol{d}^{\text{new}}$ be strictly orthogonal to $\boldsymbol{d}^{\text{old}}$. In doing so, we may revisit the same directions in our search in subsequent steps.

To avoid revisiting the same directions, we define a new search direction, $\boldsymbol{d}^{\text{new}}$, that avoids pointing in all previously visited directions. The method of conjugate gradient achieves this in a manner that avoids storing all but the current and previous directions. The means by which this is achieved is to select $\boldsymbol{d}^{\text{new}}$ to point in a very specific type of linear superposition orthogonal to the old search direction, $\boldsymbol{d}^{\text{old}}$. We do so by selecting a coefficient of superposition. Specifically, the new direction is given by

$$\boldsymbol{d}^{\text{new}} = \boldsymbol{d}' + \hat{\beta}\boldsymbol{d}^{\text{old}},$$

where $\hat{\beta}$, as we just mentioned, is selected in order to ensure that $\boldsymbol{d}'$ would have been orthogonal to the previous directions for quadratic log likelihoods. One way is via

$$\hat{\beta} = \frac{\boldsymbol{d}'\boldsymbol{d}'^{\text{T}}}{\boldsymbol{d}^{\text{old}}\boldsymbol{d}^{\text{old T}}}, \tag{3.5}$$

which is a reasonable option for locally quadratic log likelihoods.

In summary, we say that gradients are conjugate and not strictly orthogonal as in the case of gradient descent. The algorithm for *conjugate gradients* is given in Algorithm 3.4.

Algorithm 3.4 Conjugate gradient

Given $L(\boldsymbol{\theta}) = \log p(w_{1:N}|\boldsymbol{\theta}), \nabla L(\theta) = \nabla \log p(w_{1:N}|\boldsymbol{\theta})$, and an initial starting point $\hat{\boldsymbol{\theta}}^{\text{old}}$ and direction $\boldsymbol{d}^{\text{old}}$, the method of conjugate gradient iterates until convergence the following steps:

- Find $\hat{\alpha}$ by minimizing the product $\boldsymbol{d}^{\text{old}}\nabla L\left(\hat{\boldsymbol{\theta}}^{\text{old}} + \alpha\boldsymbol{d}^{\text{old}}\right)$.
- Set $\hat{\boldsymbol{\theta}}' = \hat{\boldsymbol{\theta}}^{\text{old}} + \hat{\alpha}\boldsymbol{d}^{\text{old}}$.
- Compute $\hat{\beta}$ according to Eq. (3.5).
- Set $\boldsymbol{d}^{\text{new}} = -\left(\nabla L\left(\hat{\boldsymbol{\theta}}'\right)\right) + \hat{\beta}\boldsymbol{d}^{\text{old}}$ and $\hat{\boldsymbol{\theta}}^{\text{new}} = \hat{\boldsymbol{\theta}}^{\text{old}} + \hat{\alpha}\boldsymbol{d}^{\text{new}}$.

Limitations of Optimization Methods

Questions as to whether the extrema are reached at all in nonlinear optimizations and whether they are local or global largely evade the methods presented above. Indeed, exploring multiple extrema can be attempted through heuristics: either by instantiating optimizations with parameter values neighboring the computed optima or seeding optimizations with multiple initial conditions with no guarantee of having identified the global optimum.

3.2 Observations and Associated Measurement Noise

In Chapter 2, we presented multiple dynamical forward models. The origin of uncertainty and the reason we formulated them probabilistically stems from the stochastic dynamics inherent to such systems. For example, in the systems modeled by the Langevin equation and ensuing Fokker–Planck equation, the source of uncertainty originates from the Wiener process, which, in the language of physics, is meant to model thermal agitation jostling the system.

Yet, even if the dynamics of our systems were to be deterministic, we would still require probabilistically defined likelihoods. The reason for this is simple: we have *measurement noise*. Accounting for the uncertainty inherent to the measurement process is a requirement for a full description of a likelihood and this introduces the concept of *latent variables*, which we denote by $r_{1:N}$. In full generality, these coincide with the hidden states of a system indirectly monitored through measurement $w_{1:N}$.

Indeed, instances such as Example 3.2 are too simple to model realistic scenarios because they assume no discrepancy between the measurement of a particle's position r_n and its observation w_n. Since measurement introduces some error, we now lift this assumption. That is, we make explicit the probabilistic relationship that exists between model variables and observations. While we focus mostly on processes stemming from dynamical systems, the same applies to iid processes as well.

This discussion immediately forces us to distinguish the parameters we gather in $\theta = \{\psi, \phi\}$ into two distinct categories. The first are the parameters of the *observation distribution*, or the *emission distribution* as it is often called, characterizing the measurement noise. These parameters are termed *observation parameters* or *emission parameters* and we label them ϕ. The other parameters are parameters of the generative model not involved in the observation distribution itself, which we label ψ. Of course, depending on the context our models may involve only one of the two parameter types.

In general, the observation distributions depend on the latent variables. In this setting,

$$w_n | r_n, \phi \sim \mathbb{F}_{r_n,\phi},$$

where $\mathbb{F}_{r_n,\phi}$ is the *emission distribution*, r_n is the latent state, and ϕ the emission parameters.

In the most general case, the unknowns of a model include $r_{1:N}$ and ψ, ϕ. Less generally, the measurement apparatus may have been precalibrated and, in this case, ϕ have known values. The unknowns are then reduced to $r_{1:N}$ and ψ.

Note 3.4 Generative models with measurement uncertainty

For iid random variables, a generative model accounting for measurement uncertainty reads

$$r_n|\psi \sim \mathbb{P}_{\psi},$$

$$w_n|r_n, \phi \sim \mathbb{F}_{r_n,\phi},$$

where we make explicit the dependency of each measurement on its own latent variable r_n. Similarly, for Markov random variables we have

$$r_n|r_{n-1}, \psi \sim \mathbb{P}_{r_{n-1},\psi},$$

$$w_n|r_n, \phi \sim \mathbb{F}_{r_n,\phi},$$

where we make explicit the dependency of a latent variable r_n on its predecessor r_{n-1}.

To illustrate a concrete example of a generative model for the latent variables as well as an associated observation model, Example 3.4 gives an example of a freely diffusing particle with a normal observation distribution.

Example 3.4 Gaussian observation models

We consider a freely diffusing particle whose transition probability density is described by

$$r_n|r_{n-1}, D \sim \text{Normal}(r_{n-1}, 2D\tau).$$

Here, r_n is a latent variable and $\psi = D$ is the only dynamical parameter. Assuming normal observations, we write the full generative model as follows

$$r_n|r_{n-1}, D \sim \text{Normal}(r_{n-1}, 2D\tau),$$

$$w_n|r_n, v \sim \text{Normal}(r_n, v),$$

where v is the variance of the observation distribution. Here, $\psi = v$ is the only emission parameter.

So far, we have only discussed observation models whose output, w_n, depends on an instantaneous latent variable r_n. However, if required by the measurement process, observation models can accommodate more complex scenarios.

Example 3.5 Integrative observation models

Here, we envision snapshots of a particle's position tracked by a camera with an exposure time per frame of τ_{exp}. The measurement w_n, obtained at time t_n, is the result of integration over all positions of the particle attained over the exposure time window that spans from $t_n - \tau_{\text{exp}}$ up to t_n. The measurement model then reads

$$w_n|r(\cdot), v \sim \text{Normal}\left(\frac{1}{\tau_{\text{exp}}}\int_{t_n-\tau_{\text{exp}}}^{t_n} dt\, r(t), v\right).$$

In this setting, the latent variable $r(\cdot)$ is the entire trajectory of the particle during the exposure window.

3.2.1 Completed Likelihoods

When we introduce measurement noise and the associated concept of latent variables, our likelihoods naturally grow in complexity. In this case, as we will see through examples, we must often start by specifying a joint likelihood over observations and latent variables $p(w_{1:N}, r_{1:N}|\boldsymbol{\theta})$. These joint likelihoods are called *completed likelihoods*.

The completed likelihood is obtained by considering the distributions associated with a model's measurements as well as the latent states

$$p(w_{1:N}, r_{1:N}|\boldsymbol{\theta}) = p(w_{1:N}|r_{1:N}, \boldsymbol{\theta})p(r_{1:N}|\boldsymbol{\theta}).$$

An immediate problem now arises if we attempt to maximize the completed likelihood to determine $\boldsymbol{\theta}$. What values do we ascribe to the $r_{1:N}$? These are, unfortunately, unknown.

Instead, we use the completed likelihood, $p(w_{1:N}, r_{1:N}|\boldsymbol{\theta})$, to derive the *marginal likelihood*, sometimes called the *true likelihood*, $p(w_{1:N}|\boldsymbol{\theta})$, by

$$p(w_{1:N}|\boldsymbol{\theta}) = \int dr_{1:N}\, p(w_{1:N}, r_{1:N}|\boldsymbol{\theta}).$$

The integral, over all allowed values of $r_{1:N}$, is understood as a sum when dealing with discrete latent variables.

Example 3.6 Completed and marginal likelihoods for diffusive motion

In Example 3.2, where we obtained an estimate for the diffusion coefficient, we assumed $w_n = r_n$. However, when modeling measurement noise, we introduce a joint likelihood over $w_{1:N}$ and $r_{1:N}$ that takes the form

$$\begin{aligned} &p(w_{1:N}, r_{1:N}|D, v) \\ &= p(w_N|v)p(r_N|r_{N-1}, D)\cdots p(w_2|r_2, v)p(r_2|r_1, D)p(w_1|r_1, v)p(r_1), \\ &= \left(\prod_{n=2}^{N} p(w_n|r_n, v)p(r_n|r_{n-1}, D)\right)p(w_1|r_1, v)p(r_1). \end{aligned}$$

This is the completed likelihood. This likelihood cannot be interpreted as the probability of the observation. It represents a joint distribution over the observed measurements $w_{1:N}$ and unobserved (latent) states $r_{1:N}$.

The corresponding marginal likelihood is then obtained by marginalizing $p(w_{1:N}, r_{1:N}|D, v)$ over the realizations $r_{1:N}$. In other words,

$$p(w_{1:N}|D, v) = \int dr_{1:N}\, p(w_{1:N}, r_{1:N}|D, v).$$

Maximizing the (marginal) likelihood $p(w_{1:N}|D, v)$ yields an estimate for the diffusion coefficient, $\hat{D}$, which is different from ignoring the measurement noise altogether. Indeed, we can already intuit that if our goal is to infer diffusion coefficients, D, those diffusion coefficients estimated by ignoring measurement noise will overestimate the true diffusion coefficient. Put differently, if we do not account for measurement noise, we overestimate the diffusion coefficient as both measurement noise and stochasticity in the dynamics inherent to the diffusive process positively contribute to the apparent diffusion coefficient.

Above, we discuss both completed and marginal likelihoods for a diffusing particle for which it is possible to marginalize over the positions. Though, for more general problems, such marginalizations of the completed likelihood can rarely be achieved analytically. This thought motivates our next section.

3.2.2 The EM Algorithm*

As is often the case in any inference problem, we encounter measurement noise. Within the frequentist paradigm, this implies that our goal is to maximize likelihoods, $p(w_{1:N}|\boldsymbol{\theta})$, with respect to parameters $\boldsymbol{\theta}$. In constructing completed likelihoods previously, we have had the advantage of being able to integrate over the latent variables $r_{1:N}$ and maximize the resulting marginal likelihood over $\boldsymbol{\theta}$. In general, marginalizing over the latent variable as well as maximizing over this likelihood analytically are challenging or even, most often, impossible. For this reason, here we focus on the problem of maximizing $p(w_{1:N}|\boldsymbol{\theta})$ under the assumption that marginalization over the latent variables in $p(w_{1:N}, r_{1:N}|\boldsymbol{\theta})$ is *not* analytically tractable.

The method we present to solve this problem is termed the *expectation-maximization algorithm* or EM for short. A brief motivation for this method is provided in Note 3.5.

Note 3.5 Motivation for EM

To show why it is difficult to maximize $p(w_{1:N}|\boldsymbol{\theta})$ directly, we start with

$$p(w_{1:N}|\boldsymbol{\theta}) = \sum_{r_{1:N}} p(w_{1:N}, r_{1:N}|\boldsymbol{\theta}), \tag{3.6}$$

where, for concreteness, we have assumed discrete latent variables, $r_{1:N}$, and the sum is over all allowed realizations of these latent variables.

We recall that, in order to maximize this likelihood and avoid numerical underflow, we need to maximize $\log\left(\sum_{r_{1:N}} p(w_{1:N}, r_{1:N}|\boldsymbol{\theta})\right)$. Maximizing the logarithm of the sum of a number of small terms (that may numerically underflow) is much more difficult than maximizing, say, the sum of the

* This is an advanced topic and could be skipped on a first reading.

logarithm, $\sum_{r_{1:N}} \log p(w_{1:N}, r_{1:N}|\boldsymbol{\theta})$ where each term will not typically underflow. Unfortunately, these maxima are not, in general, equivalent.

The goal of the EM algorithm is ultimately to maximize an object, defined in Eq. (3.7), expressible as a sum of logarithms with no latent variable dependence, as an approximation to maximizing over the logarithm of Eq. (3.6).

The idea behind EM is as follows: rather than maximize $\log p(w_{1:N}|\boldsymbol{\theta})$, we show that this is approximately equal to computing the expectation of $\log p(r_{1:N}, w_{1:N}|\boldsymbol{\theta})$ with respect to $p(r_{1:N}|w_{1:N}, \boldsymbol{\theta})$ (the E-step), followed by maximizing this expectation (the M-step). Overall, the EM scheme starts with an initial guess $\hat{\boldsymbol{\theta}}^{\text{old}}$ of the maximizer and iteratively improves upon this.

In particular, the E-step requires an expression for the sum or integral foreshadowing the iterative nature of EM,

$$Q_{\boldsymbol{\theta}'}(\boldsymbol{\theta}) = \sum_{r_{1:N}} \left(\log p\left(w_{1:N}, r_{1:N}|\boldsymbol{\theta}\right)\right) p\left(r_{1:N}|w_{1:N}, \boldsymbol{\theta}'\right). \tag{3.7}$$

The updated $\hat{\boldsymbol{\theta}}^{\text{new}}$ is then obtained from the M-step

$$\hat{\boldsymbol{\theta}}^{\text{new}} = \underset{\boldsymbol{\theta}}{\operatorname{argmax}}\, Q_{\hat{\boldsymbol{\theta}}^{\text{old}}}(\boldsymbol{\theta}).$$

An example (Example 3.7) and the derivation of EM follows Algorithm 3.5.

Algorithm 3.5 Expectation-Maximization

Given observations $w_{1:N}$ and an initial starting point $\hat{\boldsymbol{\theta}}^{\text{old}}$, EM iterates until convergence the following steps:

- Find

$$f_{\boldsymbol{\theta}}(r_{1:N}) = p(r_{1:N}|w_{1:N}, \boldsymbol{\theta}).$$

- Find the expectation of the complete log-likelihood

$$Q_{\boldsymbol{\theta}'}(\boldsymbol{\theta}) = \sum_{r_{1:N}} f_{\boldsymbol{\theta}'}(r_{1:N}) \log p(r_{1:N}, w_{1:N}|\boldsymbol{\theta}).$$

- Compute an improved approximation by

$$\hat{\boldsymbol{\theta}}^{\text{new}} = \underset{\boldsymbol{\theta}}{\operatorname{argmax}}\, Q_{\hat{\boldsymbol{\theta}}^{\text{old}}}(\boldsymbol{\theta}).$$

Example 3.7 EM training for the sum of two Gaussians

Consider scalar observations w_n, for $n = 1 : N$, generated by a sum of two Gaussians,

$$w_n|\boldsymbol{\theta} \sim \pi_1 \text{Normal}(\mu_1, v_1) + \pi_2 \text{Normal}(\mu_2, v_2),$$

where μ_1, μ_2 are the means and v_1, v_2 the associated variances. Here, π_1, π_2 are the probabilities of w_n stemming from the first and second component, respectively, where $\pi_1 + \pi_2 = 1$. Together, all parameters $\pi_1, \pi_2, \mu_1, \mu_2$, and v_1, v_2 are collected in $\boldsymbol{\theta}$.

In this example, we call our latent variables $s_{1:N}$. Each indicator s_n coincides with which of the two components generated the observation w_n and each s_n may take values 1 or 2 depending on whether w_n has been generated from Normal(μ_1, v_1) or Normal(μ_2, v_2), respectively. With the introduction of these indicators, our model takes the equivalent form

$$s_n|\boldsymbol{\theta} \sim \text{Categorical}_{1:2}(\pi_1, \pi_2), \tag{3.8}$$

$$w_n|s_n, \boldsymbol{\theta} \sim \text{Normal}\left(\mu_{s_n}, v_{s_n}\right). \tag{3.9}$$

The Categorical$_{1,2}(\pi_1, \pi_2)$ distribution appearing in the first equation implies that $p(s_n = 1) = \pi_1$ and $p(s_n = 2) = \pi_2$, which may be combined into a single expression

$$p(s_n|\boldsymbol{\theta}) = \pi_1^{\Delta_1(s_n)} \pi_2^{\Delta_2(s_n)},$$

where the Δs denote Kronecker deltas defined in Appendix C.

To estimate the parameters $\boldsymbol{\theta} = \{\pi_1, \pi_2, \mu_1, \mu_2, v_1, v_2\}$, as before, we seek a maximum likelihood solution, which may be obtained by applying EM to the model of Eqs. (3.8) and (3.9).

The steps involved are as follows:

- We start from some initial guess of the parameters $\boldsymbol{\theta}^{\text{old}} = \{\pi_1^{\text{old}}, \pi_2^{\text{old}}, \mu_1^{\text{old}}, \mu_2^{\text{old}}, v_1^{\text{old}}, v_2^{\text{old}}\}$.
- We then compute the conditional probability, $f_{\boldsymbol{\theta}^{\text{old}}}(s_{1:N})$, of the latent variables

$$f_{\boldsymbol{\theta}^{\text{old}}}(s_{1:N}) = p\left(s_{1:N}|w_{1:N}, \boldsymbol{\theta}^{\text{old}}\right) = \prod_{n=1}^{N} \left(\gamma_{1n}^{\text{old}}\right)^{\Delta_1(s_n)} \left(\gamma_{2n}^{\text{old}}\right)^{\Delta_2(s_n)}, \tag{*3.10*}$$

where γ_{1n}^{old} and γ_{2n}^{old} are given by

$$\gamma_{1n}^{\text{old}} = \frac{\pi_1^{\text{old}} \text{Normal}\left(w_n; \mu_1^{\text{old}}, v_1^{\text{old}}\right)}{\pi_1^{\text{old}} \text{Normal}\left(w_n; \mu_1^{\text{old}}, v_1^{\text{old}}\right) + \pi_2^{\text{old}} \text{Normal}\left(w_n; \mu_2^{\text{old}}, v_2^{\text{old}}\right)},$$

$$\gamma_{2n}^{\text{old}} = \frac{\pi_2^{\text{old}} \text{Normal}\left(w_n; \mu_2^{\text{old}}, v_2^{\text{old}}\right)}{\pi_1^{\text{old}} \text{Normal}\left(w_n; \mu_1^{\text{old}}, v_1^{\text{old}}\right) + \pi_2^{\text{old}} \text{Normal}\left(w_n; \mu_2^{\text{old}}, v_2^{\text{old}}\right)}.$$

These are essentially the probabilities of observation w_n stemming from the first and second Gaussians in the sum, respectively, under the parameter values $\boldsymbol{\theta}^{\text{old}}$.

- Next, we compute the expectation under $f_{\boldsymbol{\theta}^{\text{old}}}(s_{1:N})$ of the completed log likelihood,

$$Q_{\boldsymbol{\theta}^{\text{old}}}(\boldsymbol{\theta}) = \sum_{s_{1:N}} \log p(s_{1:N}, w_{1:N}|\boldsymbol{\theta}) f_{\boldsymbol{\theta}^{\text{old}}}(s_{1:N})$$

$$= \left(\log \pi_1 - \frac{\log v_1}{2}\right)\left(\sum_{n=1}^{N} \gamma_{1n}^{\text{old}}\right) + \left(\log \pi_2 - \frac{\log v_2}{2}\right)\left(\sum_{n=1}^{N} \gamma_{2n}^{\text{old}}\right)$$

$$- \frac{1}{2}\sum_{n=1}^{N}\left[\gamma_{1n}^{\text{old}}\frac{(w_n - \mu_1)^2}{v_1} + \gamma_{2n}^{\text{old}}\frac{(w_n - \mu_2)^2}{v_2}\right] + \text{constants}. \quad (*3.11*)$$

- Finally, our objective is to find the maximizer $\boldsymbol{\theta}^{\text{new}} = \{\pi_1^{\text{new}}, \pi_2^{\text{new}}, \mu_1^{\text{new}}, \mu_2^{\text{new}}, v_1^{\text{new}}, v_2^{\text{new}}\}$ of $Q_{\boldsymbol{\theta}^{\text{old}}}(\boldsymbol{\theta})$ under the constraint

$$\pi_1 + \pi_2 = 1.$$

For this, we form a Lagrangian

$$\mathbb{L}_{\boldsymbol{\theta}^{\text{old}}}(\boldsymbol{\theta}, \lambda) = Q_{\boldsymbol{\theta}^{\text{old}}}(\boldsymbol{\theta}) + \lambda(\pi_1 + \pi_2 - 1),$$

with multiplier λ used to enforce the constraint on the normalization of the weights. Equating the gradient of $\mathbb{L}_{\boldsymbol{\theta}^{\text{old}}}(\boldsymbol{\theta})$ to zero, we obtain

$$0 = \frac{\partial \mathbb{L}_{\boldsymbol{\theta}^{\text{old}}}(\boldsymbol{\theta}, \lambda)}{\partial \pi_1} = \frac{1}{\pi_1}\left(\sum_{n=1}^{N} \gamma_{1n}^{\text{old}}\right) + \lambda, \quad 0 = \frac{\partial \mathbb{L}_{\boldsymbol{\theta}^{\text{old}}}(\boldsymbol{\theta}, \lambda)}{\partial \pi_2} = \frac{1}{\pi_2}\left(\sum_{n=1}^{N} \gamma_{2n}^{\text{old}}\right) + \lambda,$$

which, combined with the constraint $\pi_1 + \pi_2 = 1$, yield the optimum

$$\pi_1^{\text{new}} = \frac{\sum_{n=1}^{N} \gamma_{1n}^{\text{old}}}{\sum_{n=1}^{N}\left(\gamma_{1n}^{\text{old}} + \gamma_{2n}^{\text{old}}\right)}, \qquad \pi_2^{\text{new}} = \frac{\sum_{n=1}^{N} \gamma_{2n}^{\text{old}}}{\sum_{n=1}^{N}\left(\gamma_{1n}^{\text{old}} + \gamma_{2n}^{\text{old}}\right)}.$$

From the remaining components of the Lagrangian's gradient, we obtain

$$0 = \frac{\partial \mathbb{L}_{\boldsymbol{\theta}^{\text{old}}}(\boldsymbol{\theta}, \lambda)}{\partial \mu_1} = \sum_{n=1}^{N} \gamma_{1n}^{\text{old}} \frac{(w_n - \mu_1)}{v_1},$$

$$0 = \frac{\partial \mathbb{L}_{\boldsymbol{\theta}^{\text{old}}}(\boldsymbol{\theta}, \lambda)}{\partial \mu_2} = \sum_{n=1}^{N} \gamma_{1n}^{\text{old}} \frac{(w_n - \mu_2)}{v_2},$$

$$0 = \frac{\partial \mathbb{L}_{\boldsymbol{\theta}^{\text{old}}}(\boldsymbol{\theta}, \lambda)}{\partial v_1} = \frac{1}{2v_1}\left(\frac{1}{v_1}\left(\sum_{n=1}^{N} \gamma_{1n}^{\text{old}}(w_n - \mu_1)^2\right) - \left(\sum_{n=1}^{N} \gamma_{1n}^{\text{old}}\right)\right),$$

$$0 = \frac{\partial \mathbb{L}_{\boldsymbol{\theta}^{\text{old}}}(\boldsymbol{\theta}, \lambda)}{\partial v_2} = \frac{1}{2v_2}\left(\frac{1}{v_2}\left(\sum_{n=1}^{N} \gamma_{2n}^{\text{old}}(w_n - \mu_2)^2\right) - \left(\sum_{n=1}^{N} \gamma_{2n}^{\text{old}}\right)\right),$$

which are solved to yield the remaining parameters,

$$\mu_1^{\text{new}} = \frac{\sum_{n=1}^{N} \gamma_{1n}^{\text{old}} w_n}{\sum_{n=1}^{N} \gamma_{1n}^{\text{old}}}, \qquad v_1^{\text{new}} = \frac{\sum_{n=1}^{N} \gamma_{1n}^{\text{old}}(w_n - \mu_1^{\text{new}})^2}{\sum_{n=1}^{N} \gamma_{1n}^{\text{old}}},$$

$$\mu_2^{\text{new}} = \frac{\sum_{n=1}^{N} \gamma_{2n}^{\text{old}} w_n}{\sum_{n=1}^{N} \gamma_{2n}^{\text{old}}}, \qquad v_2^{\text{new}} = \frac{\sum_{n=1}^{N} \gamma_{2n}^{\text{old}}(w_n - \mu_2^{\text{new}})^2}{\sum_{n=1}^{N} \gamma_{2n}^{\text{old}}}.$$

Why Does the EM Algorithm Work?

To illustrate the mechanism underlying the EM algorithm, we start from the factorization

$$p\left(w_{1:N}, r_{1:N}|\boldsymbol{\theta}\right) = p\left(r_{1:N}|w_{1:N}, \boldsymbol{\theta}\right) p\left(w_{1:N}|\boldsymbol{\theta}\right),$$

from which, in full generality, we obtain

$$\begin{aligned}\log p\left(w_{1:N}|\boldsymbol{\theta}\right) = & \underbrace{\sum_{r_{1:N}} \left[\log p\left(w_{1:N}, r_{1:N}|\boldsymbol{\theta}\right)\right] p\left(r_{1:N}|w_{1:N}, \boldsymbol{\theta}^{\text{old}}\right)}_{Q_{\boldsymbol{\theta}^{\text{old}}}(\boldsymbol{\theta})} \\ & - \sum_{r_{1:N}} \left[\log p\left(r_{1:N}|w_{1:N}, \boldsymbol{\theta}\right)\right] p\left(r_{1:N}|w_{1:N}, \boldsymbol{\theta}^{\text{old}}\right). \qquad (*3.12*)\end{aligned}$$

Now, for $\boldsymbol{\theta} = \boldsymbol{\theta}^{\text{old}}$ the Eq. (*3.12*) becomes

$$\begin{aligned}\log p\left(w_{1:N}|\boldsymbol{\theta}^{\text{old}}\right) = & \underbrace{\sum_{r_{1:N}} \left[\log p\left(w_{1:N}, r_{1:N}|\boldsymbol{\theta}^{\text{old}}\right)\right] p\left(r_{1:N}|w_{1:N}, \boldsymbol{\theta}^{\text{old}}\right)}_{Q_{\boldsymbol{\theta}^{\text{old}}}(\boldsymbol{\theta}^{\text{old}})} \\ & - \sum_{r_{1:N}} \left[\log p\left(r_{1:N}|w_{1:N}, \boldsymbol{\theta}^{\text{old}}\right)\right] p\left(r_{1:N}|w_{1:N}, \boldsymbol{\theta}^{\text{old}}\right).\end{aligned}$$

Subtracting $\log p\left(w_{1:N}|\boldsymbol{\theta}^{\text{old}}\right)$ from $\log p\left(w_{1:N}|\boldsymbol{\theta}\right)$, we have

$$\begin{aligned}& \log p\left(w_{1:N}|\boldsymbol{\theta}\right) - \log p\left(w_{1:N}|\boldsymbol{\theta}^{\text{old}}\right) \\ & = \underbrace{\sum_{r_{1:N}} \left[\log p\left(w_{1:N}, r_{1:N}|\boldsymbol{\theta}\right)\right] p\left(r_{1:N}|w_{1:N}, \boldsymbol{\theta}^{\text{old}}\right)}_{Q_{\boldsymbol{\theta}^{\text{old}}}(\boldsymbol{\theta})} \\ & \quad - \underbrace{\sum_{r_{1:N}} \left[\log p\left(w_{1:N}, r_{1:N}|\boldsymbol{\theta}^{\text{old}}\right)\right] p\left(r_{1:N}|w_{1:N}, \boldsymbol{\theta}^{\text{old}}\right)}_{Q_{\boldsymbol{\theta}^{\text{old}}}(\boldsymbol{\theta}^{\text{old}})} \\ & \quad + \sum_{r_{1:N}} \left[\log p\left(r_{1:N}|w_{1:N}, \boldsymbol{\theta}^{\text{old}}\right)\right] p\left(r_{1:N}|w_{1:N}, \boldsymbol{\theta}^{\text{old}}\right) \\ & \quad - \sum_{r_{1:N}} \left[\log p\left(r_{1:N}|w_{1:N}, \boldsymbol{\theta}\right)\right] p\left(r_{1:N}|w_{1:N}, \boldsymbol{\theta}^{\text{old}}\right).\end{aligned}$$

Due to Gibbs' inequality, derived in Exercise 1.16, we have

$$\begin{aligned}& \sum_{r_{1:N}} \left[\log p\left(r_{1:N}|w_{1:N}, \boldsymbol{\theta}^{\text{old}}\right)\right] p\left(r_{1:N}|w_{1:N}, \boldsymbol{\theta}^{\text{old}}\right) \\ & \quad - \sum_{r_{1:N}} \left[\log p\left(r_{1:N}|w_{1:N}, \boldsymbol{\theta}\right)\right] p\left(r_{1:N}|w_{1:N}, \boldsymbol{\theta}^{\text{old}}\right) \leq 0,\end{aligned}$$

which implies

$$\log p\left(w_{1:N}|\boldsymbol{\theta}\right) - \log p\left(w_{1:N}|\boldsymbol{\theta}^{\text{old}}\right)$$
$$\geq \underbrace{\sum_{r_{1:N}} \left[\log p\left(w_{1:N}, r_{1:N}|\boldsymbol{\theta}\right)\right] p\left(r_{1:N}|w_{1:N}, \boldsymbol{\theta}^{\text{old}}\right)}_{Q_{\boldsymbol{\theta}^{\text{old}}}(\boldsymbol{\theta})}$$
$$- \underbrace{\sum_{r_{1:N}} \left[\log p\left(w_{1:N}, r_{1:N}|\boldsymbol{\theta}^{\text{old}}\right)\right] p\left(r_{1:N}|w_{1:N}, \boldsymbol{\theta}^{\text{old}}\right)}_{Q_{\boldsymbol{\theta}^{\text{old}}}(\boldsymbol{\theta}^{\text{old}})}.$$

Next, if a $\boldsymbol{\theta}^{\text{new}}$ is selected such that $Q_{\boldsymbol{\theta}^{\text{old}}}\left(\boldsymbol{\theta}^{\text{new}}\right) \geq Q_{\boldsymbol{\theta}^{\text{old}}}\left(\boldsymbol{\theta}^{\text{old}}\right)$, we have $\log p\left(w_{1:N}|\boldsymbol{\theta}^{\text{new}}\right) \geq \log p\left(w_{1:N}|\boldsymbol{\theta}^{\text{old}}\right)$. This is guaranteed provided that

$$\boldsymbol{\theta}^{\text{new}} = \underset{\boldsymbol{\theta}}{\operatorname{argmax}}\, Q_{\boldsymbol{\theta}^{\text{old}}}\left(\boldsymbol{\theta}\right).$$

3.3 Exercise Problems

Exercise 3.1 Marginal likelihood for diffusive motion

Consider a diffusing particle in one dimension with a normal emission distribution, as we saw in Example 3.4, with known variance. Its marginal likelihood reads

$$p(w_{1:N}|D, v) = \int dr_{1:N}\, p(w_{1:N}, r_{1:N}|D, v).$$

1. Generate three data points, $w_{1:3}$, for the positions of a diffusing particle in one dimension. You can assume any initial condition you wish for r_1.
2. Obtain a diffusion coefficient estimate by maximizing the marginal likelihood.
3. Obtain another diffusion coefficient estimate analytically by maximizing a likelihood ignoring measurement noise, *i.e.*, $w_n = r_n$.
4. Explain the discrepancy you see in your two results.
5. Repeat for an arbitrary number of data points, N.
6. Reformulate the problem in three dimensions and repeat the preceding steps.

Exercise 3.2 Birth-death process likelihoods in continuous time

Start by considering a birth process alone.

1. Simulate a realization of the birth process using the Gillespie algorithm.
2. Use the results of the simulation to write down the likelihood of this sequence of realizations as they are observed in continuous time. Hint: The likelihood will be proportional to the product of terms coinciding with the probability that no event occurs over the time between events and

that an event of the type sampled occurs over the infinitesimal interval around the time at which that event occurs.
3. Maximize your likelihood to obtain the birth rate. How does the breadth of your log likelihood change as your data set grows?
4. Repeat the steps above for the birth-death process. For concreteness, use a death rate set to a tenth of the birth rate.

Exercise 3.3 Birth process likelihoods in discrete time

Consider the same setting as in Exercise 3.2, but assume that measurements are provided at regular and discrete time intervals. These time intervals can be shorter or, more interestingly, either on par or longer than the typical time it takes for a birth event to occur.

1. Simulate a realization of the birth process starting from some initial population using the Gillespie algorithm. Then create an array, $w_{1:N}$, coinciding with the population at discrete time levels. Initially, choose these time levels to be on par with the inverse birth rate.
2. Use the results of the simulation to write down the likelihood of this sequence of realizations as they are observed in discrete time. Hint: You need to solve the master equation for this portion of the problem.
3. Discuss how your likelihood here compares to your likelihood in the previous example.
4. Maximize your likelihood to obtain the birth rate.
5. What happens to your birth rate estimate as the time levels become twice, three times, and ten times as long as the inverse birth rate? Illustrate this by repeating the previous steps for these alternative cases.

Exercise 3.4 Cramér–Rao lower bound

An error bar around a maximum likelihood estimate can sometimes be approximated using the Cramér–Rao lower bound (CRLB). Briefly, under appropriate approximations the lower bound on the variance around a frequentist estimate, $\hat{\theta}$, is given by the inverse of the expectation of $-\partial^2 \log p(w_{1:N}|\theta)/\partial\theta^2$ with respect to $p(w_{1:N}|\theta)$ for a one-dimensional parameter θ. That is,

$$\text{variance}\left(\hat{\theta}\right) \geq -\frac{1}{\left\langle \frac{d^2}{d\theta^2} \log p(w_{1:N}|\theta)\right\rangle},$$

where $\langle\cdot\rangle$ denotes the expectation with respect to $p(w_{1:N}|\theta)$.

1. Compute the CRLB for N Gaussian iid realizations. On this basis, provide an intuitive interpretation for $-1/\left\langle d^2 \log p(w_{1:N}|\theta)/d\theta^2\right\rangle$ assuming a unimodal likelihood.
2. Compute the CRLB for N Beta distribution iid realizations with $\alpha = \beta = 1/2$ (see Appendix B for a definition of the Beta distribution). Note that this distribution has two maxima. Is the CRLB useful in estimating the uncertainty around the frequentist estimate? Why or why not?

Exercise 3.5 EM algorithm for two Gaussians

Implement the EM algorithm for the sum of two Gaussians in one dimension with identical variance. In other words, begin by generating synthetic data according to this model: $\left(\pi_1 e^{-\frac{(x-\mu_1)^2}{2v}} + (1-\pi_1)e^{-\frac{(x-\mu_2)^2}{2v}}\right)/\sqrt{2\pi v}$, where you have prespecified the parameters π_1, μ_1, μ_2, v. Then, implement EM in order to estimate the parameters π_1, μ_1, μ_2, v. Compare your parameter estimates from EM to the values you used to generate your data.

Exercise 3.6 Trajectory with maximum likelihood, part I

In Example 3.4 we introduced the generative model

$$r_n | r_{n-1}, D \sim \text{Normal}(r_{n-1}, 2D\tau),$$
$$w_n | r_n, v \sim \text{Normal}(r_n, v).$$

You can assume known initial conditions with position centered at the origin.

1. Given r_1, D, τ, v, and w_2, find the most likely position $\hat{r}_2$.
2. Add two more measurements $w_{3:4}$ to the known quantities; find the most likely positions $\hat{r}_{2:4}$.
3. Write down the general form of $\partial p(w_{2:N}, r_{2:N}|D, v)/\partial r_n$ for any $n = 2 : N$.
4. Simplify $\partial p(w_{2:N}, r_{2:N}|D, v)/\partial r_n = 0$ until it becomes a linear equation. We are now ready to solve a system of linear equations for the most likely trajectory $\hat{r}_{2:N}$. Write down this system of equations in matrix form.

Project 3.1 Likelihoods for exchange reactions in discrete time

Consider an exchange reaction with $\mathcal{A} \xrightarrow{\lambda_{+1}} \mathcal{B}$ and $\mathcal{B} \xrightarrow{\lambda_{-1}} \mathcal{A}$.

1. Simulate such a reaction starting from some initial population of $\mathcal{A}$ and $\mathcal{B}$ using the Gillespie algorithm. As with Exercise 3.3, create an array, $w_{1:N}$, coinciding with the population of $\mathcal{A}$s at discrete time levels.
2. Solve the master equation for the population of $\mathcal{A}$ and write down the likelihood of the sequence $w_{1:N}$.
3. Fixing λ_{+1} to the correct value, plot the logarithm of the likelihood as a function of λ_{-1}. Repeat the process fixing λ_{-1}. Verify that the likelihood logarithms are peaked at the correct values by adding a vertical line at the ground truth value in your likelihood plot.
4. Design a numerical scheme to maximize your likelihood to estimate both rates.

Project 3.2 Optical trapping

Optical trapping, a technology earning the 2018 Physics Nobel Prize, can be used to confine micron-sized colloids that would otherwise freely diffuse in solution in an approximately harmonic potential, reflecting what is sometimes called an Ornstein–Uhlenbeck process. Here, we model an optical trap in one dimension.

1. Simulate the position of a diffusing particle confined by a harmonic trap but otherwise jostled by thermal fluctuations modeled as white noise using an appropriately discretized form of the Langevin equation. That is, simulate an Ornstein–Uhlenbeck process as described in Exercise 2.7. Store the positions at regular time levels in an array $w_{1:N}$.
2. Use the results of the simulation to write down the likelihood of the sequence of realizations. Hint: You need to solve the coinciding Fokker–Planck equation.
3. Fixing the diffusion coefficient to the correct value, plot the logarithm of the likelihood as a function of trap stiffness. Repeat the process, fixing the trap stiffness, and plot the logarithm of the likelihood versus the diffusion coefficient. Verify that the likelihood logarithms are peaked at the correct values and include a vertical line in your figures at the correct ground truth values.
4. Using a numerical optimization method of your choice, maximize your log likelihood to estimate the trap stiffness and the particle's diffusion coefficient.

Project 3.3 Excited state lifetimes

It is common to estimate the amount of time a molecule spends in an excited quantum state (*i.e.*, the excited state lifetime) following a brief laser excitation pulse by measuring the time between the excitation pulse and the detection of a photon upon relaxation of the molecule back to its ground state.

Assuming the time spent in the excited state is exponential, and treating each photon arrival as iid, we may idealize the likelihood as a product of exponentials in order to estimate the exponential constant, *i.e.*, the excited state lifetime. However, this idealization is inappropriate in realistic scenarios where the laser pulse has some breadth in time and the molecule may be excited at any point over the pulse period in linear proportion to the laser intensity.

1. Begin by considering the time of arrival of the photon as being obtained from a convolution of a normal distribution with known parameters, reflecting the breadth of the laser pulse, and an exponential parametrized by an unknown lifetime. Generate synthetic data for your iid samples. To generate synthetic data, you need to first sample the time at which the molecule is excited from the normal and then add to this time the excited state lifetime sampled from the exponential. Repeat this procedure until you have a sufficient number of iid samples.
2. Derive an analytic formula for the convolution of the normal and the exponential.
3. Next, use any optimization method of your choice to infer the resulting exponential lifetime.

4. Repeat the procedure above assuming the photons are emitted from two independent molecular species with different lifetimes. Use EM to identify both lifetimes as well as the relative proportion of photons originating from one versus the other lifetime.

For the parameter values, use 1 ns as the mean of the normal and 0.1 ns as its standard deviation. Use lifetimes between 5 and 10 ns.

Project 3.4 EMCCDs and detector noise

When photons arrive at the sensors of election-multiplying charge coupled device (EMCCD) cameras, the signal generated is amplified through a sequence of election-multiplying registers ultimately giving rise to the measured output. An idealized model of an EMCCD output considers the convolution of the original Poisson photon count u with a Gamma(u/f,fG) distribution with known (precalibrated) camera parameters responsible for the signal amplification. The parameters include the excess noise factor f and the gain G and typically values are 2 and 25, respectively.

1. First, generate synthetic data similarly to Project 3.3.
2. Use EM to identify the rate of the Poisson distribution used to generate the synthetic data on one such camera pixel. Note that you need to model the photons as latent variables and that the convolution of Poisson and Gamma returns the sum of Gamma distributions.

Project 3.5 Trajectory with maximum likelihood, part II

Simulate a system as described in Exercise 3.6 for ten time steps. Given r_1, D, v, and $w_{2:10}$, find the most likely positions $\hat{r}_{2:10}$ numerically using the following schemes and compare your results.

1. Directly solve the system of linear equations you derived in Exercise 3.6.
2. Use gradient descent.
3. Use conjugate gradient.

Project 3.6 Birth-death process with FSP likelihood

Consider the birth-death process introduced in Example 2.6 with parameters λ_b and λ_d.

1. Simulate the birth-death process using the Gillespie algorithm.
2. Using the FSP algorithm of Project 2.2, approximate the solution to the master equation and use the truncated state-space to write down a likelihood for your simulated data.
3. Design a numerical scheme to maximize your likelihood of estimating both rates. How do your results change as the supplied data lengthens? How do your results change as you increase the size of your state-space in your FSP?

Additional Reading

D. S. Sivia, J. Skilling. *Data analysis: a Bayesian tutorial.* Oxford University Press, 2006.

L. Wasserman. *All of Statistics: a concise course in statistical inference.* Reprint. Springer, 2005.

S. J. Wright, B. Recht. *Optimization for data analysis.* Cambridge University Press, 2022.

A. Quarteroni, R. Sacco, F. Saleri. *Numerical Mathematics.* 2nd ed. Springer, 2006.

J. R. Shewchuk. An Introduction to the conjugate gradient method without the agonizing pain (unpublished), 1994.

J. A. Nelder, R. Mead. A simplex method for function minimization. *Computer Journal*, 7:308, 1965.

M. Tavakoli, S. Jazani, I. Sgouralis, W. Heo, K. Ishii, T. Tahara, S. Pressé. Direct photon-by-photon analysis of time-resolved pulsed excitation data using Bayesian nonparametrics. *Cell Rep. Phys. Sci.*, 1:100234, 2020.

K. B. Harpsøe, M. I. Andersen, P. Kjægaard. Bayesian photon counting with electron-multiplying charge coupled devices (EMCCDs) *Astron. Astrophys.*, 537:A50, 2012.

M. S. Robbins, B. J. Hadwen. The noise performance of electron multiplying charge-coupled devices. *IEEE Trans. Electron Dev.*, 50:1227, 2003.

B. Munsky, M. Khammash. The finite state projection algorithm for the solution of the chemical master equation. *J. Chem. Phys.*, 124:044104, 2006.

4 Bayesian Inference

By the end of this chapter, we will have presented

- *The concept of priors and posteriors*
- *Problems of model selection*
- *Graphical representations*

4.1 Modeling in Bayesian Terms

In previous chapters, we discussed common models of physical systems including observation models expressed using emission distributions. We also described how to construct likelihoods and obtain point parameter estimates via the maximum likelihood principle. This approach, often termed frequentist, is compatible with the interpretation of probabilities as frequencies of outcomes.

Within the frequentist approach, measurements and data in general are understood as random variables realized to specific values and model parameters are understood as quantities to be determined through likelihood maximization. In reality, likelihoods provide more than just parameter *point estimates*. The likelihood's curvature around its maximum tells us something about a parameter's uncertainty bound, as we briefly explored in Exercise 3.4.

However, while uncertainty bounds around maximum likelihood point estimates are helpful, these fall short of revealing the full distribution over a model's parameters. Indeed, measurement uncertainty, finite data, and intrinsic stochasticity of the underlying system probed if present, already suggest that committing to any single parameter value may be too restrictive. Instead, it appears more reasonable to seek a swathe of acceptable parameter values warranted by the data.

To satisfy this *desideratum*, *Bayesian methods* provide a sharply contrasting perspective to maximum likelihood in treating model parameters as random variables whose distribution is to be deduced from the data. Here, the effects of measurement uncertainty, finite data, and intrinsic stochasticity of the system all contribute to the distribution over parameters.

Beyond these advantages, conceptualizing parameters as random variables is also appealing for many more reasons. For instance, as we will see we sometimes recover distributions over parameters with multiple maxima.

Such features of the distribution are not otherwise captured when reporting the breadth around the maximum likelihood point estimate. Also, just as model parameters inherit the uncertainty described earlier, as we will see shortly, model structures themselves, not just their associated parameters, also inherit this uncertainty. The possibility for deducing distributions over model structures themselves and their associated parameters is detailed in later chapters in the context of our discussions on *Bayesian nonparametrics*.

4.1.1 The Posterior Distribution

Bayesian methods, the focus of this chapter, treat both data and model parameters as random variables. Of central importance in Bayesian analysis is the *posterior*, $p(\theta|w_{1:N})$. The posterior is the probability distribution over the model parameters θ after (hence, "*post*") considering the data $w_{1:N}$.

Intuitively, we think of the data as refining our knowledge of the unknown parameters distributed according to their posterior. In other words, we imagine that given more and more data, our knowledge of θ improves and the posterior sharpens around those values of the parameters that generated the data. That is, a posterior informed by $N-1$ data points, $p(\theta|w_{1:N-1})$, can be used to obtain an updated posterior $p(\theta|w_{1:N})$ as the Nth data point becomes available. Indeed, it is the likelihood, which may contain measurement models, that helps us link subsequent posteriors as we gather more data. Logically, we can write

$$p(\theta|w_{1:N}) \Longleftarrow p(\theta|w_{1:N-1}) \Longleftarrow \cdots \Longleftarrow p(\theta|w_1) \Longleftarrow \quad ?$$

The question mark arises as, in the absence of data, it is unclear how we begin updating. Put differently, what is the $p(\theta|\cdot)$ used in the updating scheme below?

$$p(\theta|w_1) \propto p(w_1|\theta)p(\theta|\cdot).$$

Here, $p(\theta|\cdot)$ represents a probability distribution reflecting our prior belief as to how the θ are distributed. It is called the *prior*. That is, the distribution obtained *prior* to the data by contrast to the posterior obtained after the data. What is more, the parameters of the prior distribution are called *hyperparameters*.

Applying Bayes' rule, we can write the posterior distribution as proportional to the product of the prior and the likelihood

$$p(\theta|w_{1:N}) = \frac{p(w_{1:N}|\theta)p(\theta|\cdot)}{p(w_{1:N})}, \qquad p(w_{1:N}) \neq 0, \tag{4.1}$$

where $p(w_{1:N})$ is obtained by completion

$$p(w_{1:N}) = \int d\theta\, p(\theta, w_{1:N}) = \int d\theta\, p(w_{1:N}|\theta)p(\theta|\cdot),$$

ensuring the normalization condition $\int d\theta\, p(\theta|w_{1:N}) = 1$. Here, as usual, the integral is understood as a sum in the case of discrete parameters.

The posterior $p(\theta|w_{1:N})$ exists only so long as the denominator $p(w_{1:N})$, sometimes called the *evidence*, is nonzero. That is, only insofar as the data at hand *can* be generated from the model we put forth.

Note 4.1 Use of proportionality in Bayesian methods

We make a note on the use of the proportionality constant here as it is widely used to simplify Bayesian calculations. We recall that the posterior is represented by $p(\theta|w_{1:N})$. This implies that the posterior is normalized over all values of θ. Thus, the proportionality constant that is missing is independent of θ and we can write

$$p(\theta|w_{1:N}) = \frac{p(w_{1:N}|\theta)p(\theta|\cdot)}{p(w_{1:N})} \propto p(w_{1:N}|\theta)p(\theta|\cdot).$$

To be clear, the proportionality constant is not constant with respect to the data $w_{1:N}$. Rather, it is constant only with respect to parameters θ.

As a simple example of how to use priors and likelihoods, we use Bernoulli trials, *i.e.*, coin flips, in Example 4.1 as an illustration.

Example 4.1 Bernoulli trials within the Bayesian paradigm

Here we return to an example involving Bernoulli trials such as a coin flip. We denote by π the probability of the first of two outcomes, say heads. The likelihood of having collected N_H heads and N_T tails with $N_H + N_T = N$ trials is then Bernoulli distributed and we may write

$$p(w_{1:N}|\pi) \propto \pi^{N_H}(1-\pi)^{N_T},$$

where the proportionality factor is found by summing $\pi^{N_H}(1-\pi)^{N_T}$ over N_H from 0 to N.

For now, we assume a prior of the form

$$p(\pi) \propto \pi^{\alpha-1}(1-\pi)^{\beta-1}, \tag{4.2}$$

where α and β are positive constants. Equation (4.2) encodes a Beta(α, β) distribution that we will revisit in greater depth in later chapters.

Despite its superficial similarity to the Bernoulli distribution, the Beta distribution is inherently different. To wit, the Beta distribution is a distribution over a continuous random variable, π, and the proportionality of Eq. (4.2) is found by integrating $\pi^{\alpha-1}(1-\pi)^{\beta-1}$ over π from 0 to 1; see Appendix B for more details. This is in sharp contrast to the Bernoulli distribution, described earlier, which is a distribution over a binary random variable.

According to Bayes' rule, Eq. (4.1), the product of the likelihood and prior yields, up to a proportionality constant independent of π, the posterior

$$p(\pi|w_{1:N}) \propto \pi^{N_H+\alpha-1}(1-\pi)^{N_T+\beta-1}.$$

While Bayesian methods return the full posterior, we may opt to maximize the posterior. The estimate for π obtained in this case is termed the *maximum a posteriori (MAP) estimate* and it follows from $\text{argmax}_\pi p(\pi|w_{1:N})$. This operation yields $\hat{\pi} = (N_H + \alpha - 1)/(N + \alpha + \beta - 2)$.

We immediately note the difference between this and the maximum of the likelihood, obtained by maximizing the likelihood over π, namely $\text{argmax}_{\pi} p(w_{1:N}|\pi)$, which yields $\hat{\pi} = N_H/N$.

Intuitively, here we see that, on account of the special mathematical form we selected for the prior, the prior added *pseudocounts* to our measurement. That is, the prior added additional $\alpha - 1$ counts to heads and $\beta - 1$ counts to tails resulting in a denominator of $N + \alpha + \beta - 2$ for the MAP estimate, in contrast to simply N from the maximum likelihood.

These results warrant a few remarks. First, even for this special choice of Bernoulli trials and Beta priors, we see that as the number of data points available grows, irrespective of our choice of α and β, our posterior eventually peaks for the same value of π as the likelihood. Next, we also see that, unlike maximum likelihood, even assuming no draws of heads after just one draw, we still ascribe some nonzero a posteriori probability of heads. Namely,

$$\hat{\pi} = \frac{\alpha}{\alpha + \beta - 1}.$$

Example 4.1 motivates a remark we now make on the importance of likelihoods. Priors essentially define the range over which parameters can be assigned nonzero posterior probability after data become available. It is reasonable to expect that, in the limit of a large N, our choice of prior becomes increasingly immaterial and the shape of the posterior is ultimately dictated by the likelihood.

This logic is illustrated in Fig. 4.1, where the effect of the prior on the values over which the posterior is distributed lessens as more data is considered. Indeed, in this simple example, for sufficiently independent observations, the likelihood's breadth eventually narrows around its putative mode. The same is not true of the prior, which is independent of the amount of data. Thus, eventually, with enough data, effects of the likelihood on the posterior dominate over the prior. This really drives home two points: we must have likelihoods that capture details of both the physical system and the observation model; there is danger in attempting parameter estimation with insufficient or poor data quality. In the latter case, our arbitrary choice of prior may deeply influence the ultimate shape of the posterior.

This discussion begs the question: How much data is enough to overcome the effects of the prior? Of course, it depends. It depends on the mathematical form for the likelihood, it depends on the quality of the data, it also depends on the "strength" of the prior. That is, how firm our prior beliefs are and how easily the likelihood can help shift the posterior away from prior biases. These topics are discussed in greater depth shortly.

Bayes' Factor

There is more to Bayes' rule than simply a mathematical device to compute posteriors. Often, we may instead be interested in comparing two hypotheses. That is, we may be interested in evaluating the posterior at two different model parameters values, say θ_1 and θ_2, given data $w_{1:N}$. We briefly explore this ratio of posteriors in Note 4.2.

Note 4.2 Bayes' factor

We consider the ratio of posteriors,

$$\frac{p(\theta = \theta_1 | w_{1:N})}{p(\theta = \theta_2 | w_{1:N})}.$$

More generally, the posterior probability over disparate parameter ranges can also be compared by integrating the posterior over different parameter ranges and taking the ratio of the ensuing probabilities.

For uniform priors, the posterior ratio reduces to a *likelihood ratio* also called a *Bayes' factor*,

$$\frac{p(w_{1:N} | \theta = \theta_1)}{p(w_{1:N} | \theta = \theta_2)},$$

which is often used to compare the relative likelihood of two parameter values.

One feature of the posterior or likelihood ratio is that it typically becomes very large or very small as more data are gathered. This feature is a consequence of how likelihoods are constructed. For instance, for iid samples, joint likelihoods are constructed from independent products of likelihoods each over a single data point. As such, this product is highly sensitive to small parameter variations. Indeed, likelihoods can quickly become vanishingly small for parameter values only slightly less optimal than others as the number of data points increases. For this reason, it is commonplace to compute logarithms of posterior or likelihood ratios. The sensitivity of likelihoods can also be a disadvantage: data point outliers, which we will discuss in Exercise 4.3, can sharply reduce likelihoods.

Indeed, it is worth mentioning that posterior or likelihood ratios are not helpful in comparing different models, only different parameter values or ranges of one model. This is because, somewhat trivially, more complex models generally fit the data better.

For example, we can imagine a likelihood for iid outcomes distributed according to a sum of normal distributions with an unknown number of normal components. The likelihood with two components will always fit the data at least as well as the simpler model. No Bayes' factor is required to ascertain this. Indeed, this is apparent from the fact that we are free to set one of the components of the two-component normal model to zero and fit just as well as we do with one component.

On the flip side, more complex models are worse at predicting the probability of future experimental outcomes. This motivation brings us to the topic of *predictive distributions*.

The Predictive Distribution

The posterior can be used to compute what is sometimes termed the *predictive distribution*, $p(w_{N+1}|w_{1:N})$. The predictive distribution assumes that data $w_{1:N}$ are known and computes the distribution over the next set of measurements, say w_{N+1}, according to $p(w_{N+1}|w_{1:N}) = \int d\theta\, p(w_{N+1}|\theta)p(\theta|w_{1:N})$.

The notion of prediction is especially relevant to model selection central to frequentist and Bayesian analysis. For example, in computing a Bayes' factor in Note 4.2, we had assumed throughout that the model structure was unchanged and that we were solely comparing ratios of posteriors or likelihoods for different parameter values. Yet all of this is under the assumption that the model structure is known.

Before discussing a Bayesian attempt at determining which model structure is best, the realm of *model selection*, we first illustrate the worsening predictive ability of complex models by discussing the predictive distribution through Example 4.2.

Example 4.2 Predictive distribution

Here, we give an example of the numerical value of the predictive distribution decreasing as the model complexity increases. We give the example of a categorical likelihood with three possible outcomes such that each iid random variable sample is realized as 1, 2, or 3. In this case, the likelihood is

$$p(w_n|\pi_{1:3}) = \pi_1^{\Delta_1(w_n)}\pi_2^{\Delta_2(w_n)}\pi_3^{\Delta_3(w_n)}. \tag{4.3}$$

Below, we introduce a prior to this 3-component categorical distribution, the Dirichlet distribution is a distribution of key importance that we will revisit on multiple occasions in future chapters,

$$p(\pi_{1:3}) = \frac{\Gamma(\alpha_1+\alpha_2+\alpha_3)}{\Gamma(\alpha_1)\Gamma(\alpha_2)\Gamma(\alpha_3)}\pi_1^{\alpha_1-1}\pi_2^{\alpha_2-1}\pi_3^{\alpha_3-1} = \text{Dirichlet}_{1:3}(\pi_{1:3};\alpha_{1:3}), \tag{4.4}$$

where $\alpha_{1:3}$ are *hyperparameters* and $\Gamma(\cdot)$ is the *gamma function*, first seen in Exercise 1.2, obtained from the normalization condition obtained under the constraint $\pi_1+\pi_2+\pi_3=1$.

On account of this particular prior choice, the posterior also takes the form of a Dirichlet distribution with hyperparameters $\alpha_{1:3}$ of the prior appearing as $\alpha_1+n_1, \alpha_2+n_2, \alpha_3+n_3$ in the posterior where n_1 are the total number of times that outcome 1 was realized in $N = n_1+n_2+n_3$ experiments. Both n_2 and n_3 are similarly defined.

Next we compute the probability that w_{N+1} is realized to outcome 1 via $p(w_{N+1}=1|w_{1:N})$. The predictive distribution reads

$$\begin{aligned} &p(w_{N+1}=1|w_{1:N}) \\ &= \int d\pi_{1:3}\, \delta_1(\pi_1+\pi_2+\pi_3)p(w_{N+1}=1|\pi_{1:3})p(\pi_{1:3}|w_{1:N}). \end{aligned} \tag{4.5}$$

Inserting Eq. (4.3) and Eq. (4.4) into Eq. (4.5), we find

$$p(w_{N+1} = 1|w_{1:N}) = \frac{\Gamma(\alpha_1 + \alpha_2 + \alpha_3 + N)}{\Gamma(\alpha_1 + n_1)\Gamma(\alpha_2 + n_2)\Gamma(\alpha_3 + n_3)}$$

$$\times \int d\pi_{1:3}\, \delta_1(\pi_1 + \pi_2 + \pi_3)\pi_1^{\alpha_1+n_1}\pi_2^{\alpha_2+n_2-1}\pi_3^{\alpha_3+n_3-1}.$$

By recognizing that the integral above is the same integral as would be required to normalize a Dirichlet distribution, we can immediately write down the integral above from the normalization of Eq. (4.4). We find that

$$p(w_{N+1} = 1|w_{1:N}) = \frac{\Gamma(\alpha_1 + \alpha_2 + \alpha_3 + N)}{\Gamma(\alpha_1 + n_1)\Gamma(\alpha_2 + n_2)\Gamma(\alpha_3 + n_3)}$$

$$\times \frac{\Gamma(\alpha_1 + n_1 + 1)\Gamma(\alpha_2 + n_2)\Gamma(\alpha_3 + n_3)}{\Gamma(\alpha_1 + \alpha_2 + \alpha_3 + N + 1)}$$

$$= \frac{\alpha_1 + n_1}{\alpha_1 + \alpha_2 + \alpha_3 + N},$$

where, in the last line, we invoked the definition of the gamma function.

In general, for the K-component categorical likelihood (and thus the K-component Dirichlet prior) we have

$$p_K(w_{N+1} = 1|w_{1:N}) = \frac{\alpha_1 + n_1}{\sum_{k=1}^K \alpha_k + N},$$

where we have made the K dependency explicit by subscripting the density. We immediately see that in comparing the predictive distribution of the two- and three-component models, we find $p_2(w_{N+1} = 1|w_{1:N}) > p_3(w_{N+1} = 1|w_{1:N})$. That is, the less complex model (*i.e.*, the two-component categorical likelihood as compared to the three-component categorical likelihood) is, as expected, strictly more predictive than that of the more complex model.

Making the dimensionality of K explicit for a K-component model, we would say that

$$p_{K'}(w_{N+1} = 1|w_{1:N}) > p_K(w_{N+1} = 1|w_{1:N})$$

for integer $K > K' \geq 2$.

On the one hand, we need models complex enough to provide a good fit to our data. On the other hand, we do not want models so complex that the predictive distribution becomes vanishingly small when evaluated for reasonable values of w_{N+1}. This is the essence of *model selection*, an important modeling challenge that will be introduced in Section 4.5.

4.1.2 Bayesian Data Analysis: The Big Picture

Before we delve into finer details of Bayesian modeling, we emphasize the big picture and key quantities.

Note 4.3 Key quantities of Bayesian modeling

For convenience, we collect in this note all key quantities of Bayesian modeling:

- The prior, $p(\theta)$.
- The likelihood, $p(w_{1:N}|\theta)$, whose maximum is termed the maximum likelihood estimator.
- The posterior, $p(\theta|w_{1:N})$, whose maximum is termed the MAP estimator.
- The evidence, $p(w_{1:N})$.
- The predictive distribution, $p(w_{N+1}|w_{1:N})$.

The breadth of the posterior or predictive distributions is often termed the *credible interval*. This is by contrast to the term *confidence interval* typically reserved to denote the breadth of distributions outside the Bayesian paradigm.

We would be remiss in our duty if we did not answer what is, almost universally, the first question to arise of any student first introduced to Bayesian methods: Aren't priors just a means of biasing results? The answer is no.

We elaborate our response, at first, through this analogy: to prove theorems in mathematics we need to start with axioms. Otherwise, we have no language in which to express our logical deductions. Similarly, when we want to compute a posterior, we need to define the domain over which the variables with unknown values, such as our parameters, are defined. Specifying this domain is the role of the prior.

Perhaps naively, we may insist on the distribution of the parameters over this domain to be flat. Yet purely uniform distributions over infinite or semi-infinite parameter spaces are not normalizable. Put differently, they are not proper distributions. To become proper distributions such priors must vanish. But if such priors are to normalize to unity, then logic dictates that they must have some positive density somewhere. As we will discuss in greater depth shortly, the idea is to select a region of positive density as diffuse, or broad, as possible. It is the duty of any Bayesian modeler to then demonstrate, often numerically, that the precise details of the priors have negligible impact on the overall shape of the posterior or to request additional data.

Note 4.4 Priors in nonparametrics

What we have said above holds for the vast majority of traditional Bayesian applications. Now, in advanced Bayesian nonparametric settings, we will find that while the effect of the prior is small it never quite vanishes. This fundamentally arises because Bayesian nonparametrics formally places priors on infinite properties of models, *i.e.*, such as placing a prior on values of a curve at any point in real space as we will see in dealing with Gaussian processes in Chapter 6. As such, any finite data never quite "swamps out" the effect of the prior on the posterior. As this is an advanced topic, we will turn to this mathematical challenge, in a problem-specific basis, in later chapters as we test the consistency of finite truncations of Bayesian nonparametric priors.

Note 4.5 summarizes the strategy for data analysis within the Bayesian paradigm.

Note 4.5 The Bayesian paradigm of data analysis

In a Bayesian problem, we follow this pattern:

- We write down the data likelihood.
- We define a domain spanned by the parameters and assign prior probability over this domain.
- We use Bayes' rule to update the prior using the likelihood and obtain the posterior distribution over the parameters.

We now end with Note 4.6 on terminology.

Note 4.6 Forward versus generative models

Now that we have introduced priors, it is a good idea to refine key terminology that we have previously introduced:

- Likelihood-based and Bayesian-based methods are *inverse methods* solving an *inverse problem*. This is the opposite of direct modeling, also introduced as forward modeling, such as computer simulations.
- The causal, probabilistic relationships between random variables required in order to simulate the data generation process is called a *generative model* or *forward model*.
- In Bayesian methods, we introduce an additional layer of probabilistic relationships with the use of priors that incorporates our random variables. Priors are not required in order to simulate the data generation process. When we refer to the causal, probabilistic relationships between random variables starting from the priors, we restrict ourselves to the use of the term *forward model* and avoid the term generative model.

4.2 The Logistics of Bayesian Formulations: Priors

We begin with a discussion of priors. Not because they are more important than likelihoods, but because they are the newest addition to our tool set. There are two types of priors: uninformative and informative. Uninformative priors intuitively meet our expectation for how priors should be distributed and, for this reason, we start with them.

4.2.1 Uninformative Priors

The simplest *uninformative prior* is inspired from Laplace's principle of insufficient reason when the set of hypotheses are complete and mutually exclusive. That is, $p(\theta)$ is independent of θ. This distribution is termed flat or uniform when speaking of a distribution over a continuous parameter over a bounded parameter range. When speaking of discrete distributions, a uniform distribution is over equiprobable a priori outcomes.

Under the assumption that $p(\theta)$ is constant over its parameter range, the posterior and likelihood are directly related,

$$p(\theta|w_{1:N}) \propto p(w_{1:N}|\theta)p(\theta) \propto p(w_{1:N}|\theta).$$

The constants of proportionality dropped are independent of θ. Therefore, the dependence of the likelihood and the posterior on θ is identical and, consequently, maximizing the posterior or maximizing the likelihood over θ results in identical parameter estimates, $\hat{\theta}$.

Example 4.3 Bernoulli trials with uninformative priors

Here we return to the simple example of Bernoulli trials. As in Example 4.1, we write down the likelihood

$$p(w_{1:N}|\pi) \propto \pi^{N_H}(1-\pi)^{N_T}$$

and assume an uninformative Beta(α, β) prior for which $\alpha = \beta = 1$. As such,

$$p(\pi) \propto \text{constant}$$

and the posterior is therefore proportional to the prior

$$p(\pi|w_{1:N}) \propto \pi^{N_H}(1-\pi)^{N_T}.$$

As intuitive as it may appear to start with flat priors, such priors may exhibit pathologies. First, flat priors can only be constructed for *bounded* continuous parameters. For example, a flat prior over the entire real line is improper, *i.e.*, it does not normalize to unity. In fact, for a parameter whose range is anywhere from $-\infty$ to $+\infty$, the flat prior cannot be normalized at all; $\int_{-\infty}^{+\infty} d\theta\, p(\theta) = \infty$. Thus $p(\theta)$ cannot be a probability distribution. Such priors are termed *improper priors* and their limitations have serious implications for more advanced Bayesian applications. Second, a flat prior over a model parameter, say θ, over a bounded interval is not quite as uninformative as it may appear as a coordinate transformation to an alternative variable, say e^{θ}, reveals that we suddenly know more about the random variable e^{θ} than we did about θ since the distribution over e^{θ} is no longer flat.

We end with Note 4.7, a mostly historical, note on other uninformative priors.

Note 4.7 Other uninformative priors

A desire to make priors invariant under continuous variable transformation motivated the development of the *Jeffreys prior*. Other concerns, in turn, motivated other uninformative priors such as those used in statistical physics, *e.g.*, the multinomial distribution and the related Shannon entropic prior or simply *entropic prior*. We do not dwell on these here though we briefly turn to entropic priors in Section 4.6 if only to explain the connection between Bayesian inference and information theory.

Fundamentally, such uninformative priors were conceived in an effort to enforce system properties or symmetries a priori irrespective of the associated computational cost of enforcing them. With the large amounts of data often used in parameter estimation today, it is no longer clear that the vanishingly small effect the prior has on the final posterior ultimately warrants the high computational cost introduced by them.

4.2.2 Informative Priors

It may appear counterintuitive to consider priors with structure. That is, one that is not completely flat over the domain of interest. Yet, as we will see shortly, informative priors have become the rule rather than the exception due to computational considerations.

A common choice of *informative prior* is directly suggested by the form of the likelihood. To see this, we reconsider how additionally available data is used to update a posterior. A new posterior, $p(\theta|w_2, w_1)$, is obtained from the old posterior, $p(\theta|w_1)$, and the likelihood as

$$p(\theta|w_2, w_1) \propto p(w_2|\theta, w_1)p(\theta|w_1).$$

In this way, the old posterior, $p(\theta|w_1)$, plays the role of the prior for the new posterior, $p(\theta|w_2, w_1)$. Intuitively, as more data become available, we expect the shape of the posterior to stop changing and be dictated by the form of the likelihood.

Now, we may further insist on the stronger condition that the prior and all future posteriors retain the same mathematical form upon consideration of every additional data point. In this case, the mathematical form for the likelihood fixes the form of the prior and we speak of prior-likelihood conjugate pairs. As shorthand, we sometimes refer to priors in such conjugate models as *conjugate priors*.

Conjugate priors can be identified, and attain analytic forms, when the likelihood belongs to the very general family of distributions, *i.e.*, the *exponential family* that we discuss in greater depth shortly.

4.2.3 Conjugate Prior-Likelihood Pairs

It turns out that the concept of conjugacy was already introduced in Example 4.1 on Bernoulli trials for Bernoulli likelihoods where we invoked the conjugate Beta distribution prior. In Example 4.4, we discuss another example of conjugacy before turning to a formal description.

Example 4.4 Determining the prior conjugate to a Poisson likelihood

To illustrate the concept of conjugacy, we consider an experiment whose likelihood for the nth event is well described by a Poisson distribution,

$$p(w_n|\mu) = \frac{\mu^{w_n}}{w_n!}e^{-\mu},$$

where w_n are nonnegative integers. Here, μ is a unitless parameter, such as a rate multiplied by an appropriate time interval, and w_n is thus the number of events sampled within that time interval.

The conjugate prior, $p(\mu)$, to the Poisson likelihood is the gamma distribution whose density reads

$$p(\mu) = \text{Gamma}(\mu; \alpha, \beta) = \frac{1}{\beta\Gamma(\alpha)}\left(\frac{\mu}{\beta}\right)^{\alpha-1} e^{-\mu/\beta}$$

and contains two hyperparameters, α and β.

To verify that the gamma distribution is indeed conjugate to the Poisson, we consider the posterior after just one observation, w_1. The posterior, expressed as the product of the likelihood and prior, reads

$$p(\mu|w_1) \propto \mu^{w_1}e^{-\mu} \times \mu^{\alpha-1}e^{-\mu/\beta} \propto \text{Gamma}\left(\mu; \alpha + w_1, \frac{1}{\frac{1}{\beta}+1}\right).$$

In other words, the posterior is also a gamma distribution, just like the prior.

By induction, after N independent measurements, $w_{1:N}$, we have

$$p(\mu|w_{1:N}) = \text{Gamma}\left(\mu; \alpha + \sum_{n=1}^{N} w_n, \frac{1}{\frac{1}{\beta}+N}\right).$$

Figure 4.1 illustrates how the posterior is dominated by the likelihood provided sufficient data and how an arbitrary choice of hyperparameters becomes less important for large enough N.

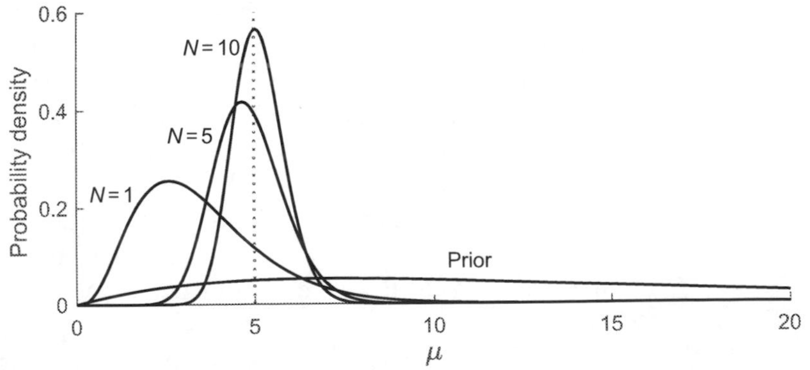

Fig. 4.1 The posterior probability sharpens as more data are accumulated. Here, we sampled data according to a Poisson distribution with $\mu = 5$, designated by the dotted line. We plotted the prior and the resulting posterior after collecting $N = 1$, then $N = 5$ and $N = 10$ points.

As we will see, values for hyperparameters in conjugate priors adopt simple physical interpretation: they provide prior estimates of parameters of the likelihood and the prior sample size, called the *conjugate prior pseudocount*, used to arrive at this estimate. The latter concept of prior sample size is especially relevant to our discussion on the *prior strength*.

On account of their mathematical structure, as we will see in subsequent chapters, conjugate priors provide direct computational advantage and, for this reason, are commonly invoked in Bayesian applications provided they can be found.

Note 4.8 Hyperparameters and hyperpriors

Whether conjugate or not, priors are probability distributions. Just like any other distribution, they have parameters on their own that we previously introduced as *hyperparameters*.

Consider hyperparameters γ, then we can write $p(\theta|\gamma)$ only if the γ themselves are random variables. Otherwise, we write $p(\theta)$. If γ are randomly distributed, then the distribution over γ is called a *hyperprior* and its parameters are called *hyperhyperparameters*, thereby establishing a hierarchy of random variable dependencies.

Putting it all together, we would say that observations probabilistically depend on latent variables, latent variables probabilistically depend on model parameters, model parameters may probabilistically depend on hyperparameters, and so forth.

4.2.4 Conjugate Priors and the Exponential Family

We now elaborate on likelihoods appearing in prior-likelihood conjugate pairs.

Likelihoods of the Exponential Family

A distribution is said to belong to a *K-parameter exponential family* if the likelihood of a single data point can be written as

$$p(w|\theta) = f(w)g(\theta)\exp\left(\sum_{k=1}^{K}\phi_k(\theta)u_k(w)\right). \tag{4.6}$$

Example 4.5 The Normal belongs to the exponential family

In this example, we rewrite a normal to make it assume the form of a K-parameter exponential family distribution. To see this explicitly, we write

$$p(w|\theta) = \frac{1}{\sqrt{2\pi v}}e^{-\frac{(w-\mu)^2}{2v}} = \left(\frac{1}{\sqrt{2\pi v}}e^{-\frac{\mu^2}{2v}}\right)e^{\frac{w\mu}{v}-\frac{w^2}{2v}},$$

which leads to the immediate identification: $f = 1$, $g = e^{-\frac{\mu^2}{2v}}/\sqrt{2\pi v}$, $K = 2$, $u_1 = w$, $\phi_1 = \mu/v$, $u_2 = w^2$, $\phi_2 = -1/2v$. Our decision to lump the factor of $-1/2$ with ϕ_2 is inconsequential.

The likelihood for N data points, assumed iid, can be written as

$$p(w_{1:N}|\theta) = \left(\prod_{n=1}^{N} f(w_n)\right) g^N(\theta) \exp\left(\sum_{k=1}^{K} \phi_k(\theta) t_k(w_{1:N})\right), \tag{4.7}$$

where $t_k(w_{1:N}) = \sum_{n=1}^{N} u_k(w_n)$. The t_k's are called *sufficient statistics* as the collection of $t_{1:K}$ encapsulates all ways of combining the data to fully specify the likelihood. For example, the exponential only has one sufficient statistic (the *point statistic* called the mean), while the normal has two; the first and second moment, or a combination thereof, including the mean and variance.

Likelihoods belonging to the exponential family are thus very special types for which the entire dataset can be reduced to point statistics which, when taken together, constitute the sufficient statistics. It is only for these special distributions that we can construct conjugate priors, leading to dramatic computational improvements.

While the structure and measurement model of any given problem may dictate complex likelihoods outside the exponential family, whenever possible seeking variable transformations or approximations under which likelihoods assume the form of distributions belonging to the exponential family is worth the effort.

Priors of the Exponential Family

Given a likelihood of the form of Eq. (4.6) or Eq. (4.7), only if the prior attains the form

$$p(\theta) \propto g^{\eta}(\theta) \exp\left(\sum_{k=1}^{K} \phi_k(\theta) \nu_k\right), \tag{4.8}$$

does the posterior read

$$p(\theta|w_{1:N}) \propto g^{\eta+N}(\theta) \exp\left(\sum_{k=1}^{K} \phi_k(\theta)(t_k(w_{1:N}) + \nu_k)\right),$$

where η and ν_k in the prior are substituted for $\eta + N$ and $\nu_k + t_k(w_{1:N})$ in the posterior, respectively.

As we now see, hyperparameters of conjugate priors have direct interpretation. Indeed, by comparison of Eq. (4.7) and Eq. (4.8) we see that the hyperparameter η plays the role of a *pseudocount* while the other, ν_k, plays the role of a *pseudoestimate* of the kth sufficient statistic, t_k. We have such a pair of hyperparameters for each sufficient statistic appearing in the likelihood.

4.2.5 Informative Priors for Normal Likelihoods

Probably the most commonly used likelihood is the normal, or Gaussian, likelihood for multiple reasons. First, it is computationally tractable and is completely defined by two parameters: the mean and variance. Secondly, it is often an excellent physical approximation for experiments. The reason for this stems from the *central limit theorem.*

Loosely, the central limit theorem states that if a random variable can be thought of as arising from the sum of multiple contributing realizations of other random variables, and provided that each of those random variables is sampled from a roughly independent distribution with finite mean and variance, then their sum is distributed according to a normal distribution. As experimental observations are often the output of multiple contributing factors, normal distributions are ubiquitous across the natural sciences.

For iid experiments, a normal likelihood takes the form

$$p(w_{1:N}|\mu, \tau^{-1}) = \text{Normal}(w_{1:N}; \mu, \tau^{-1}) = \prod_{n=1}^{N} \frac{\sqrt{\tau}}{\sqrt{2\pi}} e^{-\tau\frac{(w_n-\mu)^2}{2}}, \tag{4.9}$$

where τ, termed the *precision*, is related to the variance by $v = 1/\tau$. As we will see, it becomes convenient to start using τ more often than v from this point forward.

As the normal has two parameters, *i.e.*, two sufficient statistics, we have three cases to consider in inferring parameters for a Gaussian model. Either one of the two parameters is known, whilst our goal is then to generate a posterior over the other, or both parameters are unknown, whilst our goal is then to generate a joint posterior over both parameters. We start with the case of known mean and unknown variance (Example 4.6) and follow up immediately thereafter with the two subsequent cases (Examples 4.7 and 4.8).

Example 4.6 Conjugate prior to the normal with known mean and unknown variance

The likelihood is

$$p(w_{1:N}|\tau) \propto \tau^{N/2} \exp\left(-\frac{\tau N \nu}{2}\right), \tag{4.10}$$

where we have defined $\nu = \frac{1}{N}\sum_n^N (w_n - \mu)^2$. The prior conjugate to this likelihood, constructed using Eq. (4.8), is the gamma distribution

$$p(\tau) = \text{Gamma}(\tau; \alpha, \beta) \propto \left(\tau/\beta\right)^{\alpha-1} \exp(-\tau/\beta). \tag{4.11}$$

By comparing Eq. (4.10) and Eq. (4.11), we see upon rearrangement that the hyperparameter α is related to a pseudocount $2(\alpha - 1)$ and $\alpha\beta$ serves as the prior mean or pseudoestimate of the precision parameter.

The resulting posterior obtained by multiplying the likelihood and gamma prior also assumes the gamma form

$$p(\tau|w_{1:N}) \propto \tau^{N/2+\alpha-1} e^{-\tau/\left(N\nu/2+\beta^{-1}\right)^{-1}} \propto \text{Gamma}\left(\tau; \frac{N}{2} + \alpha, \frac{1}{\frac{1}{\beta} + N\frac{\nu}{2}}\right). \tag{4.12}$$

Example 4.7 Conjugate prior to the normal with unknown mean and known variance

We start again with the likelihood

$$p(w_{1:N}|\mu) \propto \exp\left(-\tau \frac{\sum_{n=1}^{N}(w_n - \mu)^2}{2}\right).$$

The conjugate prior, constructed from Eq. (4.8), is also a normal,

$$p(\mu) = \text{Normal}\left(\mu; \xi, \frac{1}{\psi}\right),$$

with hyperparameters ξ and ψ. The resulting posterior, obtained by multiplying prior and likelihood and completing the square to make the resulting product look like a normal in μ, is

$$p(\mu|w_{1:N}) = \text{Normal}\left(\mu; \frac{\bar{w}N\tau + \xi\psi}{N\tau + \psi}, \frac{1}{N\tau + \psi}\right),$$

where $\bar{w} = \frac{1}{N}\sum_{n=1}^{N} w_n$. We immediately see that as N becomes very large, the normal distribution mean reduces to $\bar{w}$ as expected.

Example 4.8 Conjugate prior to the normal with unknown mean and variance

We may consider a joint prior over μ and τ as being proportional to the product of a normal and a gamma probability density, yielding

$$p(\mu, \tau) = p(\mu)p(\tau) \propto \tau^{\alpha-1/2} \exp\left(-\frac{\tau}{\beta} - \psi\frac{(\mu - \xi)^2}{2}\right). \tag{4.13}$$

When multiplying Eq. (4.13) by a normal likelihood, Eq. (4.9), the posterior does not assume the form of the joint prior. For this reason, we say that the gamma distribution is only *conditionally conjugate*, assuming known mean, and that the normal distribution is similarly *conditionally conjugate* assuming known variance.

Another choice of priors over μ and τ is

$$p(\tau) = \text{Gamma}(\tau; \alpha, \beta), \qquad p(\mu) = \text{Normal}\left(\mu; \xi, \frac{1}{\psi_0\tau}\right).$$

This joint prior then takes the form

$$p(\mu, \tau) = p(\mu|\xi, \psi_0\tau)p(\tau|\alpha, \beta) \propto \tau^{\alpha-1/2} e^{-\tau\left(\frac{\psi_0(\mu-\xi)^2}{2} + \frac{1}{\beta}\right)}.$$

Here, this prior, when multiplied by a normal likelihood, Eq. (4.9), generates a posterior of the same form

$$p(\mu, \tau|w_{1:N}) \propto \tau^{N/2+\alpha-1/2} e^{-\tau\left(\frac{\psi_0(\mu-\xi)^2}{2} + \frac{1}{\beta} + \sum_{n=1}^{N}(w_n-\mu)^2\right)}. \tag{4.14}$$

4.3 EM for Posterior Maximization*

There are a number of ways from which to draw insights from posteriors. One way is by sampling posteriors using Monte Carlo methods that we explore in Chapter 5. Another one is to determine MAP estimates from posteriors using optimization methods or, alternatively, to adapt EM. The adaptation of EM for posterior maximization is given in Algorithm 4.1.

Algorithm 4.1 Expectation-Maximization

Given observations $w_{1:N}$ and an initial $\hat{\theta}^{\text{old}}$, compute an approximation of the MAP parameter values $\hat{\theta} = \text{argmax}_{\theta}\, p(\theta|w_{1:N})$ by iterating the convergence:

- Find
$$f_{\theta}(r) = p(r|w_{1:N}, \theta)$$
for appropriate latent variables r.
- Determine the expectation
$$Q_{\theta'}(\theta) = \sum_r f_{\theta'}(r) \log p(\theta|r_{1:N}, w_{1:N}).$$
- Compute new parameters values, $\hat{\theta}^{\text{new}}$, from the maximizer of $Q_{\hat{\theta}^{\text{old}}}(\theta)$.

Example 4.9 EM for the conditionally conjugate normal-gamma model

In this example we consider normally distributed scalar observations $w_{1:N}$, whose variance we wish to estimate. To accomplish this, we consider a parametrization of the normal distribution by mean μ and precision τ. We may apply independent priors on both parameters. In particular, we place a normal prior on μ and a gamma prior on τ. Thus,

$$p(\tau) = \text{Gamma}(\tau; \alpha, \beta),$$
$$p(\mu) = \text{Normal}\left(\mu; \xi, \frac{1}{\psi}\right),$$
$$p(w_n|\mu, \tau) = \text{Normal}\left(w_n; \mu, \frac{1}{\tau}\right), \qquad n = 1 : N,$$

where α, β, ξ, ψ are prespecified hyperparameters.

With this setup, the model is described by the random variables μ, τ, and the associated posterior is $p(\mu, \tau|w_{1:N})$. As we saw in Example 4.8, the model is not fully conjugate. Therefore, computational methods are required to estimate the posterior maximum with respect to τ.

* This is an advanced topic and could be skipped on a first reading.

We begin by treating μ as a latent variable r and θ of Algorithm 4.1 is understood as τ. As such, $Q_{\theta'}(\theta)$ for this problem reads

$$Q_{\tau'}(\tau) = \int d\mu\, p\left(\mu|\tau', w_{1:N}\right) \log p\left(\tau|\mu, w_{1:N}\right),$$

requiring that we compute both conditional posteriors $p\left(\mu|\tau, w_{1:N}\right)$ and $p\left(\tau|\mu, w_{1:N}\right)$. These are readily computed according to

$$p\left(\mu|\tau, w_{1:N}\right) \propto p\left(w_{1:N}|\mu, \tau\right) p\left(\mu\right),$$
$$p\left(\tau|\mu, w_{1:N}\right) \propto p\left(w_{1:N}|\mu, \tau\right) p\left(\tau\right),$$

where we can drop all constants independent of τ and μ as these do not change the location of the maximum in $Q_{\tau'}(\tau)$. The maximization of $Q_{\tau'}(\tau)$ returns an iterative equation left as an exercise; see Project 4.2.

4.4 Hierarchical Bayesian Formulations and Graphical Representations

So far, we have considered measurement output, $w_{1:N}$, latent variables, typically labeled $r_{1:N}$, parameters of the emission distribution, ϕ, and other parameters of a forward model, ψ. To be consistent with prior notation, we say that the total number of parameters $\theta_{1:K}$ is K.

In Bayesian analysis, we place priors on all parameters whose distribution is to be determined. These include all unknown elements of $\theta = \{\phi, \psi\}$. Thus a full forward model may read

$$\psi \sim \mathbb{H},$$
$$\phi \sim \mathbb{T},$$
$$r_n|\psi \sim \mathbb{P}_{\psi},$$
$$w_n|r_n, \phi \sim \mathbb{F}_{\phi_{r_n}},$$

where $\mathbb{H}$ and $\mathbb{T}$ are prior distributions. In practice, these distributions themselves may depend on parameters (hyperparameters) whose distribution, *hyperpriors*, we may also wish to specify.

As we may begin to appreciate, complex dependencies between random variables may be more conveniently represented visually to provide an immediate sense for what is otherwise encoded in forward models such as the one just above. Visual representations of forward models are called *graphical representations* and their construction must satisfy a few ground rules: every model random variable must be shown; each random variable must reside within a circle called a *node*; arrows pointing from one to another random variable indicate dependencies between the variables; finally, known random variables must be shaded.

We begin with known random variables: observations. As we see below, these are shown with time, or measurement index, progressing from left to right. On the right panel of this graphical representation, these random variables can be collected in a plate whose running index coincides with those of the random variable.

For simple cases where we do not have model latent variables, the observations themselves are directly informed by the model parameters, $\theta_{1:K}$. In this case, and under the assumption that all parameters are unknown, we have a causal arrow pointing from those parameters shown in unshaded boxes to the measurements.

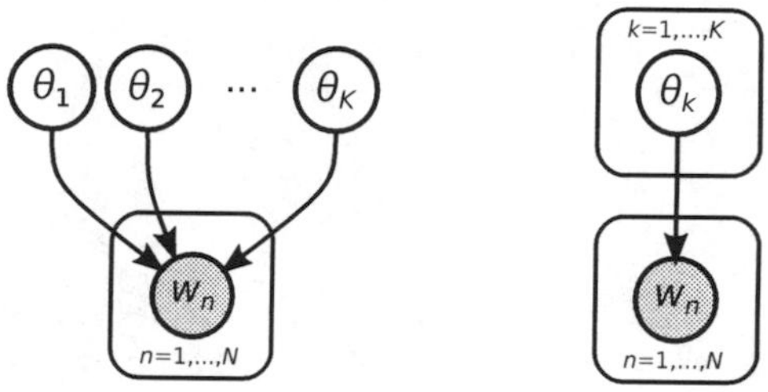

Example 4.10 Graphical model for iid normal observations

We consider three cases shown side by side for data, $w_{1:N}$, generated from Normal(μ, $1/\tau$). The cases we consider are: μ known and τ unknown; μ unknown and τ known; and μ and τ unknown.

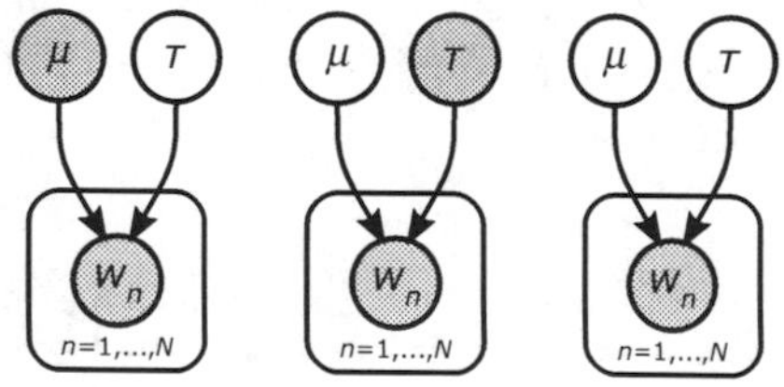

Typically, if we use conditionally conjugate priors for the mean, μ, and precision, τ, these priors are parameterized by hyperparameters. If these hyperparameters are prespecified, they are not random variables and do not appear in *nodes*.

Example 4.11 Graphical model for iid normal observations with explicit hyperparameters

We consider the case where data, $w_{1:N}$, is generated from Normal(μ, $1/\tau$). For simplicity, we only consider the case where μ and τ are unknown. If we insist on using conditionally conjugate priors, then the prior over μ is also Normal(ξ, $1/\psi$) and that over τ is Gamma(α, β).

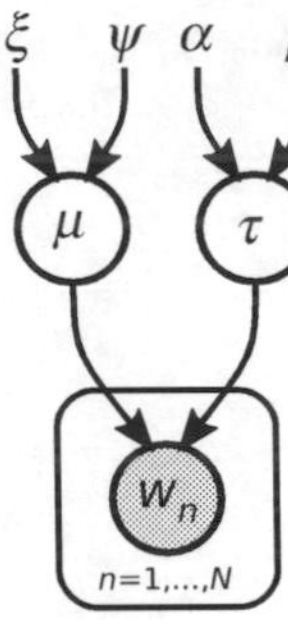

As we turn to latent variables, we start with iid observations. In the presence of latent variables, the plates can be extended to include the latent variable as the index running over the latent variable and measurement are typically the same as shown below:

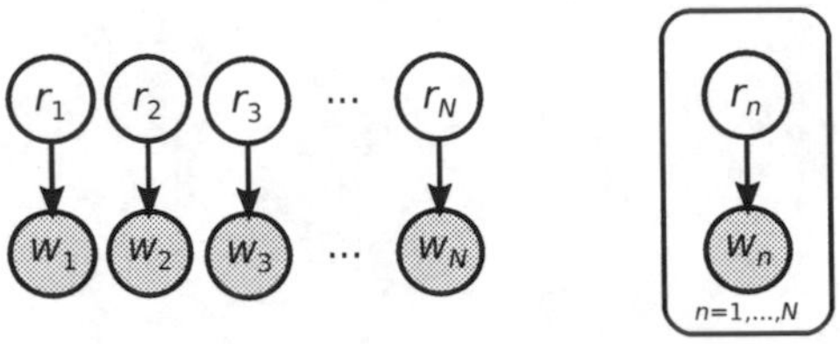

An example of this includes a sum of two normal distributions, where the latent variable generically denoted r_n, can be realized to one or two for each data point.

As we bring in latent variables, we begin discriminating between parameters of the emission distribution and other model parameters. A general graphical representation for this scenario is provided in the representation below.

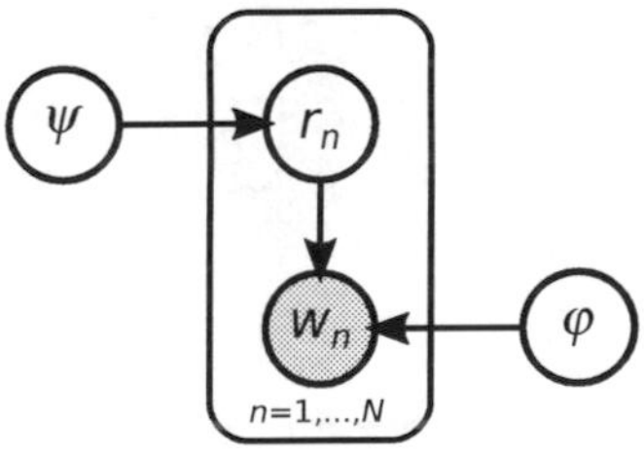

We immediately see that this graphical representation captures the forward model structure

$$
\begin{aligned}
\psi &\sim \mathbb{H}, \\
\phi &\sim \mathbb{T}, \\
r_n|\psi &\sim \mathbb{P}_{\psi}, \\
w_n|r_n, \phi &\sim \mathbb{F}_{\phi_{r_n}}.
\end{aligned}
$$

Dynamical models, *i.e.*, non-iid examples, can also be shown using graphical representations. As we will delve into greater detail on such models in later chapters, for now we simply highlight the essentials. We start with the simplest example of a dynamical model; one that satisfies the Markov property. For example, in the simple case of diffusion, *i.e.*, Brownian motion, the position at time level n causally depends on the position at previous times. An example of this is shown explicitly below:

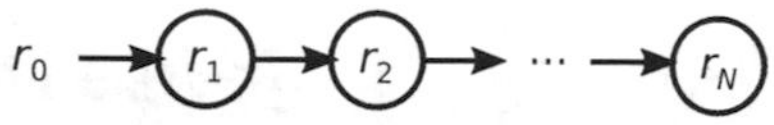

In this figure, r_0 denotes the initial condition. By modeling measurement noise explicitly, we introduce an observation coinciding with the latent variable at each time level. To be explicit, we supplement this model with a prior on the observation parameters, ϕ, and other model parameters, collectively termed ψ, including those parameters parametrizing the distribution over the initial condition, r_0, if unknown and those parameters dictating the kinetics of the system as it evolves across time levels.

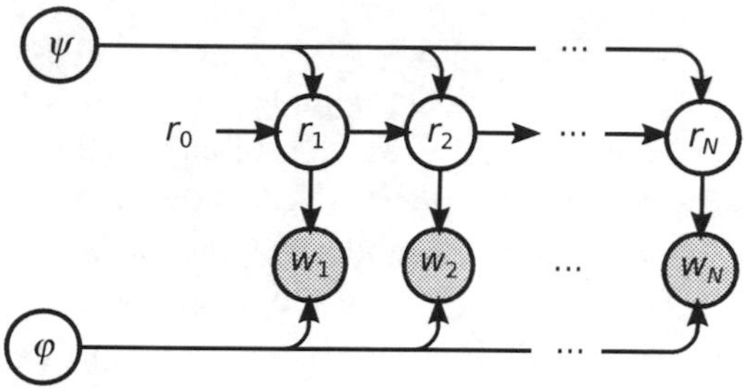

Finally, we end this section with Note 4.9 on *Markov blankets*.

Note 4.9 Markov blankets

The Markov blanket of node r includes a set of nodes that we denote $MB(r)$. The idea is that, once conditioned on all nodes contained within its Markov blanket, a node is conditionally independent of all other nodes, r',

$$p(r|MB(r), r') = p(r|MB(r)).$$

As a consequence, this implies that for two nodes, r and r', with the same Markov blanket,

$$p(r, r'|MB(r)) = p(r|MB(r))p(r'|MB(r)).$$

The Markov blanket of node r includes its parent nodes, its children, and its spouses (other parents of its children). This seems reasonable as, clearly, knowing the state of its parents informs us about the state of r. Following similar logic, knowing something about the children tells you something about its parent. What is more, by considering the effect of the spouses on the children, we can eliminate factors influencing the children not otherwise due to the instantiation of r itself.

The notion of Markov blankets, and graphical representations as a whole, can become useful in more complex inference problems. Beyond visually highlighting variable dependencies, graphical representations also make clear which node belongs within any other node's Markov blanket. Indeed, we may leverage our ability to pinpoint members of the Markov blanket to quickly simplify conditionals by excising from them those variable dependencies outside a node's Markov blanket free of burdensome derivations.

4.5 Bayesian Model Selection*

In Section 4.1.1, we introduced the problem of *model selection*. Here we turn our attention to a model selection criterion termed the *Bayesian information criterion* (BIC), sometimes also called the Schwartz information criterion (SIC).

4.5.1 The Bayesian Information Criterion

The BIC identifies a compromise between the data's goodness of fit while penalizing model complexity by identifying the most probable "model dimensionality," which is often shorthand for the number of model parameters or "components" given the data. For concreteness, we call the dimensionality of the model K and equate it to the number of model parameters in $\boldsymbol{\theta} = \theta_{1:K}$. This strategy only works for models whose complexity can be shrunk or grown. Such models, whose parameters form a subset of the parameters of a larger model, are termed *nested models*. For example a single exponential is a subset of a two-component exponential model. Such a two-component exponential model is, in turn, a subset of a model with an even larger number of components.

The basic idea underlying the BIC is to compute $p(K|w_{1:N})$, which is understood as a posterior over the number of parameters. This posterior allows us, in principle, to ascribe a probability to models of different dimensionality. In practice, as the BIC is an asymptotic result valid when $N \gg 1$, the resulting $p(K|w_{1:N})$ obtained is an approximation. This approximation may be useful in comparing relative probabilities for $p(K|w_{1:N})$ versus $p(K'|w_{1:N})$ but not in computing absolute probabilities. Alternatively, the approximate $p(K|w_{1:N})$ may simply be maximized to determine the maximum posterior value of K.

In order to compute $p(K|w_{1:N})$, we first write $p(K|w_{1:N}) = \int d\boldsymbol{\theta}\, p(\boldsymbol{\theta}, K|w_{1:N})$ where the integral is, as usual, taken to be over the entire range allowed by each parameter. The large N asymptotics creep into the calculation almost immediately as a recourse toward overcoming the intractability of the integrals over $\boldsymbol{\theta}$.

The posterior itself is constructed, as usual, from a likelihood, $p(w_{1:N}|\boldsymbol{\theta})$. In the presence of iid or, at the very least, weakly correlated observations, we are justified in rewriting the likelihood in the suggestive form

$$p(w_{1:N}|\boldsymbol{\theta}) \approx e^{N \log p_1(w_{1:N}|\boldsymbol{\theta})}, \tag{4.15}$$

* This is an advanced topic and could be skipped on a first reading.

where $p_1(w_{1:N}|\boldsymbol{\theta})$ is interpreted as a likelihood per data point, *i.e.*, $\sqrt[N]{p(w_{1:N}|\boldsymbol{\theta})}$. Strictly, the linear N scaling preceding the logarithm of Eq. (4.15) only holds for iid measurements and, as we will see, the subsequent derivation relies on the fact that any N scaling of $\log p_1(w_{1:N}|\boldsymbol{\theta})$ be strictly sublogarithmic.

Using Eq. (4.15), we now rewrite the posterior as

$$p(\boldsymbol{\theta}, K|w_{1:N}) \propto e^{N \log p_1(w_{1:N}|\boldsymbol{\theta})} \times \text{priors}, \tag{4.16}$$

where the priors, naturally independent of N, are kept intentionally vague. The subsequent integral of Eq. (4.16) over all values of $\boldsymbol{\theta}$ is now performed, asymptotically, using *Laplace's method*. That is, for N large, we recognize that the integral is dominated by that region of the integrand where $\log p_1(w_{1:N}|\boldsymbol{\theta})$ is at its maximum with respect to $\boldsymbol{\theta}$, which we call $\hat{\boldsymbol{\theta}}$. By this argument, the essence of Laplace's method, the likelihood can be approximated by a Gaussian centered at $\hat{\boldsymbol{\theta}}$. The prior over this narrow neighborhood is assumed roughly flat. In other words, we expand $\log p_1(w_{1:N}|\boldsymbol{\theta})$ around its maximum, $\hat{\boldsymbol{\theta}}$, up to quadratic order and write

$$\begin{aligned} p(K|w_{1:N}) &\propto e^{N \log p_1(w_{1:N}|\hat{\boldsymbol{\theta}})} \int d\boldsymbol{\theta}\, e^{-\frac{N}{2}(\boldsymbol{\theta}-\hat{\boldsymbol{\theta}})H\left(\log p_1(w_{1:N}|\hat{\boldsymbol{\theta}})\right)(\boldsymbol{\theta}-\hat{\boldsymbol{\theta}})^T} \\ &\propto e^{N \log p_1(w_{1:N}|\hat{\boldsymbol{\theta}})} N^{-K/2}, \end{aligned}$$

where the Hessian, H, is defined in Note 3.3 and all terms assumed constant with respect to N are lumped into the proportionality. The value of K maximizing $p(K|w_{1:N})$ now suggests an optimal number of parameters and, as we can either maximize $p(K|w_{1:N})$ or its logarithm, we opt for the latter.

Historically, the BIC is defined as -2 multiplied by the logarithm of $p(K|w_{1:N})$ and this quantity, therefore to be minimized to locate the optimal K, reads

$$\text{BIC} = -2 \log p(K|w_{1:N}) = -2 \log p(w_{1:N}|\hat{\boldsymbol{\theta}}) + K \log N, \tag{4.17}$$

where terms of order N^0 are dropped. For clarity, all prior contributions to the BIC are asymptotically vanishing and contribute to the term of order N^0, which we ignore. While our discussion above has remained abstract, we provide an example of how the BIC is used in change point detection immediately following a brief explanation as to why the BIC works just below.

4.5.2 Why the BIC Works

Why does the BIC penalize complex models or, equivalently, penalize introducing new parameters? At first sight, it appears we simply naively integrated an approximated likelihood, ignored the prior, and counterintuitively manipulated the likelihood into penalizing model parameters.

The key as to why the BIC penalizes model complexity lies in the nontrivial operation of integration itself. The more model parameters we have, the more parameters we must integrate over. Now, introducing more parameters, as we discussed in the context of Section 4.1.1, improves the fit to the data. However, this is only true of near optimal, maximum likelihood parameter values. Poorly selected parameters, especially many multiple poorly selected parameter values, can dramatically reduce a likelihood. In other words, if we integrate over all parameter values that could have been possible in high dimensional complex models, the number of "ways" of reducing the posterior eventually supersede the "ways" of increasing it. This reasoning explains why complex models, with large volumetric swathes of poorly fitting parameters, are not always preferred.

An immediate appeal of the BIC therefore lies in its ability to capture the competition between wanting to improve the goodness of fit to the data captured by the first term of Eq. (4.17), without overfitting the model, captured by the second term of of Eq. (4.17). The second term, sometimes called a "complexity penalty," scales as $\log N$ and roughly suggests that every new log order of data is deserving of a new parameter. Intuitively, this matches our expectation: the first decade of data points should certainly inform us more on the nature of the model than, say, data points appearing between $n = 1,000,000$ and $n = 1,000,010$.

Note 4.10 Warning on the BIC

While the BIC is helpful for some applications, it only makes sense so far as it is used within its domain of applicability. This includes iid, and not Markov, data. What is more, the BIC's complexity penalty with $\log N$ scaling is rigorously derived and a different dependency on N, tempting to hypothesize to either strengthen or weaken the BIC's model complexity penalty, demands an equally rigorous justification.

4.5.3 A Case Study in Change Point Detection

A creative use of the BIC lies in detecting change points in time series data. Change point algorithms locate points in the data where the statistics generate the data change. In Example 4.12, we consider data points being drawn from a normal with some (unknown) mean and variance for some portion of the trace followed by data points being drawn from a normal with a different (unknown) mean but, for simplicity only, the same (unknown) variance for all subsequent time points.

Example 4.12 Change point detection

We begin by writing down the likelihood for our problem,

$$p(w_{1:N}|K, v, \mu, j) = \prod_{k=0}^{K-1} \prod_{n=j_k}^{j_{k+1}-1} \frac{1}{\sqrt{2\pi v}} e^{-\frac{(w_n-\mu_k)^2}{2v}} = \frac{1}{(2\pi v)^{N/2}} e^{-\sum_{k=0}^{K-1} \sum_{n=j_k}^{j_{k+1}-1} \frac{(w_n-\mu_k)^2}{2v}}, \tag{4.18}$$

where j_k denotes the time level index at which a change point occurs, *i.e.*, where the change in the mean of the signal occurs. By convention, the first data point is j_0; the last is j_K. The first change point occurs at j_1.

As we have $K-1$ change points (since j_K is fixed), K means (associated to each fixed mean segment), and a variance, we have a total of $2K$ parameters. Trivially, the model maximizing this likelihood places a change point at every time level. To avoid overfitting, and since the correct values for v, $\mu_{0:K-1}$, or locations $j_{1:K-1}$ are unknown, we integrate over all these parameters. Foreshadowing the use of Laplace's method, valid for sufficiently large N, we write

$$p(K|w_{1:N}) \propto \sum_{j_{1:K-1}}{}' \int dv^{1/2}\, d\mu_{0:K-1}\, p(w_{1:N}|K, v, \mu_{0:K-1}, j_{1:K-1}),$$

where the restricted sum over change point location, denoted by the prime over the sum, excludes overlapping change points at the same point.

Unlike our earlier BIC derivation, there is no need to immediately invoke Laplace's approximation to simplify the likelihood as the integrals over the standard deviation, $v^{1/2}$, and $\mu_{0:K-1}$ can be performed analytically, though we continue treating the priors as asymptotically subdominant. Indeed, the marginalization over $\mu_{0:K-1}$ returns

$$p(K|w_{1:N}) \propto \sum_{j_{1:K-1}}{}' \int dv^{1/2} \frac{1}{v^{N/2}} \left(v^{K/2} e^{-\frac{1}{2v}S}\right), \tag{4.19}$$

where $S = n_0\hat{v}_0 + \cdots + n_{K-1}\hat{v}_{K-1}$ and

$$\hat{v}_k = \frac{1}{n_k}\sum_{n=j_k}^{j_{k+1}-1} w_n^2 - \frac{1}{n_k^2}\left(\sum_{n=j_k}^{j_{k+1}-1} w_n\right)^2,$$

where n_k counts the number of points contained in the kth step $(j_{k+1} - j_k - 1)$.

The integral over $v^{1/2}$ is also analytically performed by recognizing that the argument of Eq. (4.19) can be transformed, by variable substitution, into a gamma function integral and yields

$$\begin{aligned} p(K|w_{1:N}) &\propto \sum_{j_{1:K-1}}{}' \frac{2\pi^{-(N-K)/2}}{n_0^{1/2}\cdots n_K^{1/2}} \cdot \frac{1}{2} \cdot \left(\frac{S}{2}\right)^{-\frac{(N-K-1)}{2}} \cdot \left(\frac{N-K-3}{2}\right)! \\ &\approx \frac{2\pi^{-(N-K)/2}}{(\hat{n}_0)^{1/2}\cdots(\hat{n}_K)^{1/2}} \cdot \frac{1}{2} \cdot \left(\frac{\hat{S}}{2}\right)^{-\frac{(N-K-1)}{2}} \cdot \left(\frac{N-K-3}{2}\right)!, \end{aligned} \tag{4.20}$$

where we have approximated the intractable sum over change points by its dominant term, *i.e.*, select the location of change points that maximizes the likelihood. The quantities with hats thus coincide with those determined once the change point locations are fixed at those values maximizing the likelihood. Furthermore, it is now no longer obvious, by inspection of Eq. (4.20), that $p(K|w_{1:N})$ grows as K approaches N. To see this explicitly, we take the further simplifying assumption that all n_k are large and recover the familiar

$$\text{BIC} = -2\log p(K|w_{1:N}) = N\log\hat{S} + K\log N + \mathcal{O}\left(N^0\right) + \text{constants}, \tag{4.21}$$

where the constants capture all terms independent of K, but may otherwise depend on N.

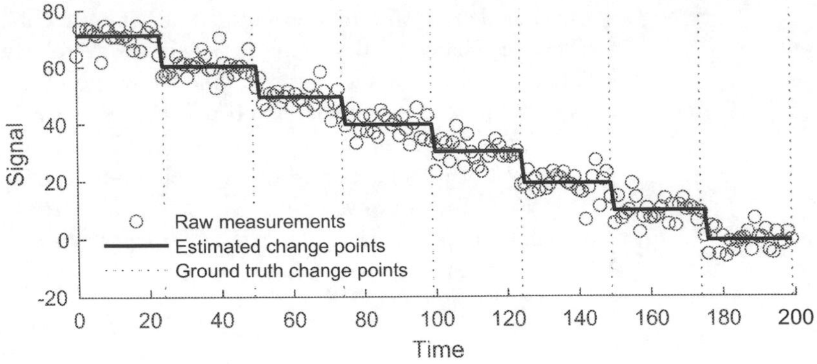

Fig. 4.2 Change point detection. Here, we generate synthetic data (dots) mimicking sharp step-like transitions. Despite the noise present, BIC identifies step times (solid line) in excellent agreement with the ground truth (dotted lines).

Figure 4.2 shows the detection of change points in synthetic data. The change point algorithm used to generate this figure is provided in Algorithm 4.2.

Algorithm 4.2 A greedy BIC change point algorithm for normal likelihoods

We start from a time trace that we assume satisfies the likelihood of Eq. (4.18). We use the BIC of Eq. (4.21) to find the change points.
Start by assuming there are no change points and compute the BIC.

- Introduce an additional change point at all allowed locations where there are no current change points.
- If the addition of a change point at any location does not reduce the BIC, as compared to the BIC with one fewer change points, stop.
- Otherwise, if a change point does reduce the BIC, find the best location (the one that minimizes the BIC) and fix the change point there and return to the first step.

Fixing a change point and identifying locations of other change points conditioned on the previous change point's location is an approximation. In principle, one should consider the locations of all change points simultaneously. The nature of this approximation, *i.e.*, the notion of making a locally optimal choice and conditioning subsequent steps of the algorithm on this choice, is the reason methods such as these are sometimes called *greedy algorithms*.

There are small issues one may quibble about regarding the BIC. What if the model that informs our likelihood can only logically increase the number of parameters by units of 3 say, imagine introducing a new normal in a model with its own associated weight, mean, and variance, while the BIC provides a maximum in K that is not a multiple of 3? Should we then round K up or down to the closest multiple of 3? In practice, a BIC equivalent should be re-derived, as we did for change points, on a case-by-case basis.

Yet, there is a much deeper challenge that remains. What do we do after having selected a K? Do we go back to square one and, armed with this K, construct a likelihood of the appropriate dimensionality and proceed onward with constructing a posterior? Somehow this feels ad hoc as we are not globally learning K and parameters. Put differently, one's optimal choice of parameters in a model of complexity K is not independent of one's choice of optimal parameters for a model of complexity $K + 1$. In other words, model parameters cannot in general by assumed independent across models. Yet, it is precisely this assumption that is made in treating parameter maximization and model dimensionality estimation as two separate, artificially disjointed steps. The ability to lift this approximation is the basis of Bayesian nonparametrics (BNPs) that we turn to shortly.

4.6 Information Theory*

We end this chapter with a brief, qualitative discussion of another paradigm of data analysis beyond frequentism, explored in the previous chapter, and Bayesianism, explored herein. This paradigm is that of *information theory* and, while developed independently of Bayesian statistics and equipped with its own language, information theory is now regarded as a special case of Bayesian inference.

The central quantity of information theory is the expression of the information content of a probability density. It is helpful here, for sake of demonstration, to initially consider a discrete probability density $p(r) = \sum_m \pi_{\rho_m} \delta_{\rho_m}(r)$ with the understanding that m runs from 1 to M. The information content contained in $p(r)$ is defined to be $I = \sum_m \pi_{\rho_m} \log_2 \pi_{\rho_m}$. To be clear, here the information, I, is interpreted as the (negative of the) average number of binary bits encoded in each symbol ρ_m whose probability of occurrence in a message consisting of a sequence of symbols is π_{ρ_m}. Often the base 2 of the logarithm, required when speaking of binary bits, is dropped especially in physics where the natural logarithm is preferred.

From this expression for the information, it is easy to see that the information is minimal when $p(r)$ is uniform and all π_{ρ_m} are set to $1/M$. In this case, $I = -\log M$. Similarly, a $p(r)$ with one π_{ρ_m} set to unity contains a maximal information of 0 when considering that $\pi_{\rho_m} \log \pi_{\rho_m}$ tends to zero as each π_{ρ_m} not approaching unity tends to zero.

Free of the language of bits encoded per symbol, loosely speaking I measures how featureless a probability density is. A featureless, flat probability density has lower information content than a sharply peaked one. Rather than speak of information, especially in the physics literature, it is quite common to speak of the *entropy* of a probability density as the

* This is an advanced topic and could be skipped on a first reading.

negative of the information, $-I$. Thus, the more featureless a probability density is, the higher its entropy. However, to avoid confusion for that portion of our audience trained in physics, we mention immediately that the entropy of the probability density is not strictly the thermodynamic entropy, though both concepts are related.

Given data, the goal is often to determine the minimal number of bits required to encode a message or, equivalently in physics, the most featureless probability density consistent with the data. To identify the ensuing $p(r)$, we minimize I, or equivalently, maximize the entropy under constraints imposed from the data using Lagrange multipliers. This procedure for reconstructing $p(r)$ in this way is termed the *maximum entropy principle* and it returns a point estimate for $p(r)$ that we call $\hat{p}(r)$. To the physicists, the thermodynamic entropy is understood as the entropy of the probability density, $-I$, when evaluated at this point estimated value, $-I(\hat{p}(r))$, up to units of Boltzmann's constant. The maximum entropy principle is of particular interest to the field of statistical physics where constraints often take the form of known averages of the means of conserved integrals of motion then used to derive point estimates for equilibrium distributions, $\hat{p}(r)$, over states of matter.

To relate the maximum entropy principle to Bayesian methods, we note immediately that constraints are none other than the logarithm of a likelihood. Indeed, constraints often take the form of a mean square difference, which is exactly the logarithm of a Normal likelihood. Similarly, the entropy can then be simply the logarithm of a prior that imposes an a priori smoothness condition on $p(r)$. Thus, $\hat{p}(r)$ is understood as a maximum a posteriori estimate under the assumption of a prior taking the form of the exponential of the entropy. The question then arises as to whether to use the exponential of the entropy as a prior on $\pi_{\rho_{1:M}}$ versus a prior conjugate, say, to the Categorical. The prior conjugate to the Categorical is the Dirichlet prior and it is often invoked as a matter of computational convenience.

While we have focused on discrete densities, it is possible to generalize the concept of information to treat continuous densities. The arguments can then, with some care in avoiding unitful arguments of the logarithm in the definition of the information, be repeated and an analog of I derived for continuous densities. An elegant starting point for the derivation of I for continuous densities begins from the so-called Shore–Johnson axioms listing *desiderata* on I. The information so derived takes any form monotonic with $I = \int dr\, p(r) \log \left(p(r)/q(r)\right)$ for continuous densities often, for convenience, simply settled upon to be I itself. Here, $q(r)$ is understood as a prior on $p(r)$. This is easy to see when we consider minimizing I subject to no constraints on data, which reveals that $\hat{p}(r) \propto q(r)$. While the form of I with the added $q(r)$ is of necessity for continuous densities, this form for I can also be written for discrete densities with the integral replaced by a sum.

Example 4.13 Signal deconvolution using maximum entropy

Here, we think of a signal, $S(t)$, as arising from the sum of exponential contributions, $S(t) = \int dr\, \pi(r) \exp(-rt)$. The goal is to deduce the continuous function $\pi(r)$ from $S(t)$. As $S(t)$ is necessarily noisy, many $\pi(r)$ are consistent with the observed $S(t)$. As such, the naive inverse Laplace transformation of $S(t)$ to obtain $\pi(r)$ is numerically unstable.

One option is to use maximum entropy to impose conditions on the smoothness of $\pi(r)$ in order to estimate a unique $\pi(r)$ from the data. Under the assumption that the noise on $S(t)$ is normally distributed with variance v, the *maximum entropy principle* suggests maximizing the following objective function in order to obtain what we understand to be a MAP estimate for $\pi(r)$,

$$\underbrace{\left(-\int dr\, \pi(r) \log\left(\pi(r)/q(r)\right)\right)}_{\text{log prior}} + \underbrace{-\left(\frac{1}{v}\left(S(t) - \int dr\, \pi(r)\exp(-rt)\right)^2\right)}_{\text{log likelihood}},$$

under the constraint that $\int dr\, \pi(r) = 1$. The maximization is then normally performed by gradient descent or related techniques.

While historically important, the above procedure is rarely used today for multiple reasons listed here in order of growing concern: while the full distribution over $\pi(r)$ can be obtained, the maximization above limits us to a MAP estimate; no conjugacy property is leveraged in selecting the prior and, as such, computational efficiency is poor; data is not always normally distributed and realistic likelihoods may further compound the already existing poor computational scaling; there is no compelling reason, other than historical, to use the prior above from which stem most computational challenges; and, as the variance of the noise is generally unknown a priori, v is treated as a tunable regularization parameter to manually strengthen or weaken the effect of the prior at will. Unfortunately, tuning v in response to data is no different than allowing the prior to depend on data, which introduces circular logic in the manipulation of probabilities and violates Bayes' theorem.

With the understanding that r may reflect a vector of parameters, say two for simplicity, which we label r_1 and r_2, we can write $p(r)$ as $p(r_1, r_2)$. If we select $q(r)$ as $p(r_1)p(r_2)$, then the information reads $I = \int dr_1 dr_2\, p(r_1, r_2) \log\left(p(r_1, r_2)/\left(p(r_1)p(r_2)\right)\right)$. Recognizing the logarithm of a ratio as the difference between the logarithm of the numerator and denominator, this information quantifies the difference in information between the joint distribution and the product of marginals. For this reason, this information it sometimes called the *mutual information*.

In an effort to estimate joint densities from data while minimizing correlations between parameters in a joint probability density, the mutual information may be minimized under constraints from data. Similar arguments used in deducing joint or conditional densities, relevant in data clustering or lossy data compression, fall under the guise of *rate-distortion theory*, where the distortion relates to the bias introduced by the data (or, in our language, the emission distribution portion of the likelihood) and the rate relates to the average rate of bits transmitted (the log prior in our language).

We conclude with the thought that, like graphical representations, information theory can help provide a new perspective to what is already encoded in Bayesian language. Indeed, information theory adds an interpretation to some Bayesian priors and probability manipulations that can otherwise be mathematically performed while adhering to Feynman's "shut up and calculate" adjuration.

4.7 Exercise Problems

Exercise 4.1 Poisson–gamma conjugacy

1. Show that the Poisson–gamma distributions form a conjugate pair.
2. Generate data sampled from a Poisson distribution following Exercise 1.23 and construct a likelihood using your generated $w_{1:N}$. Use your likelihood to construct a posterior.
3. Plot your exact posterior with 10, 100, and 1,000 data points. How does your posterior change?

Exercise 4.2 Predictive distributions

1. Compute the predictive distribution, $p(w_{N+1}|w_{1:N})$, for a Poisson likelihood with a conjugate gamma prior.
2. Simulate data, $w_{1:N}$, according to a Poisson distribution following Exercise 1.23 and plot $p(w_{N+1}|w_{1:N})$.
3. Superpose $p(w_{N+1}|w_{1:N})$ with the original Poisson distribution from which you generated your data. How do these distributions compare?

Exercise 4.3 Outliers within a normal likelihood paradigm

Consider Example 4.8 and suppose that our goal is to estimate the posterior over the mean irrespective of the value assigned to the precision. In this case, we marginalize the posterior of Eq. (4.14) over the precision

$$p(\mu|w_{1:N}) = \int_0^\infty d\tau\, p(\mu,\tau|w_{1:N})$$
$$\propto \left(1+\beta\left(\frac{\psi_0(\mu-\xi)^2}{2}+\sum_{n=1}^{N}(w_n-\mu)^2\right)\right)^{-(N/2+\alpha+1/2)}.$$

This posterior has a single mode. This is evident by first taking the logarithm of the posterior and noting that its derivative with respect to μ is a linear equation supporting only one root.

As the posterior can only accommodate one maximum, data outliers often have the effect of growing the tails of $p(\mu|w_{1:N})$ as opposed to introducing new local maxima into $p(\mu|w_{1:N})$. The question then becomes:

how can we prohibit outliers from growing the tails of the posterior? We first recognize that in constructing the conjugate prior over the mean and precision, we made important choices. One choice made was to assume that the prior over the precision of each data point was the same. Yet, this logic runs counter to that anticipated in the presence of outliers where some data points must be attributed to different, presumably larger, precisions than others. As such, we start by relaxing the constraint that all data points must have the same associated precision and write

$$p(\mu|w_{1:N}) \propto \int_0^\infty d\tau_{1:N} \prod_{n=1}^{N} p(w_n|\mu, \tau_n)p(\mu, \tau_n).$$

Under this circumstance,

$$p(\mu|w_{1:N}) \propto \prod_{n=1}^{N} \left(1 + \beta\left(\frac{\psi_0(\mu - \xi)^2}{2} + (w_n - \mu)^2\right)\right)^{-(\alpha+1/2)}.$$

To see why this posterior is indeed multimodal, we take the derivative with respect to μ of its logarithm. This derivative immediately yields a high-order polynomial in μ, suggesting multiple maxima and minima.

Now, to see how both posteriors discussed above adapt to outliers:

1. Generate iid data points, $w_{1:N}$, according to a normal distribution with occasional data points sampled from a normal distribution with a much larger precision.
2. For those data just generated, compute $p(\mu|w_{1:N})$ under the assumption that all data points are drawn from a distribution with the same precision.
3. For those data just generated, compute $p(\mu|w_{1:N})$ under the assumption that all data points are drawn from a distribution with a different precision.

Exercise 4.4 Graphical representations

In this problem, we characterize a graphical model.

1. List exhaustively every variable shown in the graphical model.
2. Sort the variables you identified into three groups: random variables with known values, random variables with unknown values, and variables that are not random.
3. For each random variable you have identified, list its Markov blanket.

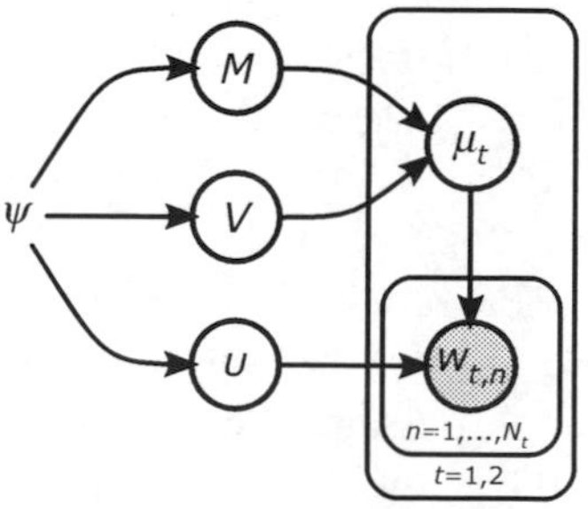

Project 4.1 Change point algorithms

Implement the BIC-based change point method laid out in Algorithm 4.2 for the simple case of a normal likelihood with identical variance assumed for all data points. In other words, begin by generating synthetic data assuming discrete steps in your data where you have prespecified the means of those steps in the signal (assume the variance associated to each step is the same). Then, implement Algorithm 4.2 to estimate the means and change point locations. Summarize your results graphically by superposing, in the same figure, the location of the ground change points with your estimates. For identical step height (call it "S") and increasing variance (call it "N"), at what value of S/N does your method start to fail for an average given number of data points associated to each step?

Project 4.2 EM for posteriors

Consider Example 4.9. Evaluate an explicit form for $Q_{\tau'}(\tau)$ and determine the maximum *a posteriori* τ estimate for synthetically generated data.

Additional Reading

P. Lee. *Bayesian statistics: an introduction*. 4th ed. Wiley, 2012.

D. S. Sivia, J. Skilling. *Data analysis: a Bayesian tutorial*. Oxford University Press, 2006.

A. Gelman, J. Carlin, H. Stern, D. Dunson, A. Vehtari, D. Rubin. Bayesian data analysis. 3rd ed. CRC Press, 2013.

P. D. Hoff. *A first course in Bayesian statistical methods*. Springer, 2009.

G. E. Schwarz. Estimating the dimension of a model. *Ann. Stat.*, 6:461, 1978.

B. Kalafut, K. Visscher. An objective, model-independent method for detection of non-uniform steps in noisy signals. *Comp. Phys. Comm.*, 179:716, 2008.

E. B. Fox. *Bayesian nonparametric learning of complex dynamical phenomena*. PhD Thesis, Dept. of Elec. Eng. & Comp. Sci. (Advisor: A. S. Willsky, J. W. Fisher III), MIT, 2019.

C. E. Shannon. A mathematical theory of communication. *Bell Sys. Tech. J.*, 27:379, 1948.

J. Shore, R. Johnson. Axiomatic derivation of the principle of maximum entropy and the principle of minimum cross-entropy. *IEEE Trans. Inf. Th.*, 26:26, 1980.

S. Pressé, K. Ghosh, J. Lee, K. A. Dill. Principles of maximum entropy and maximum caliber in statistical physics. *Rev. Mod. Phys.*, 85:1115, 2013.

U. von Toussaint, Bayesian inference in physics. *Rev. Mod. Phys.*, 83:943, 2011.

5 Computational Inference

By the end of this chapter, we will have presented

- *The Metropolis–Hastings and Gibbs algorithms*
- *Schemes for Bayesian computations*
- *Strategies to sample arbitrary distributions*

As we saw in earlier chapters, the solution of an inverse problem through Bayesian inference relies on posterior probability distributions and often requires characterizing them or, at a more quantitative level, evaluating integrals with respect to them. Except in rare cases, however, the posterior distributions involved in our problems cannot be derived analytically. Here, we describe computational methods that relax this limitation.

5.1 The Fundamentals of Statistical Computation

5.1.1 Monte Carlo Methods

Monte Carlo (MC) methods provide a wide class of algorithms for problems that can be reframed probabilistically when analytic approaches are unavailable. MC methods apply broadly, even beyond Bayesian problems. We briefly summarize their use below.

Note 5.1 What is a Monte Carlo method?

A Monte Carlo method describes a computational approach generating and using pseudorandom numbers. Such methods are mainly used to solve three wide classes of problems:

1. *Optimization*: where we seek to identify conditions leading to optimal results according to a predefined criterion.
2. *Integration*: where we seek to compute an integral of some predefined quantity.
3. *Sampling*: where we seek to obtain samples of some quantity following given predefined statistics.

Such problems arise commonly in the sciences and MC methods are employed extensively, especially to investigate cases involving complex physical systems that consist of numerous interacting components, such as idealized spins, quantum particles, molecules, cells, or agents of any kind. MC is often indispensable as it excels in efficiently handling complex interactions and high dimensionality that make alternative approaches, *i.e.*, fully deterministic approaches, practically infeasible.

MC methods solve problems attaining a probabilistic interpretation of which we have already seen some examples: the fundamental theorem of simulation in Algorithms 1.1 and 1.2 and the Gillespie simulation in Algorithm 2.1, amongst others. MC methods apply on random variables R that may be uni- or multivariate, may contain continuous or discrete components, and may gather various individual variables in a complex model. In the Bayesian context, we are mostly interested in sampling the posterior probability density, $p(r|w)$, for observations denoted by w.

To keep notation simple, throughout this chapter we denote the posterior $p(r|w)$ of interest simply as $\pi(r)$ and, since the methods we will describe also apply to cases where $\pi(r)$ may not necessarily be a posterior, but any probability density, we call $\pi(r)$ the *target density*. For clarity, we denote by Π the target's distribution. In this setup, our targeted random variable is denoted by $R \sim \Pi$ and its density is denoted $\pi(r)$.

The way MC methods work is as follows. First, we simulate a sequence $r^{(1)}, r^{(2)}, \ldots, r^{(J)}$ of iid samples of R. In other words, we obtain realizations $r^{(j)} \sim \Pi$. As we will see, we can generally perform such sampling even when an analytic formula for $\pi(r)$ is unavailable. Subsequently, we may use the sampled sequence of $r^{(j)}$ to estimate expectations

$$\langle V \rangle = \int dr\, g(r)\pi(r) \tag{5.1}$$

of any quantity $V = g(R)$ that is a function of R. Of course, if V contains discrete components, the integral in Eq. (5.1) reduces to an appropriate sum. Although we will not explicitly discriminate between integrals and sums, our description remains valid in both cases.

Since $\langle V \rangle$ in Eq. (5.1) is the mean of a transformed random variable, $V = g(R)$, we use $v^{(j)} = g\left(r^{(j)}\right)$ to compute samples of this random variable and subsequently use them to evaluate $\langle V \rangle$. Formally, this means that with MC we consider the approximation

$$\underbrace{\langle V \rangle}_{\substack{\text{desired}\\ \text{expectation}}} \approx \frac{1}{J}\sum_{j=1}^{J} v^{(j)} = \underbrace{\frac{1}{J}\sum_{j=1}^{J} g\left(r^{(j)}\right)}_{\substack{\text{sample}\\ \text{mean}}}. \tag{5.2}$$

In this approximation, the sample mean of $g\left(r^{(j)}\right)$ approximates the expectation of $\langle V \rangle$ sought. In a similar manner, when the full distribution of V is desired, this may be constructed by binning the samples $v^{(j)}$ forming a histogram instead of using them to compute only a point statistic as in Eq. (5.2).

Example 5.1 MC for the conjugate normal-gamma model

Suppose we have normally distributed observations $w_{1:N}$. Our task is to estimate the center and the spread of the underlying distribution. For this, we can consider a parametrization of the normal distribution by mean μ and precision τ. On μ and τ, we place the normal-gamma prior seen earlier in Example 4.8. Namely, our Bayesian model reads

$$\tau \sim \text{Gamma}(\alpha, \beta),$$
$$\mu|\tau \sim \text{Normal}\left(\xi, \frac{1}{\psi_0 \tau}\right), \tag{5.3}$$
$$w_n|\mu, \tau \sim \text{Normal}\left(\mu, \frac{1}{\tau}\right),$$

where $\alpha, \beta, \xi, \psi_0$ are hyperparameters that, in this example, assume known values. Within this setup, our model is described by the random variable $\boldsymbol{r} = \{\mu, \tau\}$ and the associated posterior reads

$$\pi(\boldsymbol{r}) = p(\mu, \tau|w_{1:N}).$$

Due to the factorization $p(\mu, \tau|w_{1:N}) = p(\mu|\tau, w_{1:N})p(\tau|w_{1:N})$, we may compute samples from $\pi(\boldsymbol{r})$ using *ancestral sampling*. For instance, first generating $\tau^{(j)}$ from $p(\tau|w_{1:N})$ and then generating $\mu^{(j)}$ from $p\left(\mu|\tau^{(j)}, w_{1:N}\right)$. These distributions are given by

$$p(\tau|w_{1:N}) = \text{Gamma}\left(\tau; \alpha + \frac{N}{2}, \frac{1}{\frac{1}{\beta} + \frac{1}{2}\frac{\psi_0}{\psi_0+1}\sum_{n=1}^{N}(w_n - \xi)^2}\right), \tag{*5.4*}$$

$$p\left(\mu|\tau, w_{1:N}\right) = \text{Normal}\left(\mu; \frac{\psi_0\xi + \sum_{n=1}^{N} w_n}{\psi_0 + N}, \frac{1}{\psi_0 + N}\frac{1}{\tau}\right). \tag{*5.5*}$$

Once a sufficient number, J, of samples $\boldsymbol{r}^{(j)}$ are generated through Eqs. (*5.4*) and (*5.5*), we may then apply the MC technique described above. For instance, with our MC samples, we may:

- Wish to visualize the posterior distribution of μ and τ. In this case, we may use $g(\boldsymbol{r}) = (\mu, \tau)$ with random variable $\boldsymbol{V} = (\mu, \tau)$. Here, a simple two-dimensional histogram of $\text{v}^{(j)}$ would reveal the entire shape of our posterior $p(\mu, \tau|w_{1:N})$.
- Compute the expectation of μ for which we would have to use another functional form for $g(\boldsymbol{r})$. In particular, since

$$\left(\text{Mean of } \mu\right) = \langle\mu\rangle = \int d\boldsymbol{r}\, \mu\, \pi(\boldsymbol{r}),$$

we would have to use $g(\boldsymbol{r}) = \mu$, and so the above expectation would be approximated by

$$\langle\mu\rangle \approx \frac{1}{J}\sum_{j=1}^{J}\mu^{(j)}. \tag{5.6}$$

* Reminder: The asterisks by some equations indicates that the detailed derivation can be found in Appendix F.

- Compute the variance of μ, which corresponds to the expectation

$$\left(\text{Variance of } \mu\right) = \left\langle (\mu - \langle\mu\rangle)^2 \right\rangle = \int d\boldsymbol{r} \, \left(\mu - \langle\mu\rangle\right)^2 \pi(\boldsymbol{r}).$$

In this case, we would use $g(\boldsymbol{r}) = (\mu - \langle\mu\rangle)^2$, and to obtain $\langle\mu\rangle$ in the first place, we would need to use Eq. (5.6). This choice leads to the approximation

$$\left\langle (\mu - \langle\mu\rangle)^2 \right\rangle \approx \frac{1}{J} \sum_{j=1}^{J} \left(\mu^{(j)} - \langle\mu\rangle\right)^2.$$

As an illustrative example, we consider a total of 10 observations, $w_{1:10}$, generated through Normal(5, 1). Suppose the parameters, *i.e.*, mean and variance, of the generating distribution are unknown and we need to infer them from the available observations as described above.

For simplicity, we may make some vague choices for the hyperparameters as $\alpha = 2, \beta = 1, \xi = 2, \psi_0 = 1$ and generate a total of $J = 120$ samples according to the ancestral sampling scheme of Eqs. (*5.4*) and (*5.5*). These are shown in Fig. 5.1 (upper panel). For sake of a comparison, we also show the prior over μ, τ. To obtain a more qualitative picture, Fig. 5.1 (lower panel) compares both prior $p(\mu, \tau)$ and posterior $p(\mu, \tau | w_{1:10})$ densities after the samples $\mu^{(j)}, \tau^{(j)}$ have been binned to produce a two-dimensional histogram. According to the posterior $p(\mu, \tau | w_{1:N})$, an estimate of the center of our observations is offered by the expectation of the mean μ, given by

$$\langle\mu\rangle \approx \frac{1}{J} \sum_{j=1}^{J} \mu^{(j)} = 4.50,$$

while an estimate of the spread is offered by the expectation of the standard deviation $\sigma = \sqrt{1/\tau}$, given by

$$\langle\sigma\rangle = \left\langle \sqrt{\frac{1}{\tau}} \right\rangle \approx \frac{1}{J} \sum_{j=1}^{J} \sqrt{\frac{1}{\tau^{(j)}}} = 1.04.$$

We see in Example 5.1 that once a sequence of samples $r^{(j)}$ is computed, the remaining MC choices, such as selecting the functional form of $g(r)$, are straightforward. However, unlike this example, where a computational recipe for sampling the random variable of interest R was already available, *e.g.*, Eqs. (*5.4*) and (*5.5*), generally obtaining samples from an arbitrary target $\pi(r)$ is an open problem with multiple solutions.

For this reason, we now focus the remainder of this chapter on the development of generic sampling schemes used in the simulation of sequences $r^{(j)}$. In particular, we will describe a particular class of MC methods, namely Markov chain Monte Carlo (MCMC), best suited for inference problems.

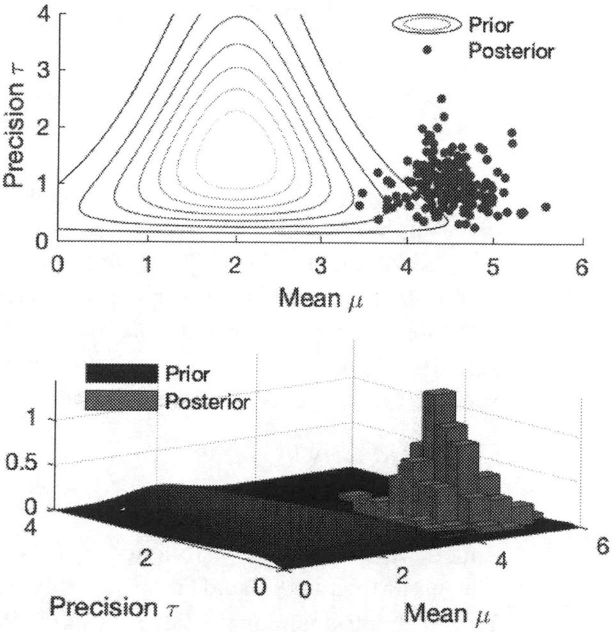

Fig. 5.1 Illustration of Example 5.1. (Top) Prior over $p(\mu, \tau)$ (contours) and MC samples $\mu^{(j)}, \tau^{(j)}$ from the posterior $p(\mu, \tau | w_{1:N})$ (dots). (Bottom) Comparison of prior $p(\mu, \tau)$ (surface) and posterior $p(\mu, \tau | w_{1:N})$ (histogram).

Note 5.2 What is a Markov chain Monte Carlo method?

MC methods are commonly used to sample complicated targets for simulation and data analysis. The idea is that, even if we are unable to evaluate the distribution of interest analytically, we generate random samples from this distribution and, as we will see shortly, use these samples to derive relevant quantities, such as point statistics and histograms.

MCMC algorithms, however, do not generate independent samples from the distribution of interest, but rather generate Markov chains of samples. This is the reason for its appellation as *Markov chain* Monte Carlo (MCMC). The main distinction between MC and MCMC is that the former generates independent samples, while, as will see shortly, the latter generates dependent samples, where dependent samples are less preferred but sometimes unavoidable.

5.1.2 Markov Chain Monte Carlo Methods

As is suggested by their name, MCMC simulations are performed in a sequential manner. Namely, each new sample $r^{(j+1)}$ of our targeted random variable $R \sim \mathbb{\Pi}$ is generated in a stochastic way *depending* on its immediate previous sample $r^{(j)}$. Formally, this means that every MCMC method relies on successive samples,

$$r^{(0)} \rightarrow r^{(1)} \rightarrow r^{(2)} \rightarrow \cdots \rightarrow r^{(j)} \rightarrow r^{(j+1)} \rightarrow \cdots \rightarrow r^{(J)},$$

forming a Markov chain. As such, our focus now shifts to the description of appropriate implementation schemes for the transitions $\cdots \rightarrow r^{(j-1)} \rightarrow r^{(j)} \rightarrow r^{(j+1)} \rightarrow \cdots$ of this chain. Similar to MC, MCMC samples $r^{(j)}$ carry the same statistical value as the target distribution itself, $\mathbb{\Pi}$, and can be generated even when we are unable to evaluate $\pi(r)$ directly as, for example, when we lack a tractable formula. However, unlike *uncorrelated* MC samples, MCMC samples are *correlated.* For this reason, to use approximations like Eq. (5.2) we need to ensure that our Markov chain satisfies certain requirements highlighted below.

Before delving into finer details, we first lay down some prerequisites. Similar to the ideas we encountered in Chapter 2, Markov chains are best studied as sequences of random variables,

$$R^{(0)} \rightarrow R^{(1)} \rightarrow R^{(2)} \rightarrow \cdots \rightarrow R^{(j)} \rightarrow R^{(j+1)} \rightarrow \cdots \rightarrow R^{(J)},$$

and from this perspective, each MCMC sample $r^{(j)}$ is a realization of a random variable $R^{(j)}$ in this sequence. To simplify our presentation from now on, instead of $R^{(j)}$ and $R^{(j+1)}$, we will use the more intuitive R^{old} and R^{new} to denote a random variable in this chain and its immediate successor. Similarly, we will be using r^{old} and r^{new} to denote realized values of these variables.

We designate a sample's value r as *feasible* when it is allowed by the target, *i.e.*, $\pi(r) > 0$. We will designate a chain as *ergodic* when its samples, in the long run, approach any feasible value, *i.e.*, as the chain grows larger, formally at the limit $J \rightarrow \infty$, some of its samples pass or come arbitrarily close to any feasible r.

Similar to any Markov chain, the transition rules leading from one sample to the next take the form of conditional probability densities $p(r^{\text{new}}|r^{\text{old}})$. These densities quantify how likely a current sample r^{old} is to move to the next r^{new} and generally say little about how the chain got into r^{old}. To avoid any notational confusion, we use $T_{r^{\text{old}}}(r^{\text{new}})$ to denote the transition densities of our MCMC chain and $\mathbb{T}_{r^{\text{old}}}$ to denote their corresponding distributions, *i.e.*, in the convention we are using from now on $T_{r^{(j)}}\left(r^{(j+1)}\right) = p\left(r^{(j+1)}|r^{(j)}\right)$ and $R^{(j+1)}|r^{(j)} \sim \mathbb{T}_{r^{(j)}}$. Additionally, we continue to only sparingly distinguish explicitly between random variables R^{old}, R^{new} and their realizations r^{old}, r^{new}, as what we mean can be deduced from the context.

As may already be clear, we will need to impose some requirements on the target $\pi(r)$. Before we list these requirements in detail, we introduce some definitions. A probability density $\pi^*(r)$ is termed *stationary* or *invariant* under the transitions $T_{r^{\text{old}}}(r^{\text{new}})$ when it satisfies the *full balance* condition

$$\pi^*(r^{\text{new}}) = \int dr^{\text{old}}\, T_{r^{\text{old}}}(r^{\text{new}}) \pi^*(r^{\text{old}}).$$

Similarly, a density $\pi^*(r)$ is termed *reversible* under $T_{r^{\text{old}}}(r^{\text{new}})$ when it satisfies the *detailed balance* condition

$$T_{r^{\text{old}}}(r^{\text{new}})\pi^*(r^{\text{old}}) = T_{r^{\text{new}}}(r^{\text{old}})\pi^*(r^{\text{new}}).$$

We now show that detailed balance implies full balance. For instance, provided r^{old} is sampled from $\pi^*(r)$, we get

$$\begin{aligned}
p(r^{\text{new}}) &= \int dr^{\text{old}}\, p(r^{\text{new}}, r^{\text{old}}) = \int dr^{\text{old}}\, p(r^{\text{new}}|r^{\text{old}})p(r^{\text{old}}) \\
&= \int dr^{\text{old}}\, T_{r^{\text{old}}}(r^{\text{new}})\pi^*(r^{\text{old}}) = \int dr^{\text{old}}\, T_{r^{\text{new}}}(r^{\text{old}})\pi^*(r^{\text{new}}) \\
&= \pi^*(r^{\text{new}}) \int dr^{\text{old}}\, T_{r^{\text{new}}}(r^{\text{old}}) \\
&= \pi^*(r^{\text{new}}).
\end{aligned}$$

Consequently, detailed balance is a *stronger* condition than full balance.

5.1.3 Monte Carlo Markov Chain Requirements

With all prerequisites established, we are now ready to see the requirements of a valid MCMC method.

Note 5.3 MCMC requirements

To generate MCMC samples $r^{(j)}$ from a target $R \sim \Pi$, we compute a Markov chain

$$r^{(0)} \to \cdots \to r^{(j-1)} \to r^{(j)} \to r^{(j+1)} \to \cdots$$

such that: (1) it has a stationary distribution; (2) this distribution is unique; and, (3) it coincides with our target.

These requirements are fairly general and, in practice, we ensure them by developing MCMC schemes generating chains under more restrictive conditions. Namely,

- *Our chain must pass from a feasible sample.* This is met as long as the initial or any subsequent sample is a feasible one.
- *Our chain must be ergodic.* This is met as long as $T_{r^{\text{old}}}(r^{\text{new}})$ allows for successive transitions between every pair of feasible values.
- *Our chain must have a stationary density.* This is met as long as $T_{r^{\text{old}}}(r^{\text{new}})$ is in full balance with some density $\pi^*(r)$. Besides ensuring that requirement (1) is met, this condition combined with initialization and ergodicity, also ensures that requirement (2) is met.
- *Our stationary density must coincide with the target.* This is met as long as $T_{r^{\text{old}}}(r^{\text{new}})$ is in detailed balance with our target $\pi(r)$. This condition ensures that requirement (3) is met.

These conditions are termed: *feasibility, irreducibility, invariance, and reversibility,* respectively.

In practice, since detailed balance implies full balance, it is *only* necessary to ensure that our MCMC scheme satisfies the feasibility, irreducibility, and reversibility conditions (*i.e.*, invariance is implied by imposing reversibility).

Generally, feasibility poses no difficulties since the initial sample $r^{(0)}$ in an MCMC scheme is specified in advance independently of the subsequent transitions and its feasibility can be readily verified. The same is also true of the irreducibility condition, which, aside from cases involving pathological transition rules $T_{r^{\text{old}}}(r^{\text{new}})$, is typically met or at least can be readily verified. However, reversibility is usually difficult to satisfy for arbitrary targets $\pi(r)$. So, in the next sections, we describe the most common MCMC schemes, or *samplers* as they are most commonly termed, for which reversibility is ensured by construction.

5.2 Basic MCMC Samplers

In this section we present two general strategies for obtaining MCMC samplers. Namely, those that rely on the *Metropolis–Hastings algorithm* and those that rely on *Gibbs sampling*. In essence, both strategies use the same theoretical foundations. However, since their implementation differs, we discuss them separately.

As we will see, in both strategies the target density $\pi(r)$ need not be normalized. So we may apply them even if the target is specified only up to a multiplicative constant, *i.e.*, a factor that *does not* depend on r, as is usually the case in Bayesian inference. To denote unnormalized targets we write $\bar{\pi}(r)$. Given $\bar{\pi}(r)$, we can readily recover the corresponding normalized one by $\pi(r) = \bar{\pi}(r)/\int dr'\, \bar{\pi}(r')$ and, for this reason, we use $\pi(r)$ and $\bar{\pi}(r)$ interchangeably.

Note 5.4 Unnormalized targets

When we operate on an unnormalized target, $\bar{\pi}(r)$, it is critical to recall that this target is associated with a probability distribution $\mathbb{\Pi}$. Accordingly, we must ensure that our $\bar{\pi}(r)$ passes a simple sanity check, namely, that it is *normalizable* in principle. This holds as long as

$$0 < \int dr\, \bar{\pi}(r) < \infty,$$

with *both inequalities* being important. A $\bar{\pi}(r)$ that fails to meet this condition is meaningless alongside any results derived from it.

Indeed, many pathologies encountered in Bayesian models originate from the usage of uniform priors over unbounded domains, such as the entire real line or only the positive half of it, for which the second inequality does not hold.

5.2.1 Metropolis–Hastings Family of Samplers

Samplers in this family can be used to generate MCMC samples $r^{(j)}$ from virtually any random variable $R \sim \mathbb{\Pi}$. These include univariate and multivariate distributions.

Metropolis–Hastings Sampler

Given a target $\bar{\pi}(r)$, to begin a Metropolis–Hastings sampler a choice for the initial sample $r^{(0)}$ must be made. This can be achieved either by assigning a fixed value or sampling a value from a specified probability distribution which need not necessarily be Π. In the later case, the chosen distribution must exclude infeasible values. Irrespective of how $r^{(0)}$ is computed, the sampler remains valid as long as $\bar{\pi}\left(r^{(0)}\right) > 0$. Next, $r^{(0)}$ is used to generate a subsequent sample, $r^{(1)}$, which, in turn, is used to generate $r^{(2)}$, and so on.

To advance the MCMC chain, the Metropolis–Hastings sampler uses a *proposal* density $p(r^{\text{prop}}|r^{\text{old}})$. For clarity, we denote the proposal density by $Q_{r^{\text{old}}}(r^{\text{prop}}) = p(r^{\text{prop}}|r^{\text{old}})$ and the associated distribution by $\mathbb{Q}_{r^{\text{old}}}$. The proposal density can be almost arbitrary since its only requirements are that the simulation of random variables $r^{\text{prop}}|r^{\text{old}} \sim \mathbb{Q}_{r^{\text{old}}}$ is *possible* and it allows the generation of *any* feasible value.

To advance from an existing sample r^{old} to the subsequent one r^{new} in the MCMC chain, the Metropolis–Hastings sampler first generates a proposal sample by $r^{\text{prop}}|r^{\text{old}} \sim \mathbb{Q}_{r^{\text{old}}}$ and then conducts a test based on the ratio

$$A_{r^{\text{old}}}(r^{\text{prop}}) = \underbrace{\frac{\bar{\pi}(r^{\text{prop}})}{\bar{\pi}(r^{\text{old}})}}_{\text{target}} \underbrace{\frac{Q_{r^{\text{prop}}}(r^{\text{old}})}{Q_{r^{\text{old}}}(r^{\text{prop}})}}_{\text{proposal}} \tag{5.7}$$

to check whether r^{prop} is *retained or discarded.* Most commonly, for the test a random number $u \sim \text{Uniform}_{[0,1]}$ is generated, then:

- If $u \leq A_{r^{\text{old}}}(r^{\text{prop}})$, the proposal is accepted and so the new sample is $r^{\text{new}} = r^{\text{prop}}$.
- If $u > A_{r^{\text{old}}}(r^{\text{prop}})$, the proposal is rejected and so the new sample is $r^{\text{new}} = r^{\text{old}}$.

Once r^{new} is obtained in this way, either through acceptance or rejection of r^{prop}, the Metropolis–Hastings sampler applies the same process again and again, until we reach the desired number of samples. In Algorithm 5.1 we summarize a computational implementation.

Algorithm 5.1 Metropolis–Hastings sampler for arbitrary targets

Given a target $\bar{\pi}(r)$, a proposal $Q_{r^{\text{old}}}(r^{\text{prop}})$, and a feasible initial sample r^{old}, the Metropolis–Hastings sampler proceeds by iterating the following:

- Generate a proposal $r^{\text{prop}} \sim \mathbb{Q}_{r^{\text{old}}}$.
- Compute the acceptance ratio $A_{r^{\text{old}}}(r^{\text{prop}})$.
- Generate $u \sim \text{Uniform}_{[0,1]}$.
- If $u < A_{r^{\text{old}}}(r^{\text{prop}})$ set $r^{\text{new}} = r^{\text{prop}}$, otherwise set $r^{\text{new}} = r^{\text{old}}$.

We emphasize that, at every iteration, whenever the proposal r^{prop} is rejected, it is necessary to maintain r^{old}. In other words, in a correct implementation of the Metropolis–Hastings sampler, following a rejection, we *must* use $r^{\text{new}} = r^{\text{old}}$. If we neglect such repetition, the sampler fails to provide correct results.

Example 5.2 Two Metropolis–Hastings schemes for the truncated normal distribution

Consider a random variable R distributed according to a normal distribution with mean μ and variance σ^2 *truncated* below 0. That is, R has a probability density given by

$$\pi(r) \propto \bar{\pi}(r) = \begin{cases} \frac{1}{\sqrt{2\pi\sigma^2}} \exp\left(-\frac{(r-\mu)^2}{2\sigma^2}\right), & r \geq 0 \\ 0, & r < 0 \end{cases}.$$

It might appear surprising but, despite its simplicity, there are no standard ways of sampling from this target. Thus, to draw samples $r^{(j)}$ from $\pi(r)$ we may develop a Metropolis–Hastings sampler.

One convenient choice of the proposal is offered by $\mathbb{Q}_{r^{\text{old}}} = \text{Normal}\left(r^{\text{old}}, \lambda^2\right)$; that is, a normal with mean on the previous sample r^{old} and a preset variance λ^2. This choice of proposal leads to the acceptance ratio

$$A_{r^{\text{old}}}(r^{\text{prop}}) = \begin{cases} \exp\left(\frac{(r^{\text{old}}-\mu)^2-(r^{\text{prop}}-\mu)^2}{2\sigma^2}\right), & r^{\text{prop}} \geq 0, \\ 0, & r^{\text{prop}} < 0. \end{cases}$$

Of course, when implementing Algorithm 5.1 we need not consider cases with $r^{\text{old}} < 0$ in the acceptance ratio since r^{old} is ensured to be positive already. If this was not true, an acceptance of an infeasible value in the previous iteration or an infeasible initialization must have occurred, neither of which is allowed.

One possible drawback of the normal proposal used above is that it may often propose negative values r^{prop}, especially if μ is close to 0, and so it may lead to considerable rejections. A different choice that avoids such unnecessary rejections is offered by a $\mathbb{Q}_{r^{\text{old}}} = \text{Gamma}(\alpha, r^{\text{old}}/\alpha)$; that is, a gamma distribution with mean on the previous sample r^{old} and a preset shape α, which is ensured to propose only positive values. This choice leads to the acceptance ratio

$$A_{r^{\text{old}}}(r^{\text{prop}}) = \begin{cases} \exp\left(\frac{(r^{\text{old}}-\mu)^2-(r^{\text{prop}}-\mu)^2}{2\sigma^2} + \alpha\left(\frac{r^{\text{prop}}}{r^{\text{old}}} - \frac{r^{\text{old}}}{r^{\text{prop}}}\right)\right)\left(\frac{r^{\text{prop}}}{r^{\text{old}}}\right)^{1-2\alpha}, & r^{\text{prop}} \geq 0, \\ 0, & r^{\text{prop}} < 0. \end{cases}$$

Figure 5.2 illustrates MCMC chains generated from both choices. Here, the target distribution is obtained with $\mu = 1$, $\sigma = 1$ and the proposals are implemented with $\lambda = \sqrt{0.2}$ and $\alpha = 4$. For both cases, sampling starts from $r^{(0)} = 1$ and continues for a total of $J = 10^5$ samples. As can be seen, independently of the proposal choice the target distribution is well sampled by both MCMC chains.

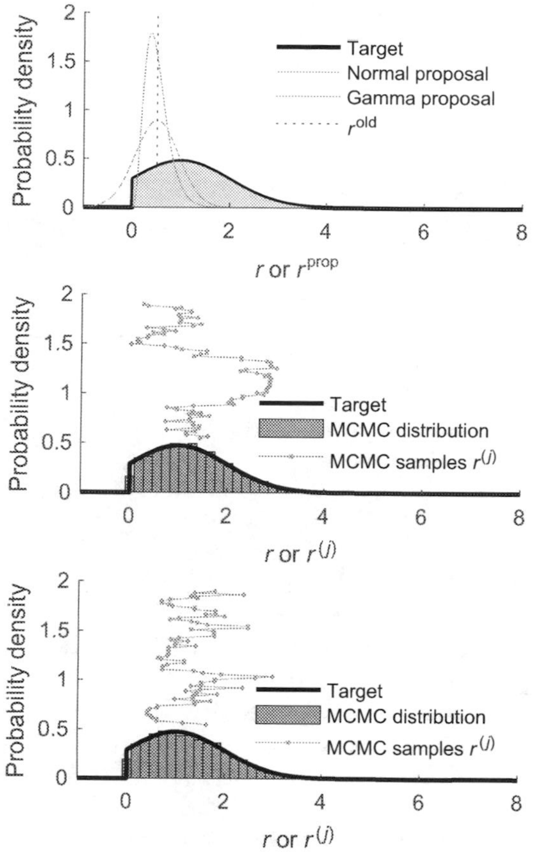

Fig. 5.2 Illustration of Example 5.2. (Top) Truncated normal target $\pi(r)$ with normal and gamma proposals $Q_{r^{old}}(r^{prop})$. (Middle) MCMC approximation of $\pi(r)$ produced by normal proposals. (Bottom) MCMC approximation of $\pi(r)$ produced by gamma proposals. Middle and bottom panels also show small segments of the generated Markov chains.

Why Does the Metropolis–Hastings Sampler Work?*

Given a target $\pi(r)$ and a proposal of choice $Q_{r^{old}}(r^{prop})$, the Metropolis–Hastings sampler ensures that the MCMC chain generated visits each feasible r with a frequency proportional to $\pi(r)$. This means that the sampler, as it passes from sample to sample, balances out two competing effects: how often the chain stays in each r; and, once leaving an r, how soon the chain returns to r. As we can deduce from the acceptance test, the ratio $A_{r^{old}}(r^{prop})$ determines whether the sampler accepts a proposal so the chain moves to a different value, or rejects the proposal and the chain retains its value. Accordingly, $A_{r^{old}}(r^{prop})$ plays an important role in setting the precise balance. Essentially, the acceptance test is a bookkeeping mechanism. Intuitively, in Eq. (5.7), the first term $\bar{\pi}(r^{prop})/\bar{\pi}(r^{old})$ balances the flow out of r^{old} and into r^{prop}, while the second term $Q_{r^{prop}}(r^{old})/Q_{r^{old}}(r^{prop})$ balances the flow of passing from r^{old} to r^{prop} and getting back from r^{prop} to r^{old}.

* This is an advanced topic and could be skipped on a first reading.

To advance a deeper understanding of the sampler, we now explain the bookkeeping mechanism in detail. Specifically, recalling Note 5.3, we verify the *reversibility* condition. For this, we explicitly show that the chain's transition rules $T_{r^{\text{old}}}(r^{\text{new}})$ are in *detailed balance* with our target $\pi(r)$.

In order first to determine the transition rules $T_{r^{\text{old}}}(r^{\text{new}})$, we need to consider all possibilities for r^{prop}. In other words, we need to complete the transition over R^{prop}. Formally, the steps are

$$
\begin{aligned}
T_{r^{\text{old}}}(r^{\text{new}}) &= p(r^{\text{new}}|r^{\text{old}}) \\
&= \int dr^{\text{prop}}\, p(r^{\text{new}}, r^{\text{prop}}|r^{\text{old}}) \\
&= \int dr^{\text{prop}}\, p(r^{\text{new}}|r^{\text{prop}}, r^{\text{old}}) p(r^{\text{prop}}|r^{\text{old}}) \\
&= \int dr^{\text{prop}}\, p(r^{\text{new}}|r^{\text{prop}}, r^{\text{old}}) Q_{r^{\text{old}}}(r^{\text{prop}}).
\end{aligned}
$$

To proceed any further, we must determine $p(r^{\text{new}}|r^{\text{prop}}, r^{\text{old}})$, which depends on the outcome of the acceptance test. Due to the way the test is performed, the probability of success $\alpha_{r^{\text{old}}}(r^{\text{prop}})$ is a function of both r^{old} and r^{prop}, and is given by

$$
\alpha_{r^{\text{old}}}(r^{\text{prop}}) = \begin{pmatrix} \text{probability of} \\ u \leq A_{r^{\text{old}}}(r^{\text{prop}}) \end{pmatrix} = \min\left(1, A_{r^{\text{old}}}(r^{\text{prop}})\right).
$$

Similarly, the probability of failure is also a function of both r^{old} and r^{prop} and is given by $1 - \alpha_{r^{\text{old}}}(r^{\text{prop}})$.

Note 5.5 Acceptance probabilities

In Exercise 5.6, we ask that the reader prove an important identity,

$$
\pi(r)\alpha_r(r') Q_r(r') = \pi(r')\alpha_{r'}(r) Q_{r'}(r),
$$

valid for any feasible r and r'. We use this identity to show that the Metropolis–Hastings sampler meets the reversibility condition.

Since there are only two options for $r^{\text{new}}|r^{\text{prop}}, r^{\text{old}}$, we reach

$$
p(r^{\text{new}}|r^{\text{prop}}, r^{\text{old}}) = \underbrace{\alpha_{r^{\text{old}}}(r^{\text{prop}})\delta_{r^{\text{prop}}}(r^{\text{new}})}_{\text{acceptance}} + \underbrace{(1 - \alpha_{r^{\text{old}}}(r^{\text{prop}}))\,\delta_{r^{\text{old}}}(r^{\text{new}})}_{\text{rejection}}.
$$

Now, with $p(r^{\text{new}}|r^{\text{prop}}, r^{\text{old}})$ derived explicitly, we may continue with the transition rule

$$
\begin{aligned}
T_{r^{\text{old}}}(r^{\text{new}}) &= \alpha_{r^{\text{old}}}(r^{\text{new}}) Q_{r^{\text{old}}}(r^{\text{new}}) \\
&\quad + \delta_{r^{\text{old}}}(r^{\text{new}}) \int dr^{\text{prop}}\, (1 - \alpha_{r^{\text{old}}}(r^{\text{prop}}))\, Q_{r^{\text{old}}}(r^{\text{prop}}).
\end{aligned}
$$

The form of $T_{r^{\text{old}}}(r^{\text{new}})$ we have derived can be used to verify that the transition rules are in detailed balance with the target. In particular, in view of Note 5.5 we obtain

$$\begin{aligned}
T_{r^{\text{old}}}(r^{\text{new}})\pi(r^{\text{old}}) &= \pi(r^{\text{old}})\alpha_{r^{\text{old}}}(r^{\text{new}})Q_{r^{\text{old}}}(r^{\text{new}}) \\
&\quad + \pi(r^{\text{old}})\delta_{r^{\text{old}}}(r^{\text{new}})\int dr^{\text{prop}}\left(1-\alpha_{r^{\text{old}}}(r^{\text{prop}})\right)Q_{r^{\text{old}}}(r^{\text{prop}}) \\
&= \pi(r^{\text{new}})\alpha_{r^{\text{new}}}(r^{\text{old}})Q_{r^{\text{new}}}(r^{\text{old}}) \\
&\quad + \pi(r^{\text{new}})\delta_{r^{\text{new}}}(r^{\text{old}})\int dr^{\text{prop}}\left(1-\alpha_{r^{\text{new}}}(r^{\text{prop}})\right)Q_{r^{\text{new}}}(r^{\text{prop}}) \\
&= T_{r^{\text{new}}}(r^{\text{old}})\pi(r^{\text{new}}).
\end{aligned}$$

Sampling of Posterior Targets

The Metropolis–Hastings sampler can be applied to yield samples from any target $\bar{\pi}(r)$. In particular, such a target may be a Bayesian model's posterior. For this special case, the acceptance ratio in Eq. (5.7) becomes

$$A_{r^{\text{old}}}(r^{\text{prop}}) = \underbrace{\frac{p(w|r^{\text{prop}})}{p(w|r^{\text{old}})}}_{\text{likelihoods}}\ \underbrace{\frac{p(r^{\text{prop}})}{p(r^{\text{old}})}}_{\text{priors}}\ \underbrace{\frac{Q_{r^{\text{prop}}}(r^{\text{old}})}{Q_{r^{\text{old}}}(r^{\text{prop}})}}_{\text{proposals}}, \tag{5.8}$$

for which we have used Bayes' rule, $\pi(r) = p(r|w) \propto p(w|r)p(r)$, to factorize the posterior in the product of the likelihood $p(w|r)$ and prior $p(r)$.

Note 5.6 Initialization of the Metropolis–Hastings sampler

For Bayesian applications, a convenient initialization is to sample $r^{(0)}$ directly from the prior. As our priors are most often described by hierarchical or generative models, their sampling can typically be carried out directly.

Often, individual terms in the acceptance ratio, especially when the supplied datasets are large, become extremely small. For such cases, to avoid numerical *underflow* the acceptance test must be performed in logarithmic space. For example, we must directly compute the ratio's logarithm,

$$L_{r^{\text{old}}}(r^{\text{prop}}) = \log\frac{p(w|r^{\text{prop}})}{p(w|r^{\text{old}})} + \log\frac{p(r^{\text{prop}})}{p(r^{\text{old}})} + \log\frac{Q_{r^{\text{prop}}}(r^{\text{old}})}{Q_{r^{\text{old}}}(r^{\text{prop}})},$$

and, after generating $u \sim \text{Uniform}_{[0,1]}$, perform the acceptance test as:

- If $\log u \leq L_{r^{\text{old}}}(r^{\text{prop}})$, the proposal is accepted and the new sample is $r^{\text{new}} = r^{\text{prop}}$.
- If $\log u > L_{r^{\text{old}}}(r^{\text{prop}})$, the proposal is rejected and the new sample is $r^{\text{new}} = r^{\text{old}}$.

Example 5.3 MCMC for the conjugate normal-gamma model

To demonstrate the application of the Metropolis–Hastings sampler in a Bayesian context, we consider the same setting as in Example 5.1. Specifically, observations $w_{1:N}$ are normally distributed and our task is to estimate the center and the spread of the generating distribution. As before, parametrizing the underlying normal by mean μ and precision τ and placing a normal-gamma prior on them, we arrive at the model

$$\tau \sim \text{Gamma}(\alpha, \beta),$$
$$\mu|\tau \sim \text{Normal}\left(\xi, \frac{1}{\psi_0 \tau}\right),$$
$$w_n|\mu, \tau \sim \text{Normal}\left(\mu, \frac{1}{\tau}\right),$$

where $\alpha, \beta, \xi, \psi_0$ are hyperparameters of known values, as before.

To generate posterior samples $\left\{\mu^{(j)}, \tau^{(j)}\right\}$, this time using a Metropolis–Hastings scheme, we may consider proposals of the form

$$\mathbb{Q}_{\mu^{\text{old}},\tau^{\text{old}}} = \text{Normal}_2\left(\left(\mu^{\text{old}}, \tau^{\text{old}}\right), \begin{pmatrix}\lambda_\mu^2, 0\\ 0, \lambda_\tau^2\end{pmatrix}\right).$$

That is, the proposals $\mu^{\text{prop}}, \tau^{\text{prop}}$ are sampled jointly from a bivariate normal with mean at the previous sample and some preset variances λ_μ^2 and λ_τ^2. With this choice, the acceptance log-ratio becomes

$$L_{\mu^{\text{old}},\tau^{\text{old}}}(\mu^{\text{prop}}, \tau^{\text{prop}}) = \begin{cases} \frac{1}{2}\sum_{n=1}^N \left(\tau^{\text{old}}(\mu^{\text{old}} - w_n)^2 - \tau^{\text{prop}}(\mu^{\text{prop}} - w_n)^2\right) \\ + \frac{N+2\alpha-1}{2}\log\frac{\tau^{\text{prop}}}{\tau^{\text{old}}} \\ + \frac{\psi_0}{2}\left(\tau^{\text{old}}(\mu^{\text{old}} - \xi)^2 - \tau^{\text{prop}}(\mu^{\text{prop}} - \xi)^2\right) \\ + \beta\left(\tau^{\text{old}} - \tau^{\text{prop}}\right), & \tau^{\text{prop}} > 0, \\ -\infty, & \tau^{\text{prop}} \le 0. \end{cases}$$

As an illustrative example, we consider a total of 10 observations $w_{1:10}$, generated through Normal(5, 1). Such observations are similar to Example 5.1. As before, we also use the same hyperparameters: $\alpha = 2, \beta = 1, \xi = 2, \psi_0 = 1$. A total of $J = 500$ samples are generated with $\lambda_\mu^2 = \lambda_\tau^2 = 1$, and shown in Fig. 5.3. As can be seen, although the sampler starts at a point $\mu^{(0)} = 1, \tau^{(0)} = 1$ at low posterior probability, the chain quickly moves towards and eventually remains near the posterior mode located around $\mu = 4.5, \tau = 1$.

The Metropolis–Hastings sampler is a powerful tool in Bayesian inference as it can be used to sample from *any* posterior irrespective of whether the prior and likelihood are conjugate. Nevertheless, its practical implementation is often challenging as the proposal $Q_{r^{\text{old}}}(r^{\text{prop}})$ used dictates the algorithm's performance. To wit, it is always possible to devise theoretically valid proposals, but a judicious choice and extensive calibration are typically critical requirements for the algorithm to sample the posterior in reasonable time. Unfortunately, with most naive choices the number of

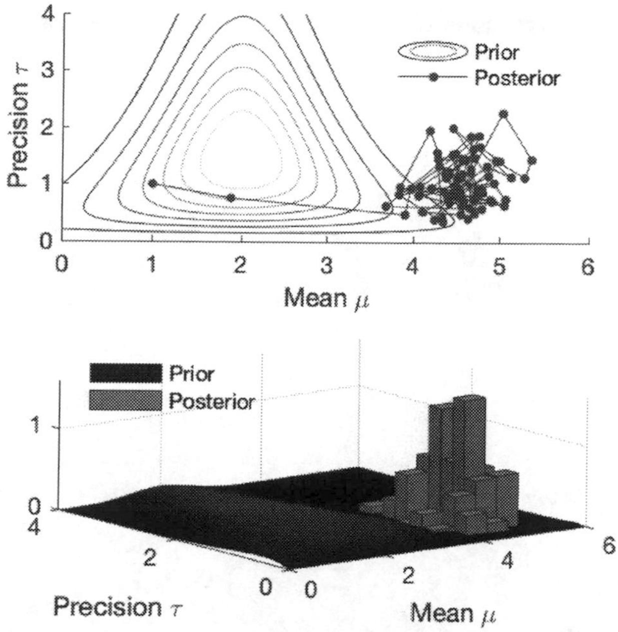

Fig. 5.3 Illustration of Example 5.3. (Top) Prior probability distribution $p(\mu, \tau)$ (contours) and MCMC samples $\left(\mu^{(j)}, \tau^{(j)}\right)$ from the posterior $p(\mu, \tau|w_{1:N})$ (dots). (Bottom) Comparison of prior $p(\mu, \tau)$ (surface) and posterior $p(\mu, \tau|w)$ (histogram).

MCMC samples, J, required to adequately sample a target may become inordinately large and, even for moderate scale applications, may involve infeasible computational cost.

Example 5.4 Choice of proposals in MCMC

To demonstrate just how important the proposal is in allowing the sampler to cover the entire target's support within few iterations, we consider a bivariate target $\pi(x, y)$ with a sum form

$$\Pi = 0.3\,\text{Normal}_2\left((-1,0), \begin{pmatrix} 0.25^2 & 0 \\ 0 & 0.25^2 \end{pmatrix}\right) + 0.3\,\text{Normal}_2\left((+1,0), \begin{pmatrix} 0.25^2 & 0 \\ 0 & 0.25^2 \end{pmatrix}\right) + 0.4\,\text{Normal}_2\left((0,+1), \begin{pmatrix} 0.25^2 & 0 \\ 0 & 0.25^2 \end{pmatrix}\right),$$

consisting of three well-separated modes. Although there exist far more efficient methods to simulate from such a target, here we consider a naive Metropolis–Hastings scheme with proposal

$$\mathbb{Q}_{x^{\text{old}},y^{\text{old}}} = \text{Normal}_2\left((x^{\text{old}}, y^{\text{old}}), \begin{pmatrix} 0.3^2 & 0 \\ 0 & 0.3^2 \end{pmatrix}\right).$$

Figure 5.4 illustrates MCMC approximations of the target produced by two different chains: a shorter chain containing $J = 5 \times 10^2$ samples and a longer

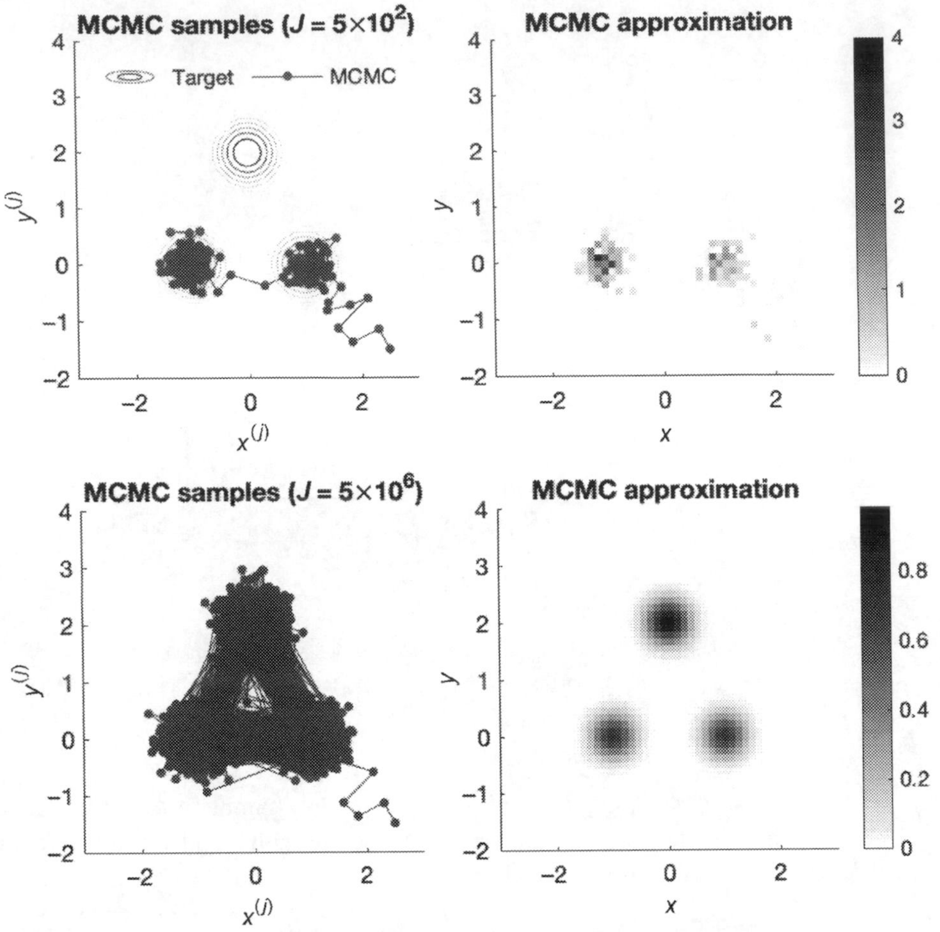

Fig. 5.4 Illustration of Example 5.4. Two Metropolis–Hastings MCMC chains are computed for the evaluation of the same target. For the chain and the corresponding histogram in the upper panels, only $J = 5 \times 10^2$ samplers are used. In contrast, the chain and the corresponding histogram in the lower panels contains $J = 5 \times 10^6$ samples. As can be seen, only the latter chain fully samples the target.

one containing $J = 5 \times 10^6$ samples. For this choice of proposal, only the latter chain samples all three modes.

For this illustrative example, each sampler iteration comes at low computational cost. As such, a large number of samples, such as $\approx 10^6$ in the second chain, can be obtained in reasonable time. However, in more complex applications, where each iteration may involve many costly operations, high numbers of samples may remain purely aspirational.

Metropolis Sampler

The Metropolis sampler is a special version of the Metropolis–Hastings sampler just seen and its implementation is identical to Algorithm 5.1. In this version of the sampler, the proposal $Q_{r^{\text{old}}}(r^{\text{prop}})$ is symmetric with respect to r^{old} and r^{prop}, *i.e.*, $Q_{r^{\text{old}}}(r^{\text{prop}}) = Q_{r^{\text{prop}}}(r^{\text{old}})$ and, due to this symmetry, the acceptance ratio simplifies to

$$A_{r^{\text{old}}}(r^{\text{prop}}) = \frac{\bar{\pi}(r^{\text{prop}})}{\bar{\pi}(r^{\text{old}})}.$$

Similarly, when the target $\bar{\pi}(r)$ is the posterior of a Bayesian model, Eq. (5.8) reduces to

$$A_{r^{\text{old}}}(r^{\text{prop}}) = \underbrace{\frac{p(w|r^{\text{prop}})}{p(w|r^{\text{old}})}}_{\text{likelihood}} \underbrace{\frac{p(r^{\text{prop}})}{p(r^{\text{old}})}}_{\text{prior}}.$$

Example 5.5 MCMC under a Cauchy prior

Consider scalar observations $w_{1:N}$ normally distributed around some unknown mean μ. For convenience, also assume that w_n and μ are scaled such that the variance is 1. In this example, the goal is to estimate μ, but instead of making the common prior choice, we use a much wider Cauchy distribution. Our entire model is

$$\mu \sim \text{Cauchy}(0, 1),$$
$$w_n|\mu \sim \text{Normal}(\mu, 1).$$

Under this model, the posterior is

$$\pi(\mu) = p(\mu|w_{1:N}) \propto \left[\prod_{n=1}^{N} p(w_n|\mu)\right] p(\mu) \propto \frac{\exp\left(-\frac{1}{2}\sum_{n=1}^{N}(w_n - \mu)^2\right)}{1 + \mu^2}.$$

Although we can easily visualize this posterior due to the analytic form above, it is still hard to compute point estimates from our posterior since we cannot analytically compute expectations with such a density. Instead, we may use a Metropolis sampler with proposals

$$\mathbb{Q}_{\mu^{\text{old}}} = \text{Normal}\left(\mu^{\text{old}}, \lambda^2\right)$$

to generate MCMC samples $\mu^{(j)}$ distributed according to $\pi(\mu)$. Once a significant number J of $\mu^{(j)}$ is obtained, using Eq. (5.2), we can compute any expectation, for example,

$$\langle\mu\rangle = \int_{-\infty}^{+\infty} d\mu\, \mu\pi(\mu) \approx \frac{1}{J}\sum_{j=1}^{J}\mu^{(j)}.$$

Figure 5.5 illustrates two MCMC chains obtained with different values of λ targeting the same posterior $\pi(\mu)$. Both chains start at $\mu^{(0)} = 1$ and include a total of $J = 200$ samples. As can be seen, a different total number of samples J is required to adequately cover the target's support for different values of λ.

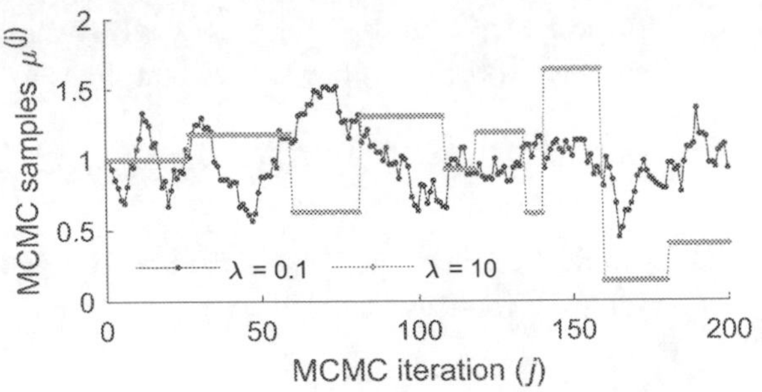

Fig. 5.5 Two Metropolis samplers on the same target. In the long run, both chains cover the entire support of the target. However, the total number of samples required differs substantially.

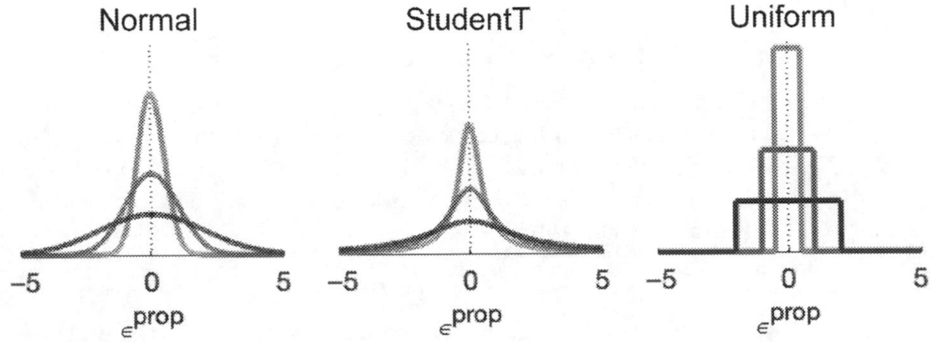

Fig. 5.6 Probability densities of common univariate Metropolis random walk perturbations.

Additive Random Walk Sampler

A straightforward approach to obtaining Metropolis proposals $\mathbb{Q}_{r^{\mathrm{old}}}$ is to consider local explorations around r^{old}. We used this idea in Examples 5.2–5.4 and here we formalize our description. A natural way to obtain such proposals is through (additive) *random walks*,

$$r^{\mathrm{prop}} = r^{\mathrm{old}} + \epsilon^{\mathrm{prop}},$$

where ϵ^{prop} is a random perturbation sampled from a density $G(\epsilon^{\mathrm{prop}})$ independent of r^{old}. For convenience, *univariate* perturbations ϵ^{prop} can be sampled from normal, StudentT, or uniform distributions. Maintaining symmetry requires $G(-\epsilon) = G(+\epsilon)$ and, for practical reasons, the perturbations need to be scaled. That is, in a unified setting convenient choices of scaling include

$$\frac{\epsilon^{\mathrm{prop}}}{\lambda} \sim \mathrm{Normal}(0,1), \qquad \frac{\epsilon^{\mathrm{prop}}}{\lambda} \sim \mathrm{StudentT}(\nu), \qquad \frac{\epsilon^{\mathrm{prop}}}{\lambda} \sim \mathrm{Uniform}_{[-1,+1]},$$

where $\lambda > 0$ is a scaling constant that characterizes the *spread* of the perturbation. In other words, λ controls the *volume* around r^{old} to be explored each time. These perturbations are demonstrated in Fig. 5.6. When a *multivariate* proposal r^{prop} is required, *e.g.*, as in Example 5.3, the univariate perturbations shown above can be readily generalized.

Note 5.7 What is *not* a Metropolis sampler?

So far we have mostly seen instances of Metropolis samplers. In most examples, the proposals $Q_{r^{\text{old}}}(r^{\text{prop}})$ have been of the special forms mentioned above. Naively, we may conclude that every distribution in these families might be suitable for a Metropolis sampler. To demonstrate that this is *not* true, we consider proposals of the form

$$\mathbb{Q}_{r^{\text{old}}} = \text{Normal}\left(r^{\text{old}}, (r^{\text{old}})^2\right).$$

Here, the spread of the proposal is not constant, *i.e.*, both the proposal's mean and variance depend on r^{old}. As a result, the proposal cannot be symmetric because, in general,

$$\text{Normal}\left(r^{\text{prop}}; r^{\text{old}}, (r^{\text{old}})^2\right) \neq \text{Normal}\left(r^{\text{old}}; r^{\text{prop}}, (r^{\text{prop}})^2\right).$$

As such, the ratio of the proposals $Q_{r^{\text{prop}}}(r^{\text{old}})/Q_{r^{\text{old}}}(r^{\text{prop}})$ does not drop from $A_{r^{\text{old}}}(r^{\text{old}})$ in Eq. (5.7) as needs to occur in a Metropolis sampler.

In most applications, it is more convenient to preselect the perturbation's density $G(\epsilon)$ and subsequently calibrate λ, which, as we show in Example 5.6, has an important effect on the resulting MCMC chain. Irrespective of calibration, however, random walk samplers can often empirically be seen to perform poorly on targets that depend on more than just a handful of variables.

Example 5.6 Scaling a MCMC random walk

We consider a univariate target $\pi(r)$ consisting of a sum of two normal densities,

$$\Pi = 0.7\,\text{Normal}(+1, 0.04) + 0.3\,\text{Normal}(-1, 0.04).$$

Although there exist more efficient choices to sample from such a target, here we consider a Metropolis random walk with uniform perturbations $\mathbb{G} = \text{Uniform}_{[-1,+1]}$. Under a scaling λ, the corresponding proposals are given by

$$\mathbb{Q}_{r^{\text{old}}} = \text{Uniform}_{[r^{\text{old}}-\lambda, r^{\text{old}}+\lambda]}.$$

Figure 5.7 shows three MCMC chains obtained with $\lambda = 0.5, 5, 50$. To facilitate a comparison, we start all chains at $r^{(0)} = 1$ and continue for up to $J = 500$ samples.

As can be seen, the chains with low λ move very slowly through the support of $\pi(r)$, while those with higher λ show higher motility. Consequently, within the first 500 samples computed, only the last two chains explored a region large enough to cover the entire support of $\pi(r)$. This is a common characteristic of random walks and it is particularly pronounced when our target consists of more than one mode separated by regions of low probability. In such cases, extremely large numbers of samples J may be required until the chain fully explores the entire posterior.

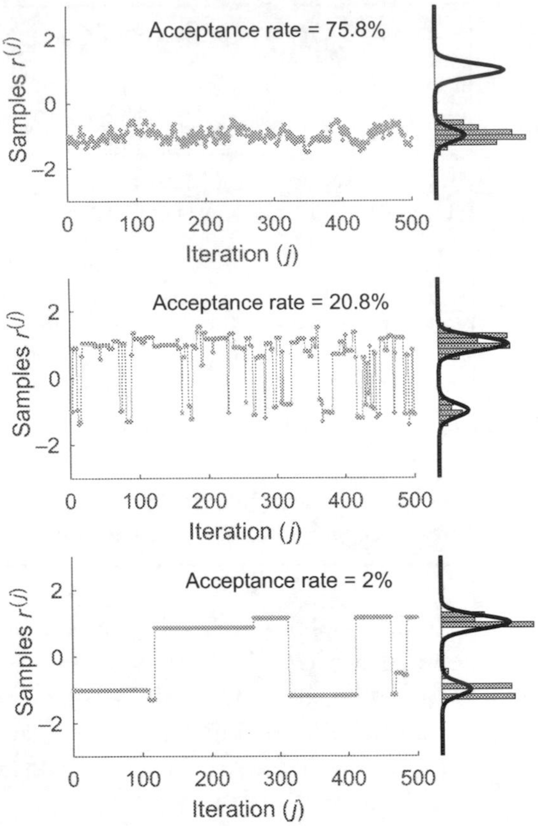

Fig. 5.7 Demonstration of the effects of a random walk's scaling λ on the resulting MCMC chain. Low λ values are associated with high acceptance rates and slowly evolving chains that may be easily trapped near a mode. Intermediate λ values are associated with fair acceptance rates and quickly evolving chains that sporadically jump among modes. High λ values are associated with low acceptance rates and chains with large but rare steps that easily jump among modes without, however, revealing much of the finer target details.

We may naively select larger values of λ and try to avoid a sampler that might be easily trapped. However, as can be seen in the last case, larger λ values tend to suffer high rejection rates, which can be deduced by the large intervals over which a chain remains constant. There are two drawbacks associated with high rejections: our chain stays at the same values often and so many samples are required in order to reveal our target's characteristics; and, as the computational cost of producing a rejected proposal is the same as an accepted proposal, large λ leads to costly samplers.

Therefore, it is not surprising that intermediate λs are most efficient as these sufficiently sample the entire target $\pi(r)$ while maintaining reasonable rejection levels.

5.2.2 Gibbs Family of Samplers

The samplers in the Gibbs family are useful only in cases where the targeted random variable is *multivariate* and partitioned into blocks of component variables. For this reason, in this section we adopt array notation, and denote the random variable of interest with $\boldsymbol{r} = r_{1:M}$, where r_m are the individual components. Of course, it is not required that each component r_m be univariate. Instead, some or all components r_m may be multivariate themselves. Accordingly, $\pi(\boldsymbol{r}) = \pi(r_{1:M}) = \pi(r_1, \ldots, r_M)$ all denote our targeted density.

The Gibbs sampler is the preferred choice when sampling a multivariate target, especially when there is a natural partitioning of the various components in groups r_m. The basic idea is that to sample $\boldsymbol{r}$, it is sufficient to sample each R_m separately. Before we describe how and why the Gibbs sampler works, we introduce some useful notation.

For simplicity, and consistency with other chapters, we use subscripts to group our variable's components,

$$r_{1:m} = (r_1, \ldots, r_m), \quad r_{m:M} = (r_m, \ldots, r_M), \quad r_{-m} = (r_1, \ldots, r_{m-1}, r_{m+1}, \ldots, r_M).$$

Specifically in r_{-m}, we use *negative* subscripts to group all components *but* r_m. Additionally, we use $\Pi^m_{r_{-m}}$ to denote the distribution of $r_m | r_{-m}$ and $\pi^m_{r_{-m}}(r_m)$ to denote its density. That is,

$$r_m | r_{-m} \sim \Pi^m_{r_{-m}}, \qquad \pi^m_{r_{-m}}(r_m) = p(r_m | r_{-m}).$$

These are termed *full conditionals* and our targeted variable $\boldsymbol{r}$ contains M of them, one for each m. For clarity, we use a superscript m to emphasize explicitly that they refer to the mth component of $\boldsymbol{r}$ and a subscript r_{-m} to emphasize that these condition on every *other* component's value.

Gibbs Sampler

Given a multivariate target $\bar{\pi}(\boldsymbol{r})$, to begin a Gibbs sampler a choice for the initial sample $\boldsymbol{r}^{(0)}$ must be made. Similar to the Metropolis–Hastings sampler, this can be achieved either by assigning a fixed value or sampling a value from some probability distribution. As before, the only requirement is that a feasible value be selected.

To advance the chain from $\boldsymbol{r}^{\text{old}} = r^{\text{old}}_{1:M}$ to $\boldsymbol{r}^{\text{new}} = r^{\text{new}}_{1:M}$, the Gibbs sampler uses M stages during which each component R_m is generated separately. In particular, the scheme proceeds by sampling each r_m conditioning on the most recent value of r_{-m} as follows:

$$\begin{array}{ll}
\text{Stage } 1 & r_1^{\text{new}} | r^{\text{old}}_{2:M} \sim \Pi^1_{r^{\text{old}}_{2:M}}, \\
\text{Stage } 2 & r_2^{\text{new}} | r_1^{\text{new}}, r^{\text{old}}_{3:M} \sim \Pi^2_{r_1^{\text{new}}, r^{\text{old}}_{3:M}}, \\
\vdots & \vdots \\
\text{Stage } M-1 & r^{\text{new}}_{M-1} | r^{\text{new}}_{1:M-2}, r^{\text{old}}_M \sim \Pi^{M-1}_{r^{\text{new}}_{1:M-2}, r^{\text{old}}_M}, \\
\text{Stage } M & r^{\text{new}}_M | r^{\text{new}}_{1:M-1} \sim \Pi^M_{r^{\text{new}}_{1:M-1}}.
\end{array}$$

That is, every component r_m^{new} is obtained by sampling from the corresponding full conditional distribution conditioned on the most recently available r_{-m}. This consists of a combination of already updated components $r_{1:m-1}^{\text{new}}$ and those that have not been updated yet $r_{m+1:M}^{\text{old}}$. For this reason, informally, we might refer to each stage as an *update*.

In Example 5.7, we provide a demonstration of the algorithm's implementation and we explicitly illustrate each of its stages. In Algorithm 5.2, we summarize a computational implementation of the entire sampling scheme.

Example 5.7 Gibbs sampling of a trivariate target

We consider a special case where $\boldsymbol{R} = (R_1, R_2, R_3)$ consists of three components. In this case, our target is $\pi(r_1, r_2, r_3) = p(r_1, r_2, r_3)$ and its full conditionals are

$$\pi^1_{r_2,r_3}(r_1) = p(r_1|r_2, r_3), \quad \pi^2_{r_1,r_3}(r_2) = p(r_2|r_1, r_3), \quad \pi^3_{r_1,r_2}(r_3) = p(r_3|r_1, r_2).$$

The Gibbs sampler uses an already computed sample $\boldsymbol{r}^{\text{old}} = (r_1^{\text{old}}, r_2^{\text{old}}, r_3^{\text{old}})$ to obtain $\boldsymbol{r}^{\text{new}} = (r_1^{\text{new}}, r_2^{\text{new}}, r_3^{\text{new}})$ by first generating

$$r_1^{\text{new}} \sim \Pi^1_{r_2^{\text{old}}, r_3^{\text{old}}},$$

and subsequently generating

$$r_2^{\text{new}} \sim \Pi^2_{r_1^{\text{new}}, r_3^{\text{old}}},$$

and finally generating

$$r_3^{\text{new}} \sim \Pi^3_{r_1^{\text{new}}, r_2^{\text{new}}}.$$

These three updates require sampling from the full conditionals $\Pi^1_{r_2,r_3}, \Pi^2_{r_1,r_3}, \Pi^3_{r_2,r_3}$, respectively, which need to be specified in advance.

Algorithm 5.2 Gibbs sampler

Given the full conditionals $\pi^m_{r_{-m}}(r_m)$ of a target and a feasible initial sample $\boldsymbol{r}^{\text{old}}$, the Gibbs sampler proceeds by iteration as follows:

- For each m from 1 up to M repeat:
  - Sample $r_m^{\text{new}} \sim \Pi^m_{r_{-m}}$ conditioning on the most recent r_{-m}.

In a Gibbs scheme, as described above, the updates of the components r_m do *not* need to take place in the given order (fixed sweep) as in Algorithm 5.2. In fact, the sampler remains valid even if the order of the updates is chosen at random (random sweep). In both versions, updating each R_m is achieved by $\Pi^m_{-r_m}$ conditioned on the most recent r_{-m}. The difference between the two versions is only in the order at which the sampler sweeps through $r_1, \ldots, r_M$.

A Gibbs scheme, either in the fixed or random sweep version, relies on the generation of samples from the full conditionals $\Pi^m_{r_{-m}}$. For this reason, it is crucial that the full conditionals be simulated directly. The choice of the partitioning of $\boldsymbol{r}$ into groups $r_{1:M}$ has a significant impact on the sampler as, essentially, it determines whether the resulting conditionals $\Pi^{m}_{r_{-m}}$ may be simulated. In Section 5.4.2, we will see an alternative scheme, however less efficient, that can be used when some of the conditionals cannot be simulated directly.

Why Does the Gibbs Sampler Work?*

Given the full conditionals $\pi^{m}_{r_{-m}}(r_m)$ of a target $\pi(\boldsymbol{r})$, the Gibbs sampler ensures that the MCMC chain generated visits each feasible $\boldsymbol{r}$ with a frequency proportional to $\pi(\boldsymbol{r})$. Similar to the Metropolis–Hastings sampler we saw earlier, this entails a sophisticated bookkeeping mechanism that balances flow in and out of each $\boldsymbol{r}$, which we illustrate in detail. Specifically, we spell out the transition rules and show that they are in balance with our target.

The Gibbs sampler advances from a sample $\boldsymbol{r}^{(j)}$ to the next one $\boldsymbol{r}^{(j+1)}$ in the MCMC chain through the successive realization of M intermediate samples,

$$\boldsymbol{R}^{(j)} = \boldsymbol{R}^{(j,0)} \rightarrow \underbrace{\boldsymbol{R}^{(j,1)}}_{\substack{\text{1st}\\ \text{update}}} \rightarrow \underbrace{\boldsymbol{R}^{(j,2)}}_{\substack{\text{2nd}\\ \text{update}}} \rightarrow \cdots \rightarrow \underbrace{\boldsymbol{R}^{(j,M)}}_{\substack{M\text{th}\\ \text{update}}} = \boldsymbol{R}^{(j+1)}.$$

The precise order of the components sampled in each update depends on the version of the sampler chosen. Nevertheless, all versions entail intermediate transition rules that each update a single component. For clarity, from now on we will denote the rule that updates the component r_m with $T^{m}_{\boldsymbol{r}^{\text{old}}}(\boldsymbol{r}^{\text{new}})$. As $T^{m}_{\boldsymbol{r}^{\text{old}}}(\boldsymbol{r}^{\text{new}})$ leaves r_{-m} unchanged, it attains the specific form

$$T^{m}_{\boldsymbol{r}^{\text{old}}}(\boldsymbol{r}^{\text{new}}) = \pi^{m}_{r^{\text{old}}_{-m}}(r^{\text{new}}_{m})\delta_{r^{\text{old}}_{-m}}(r^{\text{new}}_{-m}).$$

First, we focus on a single Gibbs update. This is equivalent to a Metropolis–Hastings scheme with proposals

$$Q_{r^{\text{old}}_{m},r^{\text{old}}_{-m}}(r^{\text{prop}}_{m}, r^{\text{prop}}_{-m}) = \pi^{m}_{r^{\text{old}}_{-m}}(r^{\text{prop}}_{m})\delta_{r^{\text{old}}_{-m}}(r^{\text{prop}}_{-m}).$$

To verify that the two schemes are indeed equivalent, we consider the acceptance ratio

$$A^{m}_{r^{\text{old}}_{m},r^{\text{old}}_{-m}}(r^{\text{prop}}_{m}, r^{\text{prop}}_{-m}) = \frac{\pi(r^{\text{prop}}_{m}, r^{\text{prop}}_{-m})}{\pi(r^{\text{old}}_{m}, r^{\text{old}}_{-m})}\frac{Q_{r^{\text{prop}}_{m},r^{\text{prop}}_{-m}}(r^{\text{old}}_{m}, r^{\text{old}}_{-m})}{Q_{r^{\text{old}}_{m},r^{\text{old}}_{-m}}(r^{\text{prop}}_{m}, r^{\text{prop}}_{-m})}$$

$$= \underbrace{\frac{\pi^{m}_{r^{\text{prop}}_{-m}}(r^{\text{prop}}_{m})}{\pi^{m}_{r^{\text{old}}_{-m}}(r^{\text{old}}_{m})}\frac{\pi^{-m}(r^{\text{prop}}_{-m})}{\pi^{-m}(r^{\text{old}}_{-m})}}_{\text{target}}\underbrace{\frac{\pi^{m}_{r^{\text{prop}}_{-m}}(r^{\text{old}}_{m})}{\pi^{m}_{r^{\text{old}}_{-m}}(r^{\text{prop}}_{m})}\frac{\delta_{r^{\text{prop}}_{-m}}(r^{\text{old}}_{-m})}{\delta_{r^{\text{old}}_{-m}}(r^{\text{prop}}_{-m})}}_{\text{proposal}}.$$

Here, we use $\pi^{-m}(r_{-m})$ to denote the probability density of R_{-m} with R_m marginalized out. As the proposal does not change r_{-m}, we have $r^{\text{prop}}_{-m} = r^{\text{old}}_{-m}$. Therefore, the acceptance ratio simplifies to $A^{m}_{r^{\text{old}}_{m},r^{\text{old}}_{-m}}$

* This is an advanced topic and could be skipped on a first reading.

$(r_m^{\text{prop}}, r_{-m}^{\text{prop}}) = 1$. Consequently, $r_m^{\text{new}} = r_m^{\text{prop}}$ and the acceptance test need not be performed since the proposal is always accepted.

We now consider the application of two successive Gibbs updates. For instance, updates m and $m+1$ in the fixed sweep version. In this case, the first update implements a transition $\boldsymbol{r}^{\text{old}} \to \boldsymbol{r}^{\text{temp}}$ to some intermediate $\boldsymbol{r}^{\text{temp}}$; while the second update implements the transition $\boldsymbol{r}^{\text{temp}} \to \boldsymbol{r}^{\text{new}}$. To derive the combined transition rule $T^{m,m+1}_{\boldsymbol{r}^{\text{old}}}(\boldsymbol{r}^{\text{new}})$, we need to consider all possibilities for the intermediate sample. In other words we need to marginalize over the intermediate variable $\boldsymbol{R}^{\text{temp}}$. Formally,

$$\begin{aligned}
T^{m,m+1}_{\boldsymbol{r}^{\text{old}}}(\boldsymbol{r}^{\text{new}}) = p\left(\boldsymbol{r}^{\text{new}}|\boldsymbol{r}^{\text{old}}\right) &= \int d\boldsymbol{r}^{\text{temp}}\, p\left(\boldsymbol{r}^{\text{new}}, \boldsymbol{r}^{\text{temp}}|\boldsymbol{r}^{\text{old}}\right) \\
&= \int d\boldsymbol{r}^{\text{temp}}\, p\left(\boldsymbol{r}^{\text{temp}}|\boldsymbol{r}^{\text{old}}\right) p\left(\boldsymbol{r}^{\text{new}}|\boldsymbol{r}^{\text{temp}}\right) \\
&= \int d\boldsymbol{r}^{\text{temp}}\, T^{m}_{\boldsymbol{r}^{\text{old}}}(\boldsymbol{r}^{\text{temp}}) T^{m+1}_{\boldsymbol{r}^{\text{temp}}}(\boldsymbol{r}^{\text{new}}).
\end{aligned}$$

With similar completions, we can derive the transition rules $T_{\boldsymbol{r}^{\text{old}}}(\boldsymbol{r}^{\text{new}})$ of the entire Gibbs scheme. These, of course, entail all M updates and the precise order they are applied depends on the particular version of the sampler.

As each Gibbs update results from a Metropolis–Hastings scheme, the associated rule $T^{m}_{\boldsymbol{r}^{\text{old}}}(\boldsymbol{r}^{\text{new}})$ is already in full balance with the target $\pi(\boldsymbol{r})$. This means that

$$\pi(\boldsymbol{r}) = \int d\boldsymbol{r}^{\text{old}}\, T^{m}_{\boldsymbol{r}^{\text{old}}}(\boldsymbol{r}) \pi(\boldsymbol{r}^{\text{old}})$$

holds for all m, which is an important result we need in order to derive the balance condition for the entire Gibbs scheme. To this end, we first consider the application of only two successive Gibbs updates, for instance m and $m+1$, as above. In this case, we have

$$\begin{aligned}
\int d\boldsymbol{r}^{\text{old}}\, T^{m,m+1}_{\boldsymbol{r}^{\text{old}}}(\boldsymbol{r}^{\text{new}}) \pi(\boldsymbol{r}^{\text{old}}) &= \int d\boldsymbol{r}^{\text{old}} \left(\int d\boldsymbol{r}\, T^{m}_{\boldsymbol{r}^{\text{old}}}(\boldsymbol{r}) T^{m+1}_{\boldsymbol{r}}(\boldsymbol{r}^{\text{new}})\right) \pi(\boldsymbol{r}^{\text{old}}) \\
&= \int d\boldsymbol{r} \left(\int d\boldsymbol{r}^{\text{old}}\, T^{m}_{\boldsymbol{r}^{\text{old}}}(\boldsymbol{r}) \pi(\boldsymbol{r}^{\text{old}})\right) T^{m+1}_{\boldsymbol{r}}(\boldsymbol{r}^{\text{new}}) \\
&= \int d\boldsymbol{r}\, \pi(\boldsymbol{r}) T^{m+1}_{\boldsymbol{r}}(\boldsymbol{r}^{\text{new}}) \\
&= \pi(\boldsymbol{r}^{\text{new}}),
\end{aligned}$$

which shows that $T^{m,m+1}_{\boldsymbol{r}^{\text{old}}}(\boldsymbol{r}^{\text{new}})$ is also in full balance with the target $\pi(\boldsymbol{r})$. We can similarly show that the transition rules $T_{\boldsymbol{r}^{\text{old}}}(\boldsymbol{r}^{\text{new}})$ of the entire Gibbs scheme are also in full balance with the target $\pi(\boldsymbol{r})$.

Of course, the full balance condition fulfilled by the Gibbs sampler is a weaker condition than detailed balance fulfilled by the Metropolis–Hastings sampler. Nevertheless, in view of Note 5.3, it is sufficient to ensure that the resulting MCMC chain is valid.

Example 5.8 When does the Gibbs sampler fail?

Consider sampling a bivariate random variable (R_1, R_2) from a target $\pi(r_1, r_2)$ that equals 1/2 whenever $-1 < r_1, r_2 < 0$, or $0 < r_1, r_2 < +1$. Essentially, this is a uniform target over two disjoint rectangular regions. Since this is a simple target, we can easily obtain the full conditionals,

$$\pi^1_{r_2}(r_1) = \begin{cases} \text{Uniform}_{[-1,0]}(r_1), & r_2 < 0, \\ \text{Uniform}_{[0,+1]}(r_1), & 0 < r_2, \end{cases}$$

$$\pi^2_{r_1}(r_2) = \begin{cases} \text{Uniform}_{[-1,0]}(r_2), & r_1 < 0, \\ \text{Uniform}_{[0,+1]}(r_2), & 0 < r_1. \end{cases}$$

For concreteness, consider a Gibbs sampler starting at some $r_1^{(0)} < 0$ and $r_2^{(0)} < 0$. This sample lies in the left region. As can be seen, the sampler remains in the same region, following the first iteration:

- Sample $r_1^{(1)}|r_2^{(0)} \sim \text{Uniform}_{[-1,0]}(r_1)$.
- Sample $r_2^{(1)}|r_1^{(1)} \sim \text{Uniform}_{[-1,0]}(r_1)$.

The same happens following the second iteration:

- Sample $r_1^{(2)}|r_2^{(1)} \sim \text{Uniform}_{[-1,0]}(r_1)$.
- Sample $r_2^{(2)}|r_1^{(2)} \sim \text{Uniform}_{[-1,0]}(r_1)$.

All subsequent iterations exhibit the same pathology. Accordingly, no matter how many iterations we perform, the sampler is unable to cross into the other region. The same also happens when the sampler is initialized with $r_1^{(0)} > 0$ and $r_2^{(0)} > 0$, in which case it remains trapped in the right region.

Recalling Note 5.3, we see that this sampler clearly meets the *feasibility* condition and, as we show above, it also meets the *invariance* condition. Nevertheless, it does *not* meet the *irreducibility* condition. In particular, because each transition $\boldsymbol{r}^{\text{old}} \to \boldsymbol{r}^{\text{new}}$ is broken down to two separate transitions, neither of which can cross into the other region, the resulting sampler is nonergodic.

Generally, Gibbs samplers encounter problems whenever the support of the target is disconnected. The same challenge persists even when the target's support is *effectively disconnected* as, for instance, when it contains ridges of very low probability. In Exercise 5.10, we show that ergodicity can be recovered by a reparametrization of the target. Such an approach suggests that reparametrization of the target offers a general remedy. However, in complex targets choosing an appropriate reparametrization may constitute an impossible task.

Sampling of Posterior Targets

Just as with Metropolis–Hastings, the Gibbs sampler holds equally well whenever our target is a Bayesian posterior $\pi(\boldsymbol{r}) = p(\boldsymbol{r}|\boldsymbol{w})$. The full conditionals in this case read

$$\pi^m_{r_{-m}}(r_m) = p(r_m | r_{-m}, \boldsymbol{w}).$$

We emphasize that, unlike a general target in which observations $\boldsymbol{w}$ are irrelevant, the full conditionals of a Bayesian posterior are also conditioned on the observations $\boldsymbol{w}$. Direct sampling from such conditionals $\pi^m_{r_{-m}}(r_m)$ is particularly convenient in Bayesian models developed in terms of conditionally conjugate distributions.

Example 5.9 MC for the conditionally conjugate normal-gamma model

In this example, we consider a variant of the model of Example 5.1. As before, suppose we have scalar, normally distributed observations $w_{1:N}$. Our task, once more, is to estimate the center and the spread of the underlying distribution. For this, we consider a similar parametrization of the normal distribution by mean μ and precision τ. However, unlike before where the priors on μ and τ were jointly specified (note the conditional on Eq. (5.3)), here we apply independent priors on the two parameters. In particular, we place a normal prior on μ and a gamma prior on τ. The entire model then reads

$$\begin{aligned} \tau &\sim \text{Gamma}(\alpha, \beta), \\ \mu &\sim \text{Normal}\left(\xi, \frac{1}{\psi}\right), \\ w_n|\mu, \tau &\sim \text{Normal}\left(\mu, \frac{1}{\tau}\right), \end{aligned} \tag{5.9}$$

where α, β, ξ, ψ are hyperparameters with known values.

With this setup, the model is described by the random variable $\boldsymbol{r} = \{\mu, \tau\}$, and the associated posterior is $\pi(\boldsymbol{r}) = p(\mu, \tau|w_{1:N})$. Despite its similarity to Example 5.1, the new model is more cumbersome. Namely, a minor change in the prior's prescription (Eq. (5.3) vs. Eq. (5.9)) has a major impact on practical applications, as the new model is not fully conjugate. Therefore, computational methods are required to characterize the induced posterior.

However, our model remains conditionally conjugate, which facilitates Gibbs sampling. For instance, a Gibbs sampler requires two stages: one to update μ and one to update τ. Suppose that $\boldsymbol{r}^{(j)} = \left\{\mu^{(j)}, \tau^{(j)}\right\}$ has been computed; the updates are:

- Sample $\mu^{(j+1)}$ from $\mu|\tau^{(j)}, w_{1:N}$.
- Sample $\tau^{(j+1)}$ from $\tau|\mu^{(j+1)}, w_{1:N}$.

Once both parameters have been updated, the new sample is $r^{(j+1)} = \left\{\mu^{(j+1)}, \tau^{(j+1)}\right\}$. The required conditionals for each update may be worked out analytically,

$$p\left(\mu|\tau, w_{1:N}\right) \propto p\left(w_{1:N}|\mu, \tau\right) p\left(\mu\right) \propto \text{Normal}\left(\mu; \frac{\psi\xi + \tau\sum_{n=1}^N w_n}{\psi + \tau N}, \frac{1}{\psi + \tau N}\right),$$

$$p\left(\tau|\mu, w_{1:N}\right) \propto p\left(w_{1:N}|\mu, \tau\right) p\left(\tau\right) \propto \text{Gamma}\left(\tau; \alpha + \frac{N}{2}, \frac{1}{\frac{1}{\beta} + \frac{1}{2}\sum_{n=1}^N \left(w_n - \mu\right)^2}\right).$$

Example 5.10 Gibbs sampling for a sum of Gaussians

As an alternative example, consider a Normal $\left(\mu_1, \sigma^2\right)$ generating observations with probability ω_1 and a Normal $\left(\mu_2, \sigma^2\right)$ generating observations with probability $\omega_2 = 1 - \omega_1$. Suppose we obtain observations $w_{1:N}$ and our task is to estimate: the locations μ_1, μ_2; their spread σ; and which of the two normal densities generated each observation w_n.

To estimate μ_1 and μ_2, we can use a common Normal$(\xi, 1/\psi)$ prior. To estimate σ, we may reparametrize the normal distributions in terms of precision $\tau = 1/\sigma^2$ and, subsequently, place a Gamma(α, β) prior on τ. Finally, for each observation w_n, we can consider an indicator variable s_n, which is 1 if w_n was generated by Normal$(\mu_1, 1/\tau)$ and 2 if w_n was generated by Normal$(\mu_2, 1/\tau)$. Thus, we simply need to estimate the values of s_n.

With this setup, our entire model is

$$\begin{aligned}
\tau &\sim \text{Gamma}(\alpha, \beta), \\
\mu_1 &\sim \text{Normal}\left(\xi, \frac{1}{\psi}\right), \\
\mu_2 &\sim \text{Normal}\left(\xi, \frac{1}{\psi}\right), \\
s_n &\sim \text{Categorical}_{1,2}(\omega_1, \omega_2), \\
w_n | s_n, \mu_1, \mu_2, \tau &\sim \text{Normal}\left(\mu_{s_n}, \frac{1}{\tau}\right),
\end{aligned}$$

and is described by the random variables $\tau, \mu_1, \mu_2, s_{1:N}$. To sample the posterior $p(\tau, \mu_1, \mu_2, s_{1:N} | w_{1:N})$, we may compute samples $\tau^{\text{new}}, \mu_1^{\text{new}}, \mu_2^{\text{new}}, s_{1:N}^{\text{new}}$ through the following Gibbs sampler:

- Generate τ^{new} by sampling from $p(\tau | \mu_1^{\text{old}}, \mu_2^{\text{old}}, s_{1:N}^{\text{old}}, w_{1:N})$.
- Generate μ_1^{new} by sampling from $p(\mu_1 | \tau^{\text{new}}, \mu_2^{\text{old}}, s_{1:N}^{\text{old}}, w_{1:N})$.
- Generate μ_2^{new} by sampling from $p(\mu_2 | \tau^{\text{new}}, \mu_1^{\text{new}}, s_{1:N}^{\text{old}}, w_{1:N})$.
- Generate $s_{1:N}^{\text{new}}$ by sampling each indicator s_n from $p(s_n | \tau^{\text{new}}, \mu_1^{\text{new}}, \mu_2^{\text{new}}, s_{1:n-1}^{\text{new}}, s_{n+1:N}^{\text{old}}, w_{1:N})$.

5.3 Processing and Interpretation of MCMC

We saw earlier that the common task in MC or MCMC is to obtain samples $r^{(1:J)}$ from a target $\pi(r)$ otherwise difficult to sample. When these samples are *independent*, such as when they are generated by running an *MC scheme,* we can directly use them in empirical approximations like Eq. (5.2). As the total number of samples J that we use increases, the approximation improves and, in the long run, empirical averages like $\frac{1}{J}\sum_{j=1}^{J} g\left(r^{(j)}\right)$ converge to their exact counterparts. Practically, this means that we can make an MC approximation as accurate as necessary by simply computing additional samples $r^{(j)}$. Therefore, when $J \gg 1$ our approximation becomes insignificant.

However, running an *MCMC scheme* yields samples that *depend* upon each other. For this reason, empirical approximations like Eq. (5.2) are generally biased and the bias depends on how strongly samples depend on one another. Fortunately, given an MCMC chain there is a sequence of standard procedures that we can follow to reduce such dependencies and, essentially, recover nearly independent MC samples.

Here, we describe some practical techniques to minimize biases introduced by MCMC sampling and improve the accuracy of our approximations. Example 5.11 provides a motivating example and we then proceed with a more formal discussion.

Example 5.11 Mixing of MC and MCMC

We consider a Gibbs sampler for a bivariate target consisting of correlated normal components $\boldsymbol{r} = (r_1, r_2)$. Specifically, the target density is

$$\pi(r_1, r_2) \propto \exp\left(-\frac{1}{2(1-\rho^2)}\left(\left(\frac{r_1-\mu_1}{\sigma_1}\right)^2 + \left(\frac{r_2-\mu_2}{\sigma_2}\right)^2 - \left(2\rho\frac{(r_1-\mu_1)(r_2-\mu_2)}{\sigma_1\sigma_2}\right)\right)\right),$$

where μ_1, μ_2 are the means, σ_1, σ_2 are the standard deviations, and ρ is the correlation coefficient.

Figure 5.8 shows an MCMC chain targeting $\pi(r_1, r_2)$. For illustrative purposes, we use $\mu_1 = 5$, $\mu_2 = 15$, $\sigma_1 = 1$, $\sigma_2 = 1$, and a high correlation coefficient $\rho = 0.95$. As our target has a standard form, we can also draw

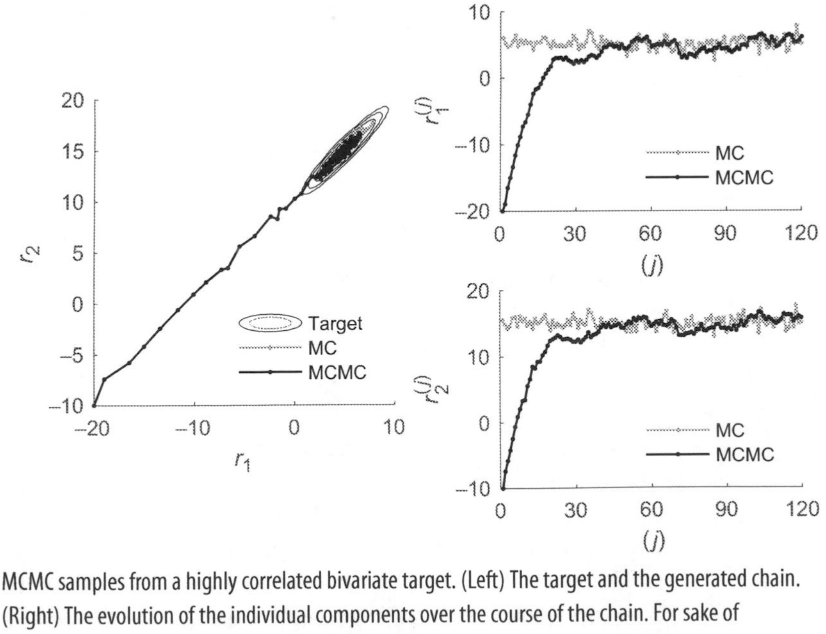

Fig. 5.8 MCMC samples from a highly correlated bivariate target. (Left) The target and the generated chain. (Right) The evolution of the individual components over the course of the chain. For sake of comparison, we also show a sequence of MC samples.

MC samples from it and, for comparison, in Fig. 5.8 we also show an MC sequence. Visual inspection of the two sampled sequences reveals three important characteristics:

- In the long run, the MC and MCMC sequences intermingle, suggesting that they both produce samples with the same statistics.
- Unlike the MC sequence, which reproduces the targeted statistics immediately, the MCMC sequence reproduces the targeted statistics only after an initial period covering the first ≈ 40 samples.
- Unlike the MC sequence where successive samples are uncorrelated, MCMC samples are deeply correlated.

These characteristics are common to MCMC chains and apply universally to the output of any MCMC method, no matter how the initial sample in the chain is selected or which transition rules are used to advance from sample to sample. For instance, a similar behavior was observed in Figs. 5.3 and 5.4, which we encountered earlier in this chapter.

These characteristics determine how well our MCMC chain approximates independent samples. Accordingly, they impact the approximations derived from $r^{(1:J)}$ and, for this reason, they determine the overall quality of our MCMC scheme. Practically, such features are influenced by the sampler used and tend to be related to each other. Typically, a good sampler performs well in all three respects and allows for reliable empirical approximations derived from chains of modest size J. When this happens, we may colloquially say that a sampling scheme has good *mixing*.

Of course, mixing is a qualitative property for which the golden standard is set by MC schemes that, due to the generation of independent samples, achieve ideal mixing. By contrast, the dependence among samples in an MCMC scheme degrade their quality and so their mixing is less than ideal. Nevertheless, as we mentioned above, given an MCMC chain, we can always improve its mixing characteristics. We will see that this entails discarding samples from the chain, which nearly always leads to dramatically improved approximations.

More formally, given an MCMC chain $r^{(0)} \rightarrow r^{(1)} \rightarrow \cdots \rightarrow r^{(J)}$, before we use its samples to derive approximations, we need to address three attributes:

- Identify whether the chain has fully explored our target and started reproducing its statistics.
- Determine the initial period until the chain starts reproducing the targeted statistics.
- Determine the lag between consecutive samples that can be considered almost independent.

Addressing these points is more or less achieved by inspection of the appropriate plots such as those shown on the left of Fig. 5.8. These are termed *trace plots* and depict how samples $r^{(j)}$ evolve over the course of the chain.

5.3.1 Assessing Convergence

Assessing whether an MCMC chain reproduces the targeted statistics is a qualitative problem. Practically, we assess whether the chain has converged to the target by inspection of the trace plots. The critical characteristic is for the chain to revisit regions already sampled rather than to keep discovering new regions and avoid previously sampled ones. For example, the MCMC chain depicted in the upper panel of Fig. 5.4 jumps from one region to the other once and never returns. This is an indication that the chain is still evolving and, when this scenario is encountered, additional samples should be produced.

Once a chain converges to the target, its trace plots fluctuate around similar values and sporadically jump back and forth between potentially separated regions. For instance, the chain depicted in the lower panel of Fig. 5.4 jumps in and out of each mode multiple times, indicating that its statistics have stabilized.

5.3.2 Burn-in Removal

Once a convergent MCMC has been identified, we need to determine the number of initial samples that it takes until convergence is achieved. Although this process is more quantitive than the assessment of convergence in the first place, it still relies on trace plots. This period can be found by pinpointing the minimum number of samples in the initial portion of the chain that need to be discarded in order for the chain's statistics to stop changing.

This phase is termed *burn-in* or *warm-up* and can be more accurately located by comparing sample statistics across successive sample batches. For example, Fig. 5.9 illustrates how the sample mean computed over three successive batches changes over the MCMC chain of Fig. 5.8. As can be seen, the mean in both trace plots over the first batch differs from the means over the second and third batches. This indicates that, for this particular chain, burn-in extends over the first batch.

The chain statistics stabilize only after burn-in, so the samples identified as belonging in this phase should be *removed* from the chain and only the remaining samples should be used for further processing. Thus, from the original chain,

$$r^{(0)} \to \cdots \to r^{(j-1)} \to r^{(j)} \to r^{(j+1)} \to \cdots \to r^{(J)},$$

after burn-in removal, what remains is a *shorter* chain,

$$r^{(j_{\min})} \to \cdots \to r^{(j-1)} \to r^{(j)} \to r^{(j+1)} \to \cdots \to r^{(J)},$$

where $j_{\min}$ marks the end of burn-in.

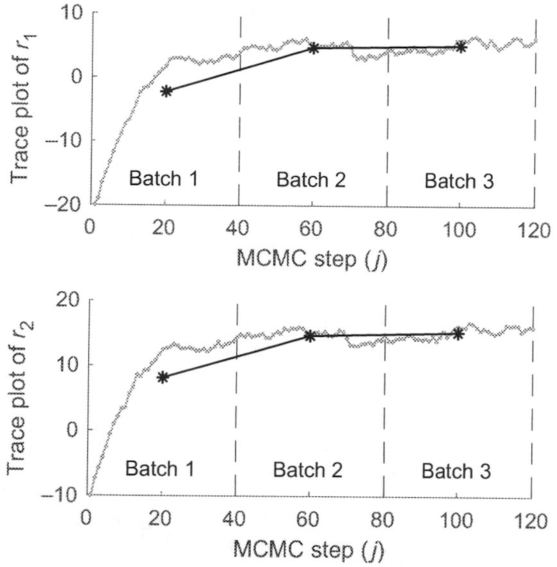

Fig. 5.9 Trace plots and corresponding batch means of the MCMC chain depicted in Fig. 5.8. As can be seen, batch statistics stabilize only after the first $\approx$ 40 samples.

5.3.3 Thinning

Finally, once we have identified a convergent MCMC chain and removed burn-in, we need to quantify the correlation between the remaining samples. As such correlations depend on the lag d separating two samples $r^{(j)}$ and $r^{(j+d)}$, this is achieved by the *autocorrelation* along the course of the chain,

$$\rho_d = \frac{\sum_{j=j_{\min}}^{J-j_{\min}+1-d}\left(r^{(j)}-\bar{r}\right)\left(r^{(j+d)}-\bar{r}\right)}{\sum_{j=j_{\min}}^{J-j_{\min}+1}\left(r^{(j)}-\bar{r}\right)^2}, \qquad d = 0, 1, \ldots, J - j_{\min},$$

where the mean is computed as $\bar{r} = \frac{1}{J-j_{\min}+1}\sum_{j=j_{\min}}^{J} r^{(j)}$.

As seen in Fig. 5.10, the autocorrelation ρ_d quantifies how tightly related successive samples are in the MCMC chain. Typically, the autocorrelation is high at small lags and decreases in a nearly exponential fashion at larger lags. Accordingly, a lag such that $\rho_d \approx 0$ is a good indication of how many samples need to be generated until our MCMC chain resembles MC sampling. In practice, of course, it is sufficient to set a lower threshold and consider the minimum lag $d_{\min}$ where $|\rho_{d_{\min}}|$ drops below this threshold.

Once $d_{\min}$ is found, a downsampling process termed *thinning*, in which we retain only one out of every $d_{\min}$ samples from our MCMC chain, can be used to recover independent samples. In summary, from the chain that remains after burn-in removal,

$$r^{(j_{\min})} \to \cdots \to r^{(j-1)} \to r^{(j)} \to r^{(j+1)} \to \cdots \to r^{(J)},$$

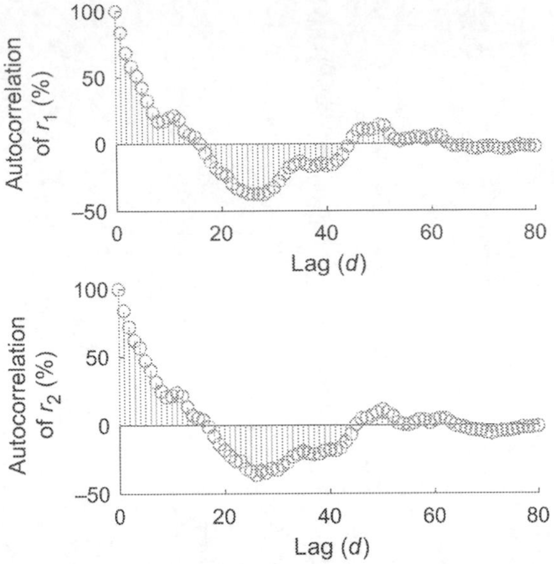

Fig. 5.10 Autocorrelation of the MCMC chain depicted in Fig. 5.8 after burn-in removal.

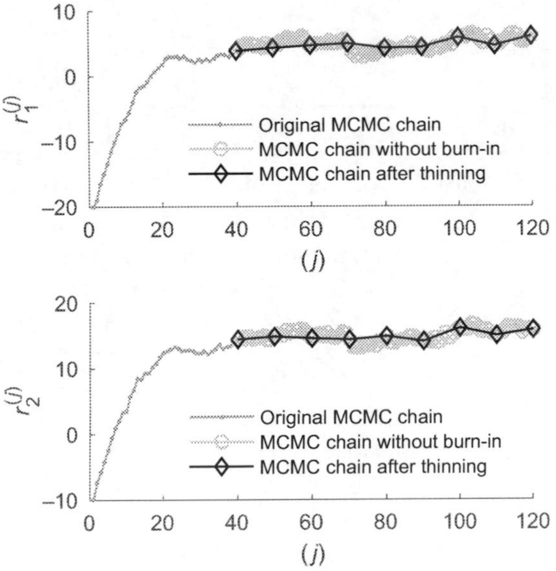

Fig. 5.11 Recovering MC samples from a given MCMC chain is achieved by discarding burn-in samples and thinning. Here, we show approximately MC samples $r^{(j_i)}$ resulting from the chain depicted in Fig. 5.8.

there remains a *thinned* chain,

$$r^{(j_1)} \to \cdots \to r^{(j_{i-1})} \to r^{(j_i)} \to r^{(j_{i+1})} \to \cdots \to r^{(j_I)},$$

where $j_i - j_{i-1} = d_{\min}$. The MCMC samples $r^{(j_i)}$ remaining in the thinned chain resample nearly iid samples drawn from our target and, essentially, can be used as if they had been generated by an MC scheme. As shown in Fig. 5.11, following the procedure above we can turn a given chain of

MCMC samples $r^{(j)}$ into MC ones $r^{(j_i)}$ on which we can safely rely to derive empirical approximations like Eq. (5.2).

5.4 Advanced MCMC Samplers*

In this section we present nonstandard samplers. Although the same theoretical foundations of Metropolis–Hastings and Gibbs samplers apply, due to subtle differences in the application context we discuss each sampler separately.

5.4.1 Multiplicative Random Walk Samplers

The additive random walk samplers we described in Section 5.2.1 are fairly easy to implement and reasonably easy to calibrate; however, their additive nature means that they exhibit similar motility over the target's support. In practice, this means that additive random walks tend to move efficiently enough only over some regions while they may move unreasonably slower over others. In particular, they rarely cross regions of low probability and so they often fail to fully explore the entire target's support. Unfortunately, whenever our targets contain multiple isolated modes, this becomes a major drawback. A convenient solution that maintains the implementation ease of additive random walks, but improves upon these drawbacks, is offered by *multiplicative* random walks. As these samplers are particularly suited for sampling *positive scalar* random variables r, below we focus on this case alone.

Proposals based on multiplicative random walks are obtained based on products

$$r^{\text{prop}} = r^{\text{old}} \epsilon^{\text{prop}},$$

where ϵ^{prop} is a random perturbation sampled from a distribution $G(\epsilon^{\text{prop}})$ independent of r^{old}. With this choice, the proposal distribution has a density $Q_{r^{\text{old}}}(r^{\text{prop}}) = G\left(r^{\text{prop}}/r^{\text{old}}\right)/r^{\text{old}}$ for which the acceptance ratio reads

$$A_{r^{\text{old}}}(r^{\text{prop}}) = \frac{\bar{\pi}(r^{\text{prop}})}{\bar{\pi}(r^{\text{old}})} \frac{r^{\text{old}}}{r^{\text{prop}}} \frac{G\left(r^{\text{old}}/r^{\text{prop}}\right)}{G\left(r^{\text{prop}}/r^{\text{old}}\right)}.$$

Convenient choices for the perturbations, such as those of Fig. 5.12, are offered by gamma or beta prime distributions,

$$\epsilon^{\text{prop}} \sim \text{Gamma}\left(\alpha, 1/\alpha\right), \qquad \epsilon^{\text{prop}} \sim \text{BetaPrime}\left(\alpha, 1+\alpha\right),$$

arranged to yield values around 1 such that we may achieve a reasonable acceptance rate simply by tuning the value of α. Unlike the additive case of

* This is an advanced topic and could be skipped on a first reading.

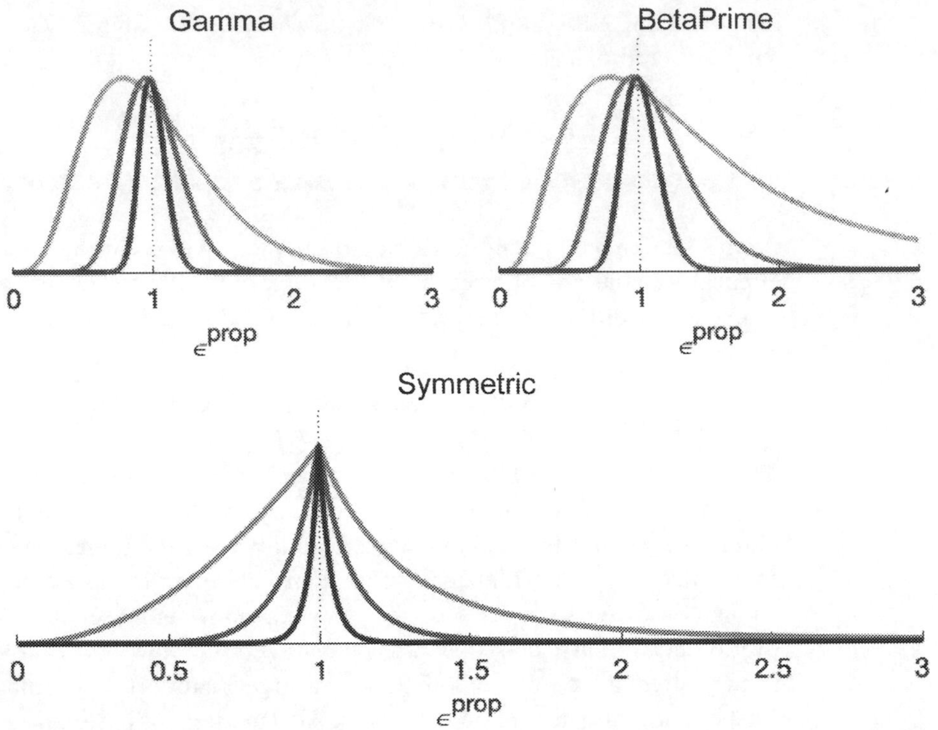

Fig. 5.12 Probability densities of common additive random walk perturbations. For clarity, all densities are normalized to a maximum of one.

Section 5.4.1, the perturbations here are unitless and do not require further scaling.

Provided, furthermore, that $G(\epsilon^{\text{prop}})$ is chosen such that the random variables ϵ^{prop} and $1/\epsilon^{\text{prop}}$ have identical densities, then the acceptance ratio simplifies to

$$A_{r^{\text{old}}}(r^{\text{prop}}) = \frac{\bar{\pi}(r^{\text{prop}})}{\bar{\pi}(r^{\text{old}})} \frac{r^{\text{prop}}}{r^{\text{old}}}.$$

In this case, we speak of a *symmetric* multiplicative random walk sampler.

Note 5.8 How do we obtain symmetric perturbations?

A symmetric multiplicative random walk requires perturbations whose density satisfies $G(\epsilon) = G(1/\epsilon)/\epsilon^2$. A convenient way of obtaining such perturbations is via a sum $G = \frac{1}{2}G_1 + \frac{1}{2}G_2$ where G_1 returns a sample from $\text{Beta}(\alpha, \beta)$ and G_2 returns the reciprocal of a sample from $\text{Beta}(\alpha, \beta)$. Such perturbations are also shown in Fig. 5.12.

Simulating $\epsilon^{\text{prop}} \sim G(\epsilon^{\text{prop}})$ from this sum can be achieved via ancestral sampling by first drawing $u \sim \text{Uniform}_{[0,1]}$ and $r \sim \text{Beta}(\alpha, \beta)$. Subsequently, if $u < 1/2$ we may set $\epsilon^{\text{prop}} = r$, otherwise $\epsilon^{\text{prop}} = 1/r$. In the typical setup, $\beta = 1$, ensuring that the perturbations concentrate around 1 and α is tuned such as to yield a reasonable acceptance rate.

5.4.2 Within Gibbs Sampling Schemes

When a multivariate random variable r is part of a complex problem, there is often a natural way to group its components $r_{1:M}$. For instance, in Example 5.10 we saw variables partitioned in three groups: tags in $r_1 = s_{1:N}$; locations in $r_2 = \mu_{1:2}$; and precision in $r_3 = \tau$. Such problem-motivated groupings often lead to intuitive Gibbs schemes. Nevertheless, in the modeling of physical systems, one or more of the full conditionals in a Gibbs scheme may be difficult or otherwise impossible to sample from directly. In this case, any draw from the full conditionals may be replaced by a Metropolis–Hastings iteration or any of its variants.

Example 5.12 Metropolis–Hastings within Gibbs

We consider the same setup as in Example 5.7. Our model variable $\boldsymbol{R} = (r_1, r_2, r_3)$ consists of three components, with target $\pi(r_1, r_2, r_3) = p(r_1, r_2, r_3)$ and full conditionals given by

$$\pi^1_{r_2,r_3}(r_1) = p(r_1|r_2, r_3), \quad \pi^2_{r_1,r_3}(r_2) = p(r_2|r_1, r_3), \quad \pi^3_{r_1,r_2}(r_3) = p(r_3|r_1, r_2).$$

As we have already seen, to advance from $\boldsymbol{r}^{\text{old}}$ to $\boldsymbol{r}^{\text{new}}$, a Gibbs scheme requires drawing successive samples

$$r_1^{\text{new}}|r_2^{\text{old}}, \quad r_3^{\text{old}} \sim \Pi^1_{r_2^{\text{old}}, r_3^{\text{old}}},$$
$$r_2^{\text{new}}|r_1^{\text{new}}, \quad r_3^{\text{old}} \sim \Pi^1_{r_1^{\text{new}}, r_3^{\text{old}}},$$
$$r_3^{\text{new}}|r_1^{\text{new}}, \quad r_2^{\text{new}} \sim \Pi^1_{r_1^{\text{new}}, r_2^{\text{new}}}.$$

Suppose that the second conditional $\Pi^2_{r_1^{\text{new}}, r_3^{\text{old}}}$ cannot be simulated, so we cannot obtain r_2^{new} directly. In this case, we may proceed with a Metropolis–Hastings step. In particular, to obtain r_2^{new}, we can first sample

$$r_2^{\text{prop}} \sim \mathbb{Q}_{r_1^{\text{new}}, r_2^{\text{old}}, r_3^{\text{old}}}$$

using a proposal distribution that may, in general, depend on r_2^{old} and potentially on the other variables as well. The acceptance ratio of this step is

$$A_{r_2^{\text{old}}}(r_2^{\text{prop}}) = \frac{\pi^2_{r_1^{\text{new}}, r_3^{\text{old}}}(r_2^{\text{prop}})}{\pi^2_{r_1^{\text{new}}, r_3^{\text{old}}}(r_2^{\text{old}})} \frac{Q_{r_1^{\text{new}}, r_2^{\text{prop}}, r_3^{\text{old}}}(r_2^{\text{old}})}{Q_{r_1^{\text{new}}, r_2^{\text{old}}, r_3^{\text{old}}}(r_2^{\text{prop}})}$$

and the acceptance test becomes

- If $u < A_{r_2^{\text{old}}}(r_2^{\text{prop}})$, the proposal is accepted and so $r_2^{\text{new}} = r_2^{\text{prop}}$.
- If $u \geq A_{r_2^{\text{old}}}(r_2^{\text{prop}})$, the proposal is rejected and so $r_2^{\text{new}} = r_2^{\text{old}}$.

Here, $u \sim \text{Uniform}_{[0,1]}$, as usual.

This strategy of obtaining nonstandard Gibbs samplers is termed *Metropolis–Hastings within Gibbs* or *Metropolis within Gibbs* depending upon the type of proposal used in the indirect sampling steps. Although the resulting schemes are valid and often easy to implement, their performance is generally poor. This is because rejections of the proposed samples lead to highly correlated MCMC chains that require long chains to adequately sample a target. To improve such a scheme, we may iterate *several times* the indirect sampling steps before proceeding. Provided the indirect sampling steps involve little computational cost, this is a viable solution for sampling complex targets.

Similar *within Gibbs* schemes can be developed even when the inner updates are carried out by more specialized samplers that may not necessarily fall in the Metropolis–Hastings family. Generally, our overall MCMC scheme remains valid as long as the sampler executing the updates is in detailed balance with the corresponding conditional target. This may be achieved even when, instead of Metropolis–Hastings, we apply appropriate auxiliary variable or Hamiltonian MCMC samplers that we see in the following sections.

5.4.3 Auxiliary Variable Samplers

Auxiliary variable samplers are more sophisticated than Metropolis–Hastings or Gibbs samplers. The main concept behind each sampler in this family is a *completion* of the target of interest with *auxiliary variables* that may lead to sampling schemes involving distributions different from the original target. Briefly, a given target $\pi(r)$ is first completed,

$$\pi(r) = \sum_{v} p(v, r), \qquad \pi(r) = \int du\, p(u, r),$$

with either discrete v or continuous u auxiliaries and, subsequently, conventional MCMC schemes are applied on the completed target $p(v, r)$ or $p(u, r)$. As the development of an auxiliary variable sampler and the choice of variables v or u are generally problem specific and almost always require an understating of the underlying structure of the problem at hand, here we illustrate only two important cases (Examples 5.13 and 5.14).

Before embarking on these important cases, we highlight the primary advantage and disadvantage of auxiliary variable samplers. As a main advantage, if chosen judiciously, auxiliary variables allow one to identify important axes along which to explore in order to efficiently sample our target. As a disadvantage, as may already be apparent, auxiliary variable samplers add complexity and necessarily involve sampling more variables leading to poor mixing of the MCMC chains.

Example 5.13 Auxiliary variables for counting data

In situations involving counting data, we are often interested in estimating rates that all contribute to a common observation. For example, in the analysis of fluorescence microscopy data, we typically have photon measurements influenced by specific and nonspecific light sources, such as fluorescent molecules and fluorescent background, respectively.

In this example, we consider a light source sending photons to our detector at a rate h and background at a rate f. Since individual photon detections, irrespective of their origin, are independent from each other, our likelihood reads

$$w_n|h,f \sim \text{Poisson}(\tau\beta(h+f)).$$

Here, τ is the integration time of the detector, β is the detector's quantum efficiency, and $w_{1:N}$ are our photon count measurements. As this likelihood does not distinguish between h and f, it is typical to also combine measurements $\bar{w}_{1:\bar{N}}$ from a different experiment counting only background photons,

$$\bar{w}_n|f \sim \text{Poisson}\left(\tau\beta f\right).$$

In this setting, our main objective is almost always to estimate the rates h,f. Accordingly, our target of interest is the posterior $p(h,f|w_{1:N},\bar{w}_{1:\bar{N}})$, which we can obtain under the priors

$$h \sim \text{Gamma}\left(H, \frac{h_{\text{ref}}}{H}\right), \qquad f \sim \text{Gamma}\left(F, \frac{f_{\text{ref}}}{F}\right),$$

where H and F are hyperparameters of the gamma distribution with h_{ref} and f_{ref} its coinciding mean.

Here, priors and likelihood do not result in a conjugate model and thus we resort to MCMC. The naive choice is to apply a Metropolis–Hastings sampler with a bivariate proposal. Nevertheless, with the aid of an auxiliary variable, we may obtain a better alternative. In particular, we consider the auxiliary variable

$$v|h,f,w_{1:N} \sim \text{Binomial}\left(\sum_{n=1}^{N} w_n, \frac{h}{h+f}\right).$$

Completing with v, our target now becomes

$$p(h,f|w_{1:N},\bar{w}_{1:\bar{N}}) = \sum_{v} p(v,h,f|w_{1:N},\bar{w}_{1:\bar{N}}),$$

and the completed target $p(v,h,f|w_{1:N},\bar{w}_{1:\bar{N}})$ can be sampled via a Gibbs scheme with three steps:

- Update v by sampling from

$$p\left(v|h,f,w_{1:N},\bar{w}_{1:\bar{N}}\right) = \text{Binomial}\left(v; W, \frac{h}{h+f}\right). \qquad (*5.10*)$$

- Update h by sampling from

$$p\left(h|v,f,w_{1:N},\bar{w}_{1:\bar{N}}\right) = \text{Gamma}\left(h; H+v, \frac{1}{\frac{H}{h_{\text{ref}}} + N\tau\beta}\right). \qquad (*5.11*)$$

- Update f by sampling from

$$p\left(f|v,h,w_{1:N},\bar{w}_{1:\bar{N}}\right) = \text{Gamma}\left(f; F+W+\bar{W}-v, \frac{1}{\frac{F}{f_{\text{ref}}} + (N+\bar{N})\tau\beta}\right). \qquad (*5.12*)$$

To simplify notation, we have abbreviated $W = \sum_{n=1}^{N} w_n$ and $\bar{W} = \sum_{n=1}^{\bar{N}} \bar{w}_n$.

Example 5.14 The slice sampler

Based on a universal completion $\bar{\pi}(r) = \int_0^{\bar{\pi}(r)} du = \int_0^1 du f(u, r)$, where

$$f(u, r) = \begin{cases} 1, & 0 < u < \bar{\pi}(r), \\ 0, & \bar{\pi}(r) \leq u, \end{cases}$$

we can develop an auxiliary random variable sampler that applies to almost any target $\bar{\pi}(r)$. Specifically, because $f(u, r)$ is essentially a joint density of u and r, we can obtain MCMC samples $u^{\text{new}}, r^{\text{new}}$ with the following Gibbs scheme:

- Sample $u|r^{\text{old}} \sim f(u, r^{\text{old}})$.
- Sample $r|u^{\text{new}} \sim f(u^{\text{new}}, r)$.

In our scheme, we use the joint $f(u, r)$ as it is proportional to the appropriate conditionals. Sampling each step is exceptionally easy as $f(u, r)$ is a uniform density and so the conditionals $f(u, r^{\text{old}})$ and $f(u^{\text{new}}, r)$ also remain uniform. More precisely, the first one is $\text{Uniform}_{[0, \bar{\pi}(r^{\text{old}})]}(u)$ while the second one is $\text{Uniform}_{H^{\text{new}}}(r)$, where H^{new} includes all feasible r with $\bar{\pi}(r) \geq u^{\text{new}}$. Each H^{new} is termed a *slice*. For this reason, this sampler is known as *slice sampler*.

5.4.4 Metropolis–Hastings Samplers with Deterministic Proposals

In the generic Metropolis–Hastings sampler that we discussed in Section 5.2.1, we saw that each proposal requires the generation of a random variable. However, this is not always necessary. In fact, we can develop valid Metropolis–Hastings samplers with *deterministic proposals* as well. To see how this is possible, we start with a target $\pi(r)$ on which we initially apply a Metropolis–Hastings scheme with proposals

$$Q^{\sigma}_{r^{\text{old}}}(r^{\text{prop}}) = \text{Normal}\left(r^{prop}; F(r^{\text{old}}), \sigma^2\right)$$

that depend on a positive real constant σ. For simplicity, we assume that our targeted random variable r is univariate. In this setting, we obtain a deterministic proposal as $\sigma \to 0$ since, in this case, $\text{Normal}\left(F(r^{\text{old}}), \sigma^2\right) \to \delta_{F(r^{\text{old}})}$. Below, we highlight the precise steps involved.

To facilitate the presentation that follows, we obtain $r^{\text{prop}}|r^{\text{old}} \sim \text{Normal}\left(F(r^{\text{old}}), \sigma^2\right)$ via additive perturbations $r^{\text{prop}} = F(r^{\text{old}}) + \sigma\epsilon$, where $\epsilon \sim \text{Normal}(0, 1)$ is independent of r^{prop}. Unlike the random walks seen earlier, where the proposals are concentrated around r^{old}, here we invoke a function $F(r^{\text{old}})$ enforcing proposals towards a position different from r^{old}.

Under our proposals, the acceptance ratio of Eq. (5.7) also depends on σ and is given by

$$A^{\sigma}_{r^{\text{old}}}(r^{\text{prop}}) = \frac{\bar{\pi}(r^{\text{prop}})}{\bar{\pi}(r^{\text{old}})} \exp\left(-\frac{1}{2}\left(\frac{r^{\text{old}} - F\left(F(r^{\text{old}}) + \sigma\epsilon\right)}{\sigma}\right)^2 + \frac{\epsilon^2}{2}\right).$$

Provided that $F(r)$ is a *continuous* function, we may simplify this ratio by considering proposals $Q^{\sigma}_{r^{\text{old}}}(r^{\text{prop}})$ increasingly concentrated around $F(r^{\text{old}})$. Formally, we may apply a limit $\sigma \rightarrow 0$ and obtain $r^{\text{old}} - F\left(F(r^{\text{old}}) + \sigma\epsilon\right) \rightarrow r^{\text{old}} - F\left(F(r^{\text{old}})\right)$.

In general, $r \neq F(F(r))$, and so $A^{\sigma}_{r^{\text{old}}}(r^{\text{prop}}) \rightarrow 0$, indicating that as our proposals become less probabilistic, our sampler becomes less likely to accept the proposed r^{prop} and, eventually, precisely at the deterministic limit where $\sigma = 0$, our sampler always rejects it. Nevertheless, provided that $F(r)$ is an *involution* function, *e.g.*, $r = F(F(r))$, we may end up with a nonzero limiting acceptance rate. This is because, in this case, we encounter an apparent indeterminate 0/0 limit that we can resolve under appropriate conditions.

In particular, provided $F(r)$ is also a *differentiable* function, we may use L'Hospital's rule to compute the limiting $A^{\sigma}_{r^{\text{old}}}(r^{\text{prop}})$ as

$$A^{\sigma}_{r^{\text{old}}}(r^{\text{prop}}) \rightarrow \frac{\bar{\pi}(r^{\text{prop}})}{\bar{\pi}(r^{\text{old}})} \exp\left(\frac{\epsilon^2}{2}\left[1 - \left(F'\left(F(r^{\text{old}})\right)\right)^2\right]\right)$$

in terms of the derivative $F'(r)$ of $F(r)$. This limiting $A^{\sigma}_{r^{\text{old}}}(r^{\text{prop}})$ is nonzero, indicating that our sampler is able, at least in principle, to accept a deterministic move from r^{old} to $r^{\text{prop}} = F(r^{\text{old}})$.

Finally, provided $F(r)$ is a function that *preserves lengths, i.e.*, $|F'(r)| = 1$, our limiting $A^{\sigma}_{r^{\text{old}}}(r^{\text{prop}})$ simplifies even further and takes the Metropolis form $A_{r^{\text{old}}}(r^{\text{prop}}) = \bar{\pi}(r^{\text{prop}})/\bar{\pi}(r^{\text{old}})$.

Gathering everything together, when our deterministic proposals $r^{\text{prop}} = F(r^{\text{old}})$ meet all of the above conditions, they lead to a valid Metropolis sampler. This is not only true of scalar random variables, but extends to the multi-dimensional case as well.

> **Note 5.9** Deterministic proposals
>
> To use deterministic proposals $r^{\text{prop}} = F(r^{\text{old}})$ in a Metropolis–Hastings scheme, we need to ensure that the function $F(r)$ is:
>
> 1. *Continuous* and *differentiable*.
> 2. *Involution, i.e.*, $F(F(r)) = r$.
> 3. *Volume preserving, i.e.*, $|J_{r \rightarrow F(r)}| = 1$ where $J_{r \rightarrow F(r)}$ is the Jacobian.
>
> Under these conditions, the acceptance ratio reduces to a Metropolis form.

In practice, samplers with naive deterministic proposals are of little interest themselves as they are generally nonergodic. At best, they alternate samples between the values of r^{old} and r^{new} and we need additional samplers or auxiliary variables to achieve ergodicity. Nevertheless, samplers with sophisticated deterministic proposals can be quite useful, especially when $F(r)$ is adapted to the target $\bar{\pi}(r)$. For instance, the Hamiltonian

MCMC samplers, which we describe in the next section, are essentially particular examples of deterministic samplers for which the proposals are specialized to the targets at hand.

5.4.5 Hamiltonian MCMC Samplers

Samplers in the Hamiltonian family apply only on *continuous* targeted variables and are best suited for multidimensional problems. What makes them particularly powerful in this setting is that, unlike Gibbs samplers, they are able to update all variables of a *high dimensional target* simultaneously, while, unlike Metropolis–Hastings samplers, maintaining a *high acceptance rate.* Although they are more complex to develop and calibrate, Hamiltonian MCMC samplers typically outperform other MCMC schemes that may also apply. To illustrate the sampling mechanism, first we consider in detail a univariate target and then describe the multivariate case, focusing on implementation details separately.

For convenience, in this section only we adopt the column-vector convention and denote all vectors with columns.

Univariate Target

Following our usual convention, we consider a univariate target with density $\bar{\pi}(r)$ that, generally, need not be normalized. At its core, a Hamiltonian sampler is an *auxiliary variable sampler* that, first, completes the target

$$\pi(r) = \int du\, p(u, r),$$

with a continuous variable u and, subsequently, samples the completed target $p(u, r)$ via a *Gibbs scheme* with the following two steps:

- Update u by sampling from $p(u|r)$.
- Update u, r jointly by sampling from $p(u, r)$.

Commonly, the auxiliary variable u is chosen to allow for direct sampling in the first Gibbs update. The second Gibbs update, however, is performed via a Metropolis–Hastings scheme that has *deterministic proposals* of the form

$$\begin{pmatrix} u^{\text{prop}} \\ r^{\text{prop}} \end{pmatrix} = F \begin{pmatrix} u^{\text{old}} \\ r^{\text{old}} \end{pmatrix}.$$

According to Note 5.9, such proposals must stem from a function $F(\cdot)$ that is involution and volume preserving. One way to fulfill these conditions is by invoking Hamiltonian dynamics.

In particular, a volume-preserving proposal is obtained by applying L successive *symplectic integration* steps to evolve along a trajectory that starts on $u^{\text{old}}, r^{\text{old}}$. This proposal becomes an involution provided a *reversion* step is applied on the terminal point of the trajectory flipping the sign of u. Schematically, the proposal mechanism is encoded in the sequence

$$\begin{pmatrix} u^{\text{old}} \\ r^{\text{old}} \end{pmatrix} = \begin{pmatrix} u_0^{\text{temp}} \\ r_0^{\text{temp}} \end{pmatrix} \xrightarrow[H]{\text{integrate}} \begin{pmatrix} u_1^{\text{temp}} \\ r_1^{\text{temp}} \end{pmatrix} \xrightarrow[H]{\text{integrate}}$$

$$\cdots \xrightarrow[H]{\text{integrate}} \begin{pmatrix} u_L^{\text{temp}} \\ r_L^{\text{temp}} \end{pmatrix} \xrightarrow[R]{\text{reverse}} \begin{pmatrix} -u_L^{\text{temp}} \\ +r_L^{\text{temp}} \end{pmatrix} = \begin{pmatrix} u^{\text{prop}} \\ r^{\text{prop}} \end{pmatrix},$$

with the operations H and R defined in Note 5.10.

Note 5.10 Operators

Both the reversion and integration steps in this sequence correspond to transformations:

$$\begin{pmatrix} u' \\ r' \end{pmatrix} = R \begin{pmatrix} u \\ r \end{pmatrix}, \qquad \begin{pmatrix} u' \\ r' \end{pmatrix} = H \begin{pmatrix} u \\ r \end{pmatrix}.$$

For instance, reversions have a simple form given by the linear operator

$$R = \begin{bmatrix} -1 & 0 \\ 0 & +1 \end{bmatrix}.$$

By contrast, integrations are problem specific and generally their operator H is nonlinear. Nevertheless, by their definition, all symplectic integrators satisfy a *characteristic property*

$$J^T_{(u,r)\mapsto H(u,r)} \begin{bmatrix} 0 & +1 \\ -1 & 0 \end{bmatrix} J_{(u,r)\mapsto H(u,r)} = \begin{bmatrix} 0 & +1 \\ -1 & 0 \end{bmatrix}.$$

Here, $J_{(u,r)\mapsto H(u,r)}$ is the Jacobian of the transformation and given by

$$J_{(u,r)\mapsto(u',r')} = \begin{bmatrix} \partial u'/\partial u & \partial u'/\partial r \\ \partial r'/\partial u & \partial r'/\partial r \end{bmatrix}.$$

Under this property, we can verify that the resulting proposal is indeed a volume preserving involution, as required in Note 5.9.

Symplectic operators are obtained by first specifying a Hamiltonian $\mathcal{H}(u, r)$ of an analogous dynamical system with *position* $r(t)$ and *momentum* $u(t)$ and then properly discretizing its canonical equations of motion

$$\frac{du}{dt} = -\frac{\partial \mathcal{H}(u, r)}{\partial r}, \qquad \frac{dr}{dt} = +\frac{\partial \mathcal{H}(u, r)}{\partial u}.$$

For instance, the popular *leapfrog* integrator assumes a separable Hamiltonian,

$$\mathcal{H}(u, r) = \frac{u^2}{2\mu} + \mathcal{U}(r),$$

with mass μ and potential $\mathcal{U}(r)$ for which the canonical equations read

$$\frac{du}{dt} = -\frac{d}{dr}\mathcal{U}(r), \qquad \frac{dr}{dt} = \frac{u}{\mu}.$$

Subsequently, the equations of motion are *discretized symplectically* over a time interval of duration τ via the expansions

$$u(t+\tau/2) \approx u(t) - \frac{\tau}{2}\frac{d}{dr}\mathcal{U}(r(t)),$$

$$r(t+\tau) \approx r(t) + \frac{\tau}{\mu}u(t+\tau/2),$$

$$u(t+\tau) \approx u(t+\tau/2) - \frac{\tau}{2}\frac{d}{dr}\mathcal{U}(r(t+\tau)).$$

Ignoring the approximations, these equations are combined leading to the integrator

$$\begin{pmatrix} u' \\ r' \end{pmatrix} = \begin{pmatrix} u \\ r \end{pmatrix} - \frac{1}{2}\begin{pmatrix} G(r)+G(r') \\ \frac{\tau}{\mu}G(r) - 2\frac{\tau}{\mu}u \end{pmatrix}, \qquad G(r) = \tau\frac{d}{dr}\mathcal{U}(r).$$

The proposal we obtain under these choices leads to an acceptance ratio

$$A_{u^{\text{old}},r^{\text{old}}}\left(u^{\text{prop}}, r^{\text{prop}}\right) = \frac{p\left(u^{\text{prop}}, r^{\text{prop}}\right)}{p\left(u^{\text{old}}, r^{\text{old}}\right)} = \frac{p\left(u^{\text{prop}}|r^{\text{prop}}\right)}{p\left(u^{\text{old}}|r^{\text{prop}}\right)}\frac{\bar{\pi}\left(r^{\text{prop}}\right)}{\bar{\pi}\left(r^{\text{old}}\right)},$$

which can be evaluated even when our target remains unnormalized.

Although the proposal function $F(\cdot)$ that results from any choice of symplectic integrator is valid, the performance of the sampler is considerably improved when the integrators H are *specialized* to the target of interest. For this reason, most commonly the Hamiltonian of the analogous system is chosen to match the completed target using

$$\mathcal{H}(u,r) = -\frac{\log p(u,r)}{\beta} = -\frac{\log p(u|r)}{\beta} - \frac{\log \bar{\pi}(r)}{\beta},$$

where β is an adjustable parameter. In this special case, the acceptance ratio takes the form

$$A_{u^{\text{old}},r^{\text{old}}}\left(u^{\text{prop}}, r^{\text{prop}}\right) = e^{\beta\mathcal{H}(u^{\text{old}},r^{\text{old}}) - \beta\mathcal{H}(u^{\text{prop}},r^{\text{prop}})}.$$

The development of a Hamiltonian sampler can be further simplified if the auxiliary variable u chosen is independent of the targeted one r and, in addition, chosen via $u \sim \text{Normal}(0, \mu/\beta)$, where μ is another adjustable parameter. In this case, the acceptance ratio takes the Metropolis form $A_{u^{\text{old}},r^{\text{old}}}\left(u^{\text{prop}}, r^{\text{prop}}\right) = \bar{\pi}(r^{\text{prop}})/\bar{\pi}(r^{\text{old}})$ of our original target and, since this ratio does not depend on the auxiliary variable, the reversion step can be neglected. With this particular auxiliary choice, the resulting Hamiltonian is identical to the Hamiltonian underlying the leapfrog integrator with mass μ and potential $\mathcal{U}(r) = -\frac{\log \bar{\pi}(r)}{\beta}$.

Example 5.15 Hamiltonian MCMC

In this example, we consider a univariate target $\pi(r)$ as shown in Fig. 5.13. To develop a Hamiltonian sampler, we complete with an auxiliary momentum $u \sim \text{Normal}\left(0, \mu/\beta\right)$. The resulting completed target $p(u, r)$ is also shown in Fig. 5.13.

Once an auxiliary momentum u^{old} is obtained, our sampler attempts a joint proposal that results by sliding along the trajectories of an analogous Hamiltonian $\mathcal{H}(u,r) = -\frac{\log p(u,r)}{\beta}$. In this particular example, the Hamiltonian takes the form $\mathcal{H}(u,r) = \frac{u^2}{2\mu} - \frac{\log \pi(r)}{\beta}$.

Figure 5.13 shows such a trajectory and highlights its initial and terminal ends as well as the proposal $u^{\text{prop}}, r^{\text{prop}}$ resulting after reversion of the terminal momentum. Due to the symplectic integration and the reversion, a trajectory starting at $u^{\text{prop}}, r^{\text{prop}}$ traces the opposite direction returning to $u^{\text{old}}, r^{\text{old}}$, indicating that, indeed, our proposal scheme is an involution. Finally, because the Hamiltonian $\mathcal{H}(u,r)$ matches our completed target, our trajectories effectively preserve the value of $p(u,r)$ resulting in a high acceptance ratio.

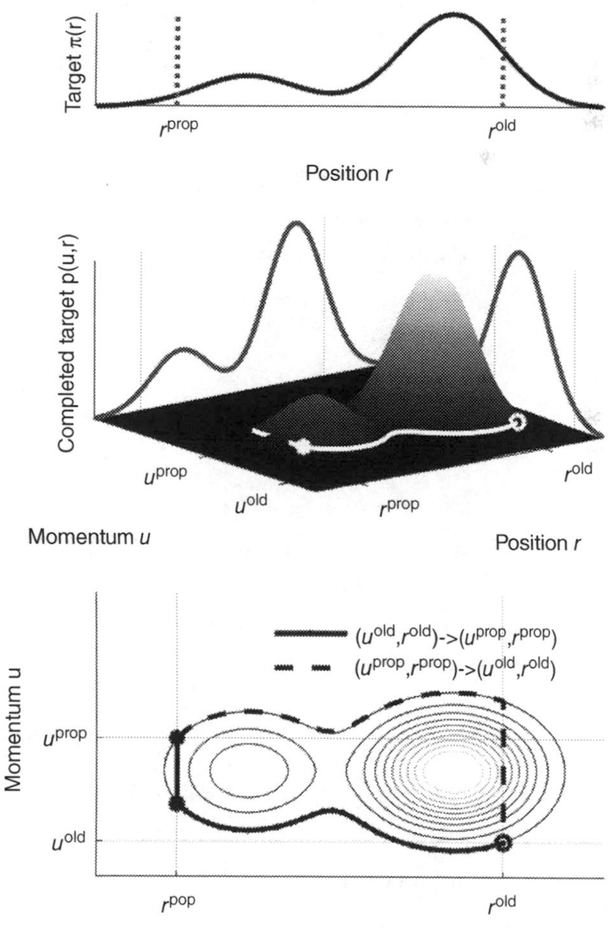

Fig. 5.13 Hamiltonian MCMC. (Top) Our target $\pi(r)$. (Middle and Bottom) The resulting completed target $p(u,r)$ along with a Hamiltonian trajectory. For comparison, the bottom panel shows also the reverse trajectory that starts at $u^{\text{prop}}, r^{\text{prop}}$ and terminates at $u^{\text{old}}, r^{\text{old}}$.

Multivariate Target

In the multidimensional case, our targeted density is $\bar{\pi}(\boldsymbol{r})$ and our random variable $\boldsymbol{r} = r_{1:M}$ consists of M scalar components r_m. A Hamiltonian sampler proceeds via the completion

$$\pi(\boldsymbol{r}) = \int d\boldsymbol{u}\, p(\boldsymbol{u}, \boldsymbol{r}),$$

with a multidimensional auxiliary variable $\boldsymbol{u} = u_{1:M}$ that also consists of M scalar components u_m. Similar to the univariate case, the completed target is sampled via two Gibbs updates:

- Update $\boldsymbol{u}$ by sampling from $p(\boldsymbol{u}|\boldsymbol{r})$.
- Update $\boldsymbol{u}, \boldsymbol{r}$ jointly by sampling from $p(\boldsymbol{u}, \boldsymbol{r})$.

As before, a choice of auxiliaries for the first step and a deterministic proposal for the second step need to be made. An auxiliary choice that allows for fast sampling, especially in a highly multivariate setting, is

$$u_m \sim \text{Normal}\left(0, \mu_m/\beta\right), \qquad m = 1 : M.$$

Volume-preserving involutions for the proposals are also obtained by invoking Hamiltonian dynamics followed by a reversion in a sequential scheme similar to the univariate case. Adopting block-matrix notation, the reversion in this proposal scheme still corresponds to a linear operator,

$$\begin{pmatrix} \boldsymbol{u}' \\ \boldsymbol{r}' \end{pmatrix} = R \begin{pmatrix} \boldsymbol{u} \\ \boldsymbol{r} \end{pmatrix}, \qquad R = \begin{bmatrix} -\mathbb{1} & 0 \\ 0 & +\mathbb{1} \end{bmatrix},$$

and the characteristic property of the symplectic integrator reads

$$J^T_{(\boldsymbol{u},\boldsymbol{r})\mapsto H(\boldsymbol{u},\boldsymbol{r})} \begin{bmatrix} 0 & +\mathbb{1} \\ -\mathbb{1} & 0 \end{bmatrix} J_{(\boldsymbol{u},\boldsymbol{r})\mapsto H(\boldsymbol{u},\boldsymbol{r})} = \begin{bmatrix} 0 & +\mathbb{1} \\ -\mathbb{1} & 0 \end{bmatrix}.$$

Here, $\mathbb{1}$ stands for the identity matrix of size M. The analogous Hamiltonian remains separable and reads

$$\mathcal{H}(\boldsymbol{u}, \boldsymbol{r}) = \frac{1}{2} \sum_{m=1}^{M} \frac{u_m^2}{\mu_m} + \mathcal{U}(\boldsymbol{r}), \qquad \mathcal{U}(\boldsymbol{r}) = -\frac{\log \bar{\pi}(\boldsymbol{r})}{\beta}.$$

Under this Hamiltonian, the resulting leapfrog integrator reads

$$\begin{pmatrix} u_1' \\ \vdots \\ u_M' \\ r_1' \\ \vdots \\ r_M' \end{pmatrix} = \begin{pmatrix} u_1 \\ \vdots \\ u_M \\ r_1 \\ \vdots \\ r_M \end{pmatrix} - \frac{1}{2} \begin{pmatrix} G_1(\boldsymbol{r}) + G_1(\boldsymbol{r}') \\ \vdots \\ G_M(\boldsymbol{r}) + G_M(\boldsymbol{r}') \\ \frac{\tau}{\mu_1} G_1(\boldsymbol{r}) - 2\frac{\tau}{\mu_1} u_1 \\ \vdots \\ \frac{\tau}{\mu_M} G_M(\boldsymbol{r}) - 2\frac{\tau}{\mu_M} u_M \end{pmatrix}, \qquad \begin{pmatrix} G_1(\boldsymbol{r}) \\ \vdots \\ G_M(\boldsymbol{r}) \end{pmatrix} = \tau \begin{pmatrix} \frac{\partial}{\partial r_1} \mathcal{U}(\boldsymbol{r}) \\ \vdots \\ \frac{\partial}{\partial r_M} \mathcal{U}(\boldsymbol{r}) \end{pmatrix}.$$

The acceptance ratio takes the Metropolis form $A_{\boldsymbol{u}^{\text{old}},\boldsymbol{r}^{\text{old}}}(\boldsymbol{u}^{\text{prop}},\boldsymbol{r}^{\text{prop}}) = \bar{\pi}(\boldsymbol{r}^{\text{prop}})/\bar{\pi}(\boldsymbol{r}^{\text{old}})$ and, since this ratio does not depend on the auxiliary variable, the reversion step can also be neglected, which leads to a simplified proposal scheme

$$\begin{pmatrix}\boldsymbol{u}^{\text{old}}\\ \boldsymbol{r}^{\text{old}}\end{pmatrix} = \begin{pmatrix}\boldsymbol{u}_0^{\text{temp}}\\ \boldsymbol{r}_0^{\text{temp}}\end{pmatrix} \xrightarrow[H]{\text{integrate}} \begin{pmatrix}\boldsymbol{u}_1^{\text{temp}}\\ \boldsymbol{r}_1^{\text{temp}}\end{pmatrix} \xrightarrow[H]{\text{integrate}} \cdots \xrightarrow[H]{\text{integrate}} \begin{pmatrix}\boldsymbol{u}_L^{\text{temp}}\\ \boldsymbol{r}_L^{\text{temp}}\end{pmatrix} = \begin{pmatrix}\boldsymbol{u}^{\text{prop}}\\ \boldsymbol{r}^{\text{prop}}\end{pmatrix}.$$

In this scheme, advancing from $\boldsymbol{u}_{\ell-1}^{\text{temp}},\boldsymbol{r}_{\ell-1}^{\text{temp}}$ to $\boldsymbol{u}_{\ell}^{\text{temp}},\boldsymbol{r}_{\ell}^{\text{temp}}$ can be broken down in three substeps:

$$\begin{aligned}
\text{v}_\ell^{\text{temp}} &= \boldsymbol{u}_{\ell-1}^{\text{temp}} - \frac{1}{2}G\left(\boldsymbol{r}_{\ell-1}^{\text{temp}}\right),\\
\boldsymbol{r}_\ell^{\text{temp}} &= \boldsymbol{r}_{\ell-1}^{\text{temp}} + \frac{\tau}{\mu}\text{v}_\ell^{\text{temp}},\\
\boldsymbol{u}_\ell^{\text{temp}} &= \text{v}_\ell^{\text{temp}} - \frac{1}{2}G\left(\boldsymbol{r}_\ell^{\text{temp}}\right).
\end{aligned}$$

That is, the three substeps involve: updating the momentum for a half-step from $\boldsymbol{u}_{\ell-1}$ to an intermediate value v_ℓ; updating the position for a full-step from $\boldsymbol{r}_{\ell-1}$ to $\boldsymbol{r}_\ell$; and, finally, updating the intermediate momentum for a half-step from v_ℓ to $\boldsymbol{u}_\ell$, respectively. A naive implementation of this marching scheme requires two separate evaluations of $G(\cdot)$, *i.e.*, one per step. However, the scheme can be implemented such that $G(\cdot)$ is interlaced and evaluated only once per step. This leads to a substantial speed up as, typically, $G(\cdot)$ is the most expensive computation. As only the initial and terminal ends of the proposal sequence are required, the entire scheme from $\boldsymbol{u}^{\text{old}},\boldsymbol{r}^{\text{old}}$ to $\boldsymbol{u}^{\text{prop}},\boldsymbol{r}^{\text{prop}}$ can be implemented efficiently as in Algorithm 5.3.

Algorithm 5.3 Hamiltonian marching scheme

Given an existing sample $\boldsymbol{u}^{\text{old}},\boldsymbol{r}^{\text{old}}$ of a Hamiltonian sampler, compute a proposal $\boldsymbol{u}^{\text{prop}},\boldsymbol{r}^{\text{prop}}$ with the steps:

- Initialize the momentum and position,
$$\boldsymbol{u}^{\text{prop}} = \boldsymbol{u}^{\text{old}},$$
$$\boldsymbol{r}^{\text{prop}} = \boldsymbol{r}^{\text{old}}.$$
- Get the gradient
$$g = \tau G(\boldsymbol{r}^{\text{prop}}).$$
- Update the momentum for the *half* step
$$\boldsymbol{u}^{\text{prop}} = \boldsymbol{u}^{\text{prop}} - \frac{g}{2}.$$
- Repeat $L-1$ times:
  - Update the position for the *full* step
$$\boldsymbol{r}^{\text{prop}} = \boldsymbol{r}^{\text{prop}} + \frac{\tau}{\mu}\boldsymbol{u}^{\text{prop}}.$$

– Get the gradient

$$\boldsymbol{g} = \tau G(\boldsymbol{r}^{\text{prop}}).$$

– Update the momentum for the *full* step

$$\boldsymbol{u}^{\text{prop}} = \boldsymbol{u}^{\text{prop}} - \boldsymbol{g}.$$

- Update the position for the *full* step

$$\boldsymbol{r}^{\text{prop}} = \boldsymbol{r}^{\text{prop}} + \frac{\tau}{\mu}\boldsymbol{u}^{\text{prop}}.$$

- Get the gradient

$$\boldsymbol{g} = \tau G(\boldsymbol{r}^{\text{prop}}).$$

- Update the momentum for the *half* step

$$\boldsymbol{u}^{\text{prop}} = \boldsymbol{u}^{\text{prop}} - \frac{\boldsymbol{g}}{2}.$$

For a complete Hamiltonian sampler, the auxiliary variables $\boldsymbol{u}^{\text{old}}$ must be sampled prior to marching and the acceptance test must be executed after marching. As the auxiliaries will be resampled in the sampler's next iteration, upon acceptance only $\boldsymbol{r}^{\text{new}}$ needs to be maintained.

5.5 Exercise Problems

Exercise 5.1 MC sampling

In the same context as in Exercise 1.18, assume that R_1, R_2, R_3 are iid $\text{Uniform}_{[-1,+1]}$ random variables and develop a Monte Carlo method to estimate the probability of the polynomial $r_1x^2 + r_2x + r_3$ having real roots.

Exercise 5.2 MC sampling of prior and posterior

In the context of Example 5.1, generate MC samples to characterize both the prior and posterior probability distributions of the parameters. For the characterization, use the generated samples to create histograms as well as to compute mean values.

Exercise 5.3 Normalization

1. Recover the normalized form $\pi(r)$ of the unnormalized target $\bar{\pi}(r)$ of Example 5.2.
2. Show analytically that $\bar{\pi}(r) = r^{-1}(1-r)^{-1}$ cannot be normalized over the interval $0 < r < 1$.

Exercise 5.4 A Metropolis–Hastings sampler for a truncated normal target

Develop a Metropolis–Hastings sampler to generate samples from a normal random variable that is truncated between r_{min} and r_{max}. Use normal and beta proposals. For the latter proposal, use a translation and a scaling such that the proposed values fall between r_{min} and r_{max}.

Exercise 5.5 A Metropolis sampler for Cauchy and Laplace targets

Develop Metropolis samplers to generate samples from a Cauchy random variable and a Laplace random variable.

Exercise 5.6 Acceptance ratios

1. Verify the identity in Note 5.5.
2. Verify the formulas for the acceptance ratios in Examples 5.2 and 5.3.

Exercise 5.7 Gibbs sampler for mixture models

In Example 5.10 we developed a Gibbs sampler for a sum of a Normal $(\mu_1, 1/\tau)$ and a Normal$(\mu_2, 1/\tau)$. These are often called *mixture models*. In the example, we assumed that the observations $w_{1:N}$ stem from the first and second components with known probabilities π_1 and $\pi_2 = 1 - \pi_1$, respectively. However, generally, the value of π_1 may be unknown and so we have to estimate it just like any other variable. In this setting, apply a beta prior on π_1 and develop a Gibbs sampler to draw posterior samples from $p(s_{1:N}, \pi_1, \mu_{1:2}, \tau | w_{1:N})$.

Exercise 5.8 A hyper-hyper-model for normal likelihoods

In Example 5.9 we applied independent normal and gamma priors to estimate the center and spread of an underlying normal distribution. In doing so, we used known values for the hyperparameters ξ, ψ, α, β. However, in many practical applications, specifying values for these hyperparameters might be difficult. In such cases, we may apply a hyper-hyper-model of the form

$$\xi \sim \text{Normal}(\eta, \zeta),$$
$$\psi \sim \text{Gamma}(\kappa, \lambda),$$
$$\alpha \sim \text{Gamma}(\gamma, \omega),$$
$$\beta \sim \text{Gamma}(\rho, \phi),$$
$$\tau | \alpha, \beta \sim \text{Gamma}(\alpha, \beta),$$

$$\mu|\xi, \psi \sim \text{Normal}\left(\xi, \frac{1}{\psi}\right),$$

$$w_n|\mu, \tau \sim \text{Normal}\left(\mu, \frac{1}{\tau}\right).$$

Assume the values of the hyper-hyperparameters $\eta, \zeta, \kappa, \lambda, \gamma, \omega, \rho, \phi$ are given and develop a Metropolis within Gibbs scheme to sample from the joint posterior $p(\mu, \tau, \alpha, \beta, \xi, \psi|w_{1:N})$. In doing so, use Gibbs updates for $\mu, \tau, \beta, \xi, \psi$ and a Metropolis update for α.

Exercise 5.9 Interpretation of MCMC

Develop a Metropolis (additive) random walk to sample from a bivariate target $\boldsymbol{r} = (r_1, r_2)$ with the density

$$\pi(\boldsymbol{r}) \propto \exp\left(-10(r_1^2 - r_2)^2 - \left(r_2 - \frac{1}{4}\right)^4\right).$$

For the perturbations that define the random walk use

$$\boldsymbol{\epsilon}^{\text{prop}} \sim \text{Normal}_2\left((0, 0), \begin{pmatrix} 1 & 0 \\ 0 & 1 \end{pmatrix}\right)$$

with three different scalings $\lambda = 0.01$, $\lambda = 1$, and $\lambda = 100$. For each case, initialize the chain at $(-1, -1)$ and generate 10^4 samples. Recover approximately MC samples for all cases.

Exercise 5.10 Ergodicity recovery of the Gibbs sampler

1. For the target $\pi(r_1, r_2)$ provided in Example 5.8, consider the following transformation of variables:

$$w_1 = \frac{r_1 + r_2}{2}, \qquad w_2 = \frac{r_1 - r_2}{2}.$$

   Develop a Gibbs sampler for the transformed random variables and show conceptually that this sampler is ergodic.
2. In the sampler of Example 5.8, consider a third Gibbs step that samples r_1, r_2 by updating jointly from the target $\pi(r_1, r_2)$ via a deterministic Metropolis–Hastings step with proposals

$$F(r_1, r_2) = (-r_1, -r_2)$$

   and show conceptually that the resulting sampler is ergodic.

Exercise 5.11 Multiplicative random walk for multimodal targets

Develop a multiplicative random walk sampler to sample a multi-modal target consisting of a sum of gamma distributions

$$\pi(r) = \sum_{m=1}^{M} \omega_m \, \text{Gamma}\left(\eta^2, \frac{\psi_m}{\eta^2}\right)$$

all having means and variances that result in the same signal-to-noise (S/N) ratio, namely η. For your sampler use symmetric proposals and for concreteness consider $M = 5$ modes of equal weight $\omega_m = 1/5$ located over $\psi_m = m$ at a signal-to-noise ratio $\eta = 50$.

Project 5.1 Auxiliary variables for counting data

In Example 5.13 we saw how an auxiliary variable sampler can be used to characterize a fluorescent microscopy model with two unknown photon emission rates. In this exercise, we develop an auxiliary variable sampler for general models that may model more than two photon sources and so they may contain more than two rates. Specifically, our model consists of

$$\lambda_m \sim \text{Gamma}\left(\phi_m, \frac{\zeta_m}{\phi_m}\right), \qquad m = 1:M,$$

$$w_n | \lambda_{1:M} \sim \text{Poisson}\left(\sum_{m=1}^{M} \lambda_m\right), \qquad n = 1:N.$$

Here, $w_{1:N}$ are the measurements, $\lambda_{1:M}$ are parameters with unknown values, and $\phi_{1:M}, \zeta_{1:M}$ are hyperparameters with known values. Complete the model with an auxiliary variable

$$v | \lambda_{1:M}, w_{1:N} \sim \text{Multinomial}\left(\sum_{n=1}^{N} w_n, \left(\frac{\lambda_1}{\sum_{m=1}^{M} \lambda_m}, \dots, \frac{\lambda_M}{\sum_{m=1}^{M} \lambda_m}\right)\right)$$

and describe a Gibbs sampler for the completed posterior $p(v, \lambda_{1:M} | w_{1:N})$.

Project 5.2 The banana distribution

The *banana distribution* is a probability distribution with density over real variables (x, y) given by

$$\pi(x, y) \propto \exp\left(-10\left(x^2 - y\right)^2 - \left(y - \frac{1}{4}\right)^4\right).$$

1. Develop a *Metropolis–Hastings* algorithm that generates MCMC samples from $\pi(x, y)$.
2. Implement the algorithm of the previous step.
3. Verify that your implementation of the previous step generates samples with the correct statistics.
4. Develop a *Hamiltonian* algorithm that generates MCMC samples from $\pi(x, y)$.
5. Implement the algorithm of the previous step.
6. Verify that your implementation of the previous step generates samples with the correct statistics.

Additional Reading

C. P. Robert, G. Casella. *Introducing Monte Carlo methods with R*. Springer, 2009.

C. P. Robert, G. Casella. *Monte Carlo statistical methods*. 2nd ed. Springer, 2010.

J. S. Liu. *Monte Carlo strategies in scientific computing*. Springer, 2002.

S. Brooks, A. Gelman, G. Jones, X. L. Meng. *Handbook of Markov chain Monte Carlo*. Chapman & Hall/CRC, 2011.

A. E. Gelfand, A. F. M. Smith. Sampling-based approaches to calculating marginal densities. *J. Am. Stat. Assoc.*, 85:398, 1990.

S. Dutta. Multiplicative random walk Metropolis–Hastings on the real line. *Sankhya B*, 74:315, 2012.

R. Neal. Slice sampling. *Ann. Stat.*, 31:3:705, 2003.

S. Sharma. Markov chain Monte Carlo methods for Bayesian data analysis in astronomy. *Ann. Rev. Astron. Astrophys.*, 55:213, 2017.

PART II

STATISTICAL MODELS

6 Regression Models

By the end of this chapter, we will have presented

- *The regression problem*
- *Gaussian processes*
- *Beta Bernoulli processes*

In the previous chapters, we focused on estimating parameters of a finite number. As we will see, these approaches are useful in *parametric regression* where the data are fitted with a parametric curve, *i.e.*, a curve whose form is parametrized by a finite number of parameters. Common examples of such curves include linear and quadratic polynomials. In this chapter, we will see that parametric regression eventually helps us motivate more powerful *nonparametric regression* frameworks where, formally, infinite numbers of parameters are estimated, such as arbitrary continuous curves defined at every point on the real line.

6.1 The Regression Problem

We begin with the definition of the regression problem.

Note 6.1 The regression problem

Often, we encounter datasets consisting of pairs $w_n = (x_n, y_n)$ where x_n is an *input* value and y_n is the corresponding *output* value. Occasionally, we may also encounter datasets consisting of triads x_n, y_n, v_n where our input/output pair is augmented with an uncertainty such as where v_n is the variance. The goal of regression is to identify the best fit curve passing through the data.

We start in this section with parametric regression, where we assume a mathematical form for the curve. The most common form of parametric regression is, invariably, *linear regression* where the curve is assumed to be a straight line with slope and intercept treated as parameters to be determined by the data.

Example 6.1 Linear regression

Suppose we have data $y_{1:N}$ expected to satisfy the linear form $f(x_n) = \alpha x_n + \beta$, where α and β, the slope and intercept, are parameters to be estimated. Here, x_n is the independent variable and our data, w_n, coincides with the associated random variable at that location, y_n.

In linear regression, we commonly assume the following generative model:

$$y_n = f(x_n) + \epsilon_n, \qquad \epsilon_n \sim \text{Normal}(0, v_n),$$

with y_n corrupted by normal measurement noise. Under this generative model, our likelihood reads

$$p(y_{1:N}|\alpha, \beta) = \prod_{n=1}^{N} p(y_n|\alpha, \beta) \propto \prod_{n=1}^{N} e^{-\frac{1}{2v_n}(y_n-(\alpha x_n+\beta))^2}.$$

The principle of maximum likelihood here reduces to what is sometimes termed χ^2-*minimization*, *i.e.*, the minimization of $\sum_{n=1}^{N}(y_n - (\alpha x_n + \beta))^2/v_n$ with respect to both α and β.

Generally, in fitting data, it is often assumed that v_n is the same for all data points and thus set to v. In this circumstance, χ^2-minimization leads to the *least squares estimator*. Indeed, least squares estimators for the slope and intercept are obtained as follows:

$$\frac{d}{d\alpha}\sum_{n=1}^{N}(y_n - (\alpha x_n + \beta))^2/v = 0, \qquad \frac{d}{d\beta}\sum_{n=1}^{N}(y_n - (\alpha x_n + \beta))^2/v = 0.$$

Evaluating these derivatives immediately results in coupled equations for the estimators $\hat{\alpha}$ and $\hat{\beta}$,

$$\hat{\alpha} = \frac{\sum_{n=1}^{N}\left(y_n - \hat{\beta}\right)x_n}{\sum_{n=1}^{N} x_n^2}, \qquad \hat{\beta} = \sum_{n=1}^{N}(y_n - \hat{\alpha}x_n),$$

which can be readily solved in order to obtain the values of both $\hat{\alpha}$ and $\hat{\beta}$.

In Example 6.1, the least squares estimates for the slope and intercept first required that we assume the data could be fit by a straight line. However, even if a straight line fit were warranted, we are still left with the possibility that, while the least square estimator returned one value for $\hat{\alpha}$ and $\hat{\beta}$, there are a range of values for the parameters α and β acceptable within error. That is, many choices of α and β are acceptable within the known variance. Thus, if for each n we allow $(y_n - (\alpha x_n + \beta))^2$ to deviate from zero by as much as v, then all acceptable values of α and β for which the following inequality holds are acceptable:

$$\sum_{n=1}^{N}(y_n - (\alpha x_n + \beta))^2 \lesssim Nv.$$

Indeed, the range of acceptable parameter combinations only grows with increasing variance. Similarly, the ambiguity arising from the growing range of acceptable parameters only becomes more acute in nonlinear regression problems such as quadratic $f(x)$.

If, following the frequentist paradigm, we insist on reporting parameter point estimates obtained from χ^2-minimization for curves parametrized with K parameters, we may eventually overfit the data. That is, we may obtain fine-tuned values for the K parameters with a resulting χ^2, *i.e.*, the squared deviation, falling well below Nv.

To avoid overfitting, it is common practice to invoke additional criteria, *i.e.*, penalties, just as we described earlier in discussing signal deconvolution in the context of information theory in Section 4.6. For example, an alternative to minimizing a χ^2 is to minimize the *objective function*

$$\underbrace{\sum_{n=1}^{N} \frac{(y_n - f(x_n))^2}{v_n}}_{\chi^2 \text{ term}} + \underbrace{\lambda_1 \int dx \, \big(f(x) - \mu(x)\big)^2 + \lambda_2 \int dx \left| \frac{d^2 f(x)}{dx^2} \right|^2}_{\text{penalty terms}} . \tag{6.1}$$

That is, we minimize a χ^2 subject to a constraint on matching a predefined curve $\mu(\cdot)$ and minimizing the curvature squared of the continuous function. The parameters λ_1 and λ_2, sometimes termed *regularization parameters* or simply regularizers, fix the relative weight placed on fitting the data, χ^2 term, versus the second and third penalty terms.

Depending on the nature of $f(\cdot)$, most often the integrals in our objective function cannot be computed analytically. Instead, we need to consider a grid of points over which to discretize our curve. This results in an objective function

$$\sum_{n=1}^{N} \frac{(y_n - f(x_n))^2}{v_n} + \bar{\lambda}_1 \sum_{n=1}^{N} \big(f(x_n) - \mu(x_n)\big)^2 + \bar{\lambda}_2 \sum_{n=1}^{N} \left(f(x_{n+1}) + f(x_{n-1}) - 2f(x_n)\right)^2 ,$$

where the Δx appearing in the discretization of the integrals and derivatives has been absorbed in the newly redefined regularization parameters, $\bar{\lambda}_1$ and $\bar{\lambda}_2$. The estimation of the continuous curve is now reduced to estimating the values of the curve only on those points at which it has been discretized.

While the data may be measured at specific *input points*, the grid over which the curve is discretized is not limited to those points. Indeed, we can interpolate values of the curve at *test points* between the measured input points and extrapolate beyond existing input points. For this reason, we now distinguish between measured or input points, measured at locations $x^{\#}$, and test points, where the function is discretized at locations $x^{\star}$. Collectively, we use x to refer to both sets of points, $x = (x^{\#}, x^{\star})$, and rewrite our objective function as follows:

$$\sum_{n=1}^{N^{\#}} \frac{(y_n - f(x_n^{\#}))^2}{v_n} + \bar{\lambda}_1 \sum_{n=1}^{N} \big(f(x_n) - \mu(x_n)\big)^2 + \bar{\lambda}_2 \sum_{n=1}^{N} \left(f(x_{n+1}) + f(x_{n-1}) - 2f(x_n)\right)^2 . \tag{6.2}$$

Here, $N^\sharp$ is the number of input points while the total number of points, N, is the sum of the input, $N^\sharp$, and test, $N^\star$, points.

For consistency, we superscript all quantities related to input points with sharps, *i.e.*, $\sharp$. We note that the first term of Eq. (6.2) depends on input points, while the second and third terms depend on all points. Furthermore, in writing Eq. (6.2) we opted for a particular way in which to discretize the second derivative in the penalty. This choice is somewhat arbitrary and, more generally, we may rewrite Eq. (6.2), as

$$\sum_{n=1}^{N^\sharp} \frac{(y_n - f(x_n^\sharp))^2}{v_n} + \sum_{n=1}^{N} f(x_n) B_n + \sum_{n=1}^{N} \sum_{m=1}^{N} f(x_n) A_{n,m} f(x_m), \tag{6.3}$$

up to additive constants independent of $f(\cdot)$, where B_n and $A_{n,m}$ are newly introduced constants.

While the form of Eq. (6.2) dictates the form of the matrix A in Eq. (6.3), the nature of the matrix introduced in Eq. (6.3) can accommodate penalties that go beyond, say, the penalties introduced in Eq. (6.1). For example, a penalty on the integral of the square of the gradient of $f(\cdot)$ can also be accommodated, when discretized, by an A albeit one with a different structure. Indeed, we can think of higher derivatives of $f(\cdot)$ appearing in the penalty integral as roughly imposing more diagonal bands adjacent to the main diagonal of A.

Up to additive constants, it is possible to rewrite Eq. (6.3) more conveniently

$$\sum_{n=1}^{N^\sharp} \frac{(y_n - f(x_n^\sharp))^2}{v_n} + \frac{1}{2} \left(f - \mu\right) C^{-1} \left(f - \mu\right)^T, \tag{6.4}$$

where $f = (f^\sharp, f^\star)$, $f^\sharp = \left(f(x_1^\sharp), \ldots, f(x_{N^\sharp}^\sharp)\right)$, and $f^\star = \left(f(x_1^\star), \ldots, f(x_{N^\star}^\star)\right)$ with μ similarly defined. Here, C is called a *covariance matrix* and is of size size $N \times N$.

Equation (6.4) reveals that our objective function may be thought of as the negative logarithm of a multivariate normal posterior. Under this scenario, the first term of Eq. (6.4) is related to a log likelihood while the second term is related to a log prior.

Note 6.2 Multivariate normal prior

Consider any finite grid $x_{1:M}$ of points. The values of a curve at those points $f = (f(x_1), \ldots, f(x_M))$ can be thought of as forming a random vector on which we apply a multivariate normal prior,

$$f \sim \text{Normal}_M(\mu, C). \tag{6.5}$$

As first introduced in Example 1.11, the parameters of this prior depend upon our grid $x_{1:M}$ and are given by

$$\mu = \left(\mu(x_1), \ldots, \mu(x_M)\right), \qquad C = \begin{pmatrix} C(x_1, x_1) & \cdots & C(x_1, x_M) \\ \vdots & \ddots & \vdots \\ C(x_M, x_1) & \cdots & C(x_M, x_M) \end{pmatrix}.$$

Borrowing notation first introduced in Section 2.5.2 and defining the increment $\Delta f(x) = f(x) - \mu(x)$, we now see how to interpret the elements of C as the expectation of $\Delta f(x_n)\Delta f(x_m)$ with respect to the multivariate normal distribution. That is, $C(x_n, x_m) = \langle \Delta f(x_n)\Delta f(x_m)\rangle$. From this expression, we see that the covariance captures the degree of correlation between points in the curve. Put differently, the covariance we specify in a multivariate normal prior imposes smoothness conditions on values of the curve discretized on the grid, $x_{1:M}$. For example, the slower the variance decays between distal points x_n and x_m, the smoother the variation between increments.

6.2 Nonparametric Regression in Continuous Space: Gaussian Process

We now turn to questions of *continuous regression*. A curve, $f(\cdot)$, consisting of an infinite array of values modeled as random variables at every point on the real line is a stochastic process. The prior placed on the infinite array of random variables, or in our case specifically the curve $f(\cdot)$, is called a *process prior*. Motivated by Eq. (6.5), we apply an infinite-dimensional multivariate normal prior on $f(\cdot)$. For this reason, we call $f(\cdot)$ a *Gaussian process* and its prior a *Gaussian process prior*, and we denote the resulting distribution by GaussianP.

Note 6.3 GaussianP prior

The GaussianP prior was first defined in Example 1.11 as $f(\cdot) \sim \text{GaussianP}_S\left(\mu(\cdot), C(\cdot,\cdot)\right)$. As S, the domain of the GaussianP, is always the real line in this chapter, we drop the subscript from hereon in. In shorthand, we write

$$f(\cdot) \sim \text{GaussianP}\left(\mu(\cdot), C(\cdot,\cdot)\right),$$

which, as we repeat from Example 1.11, captures the following features:

- The realizations from sampling are functions $f(\cdot)$ that, to any point x, are assigned the real number $f(x)$.
- The parameter $\mu(\cdot)$ is a function that, to any point x, assigns a real number $\mu(x)$.
- The parameter $C(\cdot,\cdot)$ is a function that, to any points x and x', assigns a nonnegative real number $C(x, x')$.
- For any finite subset of the real line, such as a grid $x_{1:M}$, the values $(f(x_1), \ldots, f(x_M))$ form a random vector.

- The random vector $(f(x_1),\ldots,f(x_M))$ follows a multivariate normal distribution given in Eq. (6.5).

We discriminate here between the covariance function $C(\cdot,\cdot)$, also called a *covariance kernel*, and the *covariance matrix* C consisting of elements composed of the covariance kernel evaluated at specific points.

As we may generally be interested in placing a prior on a finite number of test points, $f^\star$, of the full $f(\cdot)$ alongside input points, $f^\#$, we define the following multivariate normal as our joint prior on both grids:

$$(f^\star, f^\#) \sim \text{Normal}_{N^\star+N^\#}\left((\mu^\star,\mu^\#), \begin{pmatrix} C^{\star\star} & C^{*\#} \\ C^{\#*} & C^{\#\#} \end{pmatrix}\right).$$

In practical applications, we ignore the case where a test point and input point coincide. Typical choices for the GaussianP mean, $\mu(\cdot)$, include featureless flat curves such as the special case where $\mu(x) = 0$ for all x. By contrast, the covariance requires some finessing.

6.2.1 Covariance Kernel

The choice of $C(\cdot,\cdot)$ evaluated at any grid points, say x and x', for which we write $C(x,x')$, can be any bivariate function provided the resulting covariance matrix (obtained by evaluating the covariance kernel at specific points) be symmetric and attain positive eigenvalues, *i.e.*, that the resulting matrix be positive definite for all choices of x and x'. Of the many choices allowed for $C(x,x')$, translationally symmetric kernels for which $C(x,x') = C(|x-x'|)$ often play an important role. For example, translationally symmetric kernels, often termed *stationary*, which decay monotonically with distance between points x and x', allow us to specify a scale over which the curve retains smoothness.

Note 6.4 The squared exponential kernel

The most common choice of covariance kernel is the *squared exponential*, which is given by

$$C(x,x') = \lambda^2 \exp\left(-\frac{|x-x'|^2}{\ell^2}\right).$$

Here, ℓ sets a decay scale while λ^2 sets a correlation strength. Samples of $f(\cdot)$ are shown in Fig. 6.1 for various choices of ℓ and λ.

As ℓ tends to a small number, the values of $f(x)$ and $f(x')$ become increasingly decorrelated. In this limit, the covariance takes this form

$$C(x,x') = \lambda^2\delta_0(x-x'),$$

which models time-decorrelated, *white noise* as a prior. Indeed, white noise is an idealization which, at some level, runs counter to the purpose of GaussianP priors, which impose smoothness conditions for avoiding overfitting.

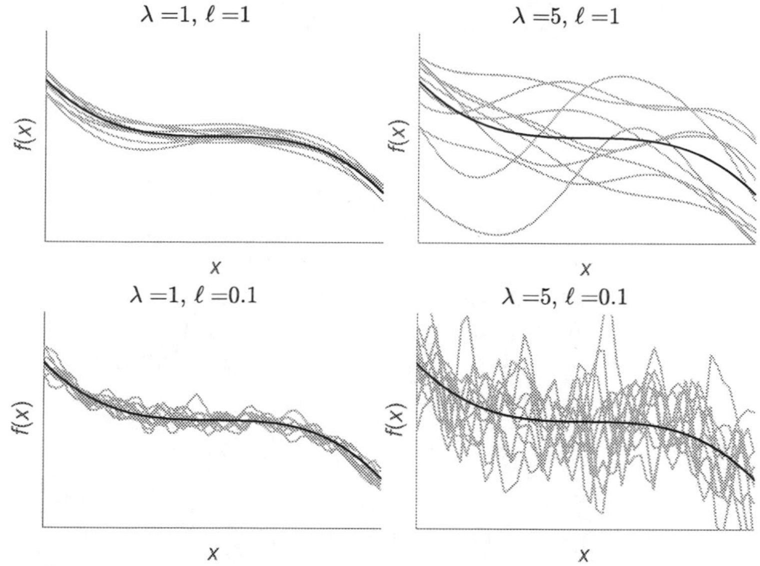

Fig. 6.1 Here, we explore the effects of ℓ and λ under the assumption of a squared exponential kernel on samples of our GaussianP.

> **Note 6.5** Periodic kernel
>
> Ringing, wave-type motion, and dampened oscillations are common phenomena in physical applications and often modeled by simple Drude or Lorenz oscillator models. For such data, it is often convenient to invoke *locally periodic kernels* such as
>
> $$C(x, x') = \lambda^2 \exp\left(-\frac{2}{\ell^2}\sin^2\left(\pi\frac{|x-x'|}{\tau}\right)\right),$$
>
> where ℓ and λ play the same role as in Note 6.4 and τ is the period.

Brownian Trajectories as a GaussianP*

To sample curves, $f(\cdot)$, coinciding with Brownian trajectories using GaussianP, we must derive a covariance consistent with Brownian motion. To model this, we consider a particle undergoing Brownian motion starting from the origin whose position after time t has elapsed is $x(t)$.

In this section, our $x(t)$ plays the role of $f(x)$. With this relabeling, the solution to the diffusion equation with open boundaries reads

$$p(x(t)|\mu(t), D) = \frac{1}{(4\pi Dt)^{1/2}} e^{-\frac{(x(t)-\mu(t))^2}{4Dt}}. \tag{6.6}$$

The particle displacement at t, averaged over $p(x(t)|\mu(t), D)$, is

$$\langle x(t)\rangle = \mu(t).$$

To construct the covariance kernel, we consider $\langle (x(t)-\mu(t))(x(t')-\mu(t'))\rangle$, which we rewrite using increments, first introduced in Section 2.5.2, as $\langle \Delta x(t)\Delta x(t')\rangle$.

* This is an advanced topic and could be skipped on a first reading.

For now, we consider the case where $t' \geq t$ and write

$$C(t,t') = \langle \Delta x(t)\Delta x(t')\rangle = \langle \Delta x(t)\Delta x(t)\rangle + \langle \Delta x(t)(\Delta x(t') - \Delta x(t))\rangle = 2Dt. \tag{6.7}$$

The result of $C(t, t') = 2Dt$ was obtained by recognizing that the first term of the right-hand side of Eq. (6.7) is $2Dt$ from the definition of Eq. (6.6). By contrast, the second term of the right-hand side of Eq. (6.7) is always zero as either $t' = t$, for which the term trivially reduces to 0, or $t' > t$, for which the increment $\Delta x(t') - \Delta x(t)$ (also sampled from a normal) is independent of the first increment $\Delta x(t)$. Importantly, we note that while $\langle \Delta x(t)(\Delta x(t') - \Delta x(t))\rangle$ is zero, $\langle \Delta x(t')(\Delta x(t') - \Delta x(t))\rangle$ is not zero as both increments here overlap. Similarly, when $t' < t$, we recover

$$C(t,t') = \langle \Delta x(t)\Delta x(t')\rangle = \langle \Delta x(t')\Delta x(t')\rangle + \langle \Delta x(t')(\Delta x(t) - \Delta x(t'))\rangle = 2Dt'.$$

Putting it all together, a Brownian GaussianP prior has the covariance kernel $C(t, t') = 2D\min(t, t')$.

Example 6.2 Brownian bridge

Suppose now that both ends of the Brownian motion are fixed to μ_0 at $t = 0$ and $t = T$. Such a motion is called a *Brownian bridge*. That is, $x(0) = x(T) = \mu_0$. Naturally, treating $x(t)$ as a Brownian motion will not satisfy this condition. For this reason, we define a new function, $\phi(t)$, such that

$$\phi(t) = \left(x(t) - \mu_0\right) - \frac{t}{T}\left(x(T) - \mu_0\right),$$

where $x(\cdot)$ is a Brownian motion with $x(0) = \mu_0$. By virtue of this definition, $\phi(t)$ satisfies these boundary conditions, $\phi(0) = \phi(T) = 0$.

As both $x(t)$ and $x(T)$ are normally distributed random variables, then the sum of two normal random variables, namely $\phi(t)$, is also a normal random variable. The mean of $\phi(t)$ taken with respect to its distribution is zero as both $x(t)$ and $x(T)$ have mean μ_0.

Next we can ask about the covariance kernel of $\phi(t)$, which we compute similarly to how we computed the covariance for the case of Brownian motion,

$$C(t,t') = 2D\left(\min(t,t') - \frac{tt'}{T}\right). \tag{*6.8*}$$

6.2.2 Sampling the GaussianP

So far, we have discussed covariance kernels and shown sample curves, technically values of the curves sampled at test points, generated using various kernels. Here, we discuss how such curves can be generated in the first place.

* Reminder: The asterisks by some equations indicates that the detailed derivation can be found in Appendix F.

Before turning to a concrete algorithm, we first make a note on the technical impossibility of sampling continuous curves given that GaussianP themselves cannot be ascribed probability densities. On the other hand, it remains possible to sample very good approximations to GaussianP, with a high density of test points, using multivariate normal distributions.

Thus, in preparation for our discussion on sampling from multivariate normal distributions, we begin by first sampling a random variable f at an arbitrary test point from a univariate normal with mean μ and variance c for which

$$f = \mu + \epsilon\sqrt{c}, \qquad \epsilon \sim \text{Normal}(0, 1).$$

The generalization to the multivariate normal with dimension M now immediately follows from above:

$$f = \mu + \epsilon C^{1/2}, \qquad \epsilon \sim \text{Normal}_M(0, \mathbb{1}), \tag{6.9}$$

where $C^{1/2}$ is any square matrix of dimension M that satisfies $(C^{1/2})^T C^{1/2} = C$ and $\mathbb{1}$ is the identity matrix.

Note 6.6 How to obtain $C^{1/2}$

The matrix $C^{1/2}$ we need in Eq. (6.9) can be obtained in at least two ways:

- First, we may write $C = R^T R$ and obtain $C^{1/2} = R$ by *Cholesky decomposition*. To compute the Cholesky factor R, we may use the Cholesky–Banachiewicz algorithm, which relies on solving coupled quadratic equations obtained by setting equal, element-wise, the known elements of C to the elements of $R^T R$ to be determined.
- An alternative is the *singular value decomposition* (SVD) of C. Briefly, since the covariance matrix is a square positive definite matrix, we can decompose it as $C = U^T S U$, where S is a diagonal matrix with elements coinciding with the eigenvalues of C while the columns of U are C's corresponding eigenvectors. In this case, we can rewrite C as the product of two matrices, $U^T S^{1/2}$ and $S^{1/2} U$ from which it follows that we can use $C^{1/2} = S^{1/2} U$.

These two are complementary to each other: Cholesky decomposition is computed faster than SVD; however, it may fail due to numerical round-off. In contrast, SVD is computed slower than Cholesky, but it remains numerically stable.

6.2.3 GaussianP Priors and Posteriors

In order to estimate continuous curves through data, we previously spoke of modeling them as random variables and alluded to applying GaussianP priors. Under the assumption that the likelihood of our data is normal, and the prior is GaussianP, as in

$$f(\cdot) \sim \text{GaussianP}(\mu(\cdot), C(\cdot,\cdot)),$$
$$y_n|f(\cdot) \sim \text{Normal}(f(x_n^\sharp), v_n),$$

we also find that the posterior over $f(\cdot)$ assumes a GaussianP form.

It is not possible to show this explicitly by application of Bayes' theorem as the prior does not assume a density. Rather, our proof follows from a discretization with $N^\sharp$ input points and $N^\star$ test points. For simplicity alone, dropping the n subscript on v, we write in shorthand,

$$(f^\star, f^\sharp) \sim \text{Normal}_{N^\star+N^\sharp}\left((\mu^\star, \mu^\sharp), \begin{pmatrix} C^{\star\star} & C^{*\sharp} \\ C^{\sharp *} & C^{\sharp\sharp} \end{pmatrix}\right),$$
$$y_n|f^\sharp \sim \text{Normal}_{N^\sharp}(f^\sharp, v\mathbb{1}).$$

The posterior over the test points, $p(f^\star|y)$ with $y_{1:N}$ abbreviated as y, is the object of interest. To compute this, we marginalize the joint distribution, which depends on $f^\sharp$ and $f^\star$, over $f^\sharp$ as

$$p(f^\star|y) = \int df^\sharp p(f^\star, f^\sharp|y) = \int df^\sharp\, p(f^\star|f^\sharp)p(f^\sharp|y), \qquad (*6.10*)$$

where the integral is understood to be over all allowed values of $f^\sharp$. An explicit evaluation of this yields

$$p(f^\star|y) = \text{Normal}_{N^\star}(f^\star; \tilde{\mu}, \tilde{C}), \qquad (*6.11*)$$

where the new parameters are

$$\tilde{\mu} = \mu^\star + C^{*\sharp}(v\mathbb{1} + C^{\sharp\sharp})^{-1}\left(y - \mu^\sharp\right), \quad \tilde{C} = C^{\star\star} - C^{*\sharp}(C^{\sharp\sharp} + v\mathbb{1})^{-1}C^{\sharp *}.$$

Now, sampling from a GaussianP posterior reveals more than just curves $f(\cdot)$. Indeed, it provides credible intervals as shown in Fig. 6.2 or statistics

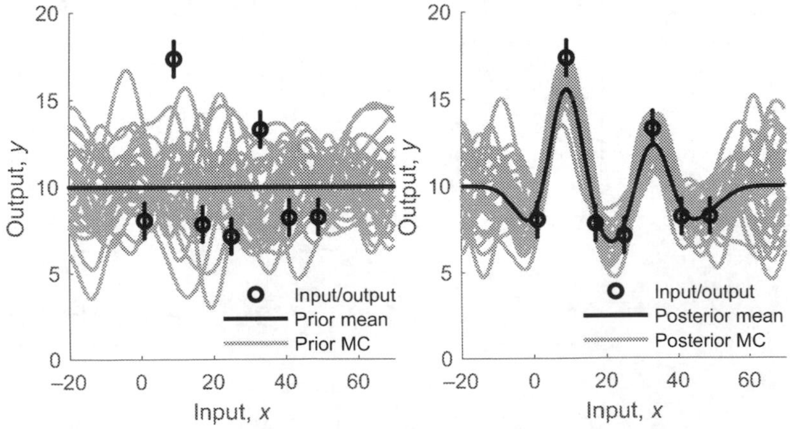

Fig. 6.2 Here, we show MC samples drawn from a GaussianP prior and posterior. We note the regions of growing credible intervals (left panel) where the data, shown by black dots, is sparser.

such as cross-correlations between two points obtained from multiple, J, samples,

$$\left\langle \Delta f^{(j)}(x)\Delta f^{(j)}(x')\right\rangle \approx \frac{1}{J}\sum_{j=1}^{J}\Delta f^{(j)}(x)\Delta f^{(j)}(x').$$

As seen in Fig. 6.2, expectedly, the credible intervals grow in regions unconstrained by data but only up to some point. That is, the credible interval does not diverge in regions infinitely far away from data. Rather, their maximum width remains bounded by that of the prior in regions where there are no data.

6.2.4 Predictive Distribution and Inducing Points

As an introduction to approximate methods, we briefly introduce how to predict the value of $f(\cdot)$ evaluated at $f^\dagger$ via $p(f^\dagger|y)$, where $f^\dagger$ may be any subset of test points omitted from the original analysis. Here, $p(f^\dagger|y)$ is often termed the *predictive distribution* as it provides a distribution over $f^\dagger$ given the data.

To compute the predictive distribution, we proceed as $p(f^\dagger|y) = \int df^\star\, p(f^\dagger|f^\star)p(f^\star|y)$. This integral is of the same form as Eq. (*6.10*) and its mean and covariance can be directly read off from $\tilde{\mu}$ and $\tilde{C}$.

Of particular interest is the distribution characterized by $p(f^\dagger|f^\star)$, which we can explicitly compute from $p(f^\dagger,f^\star)/p(f^*)$ and subsequently recompleting the squares. Namely,

$$p(f^\dagger|f^\star) = \frac{p(f^\dagger,f^\star)}{p(f^*)} = \text{Normal}_{N^\dagger}\left(f^\dagger;\mu^\dagger - C^{\dagger *}(C^{\star\star})^{-1}\left(f^\star - \mu^\star\right),\right.$$
$$\left. C^{\dagger\dagger} - C^{\dagger *}(C^{\star\star})^{-1}C^{*\dagger}\right),$$

where the dimensionality $N^\dagger$ is dictated by the size of $f^\dagger$. The mean of the last expression is itself interesting. For instance, if $f^\star$ is on a coarse grid, then we can obtain an interpolation on a finer grid $f^\dagger$, where we set all means ($\mu^\dagger$, $\mu^\star$) to zero for simplicity, through the deterministic transformation

$$f^\dagger = C^{\dagger *}(C^{\star\star})^{-1}f^\star. \tag{6.12}$$

The latter is an important observation that we will make use of in discussing *inducing point methods*.

Inducing Point Methods*

One challenge with computing a GaussianP variance lies in the inversion of an $N^\sharp \times N^\sharp$ matrix. This operation alone scales as $(N^\sharp)^3$ using standard numerical techniques. The cost of matrix inversion is further compounded by the cost associated with taking the square root of C (*e.g.*, by Cholesky decomposition) in order to sample curves, $f^\star$, if necessary. Moreover, the

* This is an advanced topic and could be skipped on a first reading.

data itself may be plentiful and redundant. As such, surplus data really only increases the computational cost without necessarily changing much in our knowledge of $f(\cdot)$.

One may be tempted to deal with this computational bottleneck by eliminating perceived redundant data points; perhaps removing all but every ten data points or eliminating data points in apparently highly sampled regions. This strategy is fraught with bias as this could result, by chance, in the elimination of data in regions where $f(\cdot)$ varies atypically. For this reason, *structured kernel* approaches, whose sole objective is to reduce correlations amongst $f(\cdot)$, are invoked.

> **Note 6.7** Warning on GaussianP approximations
>
> All GaussianP approximations reduce the computational burden, one way or another, by practically breaking or perturbing the correlations along $f(\cdot)$. Thus, they are essentially interpolation methods on a coarser grid and, for this reason, approximations should be used and interpreted cautiously.

One way to reduce computational cost is to consider our unknown function at test points called *inducing points*, f^u. We may then use the data to extrapolate values computed at f^u to a finer grid using a transformation identical to that introduced in Eq. (6.12). By analogy to Eq. (6.12), we write the likelihood and prior on the value of the function at the location of the inducing points assuming zero prior mean for simplicity, as

$$f^u \sim \text{Normal}_{N^u}\left(0, C^{uu}\right), \tag{6.13}$$

$$y|f^u \sim \text{Normal}_{N^\sharp}\left(C^{\sharp u}(C^{uu})^{-1}f^u, v\mathbb{1}\right). \tag{6.14}$$

The resulting posterior then readily follows:

$$p(f^u|y) \propto p(y|f^u)p(f^u) \propto \text{Normal}_{N^u}\left(\tilde{\mu}, \tilde{C}\right), \tag{*6.15*}$$

where

$$\tilde{\mu} = \left((C^{uu})^{-1} + M\right)^{-1} b, \qquad \tilde{C} = \left((C^{uu})^{-1} + M\right)^{-1},$$

and

$$b = v^{-1}(C^{uu})^{-1}C^{u\sharp}y, \qquad M = v^{-1}(C^{uu})^{-1}C^{u\sharp}C^{\sharp u}(C^{uu})^{-1}.$$

We note that the largest matrix inverted here is an $N^u \times N^u$. Having now established f^u on a limited grid, we are free to extrapolate onto a finer grid by means of Eq. (6.12).

6.2.5 GaussianP Regression with Nonconjugate Likelihoods*

So far we have only considered normal likelihoods for which we are assured conjugacy with the GaussianP prior. However, experimental

* This is an advanced topic and could be skipped on a first reading.

constraints may impose upon us likelihoods neither conjugate with GaussianP priors nor any other priors for that matter. At this point, we may retain GaussianP priors for the flexibility they provide in manipulating the structure of their kernel. It naturally follows that to sample from the resulting posterior, we must resort to computational methods and, for this reason, we now consider three families of samplers: Metropolis–Hastings, Gibbs, and auxiliary variable samplers.

Note 6.8 GaussianP for strictly positive data

When we invoke GaussianP priors, we assume that curves that dip into negative regions of $f^\star$ are always assigned some posterior weight. Yet some types of data, and their eventual fits, may be strictly positive, *e.g.*, the concentration profile of a chemical species in space, a refractive index map across a material slab, a temperature (in Kelvin) as a function of altitude. One way to avoid sampling negative values for strictly positive physical quantities is to retain the GaussianP prior on $f^\star$ but replace $f^\star$ in the likelihood with $e^{f^\star}$ which is always strictly greater than zero. This transformation results in loss of conjugacy and suggests the use of one of the three families of samplers explored below.

Metropolis–Hastings Samplers

We first imagine defining a grid of test points over which MCMC is performed. The grid is not allowed to change during sampling. Upon selecting a grid, we introduce the usual Metropolis–Hastings acceptance ratio, which reads

$$A_{f^{\text{old}}}(f^{\text{prop}}) = \underbrace{\frac{p(y_{1:N^\sharp}|f^{\text{prop}})p(f^{\text{prop}})}{p(y_{1:N^\sharp}|f^{\text{old}})p(f^{\text{old}})}}_{\text{target}} \underbrace{\frac{Q_{f^{\text{prop}}}(f^{\text{old}})}{Q_{f^{\text{old}}}(f^{\text{prop}})}}_{\text{proposal}} \tag{6.16}$$

and where $f = (f^\sharp, f^\star)$ with the understanding that the likelihood does not explicitly depend on $f^\star$. The challenge is now to find a proposal distribution, Q, that results in good MCMC mixing.

One choice of proposal distribution is the GaussianP prior itself for which Eq. (6.16) immediately simplifies to

$$A_{f^{\text{old}}}(f^{\text{prop}}) = \frac{p(y_{1:N^\sharp}|f^{\text{prop}})}{p(y_{1:N^\sharp}|f^{\text{old}})}.$$

Having a proposal function be a multivariate normal with a covariance identical to that of the prior may be advantageous as samples from the proposal distribution now have similar smoothness to the overall posterior informed, in part, by the prior.

Yet, as it stands, varying all points in f^{prop} simultaneously may lead to a low acceptance rate as new points must be proposed and accepted or rejected *altogether*. Furthermore, the acceptance rate may also be low as the prior does not incorporate trends in $f(\cdot)$ imposed from the data. To correct for this, we may opt for a Markov chain structure by invoking the previously accepted f, namely f^{old}, as the proposal distribution mean.

In this case, a GaussianP proposal centered at f^{old} is symmetric in f^{old} and f^{prop} and, as a result, the acceptance ratio simplifies as follows

$$A_{f^{\text{old}}}(f^{\text{prop}}) = \frac{p(y_{1:N^\sharp}|f^{\text{prop}})p(f^{\text{prop}})}{p(y_{1:N^\sharp}|f^{\text{old}})p(f^{\text{old}})}.$$

Example 6.3 Metropolis–Hastings sampler for nonconjugate likelihood GaussianP prior pair

We consider a Metropolis random walk proposal given by $f^{\text{prop}} = f^{\text{old}} + v$, where $v \sim \text{Normal}_{N^\sharp+N^\star}(0, C)$ and the covariance matrix is of dimension $N^\sharp + N^\star$. From this it follows that $Q_{f^{\text{prop}}}(f^{\text{old}}) = Q_{f^{\text{old}}}(f^{\text{prop}})$ and, under the assumption of a GaussianP prior, the explicit form of the acceptance ratio reads

$$A_{f^{\text{old}}}(f^{\text{prop}}) = \frac{p(y_{1:N^\sharp}|f^{\text{prop}})}{p(y_{1:N^\sharp}|f^{\text{old}})} \frac{\text{Normal}(f^{\text{prop}}; 0, C)}{\text{Normal}(f^{\text{old}}; 0, C)}.$$

While retaining the advantage of using the previously accepted f as proposal mean, it is also possible to avoid high rejection rates by proposing a change in a single point of f. To achieve this, while respecting the structure imposed by the covariance between points on f, we may use a *Gibbs-like sampler*. That is, first pick a point, n, at random, then sample a value for f_n from the prior $p(f_n|f_{-n})$. The challenge here is that samples may be tightly correlated and, as such, convergence may be slow.

In order to increase the acceptance rate and keep correlation between samples at a minimum, we may compromise between both solutions of sampling full curves versus sampling one point by opting for a *blocked Gibbs-like sampler* where we partition all points into K sets of points with individual sets indexed k, irrespective of whether those points defined within a block are considered contiguous or not.

In this case, we may use a GaussianP prior to construct a, typically symmetric, proposal distribution $p(f_k^{\text{prop}}|f_{-k}^{\text{old}})$. For a symmetric proposal, our acceptance ratio then reads

$$A_{f^{\text{old}}}(f^{\text{prop}}) = \frac{p(y_{1:N^\sharp}|f_k^{\text{prop}}, f_{-k}^{\text{old}})}{p(y_{1:N^\sharp}|f_k^{\text{old}}, f_{-k}^{\text{old}})} \frac{p(f_k^{\text{prop}}, f_{-k}^{\text{old}})}{p(f_k^{\text{old}}, f_{-k}^{\text{old}})}.$$

We end with a note on the difficulties of plentiful data within the GaussianP paradigm. Plentiful data already causes difficulties when sampling $f(\cdot)$ from a GaussianP posterior let alone a posterior generated from a nonconjugate likelihood-GaussianP pair as part of a Metropolis–Hastings sampler where not all samples are accepted. It is indeed in such cases that inducing point methods help mitigate the computational cost of Metropolis–Hastings.

Elliptical Slice Samplers

In an attempt to overcome the limitations of the naive Metropolis–Hastings algorithms we described above, we now discuss a more specialized strategy. The main idea we exploit here is to re-parametrize our regression problem in order to obtain conditional posteriors and then perform Gibbs sampling.

Example 6.4 Re-parametrization of a normal prior

In this example, we consider a normal random variable F that may be obtained from two different models. Namely, a simple model consisting of

$$F \sim \text{Normal}(0, c)$$

and another, more complex one, consisting of

$$V_0 \sim \text{Normal}(0, c), \qquad V_1 \sim \text{Normal}(0, c),$$
$$\theta \sim \text{Uniform}_{[0,2\pi]}, \qquad F = V_0 \sin\theta + V_1 \cos\theta.$$

A transformation of random variables from (V_0, V_1, θ) to F reveals that, indeed, the statistics of F in the first model are the same as the induced statistics of F in the second model. Accordingly, provided a likelihood $p(w|F)$, we may use either model to form a prior and reach the same posterior. Under the first model, the posterior is $p(F|w)$; while under the second model our equivalent posterior is $p(V_0, V_1, \theta|w)$.

Exactly the same approach applies for a multivariate random variable as in the case of Gaussian regression where

$$f \sim \text{Normal}_N(0, C),$$

which can be re-parametrized as

$$v_0 \sim \text{Normal}(0, C), \qquad v_1 \sim \text{Normal}(0, C),$$
$$\theta \sim \text{Uniform}_{[0,2\pi]}, \qquad f = v_0 \sin\theta + v_1 \cos\theta.$$

Following the re-parametrization of Example 6.4, we consider a mathematically equivalent prior for our f consisting of

$$v_0 \sim \text{Normal}_{N^\sharp+N^\star}(0, C),$$
$$v_1 \sim \text{Normal}_{N^\sharp+N^\star}(0, C),$$
$$\theta \sim \text{Uniform}_{[0,2\pi]},$$

where $f = v_0 \sin\theta + v_1 \cos\theta$, and seek to sample $p(v_0, v_1, \theta|y_{1:N^\sharp})$ instead of $p(f|y_{1:N^\sharp})$.

Our new target may then be completed,

$$p(v_0, v_1, \theta|y_{1:N^\sharp}) = \int dv\, p(v, v_0, v_1, \theta|y_{1:N^\sharp}),$$

with an *auxiliary* variable $v \sim \text{Normal}_{N^\sharp+N^\star}(0, C)$, independent from $v_0, v_1, \theta, y_{1:N^\sharp}$. A Gibbs sampler for the completed target $p(v, v_0, v_1, \theta|y_{1:N^\sharp})$ may now be developed with the following three steps:

- Update v by sampling from $p(v|v_0, v_1, \theta, y_{1:N^\sharp})$.
- Update jointly v_0, v_1, θ by sampling from $p(v_0, v_1, \theta|v, y_{1:N^\sharp})$.
- Update θ by sampling from $p(\theta|v, v_0, v_1, y_{1:N^\sharp})$.

As the conditional of the first update reduces to $p(v|v_0, v_1, \theta, y_{1:N^\sharp}) = \text{Normal}_{N^\sharp+N^\star}(v; 0, C)$, it can be sampled directly. By contrast, the two remaining conditionals of the subsequent steps require *within Gibbs* schemes as they are not formed by conjugate prior-likelihood pairs.

To sample from $p(v_0, v_1, \theta|v, y_{1:N^\sharp})$ in the *second update*, one choice is to invoke Metropolis–Hastings proposals with the particular form

$$\theta^{\text{prop}} \sim \text{Uniform}_{[0,2\pi]},$$

$$v_0^{\text{prop}}|v_0^{\text{old}}, v_1^{\text{old}}, \theta^{\text{old}}, \theta^{\text{prop}} \sim \delta_{\left(v_0^{\text{old}} \sin\theta^{\text{old}} + v_1^{\text{old}} \cos\theta^{\text{old}}\right) \sin\theta^{\text{prop}} + v \cos\theta^{\text{prop}}},$$

$$v_1^{\text{prop}}|v_0^{\text{old}}, v_1^{\text{old}}, \theta^{\text{old}}, \theta^{\text{prop}} \sim \delta_{\left(v_0^{\text{old}} \sin\theta^{\text{old}} + v_1^{\text{old}} \cos\theta^{\text{old}}\right) \cos\theta^{\text{prop}} - v \sin\theta^{\text{prop}}}.$$

This proposal has little associated computational cost and can be easily implemented. Furthermore, it results in an acceptance ratio of 1 irrespective of the values generated.

Finally, the *third update* requires sampling from $p(\theta|v, v_0, v_1, y_{1:N^\sharp})$. Although this conditional reduces to $p(\theta|v_0, v_1, y_{1:N^\sharp})$, due to lack of conjugacy it requires another Metropolis–Hastings scheme. A convenient solution for this is offered by the slice sampler of Example 5.14 that can take advantage of the fact that θ takes only scalar values that are, furthermore, bounded by 0 and 2π.

Since our overall Gibbs sampler always maintains variables satisfying the equality $f = v_0 \sin\theta + v_1 \cos\theta$, it can be compactly implemented as in Algorithm 6.1. Furthermore, because when considered for all possible values of θ this equality defines an ellipse in the real space of $N^\sharp + N^\star$ dimensions, it is termed the *elliptical slice sampler*.

Algorithm 6.1 Slice sampler within Gibbs for GaussianP

Given an initial f^{old}, the method iterates the following until convergence:

- Sample $\theta^{\text{old}} \sim \text{Uniform}_{[0,2\pi]}$.
- Sample $v \sim \text{Normal}_{N^\sharp+N^\star}(0, C)$.
- Set $v_0' = f^{\text{old}} \sin\theta^{\text{old}} + v \cos\theta^{\text{old}}$ and $v_1 = f^{\text{old}} \cos\theta^{\text{old}} - v \sin\theta^{\text{old}}$.
- Sample θ^{new} from $p(\theta|v_0, v_1, w_{1:N})$ using the slice sampler of Example 5.14.
- Set $f^{\text{new}} = v_0 \sin\theta^{\text{new}} + v_1 \cos\theta^{\text{new}}$.

6.3 Nonparametric Regression in Discrete Space: Beta Process Bernoulli Process*

We have previously discussed questions of regression using continuous curves or, in higher dimensions, continuous surfaces or volumes. Here, we

* This is an advanced topic and could be skipped on a first reading.

focus on *discrete regression problems*. We take, for example, the question of a two-dimensional surface covered in blips or bumps each modeled as Gaussian. Equivalently, we consider a time series with discrete jumps between otherwise flat regions similar to that considered in our discussion of change point analysis. These problems beg the question: *How many Gaussian blips or discrete jumps are required to fit our data?* In Algorithm 4.2 we invoked approximate model selection and greedy algorithms to help in identifying a model. Here, instead, we wish to operate within a Bayesian paradigm in order to obtain full posterior distributions over the number of discrete elements required to fit the data as well as parameters associated to these discrete elements; *e.g.*, the mean and standard deviation of each Gaussian blip.

To answer such questions, we turn to *beta process* and *Bernoulli process priors*, often simply abbreviated BetaBernP *priors*. Just as we introduced a one-dimensional normal in an effort to motivate the GaussianP prior, we start with a Bernoulli random variable

$$b \sim \text{Bernoulli}\,(\pi),$$

where π is interpreted as the probability of *success* and lies between zero and one. As this probability may be unknown, it is often convenient to use a beta prior, conjugate to the Bernoulli, on π and write

$$\pi \sim \text{Beta}\left(\alpha, \beta\right),$$

where α and β are beta distribution hyperparameters.

The above formulation is helpful if we wish to determine whether a discrete element is present ($b = 1$) or not ($b = 0$) as we attempt, for example, to fit a profile. Indeed, the above reasoning can even be generalized to determine how many discrete components are responsible for an observed profile such as the number of fluorescent molecules or bright galaxies in an image. For this, we need to develop the formalism of the BetaBernP. Pedagogically, it is perhaps easiest to begin, as we did with the GaussianP prior, from a finite limit.

A finite expression of the BetaBernP reads as follows

$$\pi_{\sigma_m} \sim \text{Beta}\left(\alpha, \beta\right), \qquad m = 1 : M,$$

$$b_{\sigma_m} | \pi_{\sigma_m} \sim \text{Bernoulli}\,(\pi_{\sigma_m}), \qquad m = 1 : M,$$

where M is the number of discrete elements. The densities of both beta and Bernoulli distributions assume the explicit forms

$$p\,(\pi_{\sigma_m}) = \text{Beta}\left(\pi_{\sigma_m}; \alpha, \beta\right) = \frac{\Gamma\left(\alpha + \beta\right)}{\Gamma\,(\alpha)\,\Gamma\left(\beta\right)} (\pi_{\sigma_m})^{\alpha - 1}\,(1 - \pi_{\sigma_m})^{\beta - 1},$$

$$p\,(b_{\sigma_m} | \pi_{\sigma_m}) = \text{Bernoulli}\,(b_{\sigma_m}; \pi_{\sigma_m}) = (\pi_{\sigma_m})^{b_{\sigma m}}\,(1 - \pi_{\sigma_m})^{1 - b_{\sigma m}},$$

as detailed in Appendix B. Example 6.5 shows one example of a discrete regression problem exploring how the BetaBernP may be used.

Example 6.5 BetaBernP in a regression problem

In this example of discrete regression, we try to determine the number of elements warranted by the data. For instance, we consider light emitters with a mean photon emission rate of μ_{mol} each. Based on this, we write

$$\begin{aligned} \pi_{\sigma_m} &\sim \text{Beta}\left(\alpha\frac{1}{M}, \beta\frac{M-1}{M}\right), & m &= 1:M, \\ b_{\sigma_m}|\pi_{\sigma_m} &\sim \text{Bernoulli}\left(\pi_{\sigma_m}\right), & m &= 1:M, \\ w_n|b_{\sigma_{1:M}} &\sim \text{Poisson}\left(\mu_{\text{mol}}\sum_{m=1}^{M} b_{\sigma_m}\right), & n &= 1:N. \end{aligned}$$

The particular choice of hyperparameters of the beta distribution, *e.g.*, α set to α/M, is discussed later. Here, the indicators b_{σ_m} coinciding with each emitter encodes whether the emitter is active and emits photons or is otherwise nonemitting (or does not exist at all). This variable is often called a *load*.

Assuming that μ_{mol} is known, we have two sets of unknowns, b_{σ_m} and π_{σ_m} for $m = 1 : M$, which we abbreviate as $b_{\sigma_{1:M}}$ and $\pi_{\sigma_{1:M}}$. These can be sampled in the following Gibbs sampling scheme starting from feasible values of $b_{\sigma_{1:M}}$ and $\pi_{\sigma_{1:M}}$:

- Sample $\pi_{\sigma_m}^{\text{new}}$ from $p(\pi_{\sigma_m}|b_{\sigma_m})$ for each $m = 1 : M$.
- Sample $b_{\sigma_{1:M}}^{\text{new}}$ from $p(b_{\sigma_{1:M}}|\pi_{\sigma_{1:M}}, w_{1:N})$.

Each $\pi_{\sigma_m}^{\text{new}}$ can be directly sampled from

$$p(\pi_{\sigma_m}^{\text{new}}|b_{\sigma_m}^{\text{old}}) = \text{Beta}\left(\pi_{\sigma_m}^{\text{new}}; \alpha', \beta'\right), \tag{*6.17*}$$

where $\alpha' = \alpha\frac{1}{M} + b_{\sigma_m}^{\text{old}}$ and $\beta' = \beta\frac{M-1}{M} - b_{\sigma_m}^{\text{old}} + 1$.

To sample the $b_{\sigma_{1:M}}^{\text{new}}$, as we lack conjugacy, we may employ the Metropolis algorithm and select to sample all at once. For the proposal probability distribution, we may use the prior for which the acceptance ratio immediately reduces to a likelihood ratio,

$$\begin{aligned} A_{b_{\sigma_{1:M}}^{\text{old}}}(b_{\sigma_{1:M}}^{\text{prop}}) &= \prod_{n=1}^{N} \frac{\text{Poisson}\left(w_n; \mu_{\text{mol}}\sum_{m=1}^{M} b_{\sigma_m}^{\text{prop}}\right)}{\text{Poisson}\left(w_n; \mu_{\text{mol}}\sum_{m=1}^{M} b_{\sigma_m}^{\text{old}}\right)} \\ &= \prod_{n=1}^{N} \left(\frac{\sum_{m=1}^{M} b_{\sigma_m}^{\text{prop}}}{\sum_{m=1}^{M} b_{\sigma_m}^{\text{old}}}\right)^{w_n} \exp\left(\mu_{\text{mol}}\sum_{m=1}^{M}\left(b_{\sigma_m}^{\text{old}} - b_{\sigma_m}^{\text{prop}}\right)\right). \end{aligned}$$

One challenge with introducing discrete elements to explain features in our data, such as the number of discrete steps in a signal, is that it may become difficult to avoid overfitting the data. Put differently, we need to ensure that as we include an infinite number of discrete elements in our BetaBernP, the number of active elements warranted by the data does not itself diverge. This condition forces upon us restrictions on hyperparameters that we consider below.

6.3.1 Posterior Convergence of the BetaBernP

Formally, the BetaBernP is understood to include an infinite number of discrete elements. The question then arises, as we allow M to become infinite, how we keep the number of active elements, $B = \sum_{m=1}^{M} b_{\sigma_m}$ equal to the sum of b's realized to unity, from diverging. As we will now show, this can be assured by a judicious choice of prior hyperparameters.

Note 6.9 Picking hyperparameters

Formally, our goal is to identify the prior distribution over B and assure ourselves that this distribution has finite mean as M tends to infinity. To do this, we first construct the distribution over $p(b_{\sigma_m})$ by marginalization

$$p\left(b_{\sigma_m}\right) = \int_0^1 d\pi_{\sigma_m}\, p\left(b_{\sigma_m}|\pi_{\sigma_m}\right) p\left(\pi_{\sigma_m}\right) = \begin{cases} \frac{\alpha}{\alpha+\beta}, & b_{\sigma_m} = 1, \\ \frac{\beta}{\alpha+\beta}, & b_{\sigma_m} = 0. \end{cases} \tag{*6.18*}$$

It follows that if b_{σ_m} is Bernoulli distributed, then $B = \sum_{m=1}^{M} b_{\sigma_m}$ follows a Binomial distribution,

$$p(B) = \text{Binomial}\left(B; M, \frac{\alpha}{\alpha+\beta}\right).$$

As $M \to \infty$, we expect the mean of the resulting Binomial to converge and satisfy the condition

$$B_{\text{mean}} = \lim_{M\to\infty} M\frac{\alpha}{\alpha+\beta} < \infty.$$

This now implies that the parameters of the beta distribution, α and β, cannot be independent of M, though there remain many choices for α and β that satisfy this condition. One such choice for the dependency of the beta hyperparameters on M is used in Example 6.5.

Marginalizing Out Beta Variables

It follows from the marginalization of π_{σ_m} in Eq. (*6.18*) that the beta prior can be eliminated altogether, from which we arrive at

$$b_{\sigma_m} \sim \text{Bernoulli}\left(\frac{\alpha}{\alpha+(M-1)\beta}\right),$$

where we used $\alpha/M, \beta\,(1-1/M)$ as the hyperparameters of the beta prior. This choice of Bernoulli prior and hyperparameter on $b_{\sigma_{1:M}}$ now lend themselves to a simpler alternative to the discrete regression problem. In Example 6.6, we briefly reformulate Example 6.5 with this new prior.

Example 6.6 Eliminating the beta prior for a regression problem

The simplified model from Example 6.5 reads

$$b_{\sigma_m} \sim \text{Bernoulli}\left(\frac{\alpha}{\alpha + (M-1)\beta}\right), \qquad m = 1 : M,$$

$$w_n | b_{\sigma_{1:M}} \sim \text{Poisson}\left(\mu_{\text{mol}} \sum_{m=1}^{M} b_{\sigma_m}\right), \qquad n = 1 : N,$$

thereby eliminating $\pi_{\sigma_{1:M}}$. Put differently, we have reduced our posterior of Example 6.5 from $p(b_{\sigma_{1:M}}, \pi_{\sigma_{1:M}} | w_{1:N})$ to $p(b_{\sigma_{1:M}} | w_{1:N})$, thereby improving mixing.

6.4 Exercise Problems

Exercise 6.1 Modeling experimental replicates with variable noise

A recurring problem in modeling experimental data is to deal with cases where day-to-day fluctuations in conditions directly impact the parameters of the experimental instrument's emission distribution (such as the variance of a normal model). Despite these complications, it is often reasonable to assume that the physical laws dictating a system's behavior under study remain unchanged. The question then naturally arises: How do we use all data to learn properties of the system when conditions under which the data are collected exhibit irreproducibility?

Here, we reconsider a problem of linear regression where we commonly assume the generative model

$$y_n = f(x_n) + \epsilon_n, \qquad \epsilon_n \sim \text{Normal}(0, 1/\tau),$$

where $f(x_n) = \alpha x_n + \beta$ and (α, β), the slope and intercept, are parameters to be estimated. Each measurement y_n is corrupted by normal noise with a common precision, τ.

In this model, the likelihood of one replicate containing N data points reads

$$p(y_{1:N} | \alpha, \beta) = \prod_{n=1}^{N} p(y_n | \alpha, \beta) \propto \prod_{n=1}^{N} e^{-\frac{\tau}{2}(y_n - (\alpha x_n + \beta))^2}.$$

Consider three replicates containing N, N', and N'' data points. Assume that the precision for each replicate is different but that otherwise α and β are the same across all replicates.

1. Generate synthetic data under these conditions. Then, place an independent gamma prior on each precision and a prior of your choice on α and β. Develop a Gibbs sampling scheme to sample α and β as well as all the variances.
2. Histogram samples over all five quantities, after burn-in removal, and make sure to include the ground truth as vertical lines in all your histograms.

Exercise 6.2 Stream-bed profile estimation

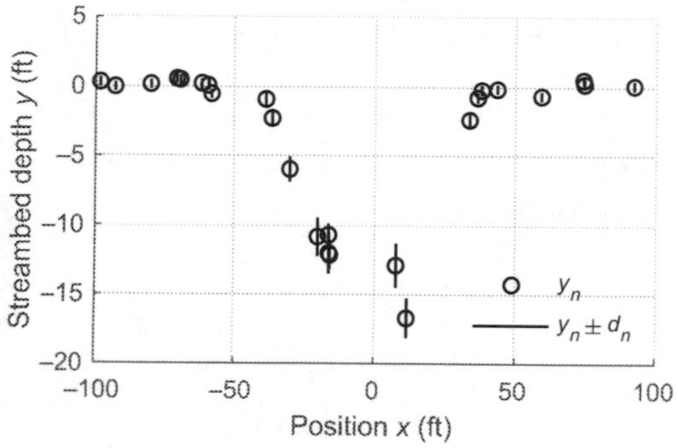

x_n (ft)	y_n (ft)	d_n (ft)
−16.33	−10.71	0.82
43.37	−0.17	0.33
−98.40	0.29	0.33
−38.91	−0.92	0.47
−69.54	0.48	0.33
−80.25	0.15	0.33
−61.76	0.21	0.33
−30.40	−5.96	0.90
−20.32	−10.84	1.39
7.64	−12.90	1.60
−15.91	−12.15	1.03
36.46	−0.77	0.38
−58.18	−0.54	0.33
74.43	0.19	0.33
−93.03	−0.04	0.33
33.56	−2.38	0.43
−16.28	−12.02	1.49
11.55	−16.70	1.44
−70.79	0.55	0.33
−59.43	0.06	0.33
59.20	−0.68	0.33
92.18	0.07	0.33
−36.73	−2.28	0.45
37.86	−0.24	0.36
74.09	0.49	0.33

The dataset shown above provides measurements of the depth, y_n, of a water channel obtained at randomly chosen positions, x_n, along its cross-section. Each depth measurement, y_n, is contaminated with additive normal noise of zero mean and standard deviation, d_n, illustrated by the error bars.

1. Formulate a Bayesian regression model employing a GaussianP to estimate the depth profile along the channel's cross-section. In the GaussianP prior, make your own choices for the mean and covariance.
2. Sample your prior depth profiles and summarize the results graphically with the mean curve shown in a thick line and all other curves shown as thin lines.
3. Repeat the above for the posterior and highlight the MAP estimate.
4. At each of the positions

$$x_A = -20, \quad x_B = -10, \quad x_C = 0, \quad x_D = +10, \quad x_E = +20 \text{ ft}$$

along the channel's cross-section, compute the posterior probability that the depth is less than 10 ft. Derive your results analytically as well as through sampling.

5. Compute the posterior probability that the depth is less than 10 ft in all positions x_A, x_B, x_C, x_D, x_E simultaneously. Derive your results analytically and by sampling.

Exercise 6.3 GaussianP on a simple polynomial curve

Simulate a polynomial curve with normal noise. Implement the GaussianP to learn the shape of the underlying curve. Use a zero mean GaussianP prior with a squared exponential covariance kernel. To present your results, plot the posterior mean curve estimate in a thick line and all other curves sampled from your posterior as thin lines on a grid of test points. Superpose the ground truth profile in a different color.

Exercise 6.4 Inducing point methods

Repeat Exercise 6.3 but with fewer, say half, as many test points now treated as inducing points. Interpolate between the inducing point values on a finer grid using the deterministic transformation discussed in Section 6.2.4. To present your results, plot the interpolated mean curve estimate in a thick line and superpose the ground truth profile in a different color. On your figure, superpose other mean curves obtained by interpolating the curve using even sparser inducing points in your analysis.

Exercise 6.5 Metropolis–Hastings for nonconjugate GaussianP models

Following the discussion in Note 6.8, generate a one-dimensional purely positive profile and add a small level of gamma noise at selected points $x_n^{\sharp}$. The goal is to estimate your profile.

Place a GaussianP prior on $f(\cdot)$, with a mean and covariance of your choosing, while expressing the profile using $e^{f(\cdot)}$ in the likelihood. Use Example 6.3 to sample from your posterior. Illustrate your result graphically by superposing the mean sampled profile as a thick black line and all other samples as thin black lines after removal of burn-in. Also, superpose the ground truth profile in a different color.

Repeat the procedure for the Gibbs-like sampler provided in the discussion below Note 6.8 as well as the blocked Gibbs-like sampler.

How does the performance of samplers change as the magnitude of noise varies? To answer the latter, repeat the procedure for all three samplers but with data generated at higher noise levels.

Exercise 6.6 Elliptical slice sampler for the spatially inhomogeneous Poisson distribution

Realistic likelihoods may deviate from the normal form, thereby breaking conjugacy with the GaussianP prior. Here, we consider a spatially inhomogeneous Poisson distribution, *i.e.*, a Poisson distribution with a spatially dependent mean $\mu(\cdot)$ along one dimension in space. The exact spatial dependence of the mean is unknown and to be determined.

1. Start by specifying a form for the Poisson mean in space and sample from the Poisson distribution according to its mean on a regular grid of locations in one dimension.
2. Place a GaussianP prior on the logarithm of the mean and use the elliptical slice sampler to sample the mean curve along a grid of test points.
3. Report the MAP sample of your Poisson mean on these test points and compare to the ground truth.
4. Repeat this exercise for a Poisson distribution with a spatially dependent mean along two dimensions of space. In doing so, use inducing points to interpolate your estimate for the mean surface.

Exercise 6.7 GaussianP on Langevin dynamics

When a particle undergoes Brownian-like motion in a potential, $U(x)$, its motion may be described by Langevin dynamics,

$$x_{n+1}|x_n, U \sim \text{Normal}\left(x_n + \frac{\tau}{\zeta}f(x_n), \frac{2\tau k_B T}{\zeta}\right),$$

where τ is the time step, ζ is the friction coefficient, $k_B T$ is the temperature multiplied by Boltzmann's constant, and $f(\cdot) = -U'(\cdot)$ where $U'(\cdot)$ is the derivative of $U(\cdot)$.

1. Simulate the trajectory of a particle within a harmonic potential well described by $U(x) = kx^2/2$.
2. Applying a GaussianP prior on the force, $f(\cdot)$, derive a posterior given a trajectory $x_{1:N}$. Hint: The posterior should be another GaussianP.
3. Find the MAP values for your force along a grid of test points. Numerically integrate the force to get an estimate for the potential, $U(\cdot)$, at each grid point. Plot your mean potential as a solid line and other samples as thin lines alongside the ground truth.
4. Normally, we would expect to poorly estimate the potential in regions with limited data, such as regions of high potential. If this is the case, why does the sample variance not grow monotonically as you move away from regions of lower potential visited by the particle?
5. Repeat all of the above steps for a quartic potential with two minima separated by a barrier. Hint: Use a low enough barrier in order to allow the particle to visit both wells over the course of your simulated trajectory that you will use as synthetic data.

Exercise 6.8 BetaBernP in estimating molecule numbers

Here we will repeat Example 6.5 by first generating synthetic data following the provided generative model. Next, use the Gibbs sampling scheme provided in order to sample both $\pi_{\sigma_{1:M}}$ and $b_{\sigma_{1:M}}$. Illustrate your results by showing the estimated active loads, B, as a function of MCMC iterations. Make sure to include the ground truth number of active loads as a horizontal line. Perform this procedure for 2, 5, and 10 active loads.

Repeat the procedure above by now treating μ_{mol} as an unknown and placing a conjugate gamma prior on μ_{mol}. Illustrate your results by showing the estimated active loads, B, as well as μ_{mol} as a function of MCMC iterations (make sure to include ground truth in your figures). Comment on the model indeterminacy encountered.

Project 6.1 Learning kernel parameters

In Note 6.5 we discussed locally periodic kernels relevant in modeling dampened oscillations. Dampened oscillations are a recurring pattern of nature appearing in models of molecular dipoles treated as harmonically bound charges fluorescing near surfaces (R. R. Chance, A. Prock, R. Silbey. Molecular fluorescence and energy transfer near interfaces. *Adv. Chem. Phys.* 37:1, 1978). Oscillations are equally relevant in describing bacterial spatial and temporal patterns including gene expression, cell division, and progression (*e.g.*, P. Lenz, L. Søgaard-Andersen. Temporal and spatial oscillations in bacteria. *Nat. Rev. Micro. Biol.* 9:565, 2011).

1. Begin by generating synthetic data by simulating a trajectory from an underdamped Langevin equation, Eq. (2.44), in a harmonic potential. Your generated trace should show a few tens of oscillations prior to damping.
2. Apply a GaussianP prior on your curve using the squared exponential kernel from Note 6.4. Use fixed values for λ and apply a hyperprior on ℓ. Develop an MCMC scheme in order to sample the curve alongside ℓ.
3. Plot your ℓ versus MCMC iterations.
4. Repeat the above for different values of the temperature in your underdamped Langevin equation. How does your ℓ vary as you increase temperature?

Project 6.2 Step finding with BetaBernP

A system may visit an unknown number of different states over the course of an experiment. Each observation in such an experiment can be modeled by

$$w_n = \epsilon_n + \sum_{m=1}^{M} b_{\sigma_m} h_{\sigma_m} H\left(t - t_{\sigma_m}\right), \qquad \epsilon_n \sim \text{Normal}\left(0, \frac{1}{\tau}\right),$$

where ϵ_n represents simple additive normal noise, b_{σ_m} is an indicator (or load) variable determining whether the associated step is present or absent, h_{σ_m} is a (constant) signal level associated to that step, t_{σ_m} is the time at which the mth step occurs, and $H\left(t - t_{\sigma_m}\right)$ models an idealized instantaneous detector response according to a modified Heaviside function,

$$H(t) = \begin{cases} 1, & t \leq 0, \\ 0, & t > 0. \end{cases}$$

Determining $h_{\sigma_{1:M}}$, $h_{\sigma_{1:M}}$, and $t_{\sigma_{1:M}}$ requires a Bayesian nonparametric inference as M is fundamentally unknown. Given the likelihood associated with this model,

$$w_n|\tau, b_{\sigma_{1:M}}, h_{\sigma_{1:M}}, t_{\sigma_{1:M}} \sim \text{Normal}\left(\sum_{m=1}^{M} b_{\sigma_m} h_{\sigma_m} H\left(t - t_{\sigma_m}\right), \frac{1}{\tau}\right), \tag{6.19}$$

we apply the following priors on the parameters:

$$\tau \sim \text{Gamma}\left(\phi, \frac{\tau_{\text{ref}}}{\phi}\right),$$
$$b_{\sigma_m} \sim \text{Bernoulli}\left(\frac{\gamma}{M}\right),$$
$$h_{\sigma_m} \sim \text{Normal}\left(h_{\text{ref}}, \frac{1}{\chi}\right),$$
$$t_{\sigma_m} \sim \text{Uniform}_{[t_0, t_N]},$$

with the understanding that $m = 1 : M$.

Generate synthetic data according to Eq. (6.19), then develop a Metropolis–Hastings within Gibbs scheme to learn the values of τ, $b_{\sigma_{1:M}}$, $h_{\sigma_{1:M}}$, and $t_{\sigma_{1:M}}$. Hint: Make sure to have a sufficient number of data points per step. Also make sure that the emission levels are sufficiently well separated in order for your data to be analyzable.

Report your final results by superposing your ground truth without noise, your actual data with noise faded in the background, and your MAP estimate as a thick line.

Project 6.3 Discrete regression with BetaBernP

One common question in microscopy and astronomy is to ask about the number of discrete elements present in an acquired image frame whether these be fluorescent molecules, groups of stars, or galaxies.

Indeed, photons emitted by a point source of light, such as fluorescent molecules whose emission is collected by a microscope objective, form an extended two-dimensional spatial pattern over the camera. This pattern is termed the point spread function (PSF). Under ideal conditions, this spatial pattern is described by an Airy pattern given in Project 1.1.

For concreteness, we will consider single molecules as emitters. We start from a Gaussian approximation to the Airy function in Project 1.1 for a single emitter which reads

$$f(\boldsymbol{r}) \propto \exp\left(-\frac{(\boldsymbol{r} - \boldsymbol{r}_\star)^2}{2v_{\text{PSF}}}\right),$$

where the PSF variance, v_{PSF}, is set to $\lambda^2/(2\pi^2 N_A^2)$ with the wavelength in vacuum set to $\lambda = 500$ nm and a numerical aperture of $N_A = 1.2$. For more emitters, $f(\cdot)$ will be a sum of Gaussians all assumed to have the same variance but different centers.

1. As a sanity check, compare the Gaussian approximation above to the exact Airy function of Project 1.1.

2. Generate a synthetic data set consisting of $w_{1:N}$ on a regular two-dimensional grid of designated position $\boldsymbol{r}_{1:N}$, idealizing a pixelated image. Assume the measurements are contaminated with error of the form $w_n \sim \text{Normal}(If(r_n) + b, v)$ where $f(\cdot)$ is the profile, *i.e.*, a sum of Gaussians, that we later infer from this data. Here, I is the photon emission rate multiplied by the camera exposure, *i.e.*, the intensity, set to 4,000. We also set v to 25. The background emission (not originating from the emitters and constant over all pixels) b is also set to 25. This noise distribution is intended to approximate a combination Poisson shot noise and detector noise. In generating the synthetic data, consider two emitters, three standard deviations ($3\sqrt{v_{\text{PSF}}}$) apart.
3. Use the BetaBernP to formulate a regression problem estimating the profile $f(\cdot)$, *i.e.*, that estimates the loads, as well as estimates the center of each Gaussian assuming known, precalibrated, v_{PSF}, v, I, and b. Make your own choices for your priors and hyperparameters and develop an MCMC scheme to characterize the posterior distribution of your model.
4. Illustrate your results by superposing the ground truth and MAP profile.
5. Repeat the above with three emitters, two standard deviations apart in a triangular configuration on one plane.

Project 6.4 Discrete regression with BetaBernP and pixelization

Here we repeat Project 6.3 but with the added complexity of camera/detector pixelization. We consider a 32 × 32 pixel-sized image with a measurement, w_n, over each of these pixels. Thus $w_{1:N}$ coincides with a 32 × 32 array of values. We assume each pixel is 120 nm wide. Just as before, the measurements are contaminated with error of the form $w_n \sim \text{Normal}(If(r_n) + b, v)$.

The difference with Project 6.3 is that $f(r_n)$ is itself an integral over the nth pixel area, A_n, of an underlying profile $g(\cdot)$. That is,

$$f(r_n) = \int_{A_n} g(\boldsymbol{r})d\boldsymbol{r},$$

with $g(\cdot)$ the sum of an unknown number of Gaussians of equal and known spread but with different and unknown centers.

We assume throughout known values of I, b, v, and v_{PSF}, which we set identical to Project 6.3.

1. Generate synthetic data according to the prescription above for a profile generated from two and then three emitters in the configurations specified in Project 6.3.
2. Now use a BetaBernP to formulate a regression problem estimating the profile $g(\cdot)$, *i.e.*, estimate the loads, as well as the center of each Gaussian. Make your own choices for your priors and hyperparameters and develop an MCMC scheme to characterize the posterior distribution of your model.
3. Illustrate your results by superposing the ground truth and MAP profile.

Project 6.5 Emitter localization

The optics of diffraction often prohibit us from being able to precisely pinpoint the number of closely spaced emitting objects such as molecules in an image.

Our goal here is to illustrate how far we can leverage the BetaBernP to correctly identify the number of emitters as a function of their center-to-center distance and continue to use the idealized Gaussian model from Project 6.3 with the same, known, I, b, v, and v_{PSF}, which we set identical to Project 6.3. We use the same pixelization condition as in Project 6.4 with the same size pixels.

1. Generate data assuming four emitters in the same plane sitting on the corners of a square three standard deviations ($3\sqrt{v_{\text{PSF}}}$) apart. Make sure to generate a pixelated output, as you did in Project 6.4, and, just as before, add measurement error of the form $w_n \sim \text{Normal}(If(r_n) + b, v)$.
2. Use the BetaBernP to formulate a regression problem that estimates the number of active loads. How close must the two emitters be in order for your method to fail? That is, repeat the first item but with four emitters two, and then one, standard deviation apart.
3. Realistically, the actual form for the PSF is rarely known. Not knowing the exact PSF can impact the number of loads we determine. Here, regenerate data, just as you did earlier, but with the exact Airy function PSF. Now analyze the data under the approximation of a Gaussian PSF. In this case, how close must the two emitters be in order for your method to fail?

Project 6.6 Localization with an aberrated PSF

Here, we go beyond Project 6.5 and consider more realistic forms of the PSF. According to the scalar diffraction theory of light, the PSF at location (x, y) generated by an emitter at location $(x_\star, y_\star, z_\star)$ is described by

$$\text{PSF}(x, y;\ x_\star, y_\star, z_\star, \mathcal{P})$$
$$\propto \left| \iint dk_x dk_y \left(\mathcal{P}\left(k_x, k_y\right) \exp\left[2i\pi z_\star \sqrt{(n_i/\lambda)^2 - k_x^2 - k_y^2}\right]\right)\right.$$
$$\left. \exp\left[2i\pi (k_x(x - x_\star) + k_y(y - y_\star))\right] \right|^2.$$

The integrand includes the *pupil function*, $\mathcal{P} = \mathcal{A}e^{i\Phi}$, arising from optical aberrations and a *defocus* term (the first exponential of the integrand) arising from the axial location of the emitter with respect to the in-focus imaging plane, where n_i is the refractive index set to 1.33. Here, the particle coordinates are starred. The integral is performed over the area where the wave vector components satisfy $k_x^2 + k_y^2 \leq (N_A/\lambda)^2$, where N_A is the numerical aperture of the objective lens used to collect the light.

An approximate discretization of the integral above reads

$$\mathrm{PSF}(p, q; x_\star, y_\star, z_\star, \mathcal{A}, \Phi) = \alpha_{\mathrm{PSF}} \Bigg| \sum_{\mu=1}^{N} \sum_{\nu=1}^{N} M_{\mu\nu} \mathcal{A}_{\mu\nu} \times \exp\left[i\Phi_{\mu\nu} + 2i\pi z_\star \sqrt{(n_i/\lambda)^2 - k_\mu^2 - k_\nu^2} \right] \times \exp\left[2i\pi \left(k_\mu (p - x_\star) + k_\nu (q - y_\star) \right) \right] \Bigg|^2 .$$

Here, p and q denote the locations along the x and y axes coinciding with the center of each pixel and μ and ν count the discrete Fourier modes $k_\mu, k_\nu = \frac{-1}{2a}, \frac{-1}{2a} + \frac{1}{aN}, \ldots, \frac{1}{2a} - \frac{1}{aN}, \frac{1}{2a}$ where a is the pixel dimension (assumed square and, as before, set to 120 nm) and where N is the number of pixels along each direction, 32.

Furthermore, α_{PSF} is the normalization constant (assuring the PSF is normalized over a sum over all allowed p and q) and $M_{\mu\nu}$ is

$$M_{\mu\nu} = \begin{cases} 1 & k_\mu^2 + k_\nu^2 \le \frac{N_A^2}{\lambda^2} \\ 0 & k_\mu^2 + k_\nu^2 > \frac{N_A^2}{\lambda^2} \end{cases}$$

to guarantee that the sum is only over the wave vectors within $k_\mu^2 + k_\nu^2 \le (N_A/\lambda)^2$. As such, the PSF can be obtained by a fast Fourier transform (FFT).

We continue to use the same values for I, b, v, and λ, which we set identical to Project 6.3, and set N_A to 1.2. We assume all these to be known.

1. Verify, as a sanity check, that in an aberration-free setup when $\mathcal{A}_{\mu\nu} = 1$ and $\Phi_{\mu\nu} = 0$, we obtain the resulting Airy pattern described in Project 1.1.
2. Use the above expression to simulate three aberrated PSFs located at $(x_\star, y_\star, z_\star)$ of $(7.5, -12.5, 0)$, $(-11.0, 1.0, 0)$, and $(0.75, 9.0, 0)$ over a 32×32 image frame. The positions are given in pixel units such that a decimal of .5 is in the pixel center, .0 is to the left, and the center of the frame is defined as the origin. For simplicity alone, herein, we assume $\mathcal{A}_{\mu\nu} = 1$ indicating there is no aberration in the form of attenuation in the system. Furthermore, we will assume a form of aberration arising from light propagating from the emitter along different directions focusing at different planes introducing astigmatism with $\Phi(k_\mu, k_\nu) = -2\left(4(k_\mu^2 + k_\nu^2)^2 - 3(k_\mu^2 + k_\nu^2)\right)\cos(2\theta)$, where $\theta = \arctan(k_\nu/k_\mu)$. Now assume the measurements are contaminated with error identical to that in Project 6.3.
3. Next, introduce the appropriate priors under the assumption of three emitters and learn the emitter positions by developing your own MCMC procedure. Generate two-dimensional histograms for your sampled positions (after removing burn-in) and make sure to mark the ground truth positions.
4. Now assuming that the number of emitters are unknown, use a BetaBernP to learn the number of emitters as well as the corresponding locations. As before, develop an appropriate MCMC scheme and generate

a two-dimensional histogram for your positions, after burn-in removal, and mark the ground truth positions.

Project 6.7 Learning point spread functions using the GaussianP

Starting from the Airy PSF, generate synthetic data with the same parameters (v, I, b, λ, n_α, N_A, pixel size, number of pixels) as Project 6.6 albeit with only one emitter located at the origin. Use a GaussianP prior on the PSF under known v, I, b, λ, n_α, N_A, and pixel size to develop an MCMC scheme to learn the PSF. Compare your PSF MAP estimate to your ground truth. Repeat for the astigmatic PSF provided in Project 6.6. Hint: The parameters of the kernel of the GaussianP prior may require calibration.

Explain, in words alone, the feasibility of determining an unknown PSF when the number of emitters are unknown.

Additional Reading

D. J. C. MacKay. Introduction to Gaussian processes. *Neural networks and Machine Learning. NATO ASI Series. Series F, Computer and System Sciences*, 168:133, 1998.

A. G. Wilson, H. Nickisch. Kernel interpolation for scalable structured Gaussian processes (KISS-GP). *Intl. Conf. Mach. Learn.*, PMLR, 2015.

M. K. Titsias, M. Rattray, N. D. Lawrence. Markov chain Monte Carlo algorithms for Gaussian processes. *Inference and Estimation in Probabilistic Time-Series Models*, 9:298, 2008.

I. Murray, R. Adams, D. MacKay. Elliptical slice sampling. *Proc. 13th Intl. Conf. AI & Stat. JMLR Workshop and Conference Proceedings*, 2010.

L. P. Swiler, M. Gulian, A. L. Frankel, C. Safta, J. D. Jakeman. A survey of constrained Gaussian process regression: approaches and implementation challenges. *J. Mach. Learn. Mod. Comp.*, 1, 2020.

D. Foreman-Mackey, E. Agol, S. Ambikasaran, R. Angus. Fast and scalable Gaussian process modeling with applications to astronomical time series. *Astro. J.*, 154:220, 2017.

J. S. Bryan IV, I. Sgouralis, S. Pressé. Inferring effective forces for Langevin dynamics using Gaussian processes. *J. Chem. Phys.*, 152:124106, 2020.

J. S. Bryan IV, P. Basak, J. Bechhoefer, S. Pressé. Inferring potential landscapes from noisy trajectories of particles within an optical feedback trap. *iScience*, 25.9:104731, 2022.

Z. Ghahramani, T. L. Griffiths. Infinite latent feature models and the Indian buffet process. *Adv. Neur. Inf. Proc. Syst.*, 18, 2005.

T. L. Griffiths, Z. Ghahramani. The Indian buffet process: an introduction and review. *J. Mach. Learn. Red.*, 12:4, 2011.

L. Al-Abadi, M. Zarepour. On approximations of the beta process in latent feature models: point processes approach. *Sankhya A*, 80:59, 2018.

J. Paisley, L. Carin. Nonparametric factor analysis with beta process priors. *Proc. 26th Ann. Intl. Conf. Mach. Learn. ACM*, 777, 2009.

M. Tavakoli, S. Jazani, I. Sgouralis, O. M. Shafraz, B. Donaphon, S. Sivasankar, M. Levitus, S. Pressé. Pitching single-focus confocal data analysis one photon at a time with Bayesian nonparametrics. *Phys. Rev. X*, 10:011021, 2020.

S. Jazani, L. W. Q. Xu, I. Sgouralis, D. Shepherd, S. Pressé. Computational proposal for tracking multiple molecules in a multi-focus confocal setup. *ACS Photonics*, 9.7:2489, 2022.

M. Fazel, S. Jazani, L. Scipioni, A. Vallmitjana, E. Gratton, M. A. Digman, S. Pressé. High resolution fluorescence lifetime maps from minimal photon counts. *ACS Photonics*, 9:1015, 2022.

S. Liu, E. B. Kromann, W. D. Krueger, J. Bewersdorf, K. A. Lidke. Three dimensional single molecule localization using a phase retrieved pupil function. *Opt. Express*, 21:29462, 2013.

B. Huang, W. Wang, M. Bates, X. Zhuang. Three-dimensional super-resolution imaging by stochastic optical reconstruction microscopy. *Science*, 319:810, 2008.

J. Enderlein, E. Toprak, P. R. Selvin. Polarization effect on position accuracy of fluorophore localization. *Opt. Express*, 14:8111, 2006.

M. Fazel, M. J. Wester, H. Mazloom-Farsibaf, M. Meddens, A. S. Eklund, T. Schlichthaerle, F. Schueder, R. Jungmann, K. A. Lidke. Bayesian multiple emitter fitting using reversible jump Markov chain Monte Carlo. *Scientific Rep.*, 9:1, 2019.

M. Fazel, M. J. Wester. Analysis of super-resolution single molecule localization microscopy data: a tutorial. *AIP Adv.*, 12:010701, 2022.

7 Mixture Models

> *By the end of this chapter, we will have presented*
>
> - *The clustering problem*
> - *Mixture models*
> - *Dirichlet priors*

In Chapter 3 we briefly introduced one of the simplest examples of what is sometimes called a *mixture model* (MM): a two-component Gaussian. In the context of this two-component Gaussian, in Example 5.10 we presented a Gibbs sampler to assign individual data points to specific mixture components that, depending on the context, may be more accurately termed *clusters, classes, categories,* or even *constitutive states.* This is an example of what is otherwise considered the general *data classification* problem. Within the context of Example 5.10, we also simultaneously estimated other unknowns such as the center of each Gaussian component that might be relevant to an MM. Taken together, both classification and parameter inference are themselves part of the broader *data clustering* problem.

As the flavors of questions we can ask within a clustering application are quite broad, here we focus on MMs within a Bayesian framework, which allows us to readily generalize and estimate even the probability associated to each component in an effort to shed light on the number of categories warranted by the data. Before we start, we discuss some applications to which MMs presented herein are relevant.

Many questions in the natural sciences naturally lend themselves to MM formulations where the intent is to classify data points and learn parameters distinguishing mixture components.

For instance, given a decay curve assumed to arise from the sum of many exponential components, it is common to ask, given inherent uncertainty in the data, the relative weight associated to each exponential component and to ask about the decay constant associated to each component (the parameter inference problem). Similarly, we can ask which exponential component gave rise to individual data points (the data classification problem). This question, involving exponentials, is visited in the context of a particular fluorescence lifetime imaging application in Project 7.2.

Similar questions can be formulated on the localization and assignment of point light emitters, whether assigning stars to star clusters from astronomical data or characterizing overlapping fluorescent molecules from microscopy images using mixtures. These examples can be addressed using simple, sometimes Gaussian, mixtures. Yet the MMs we explore in this chapter are not limited to exponential or normal mixtures. Indeed, mixtures can arise from the superposition of arbitrarily complicated densities as required by the underlying problem, *e.g.*, such as astigmatic point spread functions explored in Project 6.6.

Whatever the case may be, data often originates from multiple independent components and developing methods to achieve inference when distributions involved are expressed as sums of other distributions constitutes a critical tool for both static problems, as we explore in this chapter, as well as dynamic problems, as we explore in Chapter 8.

7.1 Mixture Model Formulations with Observations

In Chapter 1, we studied discrete probability distributions and we saw that their densities take the form

$$p(s) = \sum_{m=1}^{M} \pi_{\sigma_m} \delta_{\sigma_m}(s).$$

Here, s is the random variable of interest and $\sigma_{1:M}$ are its possible values. Such densities are the starting point for more general mixture distributions for which individual components are not delta distributed but rather have some associated breadth. For example, we may think of observations w as being distributed according to a *mixture distribution* characterized by a probability density of the form

$$p(w) = \sum_{m=1}^{M} \pi_{\sigma_m} f_{\sigma_m}(w), \tag{7.1}$$

where the mixture weights, which we gather in $\boldsymbol{\pi} = \pi_{\sigma_{1:M}}$, are nonnegative scalars such that

$$\sum_{m=1}^{M} \pi_{\sigma_m} = 1.$$

Each component $f_{\sigma_m}(\cdot)$ in our mixture is a probability density itself and so it must be normalized as

$$\int dw\, f_{\sigma_m}(w) = 1,$$

where it is understood that the integral is over all allowed values of w.

Note 7.1 Normalization

Using the properties listed above, we can readily verify that mixture densities are also normalized,

$$\int dw\, p(w) = \int dw \sum_{m=1}^{M} \pi_{\sigma_m} f_{\sigma_m}(w) = \sum_{m=1}^{M} \int dw\, \pi_{\sigma_m} f_{\sigma_m}(w)$$
$$= \sum_{m=1}^{M} \pi_{\sigma_m} \int dw\, f_{\sigma_m}(w) = \sum_{m=1}^{M} \pi_{\sigma_m} = 1,$$

as needed in order to form valid probability densities.

7.1.1 Representations of a Mixture Distribution

Very often, components of a mixture represent *categories* or *constitutive states* $\sigma_{1:M}$ coinciding with the allowed realization of the states s_n associated to each data point. In our language, the mixture probability density, Eq. (7.1), can therefore alternatively be represented, as seen in Chapter 3, as

$$s \sim \text{Categorical}_{\sigma_{1:M}}(\boldsymbol{\pi}),$$
$$w|s \sim \mathbb{F}_s,$$

where $\mathbb{F}_{\sigma_m}$ is a distribution with coinciding density $f_{\sigma_m}(w)$. The relationship between both representations can be made explicit by considering the completion

$$p(w) = \sum_s p(w, s) = \sum_s p(w|s)p(s)$$
$$= \sum_s f_s(w)\text{Categorical}_{\sigma_{1:M}}(s; \boldsymbol{\pi}) = \sum_{m=1}^{M} f_{\sigma_m}(w)\pi_{\sigma_m}.$$

Although mathematically equivalent, the new formulation is more convenient as it immediately lends itself to a graphical representation, Fig. 7.1, and to the ancestral sampling scheme of Algorithm 7.1.

Fig. 7.1 Graphical representation of an MM. Here, a single observation w is drawn from a mixture with weights $\boldsymbol{\pi}$.

Algorithm 7.1 Ancestral sampling for mixture distributions

Given weights π and distributions $\mathbb{F}_{\sigma_{1:M}}$, to sample from a MM:

- Generate $s \sim \text{Categorical}_{\sigma_{1:M}}(\pi_{\sigma_{1:M}})$.
- Generate $w|s \sim \mathbb{F}_s$.

We recall that the distributions $\mathbb{F}_{\sigma_m}$ are called *emission distributions* and generally depend on state-specific parameters ϕ_{σ_m}. For this reason, we express the emission distributions in a parametrized fashion, $\mathbb{F}_{\sigma_m} = \mathbb{G}_{\phi_{\sigma_m}}$, where $\mathbb{G}_\phi$, called a *mother distribution*, attains a form that depends on a generic parameter ϕ. With this convention, categories such as σ_m are associated with unique values ϕ_{σ_m} and we may write

$$w|s \sim \mathbb{G}_{\phi_s}.$$

For notational convenience, we may gather the emission parameters of an MM into an array $\boldsymbol{\phi} = \phi_{\sigma_{1:M}}$.

Note 7.2 Emission models

Generally, the mother distribution $\mathbb{G}_\phi$, or the functional form of its density $G_\phi(w)$, is motivated by physical considerations. For example, observations corrupted by Brownian noise are typically represented by normal distributions, while observations corrupted by shot noise are typically represented by Poisson distributions. At the crudest level, such distinction can be made based on whether our measurements $w|s$ are corrupted by *additive noise* that does not depend on the state variable or *multiplicative noise* that does depend on the state variable.

Nearly always, parameters ϕ have direct physical interpretation. In particular, in problems where ϕ is equal to the mean of $\mathbb{G}_\phi$, the individual emission parameters ϕ_{σ_m} are often termed *emission levels*.

7.1.2 Likelihoods for MMs

For a dataset $w_{1:N}$, our model reads

$$s_n \sim \text{Categorical}_{\sigma_{1:M}}(\boldsymbol{\pi}),$$
$$w_n|s_n \sim \mathbb{G}_{\phi_{s_n}}.$$

This model is depicted graphically in Fig. 7.2. Here, we assume that $\boldsymbol{\pi}$ and $\boldsymbol{\phi}$ attain known values and for this reason they are not random variables. We may use this MM to write down the completed likelihood,

$$p(w_{1:N}, s_{1:N}|\boldsymbol{\pi}, \boldsymbol{\phi}) = \prod_{n=1}^{N} G_{\phi_{s_n}}(w_n)\pi_{s_n}, \tag{*7.2*}$$

* Reminder: The asterisks by some equations indicates that the detailed derivation can be found in Appendix F.

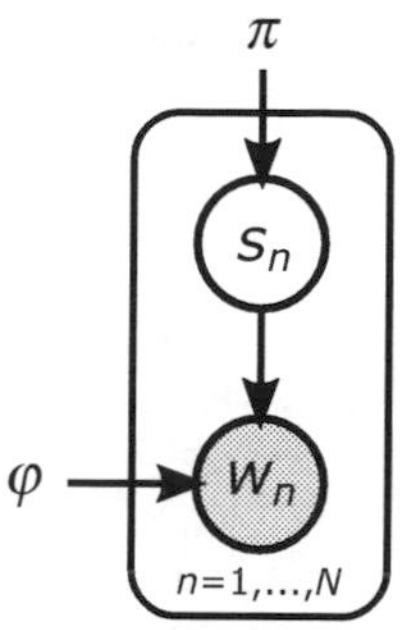

Fig. 7.2 Graphical representation of an MM with iid observations $w_{1:N}$ and known parameters π, ϕ.

and use it to compute the marginal likelihood,

$$p(w_{1:N}|\pi, \phi) = \prod_{n=1}^{N} \sum_{m=1}^{M} G_{\phi_{\sigma_m}}(w_n)\pi_{\sigma_m}. \qquad (*7.3*)$$

Note 7.3 EM and the MM

In Chapter 3, we saw that when model variables π, ϕ are unknown, we can use EM in likelihood maximization. This provided us with point estimates $\hat{\pi}, \hat{\phi}$. Yet, in more complex settings we may be interested in fully characterizing the posterior over $s_{1:N}, \pi, \phi$. For this reason, in this chapter we turn our focus to MMs within the Bayesian paradigm. Our presentation will also serve as an important starting point for the Bayesian hidden Markov model of Chapter 8.

7.1.3 State-Space Labeling and Likelihood Invariance*

Algorithms that we will present here and in the next chapter involve likelihoods $p(w_{1:N}|\pi, \phi)$ or $p(w_{1:N}, s_{1:N}|\pi, \phi)$. These likelihoods are *invariant to permutations* of the constitutive state labels. Thus, any given value of the likelihood, evaluated at specific values of π and ϕ, such as the maximum likelihood estimators, is $M!$-fold degenerate.

Example 7.1 State relabeling

To illustrate the likelihood's degeneracy, we consider a simplified MM with $N = 3$ datapoints and $M = 2$ constitutive states. The marginal likelihood then exhibits the following invariance with respect to state label permutation:

$$p\left(w_{1:3}|\pi = (\pi_\alpha, \pi_\beta), \phi = (\phi_\alpha, \phi_\beta)\right)$$
$$= p\left(w_{1:3}|\pi = (\pi_\beta, \pi_\alpha), \phi = (\phi_\beta, \phi_\alpha)\right).$$

Similarly, invariance of the completed likelihood reads

$$p\left(w_{1:3}, s_{1:3} = (\alpha, \beta, \beta)|\pi = (\pi_\alpha, \pi_\beta), \phi = (\phi_\alpha, \phi_\beta)\right)$$
$$= p\left(w_{1:3}, s_{1:3} = (\beta, \alpha, \alpha)|\pi = (\pi_\beta, \pi_\alpha), \phi = (\phi_\beta, \phi_\alpha)\right).$$

* This is an advanced topic and could be skipped on a first reading.

Both cases are produced by considering every possible permutation of the constitutive states, which for this simple example reads

$$\begin{matrix} \sigma_1 & \sigma_2 \end{matrix}$$
$$\begin{pmatrix} \alpha & \beta \\ \beta & \alpha \end{pmatrix}.$$

With a larger state-space, the total number of label permutations grows. For instance, with $M = 3$, the likelihood admits the 6-fold degeneracy

$$\begin{matrix} \sigma_1 & \sigma_2 & \sigma_2 \end{matrix}$$
$$\begin{pmatrix} \alpha & \beta & \gamma \\ \alpha & \gamma & \beta \\ \beta & \alpha & \gamma \\ \beta & \gamma & \alpha \\ \gamma & \alpha & \beta \\ \gamma & \beta & \alpha \end{pmatrix}$$

and, in general, a state-space of size M, entails an $M!$-fold degeneracy.

In other words, the MM in not identifiable in the *strict* sense where unique estimators over parameters are associated with each label m. Instead, an MM is identifiable only up to *permutations* of the state labels. As we are mostly interested in determining values for the emission parameters and probabilities of the mixture components, in practice, such nonidentifiability does not pose a problem. For example, the sequence of emission parameters $\phi_{s_{1:N}}$ attained by our model does not depend on a particular labeling and so is uniquely identifiable.

In essence, the likelihood's invariance under relabeling originates in $p(w_{1:N}|\boldsymbol{\pi}, \boldsymbol{\phi})$ or $p(w_{1:N}, s_{1:N}|\boldsymbol{\pi}, \boldsymbol{\phi})$ and occurs because emission and other model parameters are defined *only* with regards to the constitutive states σ_m and *not* with regards to their labels m. To eliminate this invariance, *additional* assumptions in an MM are required in order to establish an association between constitutive states σ_m and their labels m.

In practice, we can resolve nonidentifiability by imposing post hoc constraints in terms of the state labels. For instance, once an estimator is chosen this can be achieved by restricting the entire parameter space only to the semi-orthant on which this estimator lies. In this case, problem-specific heuristic rules are needed to impose a particular relation of the constitutive states with their assigned labels. As such, when restricting ourselves to a particular semi-orthant of the parameter space entails that from all $M!$ equally likely relabelings, we select the one lying closest to the ad hoc chosen labeling of the estimator using assignment methods such as the *Hungarian algorithm*.

7.2 MM in the Bayesian Paradigm

In its most general setting, the goal of Bayesian clustering is to obtain full joint posteriors over: states $s_{1:N}$; parameters of each component π_{σ_m}, ϕ_{σ_m};

and, most ambitiously, shed light on how many mixture components M may have given rise to the data. Here, we will begin with the first challenge of determining $s_{1:N}$, *i.e.*, estimating to which category each s_n is assigned. This constitutes the *classification task,* and initiates progress toward addressing both subsequent challenges.

7.2.1 Classification: Estimating Categories

We begin by reconsidering the emission part of our MM. In our model, the state s_n is realized to one of the allowed categories that generates datapoint w_n. On the basis of this model, we can immediately write down the model's posterior as we did in Eq. (*7.2*). This reads,

$$p(s_{1:N}|w_{1:N}) \propto p(w_{1:N}|s_{1:N})p(s_{1:N}) = \prod_{n=1}^{N} f_{s_n}(w_n)\pi_{s_n}.$$

In turn, after normalization $p(s_{1:N}|w_{1:N})$ becomes

$$p(s_{1:N}|w_{1:N}) = \prod_{n=1}^{N} \frac{f_{s_n}(w_n)\pi_{s_n}}{\sum_{m=1}^{M} f_{\sigma_m}(w_n)\pi_{\sigma_m}}. \quad (*7.4*)$$

Under known parameters π and ϕ, this becomes is product of categorical densities and can be sampled as in Algorithm 7.2.

Algorithm 7.2 Sampling MM states

Given observations $w_{1:N}$ and parameters π, ϕ, the states are independently sampled as follows:

- For $n = 1 : N$, draw

$$s_n|w_n \sim \text{Categorical}_{\sigma_{1:M}}\left(\frac{f_{\sigma_1}(w_n)\pi_{\sigma_1}}{\sum_{m=1}^{M} f_{\sigma_m}(w_n)\pi_{\sigma_m}}, \ldots, \frac{f_{\sigma_M}(w_n)\pi_{\sigma_M}}{\sum_{m=1}^{M} f_{\sigma_m}(w_n)\pi_{\sigma_m}}\right).$$

7.2.2 Mixture Weights and the Dirichlet Distribution

In many problems, both the weights, π, and the emission parameters, ϕ, are unknown. Estimating the weights directly motivates why we need the Dirichlet distribution. In such cases, a more complete model reads

$$\begin{aligned}
\phi_{\sigma_m} &\sim \mathbb{H}, \\
\pi &\sim \text{Dirichlet}_{\sigma_{1:M}}(\eta), \\
s_n|\pi &\sim \text{Categorical}_{\sigma_{1:M}}(\pi), \\
w_n|s_n, \phi &\sim \mathbb{G}_{\phi_{s_n}}.
\end{aligned}$$

In this model we have introduced two priors: independent priors on each emission parameter represented by $\mathbb{H}$ and a prior simultaneously acting on all elements of π represented by the Dirichlet distribution. In the latter, η is a hyperparameter.

This model is described by the posterior $p(s_{1:N}, \pi, \phi | w_{1:N})$. As a Gibbs sampling formalism may be invoked to sample from this posterior, it is often convenient to apply conditionally conjugate priors. For this reason, we now focus on the prior to the categorical, namely the Dirichlet distribution.

Properties of the Dirichlet Distribution

A Dirichlet random variable, $\pi_{\sigma_{1:M}}$, is an array of scalar probabilities π_{σ_m} that are simultaneously sampled,

$$\pi_{\sigma_{1:M}} \sim \mathrm{Dirichlet}_{\sigma_{1:M}}\left(\eta_{\sigma_{1:M}}\right),$$

and supported over the simplex

$$\sum_{m=1}^{M} \pi_{\sigma_m} = 1.$$

The hyperparameter of a Dirichlet distribution, $\eta_{\sigma_{1:M}}$, is another array of nonnegative weights, at least one of which is nonzero. The Dirichlet probability density is

$$\mathrm{Dirichlet}_{\sigma_{1:M}}\left(\pi_{\sigma_{1:M}}; \eta_{\sigma_{1:M}}\right) = \frac{\Gamma\left(\sum_{m=1}^{M} \eta_{\sigma_m}\right)}{\prod_{m=1}^{M} \Gamma(\eta_{\sigma_m})} \prod_{m=1}^{M} \pi_{\sigma_m}^{\eta_{\sigma_m}-1},$$

where, by normalization, we have

$$\int \cdots \int d\pi_{\sigma_{1:M}} \, \mathrm{Dirichlet}_{\sigma_{1:M}}\left(\pi_{\sigma_{1:M}}; \eta_{\sigma_{1:M}}\right) = 1.$$

Note 7.4 Conjugacy of the Dirichlet-categorical pair

We begin by considering a categorical likelihood over the states

$$p(s_{1:N} | \pi_{\sigma_{1:M}}) = \prod_{n=1}^{N} \prod_{m=1}^{M} \pi_{\sigma_m}^{\Delta_{\sigma_m}(s_n)}.$$

Conjugacy is ascertained when both $p(\pi_{\sigma_{1:M}} | s_{1:N})$ and the Dirichlet prior attain the same functional form. To verify this explicitly, we write:

$$\begin{aligned}
p(\pi_{\sigma_{1:M}} | s_{1:N}) &\propto p(s_{1:N} | \pi_{\sigma_{1:M}}) p(\pi_{\sigma_{1:M}}) \\
&= \prod_{n=1}^{N} \prod_{m=1}^{M} \pi_{\sigma_m}^{\Delta_{\sigma_m}(s_{1:N})} \pi_{\sigma_m}^{\eta_{\sigma_m}-1} \\
&\propto \mathrm{Dirichlet}_{\sigma_{1:M}}\left(\eta_{\sigma_1} + \sum_{n=1}^{N} \Delta_{\sigma_1}(s_n), \ldots, \eta_{\sigma_M} + \sum_{n=1}^{N} \Delta_{\sigma_M}(s_n)\right).
\end{aligned}$$

Example 7.2 Dirichlet random variables of size three

A Dirichlet random variable with three categories,

$$\pi_{\sigma_{1:3}} \sim \text{Dirichlet}_{\sigma_{1:3}}(\eta_{\sigma_{1:3}}),$$

is a random vector supported over the simplex $\pi_{\sigma_1} + \pi_{\sigma_2} + \pi_{\sigma_3} = 1$. The density is

$$\text{Dirichlet}_{\sigma_{1:3}}(\pi_{\sigma_{1:3}}; \eta_{\sigma_{1:3}}) = \frac{\Gamma(\eta_{\sigma_1} + \eta_{\sigma_2} + \eta_{\sigma_3})}{\Gamma(\eta_{\sigma_1})\Gamma(\eta_{\sigma_2})\Gamma(\eta_{\sigma_3})} \pi_{\sigma_1}^{\eta_{\sigma_1}-1} \pi_{\sigma_2}^{\eta_{\sigma_2}-1} \pi_{\sigma_3}^{\eta_{\sigma_3}-1}.$$

Normalization of this density reads

$$\iiint\limits_{\pi_{\sigma_1}+\pi_{\sigma_2}+\pi_{\sigma_3}=1} d\pi_{\sigma_{1:3}} \, \text{Dirichlet}_{\sigma_{1:3}}(\pi_{\sigma_{1:3}}; \eta_{\sigma_{1:3}}) = 1.$$

Gamma Random Variables and the Dirichlet Distribution

Here, we explore the relationship between gamma and Dirichlet random variables. This will allow us to show how to sample from the Dirichlet distribution and illustrate the steps involved in normalizing the Dirichlet distribution. We start (Algorithm 7.3) with how to sample from the Dirichlet.

Algorithm 7.3 Sampling Dirichlet random variables

To simulate from a Dirichlet random variable:

- First, draw gamma random variables

$$\tau_{\sigma_m} \sim \text{Gamma}(\eta_{\sigma_m}, \lambda).$$

- Then normalize

$$\pi_{\sigma_m} = \frac{\tau_{\sigma_m}}{\sum_{m'=1}^{M} \tau_{\sigma_{m'}}}.$$

For any λ chosen, the resulting $\pi_{\sigma_{1:M}}$ follows $\text{Dirichlet}_{\sigma_{1:M}}(\eta_{\sigma_{1:M}})$ statistics.

Why Does This Sampling Scheme Work?*

We begin by considering the joint distribution over iid gamma random variables,

$$p(\tau_{\sigma_{1:M}}) = \prod_{m=1}^{M} \text{Gamma}(\tau_{\sigma_{m=1}^{M}}; \eta_{\sigma_m}, \lambda) = \prod_{m=1}^{M} \left(\frac{(\tau_{\sigma_m}/\lambda)^{\eta_{\sigma_m}-1} e^{-\tau_{\sigma_m}/\lambda}}{\lambda \Gamma(\eta_{\sigma_m})} \right),$$

which we suggestively rewrite as

$$p(\tau_{\sigma_{1:M}}) = \frac{e^{-\sum_{m=1}^{M} \tau_{\sigma_m}/\lambda}}{\prod_{m=1}^{M} \lambda \Gamma(\eta_{\sigma_m})} \left(\prod_{m=1}^{M-1} \left(\frac{\tau_{\sigma_m}}{\lambda} \right)^{\eta_{\sigma_m}-1} \right) \left(\frac{\tau_{\sigma_M}}{\lambda} \right)^{\eta_{\sigma_M}-1}.$$

We now transform $\tau_{\sigma_{1:M}}$ into $\{\tau_{\sigma_{1:M-1}}/(c\lambda), c\}$ where $c = \sum_{m=1}^{M} \tau_{\sigma_m}/\lambda$. For convenience, we define $\pi_{\sigma_m} = \tau_{\sigma_m}/(c\lambda)$ and note that, under this definition,

* This is an advanced topic and could be skipped on a first reading.

$\sum_{m=1}^{M} \pi_{\sigma_m} = 1$. Under these new coordinates, and invoking the transformation rules described in Section 1.2.1, the gamma density reads

$$p(\tau_{\sigma_{1:M}}) = \lambda^M c^{M-1} p(\pi_{\sigma_{1:M-1}}, c),$$

where the domain of π_{σ_m} is restricted to the range 0 to 1 while c's domain spans the positive real line.

Marginalizing the above over all values of c, using the definition of the gamma function of Example 4.2, immediately yields the $\text{Dirichlet}_{\sigma_{1:M}}(\pi_{\sigma_{1:M}}; \eta_{\sigma_{1:M}})$ whose normalization reads

$$\int d\pi_{\sigma_{1:M-1}} \text{Dirichlet}_{\sigma_{1:M}}(\pi_{\sigma_{1:M}}; \eta_{\sigma_{1:M}}) = 1.$$

Note 7.5 An integral identity

We note that a Dirichlet distribution of size two is, by definition, the beta distribution. That is,

$$p(\pi_{\sigma_{1:2}}) = \frac{\Gamma(\eta_{\sigma_1} + \eta_{\sigma_2})}{\Gamma(\eta_{\sigma_1})\Gamma(\eta_{\sigma_2})} \pi_{\sigma_1}^{\eta_{\sigma_1}-1} \pi_{\sigma_2}^{\eta_{\sigma_2}-1},$$

from which, by normalization, it must follow that

$$\int_0^1 d\pi_{\sigma_1}\, \pi_{\sigma_1}^{\eta_{\sigma_1}-1} (1 - \pi_{\sigma_1})^{\eta_{\sigma_2}-1} = \frac{\Gamma(\eta_{\sigma_1})\Gamma(\eta_{\sigma_2})}{\Gamma(\eta_{\sigma_1} + \eta_{\sigma_2})}. \tag{7.5}$$

Marginals of the Dirichlet Distribution*

While the Dirichlet distribution itself may appear novel, it is in fact the higher-dimensional form of the beta distribution. Put differently, the marginal of the Dirichlet distribution is the beta distribution

$$\pi_{\sigma_{m'}} \sim \text{Beta}\left(\eta_{\sigma_{m'}}, \sum_{m \neq m'} \eta_{\sigma_m}\right),$$

where the restricted sum over m excludes m'. This realization stems from the *aggregate property* of the Dirichlet distribution that, more generally, states that

$$\left(\pi_{\sigma_{1:m'-1}}, \pi_{\sigma_{m'}} + \pi_{\sigma_{m'+1}}, \pi_{\sigma_{m'+2:M}}\right)$$
$$\sim \text{Dirichlet}_{\sigma_{1:m'-1},\sigma_*,\sigma_{m'+2:M}}\left(\eta_{\sigma_{1:m'-1}}, \eta_{\sigma_{m'}} + \eta_{\sigma_{m'+1}}, \eta_{\sigma_{m'+2:M}}\right).$$

Here, we use σ_* to denote the aggregation of $\sigma_{m'}$ and $\sigma_{m'+1}$ into a single superstate. That is, $\sigma_* = \sigma_{m':m'+1}$.

Note 7.6 Proof of the aggregate property

Without loss of generality, we consider aggregating the last two categories. That is, we wish to compute the distribution over $(\pi_{\sigma_{1:M-2}}, \pi_{\sigma_*})$, where $\pi_{\sigma_*} = \pi_{\sigma_{M-1}} + \pi_{\sigma_M}$, starting from $\text{Dirichlet}_{\sigma_{1:M}}(\pi_{\sigma_{1:M}}; \eta_{\sigma_{1:M}})$ and show

* This is an advanced topic and could be skipped on a first reading.

that the distribution following aggregation is equal to $\text{Dirichlet}_{\sigma_{1:M-1}}(\pi_{\sigma_{1:M-2}}, \pi_{\sigma_*}; \eta_{\sigma_{1:M-2}}, \eta_{\sigma_*})$, where $\pi_{\sigma_*} = \pi_{\sigma_{M-1}} + \pi_{\sigma_M}$ and $\eta_{\sigma_*} = \eta_{\sigma_{M-1}} + \eta_{\sigma_M}$. To do so, we consider

$$p(\pi_{\sigma_{1:M-2}}, \pi_{\sigma_*}) = \frac{\Gamma\left(\sum_{m=1}^{M} \eta_{\sigma_m}\right)}{\prod_{m=1}^{M} \Gamma(\eta_{\sigma_m})} \int_0^{\pi_{\sigma_*}} du\, u^{\eta_{\sigma_{M-1}}-1} (\pi_{\sigma_*} - u)^{\eta_{\sigma_M}-1} \left(\prod_{m=1}^{M-2} \pi_{\sigma_m}^{\eta_{\sigma_m}-1} \right).$$

To move forward, it is convenient to turn the above integral to the form of Eq. (7.5). To this end, we perform a variable rescaling in order to transform the upper bound of the integral from π_{σ_*} to 1. Upon this rescaling and invoking Eq. (7.5), we now find

$$p(\pi_{\sigma_{1:M-2}}, \pi_{\sigma_*}) = \text{Dirichlet}_{\sigma_{1:M-1}}(\pi_{\sigma_{1:M-2}}, \pi_{\sigma_*}; \eta_{\sigma_{1:M-2}}, \eta_{\sigma_*}).$$

Similarly, it follows that this property holds for any pair of states and that repeated application of the aggregate property yields the beta distribution as the final marginal.

An Alternative Dirichlet Distribution Parametrization

To help introduce an alternative parametrization of the Dirichlet distribution, it is helpful to discuss the significance of its hyperparameters.

Example 7.3 Interpreting the Dirichlet distribution hyperparameters

The Dirichlet density is

$$\text{Dirichlet}_{\sigma_{1:M}}(\pi_{\sigma_{1:M}}; \eta_{\sigma_{1:M}}) \propto \prod_{m=1}^{M} \pi_{\sigma_m}^{\eta_{\sigma_m}-1}. \tag{7.6}$$

In order to build intuition on the role of the hyperparameters $\eta_{\sigma_{1:M}}$, we consider the expectation of π_{σ_m}, which reads

$$\mathbb{E}(\pi_{\sigma_m}) = \frac{\eta_{\sigma_m}}{\sum_{m'=1}^{M} \eta_{\sigma_{m'}}}. \tag{7.7}$$

From this, we recognize the set of $\eta_{\sigma_m} / \sum_{m'=1}^{M} \eta_{\sigma_{m'}}$ as probabilities in themselves. We also see, based on the structure of the Dirichlet density from Eq. (7.6), that draws from the Dirichlet distribution are distributions whose deviations away from the mean distribution, $\mathbb{E}(\pi_{\sigma_m}) \propto \eta_{\sigma_m}$, are exponentially suppressed.

Recognizing η_{σ_m} as proportional to a probability (compare with Eq. (7.7)), we may parametrize the Dirichlet distribution, $\text{Dirichlet}_{\sigma_{1:M}}\left(\pi_{\sigma_{1:M}}; \eta_{\sigma_{1:M}}\right)$, as

$$\pi_{\sigma_{1:M}} \sim \text{Dirichlet}_{\sigma_{1:M}}\left(\alpha \beta_{\sigma_{1:M}}\right),$$

where α is a strictly positive hyperparameter and $\beta_{\sigma_{1:M}}$ is an array of scalar probabilities that also belongs in the simplex $\sum_{m=1}^{M} \beta_{\sigma_m} = 1$. Commonly, α is termed the *concentration* hyperparameter and $\beta_{\sigma_{1:M}}$ the *base distribution*. Under this re-parametrizaion, the Dirichlet density reads

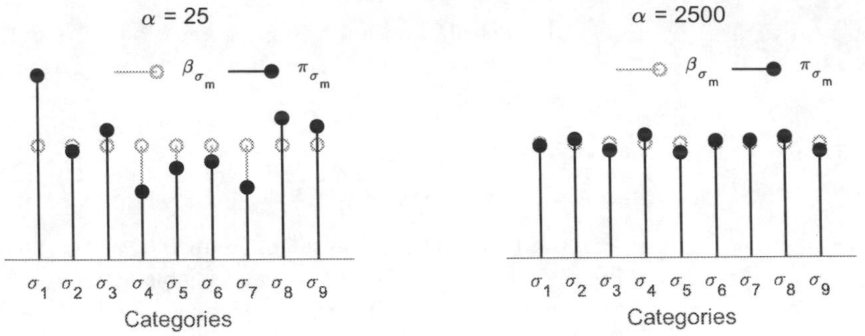

Fig. 7.3 $\text{Dirichlet}_{\sigma_{1:M}}(\alpha\beta_{1:M})$ random variables with low (left) and high (right) concentration parameter α. For illustration, here we use a size $M = 9$ and a uniform base $\beta_{\sigma_m} = 1/M$.

$$\text{Dirichlet}_{\sigma_{1:M}}\left(\pi_{\sigma_{1:M}}; \alpha\beta_{\sigma_{1:M}}\right) = \frac{\Gamma(\alpha)}{\prod_{m=1}^{M} \Gamma(\alpha\beta_{\sigma_m})} \prod_{m=1}^{M} \pi_{\sigma_m}^{\alpha\beta_{\sigma_m}-1}, \tag{7.8}$$

while both parameterizations are related by

$$\alpha = \sum_{m=1}^{M} \eta_{\sigma_m}, \qquad \beta_{\sigma_m} = \frac{\eta_{\sigma_m}}{\sum_{m=1}^{M} \eta_{\sigma_m}}, \qquad \eta_{\sigma_m} = \alpha\beta_{\sigma_m}.$$

The intuition as to why the α is termed a concentration parameter is now clear: large values for α *concentrate* draws of $\pi_{\sigma_{1:M}}$ from the Dirichlet around the base distribution, $\beta_{\sigma_{1:M}}$. Draws differing from the base are exponentially suppressed with exponential constant proportional to α; see Fig. 7.3.

> **Note 7.7** Setting elements of the base distribution to zero
>
> In writing Eq. (7.6) and Eq. (7.8), we made an implicit assumption regarding weights associated with zero η_{σ_m} or, equivalently, zero β_{σ_m}. Indeed, by definition, when η_{σ_m} is zero, we assume that the coinciding weight π_{σ_m} is zero as well such that the associated factor in the density, $\pi_{\sigma_m}^{\eta_{\sigma_m}}$, is unity in the limiting case that both π_{σ_m} and η_{σ_m} vanish.

7.2.3 Estimating Weights and Other Parameters

In order to demonstrate how we determine posteriors over weight parameters, π, we now make immediate use of the Dirichlet distribution.

Example 7.4 Learning unknown weights with a Dirichlet prior

We consider a MM with $M = 4$ components,

$$\underbrace{\text{Normal}\left(\mu_{\sigma_1}, v\right)}_{\text{category } \sigma_1} \quad \underbrace{\text{Normal}\left(\mu_{\sigma_2}, v\right)}_{\text{category } \sigma_2} \quad \underbrace{\text{Normal}\left(\mu_{\sigma_3}, v\right)}_{\text{category } \sigma_3} \quad \underbrace{\text{Normal}\left(\mu_{\sigma_4}, v\right)}_{\text{category } \sigma_4}$$

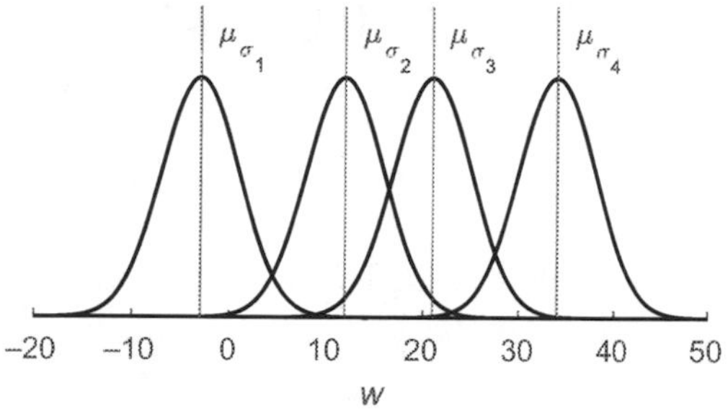

Fig. 7.4 Components of an MM with size $M = 4$.

located at $\mu_{\sigma_1} = -3$, $\mu_{\sigma_2} = 12$, $\mu_{\sigma_3} = 21$, $\mu_{\sigma_4} = 34$ with a variance of $v = 4$, as shown in Fig. 7.4. Our model reads

$$\pi \sim \text{Dirichlet}_{\sigma_{1:4}}(\alpha\boldsymbol{\beta}),$$
$$s_n|\pi \sim \text{Categorical}_{\sigma_{1:4}}(\pi),$$
$$w_n|s_n \sim \text{Normal}\left(\mu_{s_n}, v\right).$$

The hyperparameters are $\alpha = 1$ and $\boldsymbol{\beta} = (1/4, 1/4, 1/4, 1/4)$.

We may sample our posterior distribution $p(\pi|w_{1:N})$ by Gibbs sampling the joint distribution $p(s_{1:N}, \pi|w_{1:N})$ and later marginalize over $s_{1:N}$. A Gibbs sampler targeting $p(s_{1:N}, \pi|w_{1:N})$, uses the updates:

- Update s_1 by sampling from $p(s_1|s_{2:N}, \pi, w_{1:N}) = p(s_1|s_{-1}, \pi, w_{1:N})$.
- ...
- Update s_N by sampling from $p(s_N|s_{1:N-1}, \pi, w_{1:N}) = p(s_N|s_{-N}, \pi, w_{1:N})$.
- Update π by sampling from $p(\pi|s_{1:N}, w_{1:N})$.

In our sampler, each s_n is updated from

$$p(s_n|s_{-n}, \pi, w_{1:N}) \propto \text{Normal}\left(w_n; \mu_{s_n}, v\right) \text{Categorical}_{\sigma_{1:4}}(s_n; \pi)$$
$$= \frac{e^{-\frac{(w_n-\mu_{s_n})^2}{2v}}}{\sqrt{2\pi v}} \pi_{s_n} \propto e^{-\frac{(w_n-\mu_{s_n})^2}{2v}} \pi_{s_n}, \quad (*7.9*)$$

while π is updated from

$$p(\pi|s_{1:N}, w_{1:N}) = p(\pi|s_{1:N}) \propto p(s_{1:N}|\pi)p(\pi) \propto \prod_{m=1}^{4} \pi_{\sigma_m}^{\alpha\beta_{\sigma_m}+c_{\sigma_m}-1}, \quad (*7.10*)$$

where c_{σ_m} is the total count in σ_m, *i.e.*, the number of $s_n = \sigma_m$ over all $s_{1:N}$.

Determining other parameters, $\boldsymbol{\phi}$, as well (in addition to the weights) is now a matter of developing Gibbs samplers on the full posterior, $p(s_{1:N}, \pi, \boldsymbol{\phi}|w_{1:N})$, which requires the specification of $\mathbb{H}$ as in Section 7.2.2, ideally conjugate to $\mathbb{G}_{\phi_{\sigma_m}}$ if at all possible. For example, in developing the simple Gaussian mixture model of Example 5.10, the priors on the cluster centers, μ_{σ_m}, and precision, τ, could be supplemented by priors on the weights, π, following the prescription of Example 7.4 in order to construct the full posterior precisely as we laid out in Exercise 5.7.

7.3 The Infinite MM and the Dirichlet Process*

Often, we are presented with the more challenging problem of not knowing how many mixture components, M, are represented in our data and therefore not knowing how to set the dimensionality of our Dirichlet distribution. In such cases, we turn to the Bayesian nonparametric paradigm and allow M to tend to infinity where we now speak of the Dirichlet *process prior*. The idea here is akin to notions already introduced in the context of other nonparametric process priors discussed, namely the GaussianP or BetaBernP priors. As we recall, in the case of the GaussianP prior, the number of test points tended to infinity. Similarly, the number of loads tended to infinity using the BetaBernP.

For computational convenience alone, we may choose to truncate the number of arguments of the Dirichlet process prior reducing it to a high-dimensional Dirichlet distribution. Later, in Note 8.15, we will show that it is possible to avoid this approximation, though with computational cost. In other words, the idea is to show that the number of mixture components warranted by the data, say K, assuming a large enough M becomes independent of the limit we set for M. To demonstrate that our conclusions are independent of our limit, we imagine an experiment where we have drawn observations

$$\underbrace{\overbrace{c_{\sigma_1}, \ldots, c_{\sigma_K}}^{\substack{K \\ \text{categories} \\ \text{populated}}}, \overbrace{0, \ldots, 0}^{\substack{M-K \\ \text{categories} \\ \text{unpopulated}}}}_{M \text{ categories in total}}.$$

Starting from the prior $\text{Dirichlet}_{\sigma_{1:M}}(\pi_{\sigma_{1:M}}; \eta_{\sigma_{1:M}})$, the posterior then reads

$$p(\pi_{\sigma_{1:M}} | c_{\sigma_{1:K}}, 0, \ldots, 0)$$
$$= \text{Dirichlet}_{\sigma_{1:M}}(\pi_{\sigma_{1:M}}; \eta_{\sigma_1} + c_{\sigma_1}, \ldots, \eta_{\sigma_K} + c_{\sigma_K}, \eta_{\sigma_{K+1:M}}).$$

As before, we can aggregate the unsampled categories in $\sigma_* = \sigma_{K+1:M}$. Using the results of Section 7.2.2, we find

$$p(\pi_{\sigma_{1:K}}, \pi_{\sigma_*} | c_{\sigma_{1:K}})$$
$$= \text{Dirichlet}_{\sigma_{1:K},\sigma_*}\left(\pi_{\sigma_{1:K}}, \pi_{\sigma_*}; \eta_{\sigma_1} + c_{\sigma_1}, \ldots, \eta_{\sigma_K} + c_{\sigma_K}, \eta_{\sigma_*}\right),$$

where $\pi_{\sigma_*} = \sum_{m=K+1}^{M} \pi_{\sigma_m}$ and $\eta_{\sigma_*} = \sum_{m=K+1}^{M} \eta_{\sigma_m}$. We now investigate the posterior for large enough M. For simplicity, we consider a uniform base. That is, $\eta_{\sigma_m} = \alpha\beta_{\sigma_m} = \alpha/M$. Under this condition, the dependency of η_{σ_*} on M in the posterior simplifies to

* This is an advanced topic and could be skipped on a first reading.

$$\sum_{m=K+1}^{M} \eta_{\sigma_m} = \alpha \sum_{m=K+1}^{M} \beta_{\sigma_m} = \alpha \frac{M-(K+1)}{M} \approx \alpha$$

as $M \to \infty$. Putting it differently, the posterior becomes independent of M for large enough M.

While the demonstration above is valid for a Uniform base distribution, we now explore the case of a non-uniform base distribution in Note 7.8.

Note 7.8 Dirichlet posterior convergence for nonuniform base

Just as above, we consider $\alpha \sum_{m=K+1}^{M} \beta_{\sigma_m}$ with the intention of demonstrating that it is independent of M for a non-uniform base distribution.

The non-uniform base distribution we consider must have weights β_{σ_m} that begin gradually vanishing for large enough m approaching M. This condition would be reasonable for normalizable densities. Now, under these conditions,

$$\alpha \sum_{m=K+1}^{M} \beta_{\sigma_m} = \alpha \left(1 - \sum_{m=1}^{K} \beta_{\sigma_m}\right) \approx 0,$$

where the approximation holds when $\sum_{m=1}^{K} \beta_{\sigma_m}$ approaches the total normalized probability (unity).

In other words, provided $\sum_{m=1}^{K} \beta_{\sigma_m}$ approaches unity, the contribution to the posterior arising from the sum of the remaining weights excluded from the sum is negligible.

We end this section with a note on the interpretation of samples from a Dirichlet prior.

Note 7.9 Interpretation of the samples from a Dirichlet prior

We motivated nonparametrics here with a discussion pertaining to estimating the number of categories, M, when these are unknown in a clustering application. Our intuition already dictates that Bayesian methods will not provide such sharp point estimates, only probabilities in light of uncertain data. In this vein, we need to make clear that the use of Dirichlet process priors does not provide an M estimate. Rather, a Dirichlet prior, truncated for large enough M, provides the ability to construct a posterior over $\pi_{\sigma_{1:M}}$. That is, a weight over all categories, $\sigma_{1:M}$, some of which will have an associated weight, π_{σ_m}, far exceeding their expected prior value of β_{σ_m} due to the effect of the likelihood. We may then interpret the number of weights exceeding the prior expectation as coinciding with the number of categories warranted by the data.

7.4 Exercise Problems

Exercise 7.1 Gibbs sampler for Gaussian mixtures, part I

In Exercise 5.7 we developed a Gibbs sampler for a two component Gaussian MM. Here, we consider the more general model,

$$\begin{aligned}
\pi_{\sigma_{1:M}} &\sim \text{Dirichlet}_{\sigma_{1:M}}\left(\frac{\alpha}{M}, \ldots, \frac{\alpha}{M}\right), \\
s_n | \pi_{\sigma_{1:M}} &\sim \text{Categorical}_{\sigma_{1:M}}(\pi_{\sigma_{1:M}}), \\
w_n | s_n, \mu_{s_n} &\sim \text{Normal}_2\left(\mu_{s_n}, v\mathbb{1}\right).
\end{aligned}$$

Assume $M = 3$ and place the centers of the Gaussians at positions (x, y) for $\mu_{\sigma_1} = (-1, 0)$, $\mu_{\sigma_2} = (1, 0)$, and $\mu_{\sigma_3} = (0, 1)$. Set the variance to $v = 0.25^2$. Assume all of these quantities are known.

1. Using all these parameters, generate synthetic data by using an ancestral sampling scheme.
2. Develop a Gibbs sampler to draw posterior samples from $p(s_{1:N}, \pi_{\sigma_{1:M}} | w_{1:N})$ from the data generated.
3. Plot values of $\pi_{\sigma_{1:2}}$ as a function of MCMC iterations. Make sure to include ground truth as a horizontal line in your figures. Note that you only need to plot two of the three weights as the third is constrained by normalization.
4. How does your success at classification, as estimated by the deviation of the MAP estimate for your $s_{1:N}$ from the ground truth, depend on the overlap between the Gaussians of your MM? To address this question, repeat the problem by generating data with the same centers but assume a variance of $v = 0.5^2$ and then $v = 1$.

Exercise 7.2 Gibbs sampler for Gaussian mixtures, part II

Repeat Exercise 7.1 except that you are now asked to estimate the variance and centers for this more general formulation

$$\begin{aligned}
\pi_{\sigma_{1:M}} &\sim \text{Dirichlet}_{\sigma_{1:M}}\left(\frac{\alpha}{M}, \ldots, \frac{\alpha}{M}\right), \\
\tau &\sim \text{Gamma}(\alpha, \beta), \\
\mu_{\sigma_m} &\sim \text{Normal}_2\left(\xi, \frac{1}{\psi}\mathbb{1}\right), \\
s_n | \pi_{\sigma_{1:M}} &\sim \text{Categorical}_{\sigma_{1:M}}(\pi_{\sigma_{1:M}}), \\
w_n | s_n, \mu_{s_n}, \tau &\sim \text{Normal}_2\left(\mu_{s_n}, \frac{1}{\tau}\mathbb{1}\right).
\end{aligned}$$

That is:

1. Generate synthetic data (with $M = 3$, using $v = 0.25^2$ and the same means as in Exercise 7.1) by using an ancestral sampling scheme.
2. Develop a Gibbs sampler to draw posterior samples from $p(s_{1:N}, \pi_{\sigma_{1:3}}, \mu_{\sigma_{1:3}}, \tau | w_{1:N})$.

3. Plot values of $\pi_{\sigma_{1:2}}, \mu_{\sigma_{1:3}}, \tau$ as a function of MCMC iterations. Make sure to include the ground truth as a horizontal line in your figures.

Exercise 7.3 Moments of the Dirichlet

Prove Eq. (7.7). Once complete, find the variance of the Dirichlet distribution.

Project 7.1 Dirichlet process prior on Gaussian mixtures

Here, we will repeat part of Exercise 7.1 but taking the large M limit.

1. Generate your data with $M = 3$, using $v = 0.25^2$ and the same means as in Exercise 7.1 but use an approximation to the Dirichlet process prior with large M to sample $\pi_{\sigma_{1:M}}$ and $s_{1:N}$. Assume known and identical variances for all mixture components.
2. Plot $\pi_{\sigma_{1:M-1}}$ as a function of MCMC iteration for large enough M and include the ground truths as solid lines.
3. How do you reasonably interpret how many mixture components may be present from your MAP estimate?

Project 7.2 Fluorescence lifetime imaging

A common goal of quantitative fluorescence is to assess the number of molecular species, M, in a homogeneous sample, such as a well-stirred solution, or various parts of a sample, such as a heterogeneous cell. One way to achieve this is by fluorescence lifetime imaging microscopy (FLIM) where one region of a sample is pulsed with a laser. Following this sharp pulse, molecules may be excited and emit photons. The time delay between the onset of excitation and the detected photon is Δt. This delay arises because molecules spend some time in the excited state before emitting their photon. The emission rate, *i.e.*, escape rate, of the photon from the excited state, λ_{σ_m}, is unique to each species, σ_m.

Generally, the intensity of the pulse is low enough that either no photon or just one is detected. Thus, it is reasonable to assume that only a small fraction of the molecules are excited by any one pulse. Pulses are then repeated and iid data are collected in the form of $\Delta t_{1:N}$.

The forward model is given by

$$\begin{aligned}
\lambda_{\sigma_m} &\sim \text{Gamma}(\alpha, \beta), \\
\pi_{\sigma_{1:M}} &\sim \text{Dirichlet}_{\sigma_{1:M}}\left(\frac{\alpha}{M}, \dots, \frac{\alpha}{M}\right), \\
s_n|\pi_{\sigma_{1:M}} &\sim \text{Categorical}_{\sigma_{1:M}}(\pi_{\sigma_{1:M}}), \\
\Delta t_n|s_n, \lambda_{\sigma_{1:M}} &\sim \text{Exponential}(\lambda_{s_n}).
\end{aligned}$$

Reasonable rates, λ_{σ_m}, are on the order of 1–10 ns.

1. Generate data with three species, $\lambda_{\sigma_{1:3}} = \{1, 3, 5\}$ ns, and estimate $\pi_{\sigma_{1:M}}$ from your data under the assumption of large M in addition to the rates of each species.
2. Report your results as a two-dimensional histogram showing the sampled $\lambda_{\sigma_{1:M}}$ and associated $\pi_{\sigma_{1:M}}$ after burn-in removal. Include the ground truth in your histogram.
3. From your MAP estimate, how would you reasonably interpret how many chemical species are present?
4. How is $\pi_{\sigma_{1:M}}$ related to the concentration of each species under local excitation by the laser pulse?

Project 7.3 Fluorescence lifetime imaging with instrumental response function artefacts

In Project 7.2 we assumed that the pulse is almost instantaneous and the detector detects arriving photons with almost no delay. However, in reality the pulse has some duration and the detector reports each photon arrival time subject to a stochastic delay upon photon impact. To capture these effects, it is more realistic to assume that the measurement model of Project 7.2 reflects the convolution of a normal instrumental response function (IRF) and the exponential waiting time

$$\Delta t_n | s_n, \lambda_{\sigma_{1:M}} \sim \text{Normal}(\mu_{\text{IRF}}, v_{\text{IRF}}) * \text{Exponential}(\lambda_{s_n}),$$

where μ_{IRF} and v_{IRF} are precalibrated (known) means and variance capturing the effects described above. For now, suppose that $\mu_{\text{IRF}} = 3$ ns and $v_{\text{IRF}} = 1$ ns^2.

1. As before, generate data with three species. Start with $\lambda_{\sigma_{1:3}} = \{1, 5, 10\}$ ns. For large M, sample $\pi_{\sigma_{1:M}}$ from your data in addition to the rates of each species.
2. Report your results as a two-dimensional histogram showing the sampled $\lambda_{\sigma_{1:M}}$ and associated $\pi_{\sigma_{1:M}}$ after burn-in removal. Include the ground truth in your histogram.
3. How has the presence of an instrumental response function altered your ability to perform inference as compared to the idealized scenario of Project 7.2?
4. Repeat the points above by substituting the rates used in generating data for the more difficult $\lambda_{\sigma_{1:3}} = \{1, 3, 5\}$ ns.

Additional Reading

A. Gelman, J. Carlin, H. Stern, D. Dunson, A. Vehtari, D. Rubin. *Bayesian data analysis*. 3rd ed. CRC Press, 2013.

T. S. Ferguson. A Bayesian analysis of some nonparametric problems. *Ann. Stat.*, 1:209, 1973.

Y. W. Teh. Dirichlet process. *Encycl. Mach. Learn.*, 1063:280, 2010.

R. M. Neal. Markov chain sampling methods for Dirichlet process mixture models. *J. Comp. Graph. Stat.*, 9:249, 2000.

D. M. Blei, M. I. Jordan. Variational inference for Dirichlet process mixtures. *Bayesian Anal.*, 1:121, 2006.

N. L. Hjort, C. Holmes, P. Müller, S. G. Walker (eds.). *Bayesian nonparametrics*, Vol. 28. Cambridge University Press, 2010.

C. E. Antoniak. Mixtures of Dirichlet processes with applications to Bayesian nonparametric problems. *Ann. Stat.*, 2:1152, 1974.

H. Ishwaran, M. Zarepour. Exact and approximate sum representations for the Dirichlet process. *Canadian J. Stat.*, 30:269, 2002.

M. D. Escobar. Estimating normal means with a Dirichlet process prior. *J. Am. Stat. Assoc.*, 89:268, 1994.

J. M. Marin, K. Mengersen, C. P. Robert. Bayesian modelling and inference on mixtures of distributions. *Handb. Stat.*, 25:459, 2005.

M. A. Kuhn, E. D. Feigelson. Mixture models in astronomy. arXiv:1711.11101, 2017.

M. Tavakoli, S. Jazani, I. Sgouralis, W. Heo, K. Ishii, T. Tahara, S. Pressé. Direct photon-by-photon analysis of time-resolved pulsed excitation data using Bayesian nonparametrics. *Cell Rep. Phys. Sci.*, 1:100234, 2020.

8 Hidden Markov Models

By the end of this chapter, we will have presented

- *Statistical ways of modeling dynamics*
- *Fundamentals of hidden Markov models*
- *Specialized computational algorithms*

In this chapter we are exclusively concerned with modeling *time-dependent* measurements. We revisit some of the systems introduced in Chapter 2 and present, in a unified framework, several methods to combine dynamics and observation likelihoods. It will become apparent soon that computational tractability is by no means guaranteed in time-dependent problems and often we need to consider specialized algorithms that build upon or extend those of Chapter 5. For this reason, we also present appropriate computational methods that can be used to train the resulting models in an efficient manner. In this chapter, we focus on modeling *discrete systems* evolving in *discrete time*; while we present more general systems in the subsequent chapters.

8.1 Introduction

Throughout this chapter, we consider a system that may access a number of discrete states similar to the systems seen in Section 2.4. For convenience, throughout this chapter we denote the constitutive states of the system by σ_m, and use numerical labels $m = 1 : M$ to distinguish between them. The number of different states, M, that the system may occupy depends upon the problem at hand.

When such a system evolves in time, these fundamental questions arise: *"What is the sequence of successive states the system occupies across time?"* and *"What are the properties of the states occupied across time?"* To help formulate our questions more precisely, we consider ordered time levels t_n, indexed $n = 1 : N$, and use s_n to denote the state occupied by the system at t_n. That is, for a given n, the passing state s_n takes its value from the constitutive states $\sigma_{1:M}$. Thus, our questions about the system at hand can be answered by estimating the trajectory $s_{1:N}$ and the properties of each σ_m.

Note 8.1 Label and index conventions

Just as with the systems we encountered in Note 2.5, only the labeled states σ_m carry meaning, while the m labels themselves are otherwise only an arbitrary index. Such distinction does not carry over to the time level indices, n. By convention, our time levels are ordered $t_{n-1} < t_n$ indicating that, contrary to the m labels, our n indices carry information.

A critical aspect of modeling time-evolving systems is to recognize that states are not directly observed. Rather, only a version of them corrupted by measurement noise is typically assessed experimentally. Thus, whenever a system occupies a state σ_m, it generates observations according to a probability distribution $\mathbb{F}_{\sigma_m}$, or its associated density $F_{\sigma_m}(w)$, unique to σ_m.

To derive a concise formulation incorporating measurement noise, we will assume the case where only *one* observation, denoted by w_n, is gathered per time level t_n. In other words, our *assessment rule* reads

$$w_n|s_n \sim \mathbb{F}_{s_n}, \tag{8.1}$$

with the understanding that each observation w_n may consist of more than one scalar quantity, *i.e.*, our individual observations may be array-valued.

As we saw in Section 7.1.1, a more convenient way of representing Eq. (8.1) is through a mother distribution $\mathbb{G}_\phi$ with state specific parameters ϕ_{σ_m}. In this case, our assessment rules take the form

$$w_n|s_n, \boldsymbol{\phi} \sim \mathbb{G}_{\phi_{\sigma_m}}. \tag{8.2}$$

Here, for convenience, we use $\boldsymbol{\phi}$ to gather all emission parameters $\phi_{\sigma_{1:M}}$. Equation (8.1) or (8.2) provides a means to incorporate measurement and, unlike in Chapter 7, the *order in which observations are made* provides important information on *dynamics*, including, say, the probability of transitioning to particular states σ_m at subsequent time levels.

We saw in Chapter 2 that dynamics for systems with discrete state-spaces evolving in discrete time are best described by assigning appropriate probability distributions on the passing states s_1 and $s_n|s_{n-1}$ dictating the *initialization* and *transition rules*. Next, we discuss some modeling options to consider when selecting such distributions.

8.2 The Hidden Markov Model

8.2.1 Modeling Dynamics

From the modeling point of view, the simplest and often most convenient way to incorporate dynamics into an observation model is to adopt transition probabilities between any pair σ_m and $\sigma_{m'}$ of states in the system's state-space. That is, such formulations generally lead to intuitive and computationally tractable problems.

In general, our system's transitions need not be reversible. As such, we may adopt different probabilities for transitions $\sigma_m \to \sigma_{m'}$ and $\sigma_{m'} \to \sigma_m$. In the most general case, we denote by $\pi_{\sigma_m \to \sigma_{m'}}$ the probability of the system starting at σ_m and, within *one* time step, transitioning to $\sigma_{m'}$. In this setup, some $\pi_{\sigma_m \to \sigma_{m'}}$ can be zero, indicating that the system cannot undergo transitions $\sigma_m \to \sigma_{m'}$ in a single step.

To facilitate the presentation that follows, we gather all transition probabilities out of the same state σ_m into an array $\boldsymbol{\pi}_{\sigma_m} = [\pi_{\sigma_m \to \sigma_1}, \pi_{\sigma_m \to \sigma_2}, \ldots, \pi_{\sigma_m \to \sigma_M}]$. Since once the system departs from any σ_m, it necessarily lands somewhere *within* the state-space $\sigma_{1:M}$, the individual transition probabilities assigned must satisfy $\sum_{m=1}^{M} \pi_{\sigma_m \to \sigma_{m'}} = 1$. Consequently, each $\boldsymbol{\pi}_{\sigma_m}$ is, in fact, a *probability* array.

Note 8.2 Transition probability matrix

To simplify the notation, we tabulate the transition probabilities into

$$\begin{array}{c} \\ \sigma_1 \\ \sigma_2 \\ \vdots \\ \sigma_M \end{array} \begin{array}{c} \begin{array}{cccc} \sigma_1 & \sigma_2 & \cdots & \sigma_M \end{array} \\ \begin{bmatrix} \pi_{\sigma_1 \to \sigma_1} & \pi_{\sigma_1 \to \sigma_2} & \cdots & \pi_{\sigma_1 \to \sigma_M} \\ \pi_{\sigma_2 \to \sigma_1} & \pi_{\sigma_2 \to \sigma_2} & \cdots & \pi_{\sigma_2 \to \sigma_M} \\ \vdots & \vdots & \ddots & \vdots \\ \pi_{\sigma_M \to \sigma_1} & \pi_{\sigma_M \to \sigma_2} & \cdots & \pi_{\sigma_M \to \sigma_M} \end{bmatrix} \end{array} = \begin{bmatrix} \boldsymbol{\pi}_{\sigma_1} \\ \boldsymbol{\pi}_{\sigma_2} \\ \vdots \\ \boldsymbol{\pi}_{\sigma_M} \end{bmatrix} = \mathbf{\Pi}.$$

This matrix is similar to the transition probability matrices we encountered in Chapter 2.

Under this formulation, dynamics are represented generically by the transition rules

$$s_n | s_{n-1} \sim \text{Categorical}_{\sigma_{1:M}}\left(\boldsymbol{\pi}_{s_{n-1}}\right). \tag{8.3}$$

The system's initial state s_1 is not included in Eq. (8.3) as there is no predecessor passing state. To complete our formulation, we need to adopt separate probabilities for s_1, which we denote by $\boldsymbol{\rho} = [\rho_{\sigma_1}, \rho_{\sigma_2}, \ldots, \rho_{\sigma_M}]$. As such, Eq. (8.3) is combined with the initialization rule

$$s_1 \sim \text{Categorical}_{\sigma_{1:M}}(\boldsymbol{\rho}),$$

thereby completing the description of the system's dynamics.

Note 8.3 Deterministic initialization

When the initial state of our dynamical system is specified deterministically, we may still maintain the same formulation by simply setting $\rho_{\sigma_m} = 1$ for the constitutive state σ_m from which the system is initialized and, thus, $\rho_{\sigma_{m'}} = 0$ for every other state. For example, for a system initialized at σ_2, the initial probabilities are $\boldsymbol{\rho} = [0, 1, 0, \ldots, 0]$.

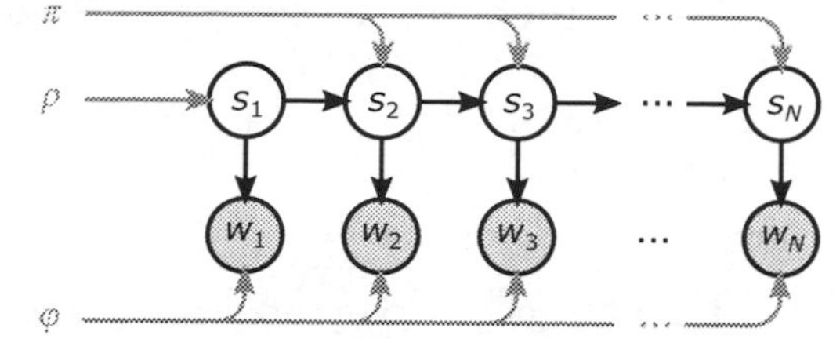

Fig. 8.1 Graphical representation of an HMM. Here, the parameters ρ, $\mathbf{\Pi}$, and ϕ are assumed known.

8.2.2 Modeling Overview

The statistical model we have described so far is depicted graphically in Fig. 8.1 and is summarized as

$$s_1|\rho \sim \text{Categorical}_{\sigma_{1:M}}(\rho), \tag{8.4}$$

$$s_n|s_{n-1}, \mathbf{\Pi} \sim \text{Categorical}_{\sigma_{1:M}}\left(\pi_{s_{n-1}}\right), \qquad n = 2:N, \tag{8.5}$$

$$w_n|s_n, \phi \sim \mathbb{G}_{\phi_{s_n}}, \qquad n = 1:N, \tag{8.6}$$

where, for clarity, we emphasize the dependencies upon the parameters $\rho, \mathbf{\Pi}, \phi$ by conditioning explicitly upon them. The three equations model the initialization, transitions, and observations of the system under study, respectively, and combined with a clear specification of the state-space $\sigma_{1:M}$, provide a complete description of our problem.

The model just defined is termed the *hidden Markov model* (HMM). It contains two sets of parameters: dynamical, $\rho, \mathbf{\Pi}$, and observational, ϕ. From the inference point of view, the trajectory $s_{1:N}$ gathers latent (*i.e.*, hidden) variables; $w_{1:N}$ gathers measurements; and $\rho, \mathbf{\Pi}, \phi$ gather parameters that, depending on context, may either be known or unknown. The dependencies among variables are illustrated in Fig. 8.1.

The HMM's formulation in Eqs. (8.4) to (8.6) is very general and, for this reason, is one of the most widely used models in time series analysis. Since it is already formulated in generative form, when tackling direct problems it is straightforward to use this model for the simulation of synthetic measurements $w_{1:N}$ via ancestral sampling; see Algorithm 1.3. However, as we will see shortly, the HMM is mostly useful in tackling inverse problems. In particular, an inverse formulation of the HMM can be used to shed light on the following questions:

1. Given observations $w_{1:N}$ and parameter values $\rho, \mathbf{\Pi}, \phi$, what is the likelihood of $w_{1:N}$?
2. Given observations $w_{1:N}$ and parameter values $\rho, \mathbf{\Pi}, \phi$, what are the passing states $s_{1:N}$?
3. Given observations $w_{1:N}$, what are the values of the parameters $\rho, \mathbf{\Pi}, \phi$?

These questions are commonly referred to as: *evaluation*, *decoding*, and *estimation*, respectively. To answer them, we can follow two complementary routes: frequentist and Bayesian. We describe these separately in the subsequent sections.

Note 8.4 Time indexing and missing observations

With the indexing convention adopted, we designate $n = 1$ to be the *earliest* time level associated with an observation and $n = N$ with the *latest* one. Further, we assume that *every* intermediate time level $n = 2, \ldots, N-1$ is also associated with an observation. This is a measurement-centric convention in the sense that the timing schedule of the measurement acquisition protocol determines the precise structure of the hidden state sequence.

Occasionally, we might encounter situations where we need to incorporate into our formulation time levels *without* observations, for example when modeling measurements collected at *irregular times*. In such circumstances, we may generalize our formulation in at least two possible ways:

1. Use the same indexing convention with precisely one observation per time level and adopt *time dependent kinetics*, for example by explicitly requiring transition probabilities $\pi_{n,\sigma_m \to \sigma_{m'}}$ that may change across time levels.
2. Use an indexing scheme with *redundant time levels*. In particular, we may choose to maintain a hidden state sequence at a finer, but regular, time spacing and associate only some of the passing states with the observations while leaving the others unassociated.

The theory we present next can readily accommodate both these cases with minor modification.

8.3 The Hidden Markov Model in the Frequentist Paradigm

The concepts of this section are direct extensions of Chapter 2. As we have already introduced them, here we treat mostly computational aspects specifically tailored to HMMs.

8.3.1 Evaluation of the Likelihood

Evaluation of an HMM requires the computation of the (marginal) likelihood,

$$p\left(w_{1:N} | \boldsymbol{\rho}, \boldsymbol{\Pi}, \boldsymbol{\phi}\right) = \sum_{s_{1:N}} p\left(w_{1:N}, s_{1:N} | \boldsymbol{\rho}, \boldsymbol{\Pi}, \boldsymbol{\phi}\right),$$

with the sum taken over every possible state sequence $s_{1:N}$. Naive evaluation of this enormous sum, where a term $p(w_{1:N}, s_{1:N}|\boldsymbol{\rho}, \boldsymbol{\Pi}, \boldsymbol{\phi}) = p(w_{1:N}|s_{1:N}, \boldsymbol{\phi})\, p(s_{1:N}|\boldsymbol{\rho}, \boldsymbol{\Pi})$ is computed for each possible trajectory $s_{1:N}$ and summed, requires the evaluation and addition of M^N terms. This is prohibitively large even for small problems. Instead, below, we describe a particular computational scheme, termed *filtering*, scaling as M^2N.

Instead of completing over the entire state sequence $s_{1:N}$, the computation of the likelihood is achieved most efficiently by completing first only with respect to the terminal passing state s_N, as follows:

$$p(w_{1:N}|\boldsymbol{\rho}, \boldsymbol{\Pi}, \boldsymbol{\phi}) = \sum_{s_N} p(w_{1:N}, s_N|\boldsymbol{\rho}, \boldsymbol{\Pi}, \boldsymbol{\phi}) = \sum_{s_N} \mathcal{A}_N(s_N). \qquad (8.7)$$

This sum is readily computed so long as $\mathcal{A}_N(s_N)$, called *forward variables*, are available for all possible values of s_N. Written explicitly, these are $\mathcal{A}_N(\sigma_1), \mathcal{A}_N(\sigma_2), \ldots, \mathcal{A}_N(\sigma_M)$. We can define forward variables, more generally, for any time level by the joint distributions

$$\mathcal{A}_n(s_n) = p(w_{1:n}, s_n|\boldsymbol{\rho}, \boldsymbol{\Pi}, \boldsymbol{\phi}) \qquad (8.8)$$

and compute them recursively. Our recursion relies on

$$\mathcal{A}_n(s_n) = G_{\phi_{s_n}}(w_n) \sum_{s_{n-1}} \pi_{s_{n-1}\to s_n} \mathcal{A}_{n-1}(s_{n-1}), \qquad n = 2:N, \qquad (*8.9*)$$

and requires the initial condition $\mathcal{A}_1(s_1)$ to iterate forward. As a direct consequence of the definition of $\mathcal{A}_1(s_1)$, the initial condition is

$$\mathcal{A}_1(s_1) = G_{\phi_{s_1}}(w_1)\, \rho_{s_1}. \qquad (*8.10*)$$

The steps involved are summarized in Algorithm 8.1.

Algorithm 8.1 Forward recursion for the HMM (unstable version)

Given observations $w_{1:N}$ and parameters $\boldsymbol{\rho}, \boldsymbol{\Pi}, \boldsymbol{\phi}$, the forward terms $\mathcal{A}_n(\sigma_m)$ for each time level n and each state σ_m are computed as follows:

- At $n = 1$, initialize with

$$\mathcal{A}_1(\sigma_m) = G_{\phi_{\sigma_m}}(w_1)\rho_{\sigma_m}.$$

- For $n = 2:N$, compute recursively

$$\mathcal{A}_n(\sigma_m) = G_{\phi_{\sigma_m}}(w_n) \sum_{m'=1}^{M} \pi_{\sigma_{m'}\to\sigma_m} \mathcal{A}_{n-1}(\sigma_{m'}).$$

* Reminder: The asterisks by some equations indicates that the detailed derivation can be found in Appendix F.

Upon completion, the algorithm yields every $\mathcal{A}_n(\sigma_m)$, which may be tabulated as follows

$$\begin{array}{c} \\ t_1 \\ t_2 \\ \vdots \\ t_n \\ \vdots \\ t_N \end{array} \begin{array}{c} \begin{array}{cccc} \sigma_1 & \sigma_2 & \cdots & \sigma_M \end{array} \\ \left[\begin{array}{cccc} \mathcal{A}_1(\sigma_1) & \mathcal{A}_1(\sigma_2) & \cdots & \mathcal{A}_1(\sigma_M) \\ \mathcal{A}_2(\sigma_1) & \mathcal{A}_2(\sigma_2) & \cdots & \mathcal{A}_2(\sigma_M) \\ \vdots & \vdots & \ddots & \vdots \\ \mathcal{A}_n(\sigma_1) & \mathcal{A}_n(\sigma_2) & \cdots & \mathcal{A}_n(\sigma_M) \\ \vdots & \vdots & \ddots & \vdots \\ \mathcal{A}_N(\sigma_1) & \mathcal{A}_N(\sigma_2) & \cdots & \mathcal{A}_N(\sigma_M) \end{array}\right] \end{array} \begin{array}{c} \\ \mathcal{A}_1(s_1) \\ \mathcal{A}_2(s_2) \\ \vdots \\ \mathcal{A}_n(s_n) \\ \vdots \\ \mathcal{A}_N(s_N). \end{array}$$

Note 8.5 Vectorization

Gathering the forward terms of the same time level in *row* arrays,

$$\mathbb{A}_n = \left[\mathcal{A}_n(\sigma_1) \quad \mathcal{A}_n(\sigma_2) \quad \cdots \quad \mathcal{A}_n(\sigma_M)\right],$$

and similarly for the likelihood terms,

$$\mathbb{\Gamma}_n = \left[G_{\phi_{\sigma_1}}(w_n) \quad G_{\phi_{\sigma_2}}(w_n) \quad \cdots \quad G_{\phi_{\sigma_M}}(w_n)\right],$$

the filtering recursions can be executed in vectorized form as

$$\mathbb{A}_1 = \mathbb{\Gamma}_n \odot \boldsymbol{\rho},$$
$$\mathbb{A}_n = \mathbb{\Gamma}_n \odot (\mathbb{A}_{n-1}\boldsymbol{\Pi}), \qquad n = 2:N,$$

where $\odot$ denotes the Hadamard (element-wise) product. If, instead of a row array, we represent $\mathbb{\Gamma}_n$ as a diagonal matrix $\mathbb{D}_{\mathbb{\Gamma}_n}$, then the filtering recursions take a more conventional form,

$$\mathbb{A}_1 = \boldsymbol{\rho}\, \mathbb{D}_{\mathbb{\Gamma}_n},$$
$$\mathbb{A}_n = (\mathbb{A}_{n-1}\boldsymbol{\Pi})\, \mathbb{D}_{\mathbb{\Gamma}_n}, \qquad n = 2:N,$$

that only use ordinary matrix-vector operations. From these two sets of filtering equations, the first is preferred for computational implementations while the second is more convenient in theoretical derivations.

8.3.2 Decoding the State Sequence

Decoding an HMM may be achieved in at least two meaningful ways. Depending on the problem specifics, we might be interested in finding a single passing state s_n^* maximizing the marginal $p(s_n|w_{1:N}, \boldsymbol{\rho}, \boldsymbol{\Pi}, \boldsymbol{\phi})$, or in finding the sequence $s^{\#}_{1:N}$ maximizing the joint $p(s_{1:N}|w_{1:N}, \boldsymbol{\rho}, \mathbf{\Pi}, \boldsymbol{\phi})$. Generally, individual states s^*_n are useful in problems where optimal passing states for a *particular time level* are sought. By contrast, $s^{\#}_{1:N}$ is useful in problems where the optimal trajectory over the *entire time course* is sought.

Note 8.6 Marginal and joint state sequences

Collecting passing states s^*_n across all time levels, we may form a state sequence $s^*_{1:N}$. This sequence, however, must be used with caution as it might violate the kinetics in $\mathbf{\Pi}$. In particular, since $s^*_{1:N}$ considers each s^*_n *irrespective* of s^*_{n-1}, it may very well contain transitions $s^*_{n-1} \to s^*_n$ coinciding with forbidden probabilities, *i.e.*, $\pi_{s^*_{n-1} \to s^*_n} = 0$. In contrast, $s^{\#}_{1:N}$ is *guaranteed* to obey the kinetics in $\mathbf{\Pi}$ as any sequence containing prohibited transitions $\pi_{s^{\#}_{n-1} \to s^{\#}_n}$ is, as we will see, excluded by construction.

Marginal Decoding

To obtain each s^*_n, termed *marginal decoding*, we need to compute $p(s_n|w_{1:N}, \boldsymbol{\rho}, \mathbf{\Pi}, \boldsymbol{\phi})$. This can be efficiently achieved using the $\mathcal{A}_n(s_n)$ variables through the relations

$$p(s_N|w_{1:N}, \boldsymbol{\rho}, \mathbf{\Pi}, \boldsymbol{\phi}) \propto \mathcal{A}_N(s_N), \qquad (*8.11*)$$

$$p(s_n|w_{1:N}, \boldsymbol{\rho}, \mathbf{\Pi}, \boldsymbol{\phi}) \propto \mathcal{A}_n(s_n)\mathcal{B}_n(s_n), \qquad n = 1 : N-1, \qquad (*8.12*)$$

where we define *backward variables* by

$$\mathcal{B}_n(s_n) = p\left(w_{n+1:N}|s_n, \mathbf{\Pi}, \boldsymbol{\phi}\right), \qquad n = 1 : N-1. \qquad (8.13)$$

In both Eq. (*8.11*) and Eq. (*8.12*), the missing proportionality constants do not affect the maximization of $p(s_n|w_{1:N}, \boldsymbol{\rho}, \mathbf{\Pi}, \boldsymbol{\phi})$ and, as such, need not be computed. Given $\mathcal{A}_n(s_n)$ and $\mathcal{B}_n(s_n)$, the sequence $s^*_{1:N}$ is readily computed by

$$s^*_n = \underset{\sigma_m}{\operatorname{argmax}}\, \mathcal{A}_n(\sigma_m)\mathcal{B}_n(\sigma_m).$$

The terms $\mathcal{B}_n(s_n)$, similarly to $\mathcal{A}_n(s_n)$, may be computed recursively. In this case, the recursion relies on

$$\mathcal{B}_n(s_n) = \sum_{s_{n+1}} \mathcal{B}_{n+1}(s_{n+1}) G_{\phi_{s_{n+1}}}(w_{n+1})\, \pi_{s_n \to s_{n+1}}, \qquad n = N-1 : 1, \qquad (*8.14*)$$

and requires the final condition $\mathcal{B}_N(s_N)$ to iterate backward. By comparing Eq. (*8.11*) and Eq. (*8.12*), we fulfill the terminal condition, conventionally, by setting

$$\mathcal{B}_N(s_N) = 1.$$

The steps involved are summarized in Algorithm 8.2.

Algorithm 8.2 Backward recursion for HMM (unstable version)

Given observations $w_{1:N}$ and parameters $\mathbf{\Pi}$, $\boldsymbol{\phi}$, the backward terms $\mathcal{B}_n(\sigma_m)$ for each time level n and each state σ_m are computed as follows:

- At $n = N$, initialize by

$$\mathcal{B}_N(\sigma_m) = 1.$$

- For $n = N - 1 : 1$, compute recursively

$$\mathcal{B}_n(\sigma_m) = \sum_{m'=1}^{M} \mathcal{B}_{n+1}(\sigma_{m'}) G_{\phi_{\sigma_{m'}}}(w_{n+1})\, \pi_{\sigma_m \to \sigma_{m'}}.$$

Upon completion, the algorithm yields every $\mathcal{B}_n(\sigma_m)$, which may be tabulated as follows

$$\begin{array}{c} \\ t_1 \\ t_2 \\ \vdots \\ t_n \\ \vdots \\ t_N \end{array}
\begin{array}{c} \begin{array}{cccc} \sigma_1 & \sigma_2 & \cdots & \sigma_M \end{array} \\ \begin{bmatrix} \mathcal{B}_1(\sigma_1) & \mathcal{B}_1(\sigma_2) & \cdots & \mathcal{B}_1(\sigma_M) \\ \mathcal{B}_2(\sigma_1) & \mathcal{B}_2(\sigma_2) & \cdots & \mathcal{B}_2(\sigma_M) \\ \vdots & \vdots & \ddots & \vdots \\ \mathcal{B}_n(\sigma_1) & \mathcal{B}_n(\sigma_2) & \cdots & \mathcal{B}_n(\sigma_M) \\ \vdots & \vdots & \ddots & \vdots \\ \mathcal{B}_N(\sigma_1) & \mathcal{B}_N(\sigma_2) & \cdots & \mathcal{B}_N(\sigma_M) \end{bmatrix} \end{array}
\begin{array}{c} \\ \mathcal{B}_1(s_1) \\ \mathcal{B}_2(s_2) \\ \vdots \\ \mathcal{B}_n(s_n) \\ \vdots \\ \mathcal{B}_N(s_N). \end{array}$$

Joint Decoding

The computation of $s^{\sharp}_{1:N}$, termed *joint decoding*, relies on the factorization

$$p(s_{1:N}|w_{1:N}, \boldsymbol{\rho}, \mathbf{\Pi}, \boldsymbol{\phi}) = p(s_N|w_{1:N}, \boldsymbol{\rho}, \mathbf{\Pi}, \boldsymbol{\phi}) \prod_{n=1}^{N-1} p(s_n|s_{n+1:N}, w_{1:N}, \boldsymbol{\rho}, \mathbf{\Pi}, \boldsymbol{\phi}),$$

which implies that the maximizer $s^{\sharp}_{1:N}$ can be computed by the maximizers $s^{\sharp}_n$ of the individual factors $p(s_N|w_{1:N}, \boldsymbol{\rho}, \mathbf{\Pi}, \boldsymbol{\phi})$ and $p(s_n|s^{\sharp}_{n+1:N}, w_{1:N}, \boldsymbol{\rho}, \mathbf{\Pi}, \boldsymbol{\phi})$. Given $\mathcal{A}_n(s_n)$, maximization of each factor can be simplified through the relations

$$p(s_N|w_{1:N}, \boldsymbol{\rho}, \mathbf{\Pi}, \boldsymbol{\phi}) \propto \mathcal{A}_N(s_N), \tag{8.15}$$

$$p(s_n|s_{n+1:N}, w_{1:N}, \boldsymbol{\rho}, \mathbf{\Pi}, \boldsymbol{\phi}) \propto \mathcal{A}_n(s_n)\pi_{s_n \to s_{n+1}}, \quad n = 1 : N - 1. \tag{*8.16*}$$

The steps involved, known as *Viterbi recursion*, are summarized in Algorithm 8.3.

Algorithm 8.3 Viterbi recursion for HMM

Given observations $w_{1:N}$, kinetic parameters $\mathbf{\Pi}$, and every $\mathcal{A}_n(\sigma_m)$, the Viterbi sequence $s^{\#}_{1:N}$ is computed as follows:

- At $n = N$, initialize by

$$s^{\#}_N = \underset{\sigma_m}{\text{argmax}}\, \mathcal{A}_N(\sigma_m).$$

- For $n = N - 1 : 1$, compute recursively

$$s^{\#}_n = \underset{\sigma_m}{\text{argmax}}\, \mathcal{A}_n(\sigma_m)\pi_{\sigma_m \to s^{\#}_{n+1}}.$$

8.3.3 Estimation of the Parameters

Estimation of an HMM seeks the maximizer of the likelihood,

$$\{\boldsymbol{\rho}^*, \mathbf{\Pi}^*, \boldsymbol{\phi}^*\} = \underset{\boldsymbol{\rho},\mathbf{\Pi},\boldsymbol{\phi}}{\text{argmax}}\, p\left(w_{1:N}|\boldsymbol{\rho}, \mathbf{\Pi}, \boldsymbol{\phi}\right).$$

When completing the likelihood $p\left(w_{1:N}|\boldsymbol{\rho}, \mathbf{\Pi}, \boldsymbol{\phi}\right)$ with the state sequence $s_{1:N}$, for instance, as

$$p\left(w_{1:N}|\boldsymbol{\rho}, \mathbf{\Pi}, \boldsymbol{\phi}\right) = \sum_{s_{1:N}} p\left(s_{1:N}, w_{1:N}|\boldsymbol{\rho}, \mathbf{\Pi}, \boldsymbol{\phi}\right),$$

we may perform this maximization with an EM procedure, where we iterate between an expectation (E) step and a maximization (M) step, similar to Section 3.2.2. The entire procedure, adapted to the HMM, is known as the *Baum–Welch algorithm* and the steps involved are summarized in Algorithm 8.4. In the next sections, we describe the steps of Algorithm 8.4 in detail.

Algorithm 8.4 Baum–Welch algorithm

Given observations $w_{1:N}$ and an initial guess for the model parameters $\boldsymbol{\rho}, \mathbf{\Pi}, \boldsymbol{\phi}$, the Baum–Welch algorithm computes successively improved approximations of the maximizer of $p\left(w_{1:N}|\boldsymbol{\rho}, \mathbf{\Pi}, \boldsymbol{\phi}\right)$ by repeating the following steps:

- E-step:
  - Use $\boldsymbol{\rho}, \mathbf{\Pi}, \boldsymbol{\phi}$ to compute $\mathcal{A}_n(\sigma_m)$ and $\mathcal{B}_n(\sigma_m)$.
  - Use $\mathcal{A}_n(\sigma_m)$ and $\mathcal{B}_n(\sigma_m)$ to compute

$$\zeta_n(\sigma_m) = \mathcal{A}_n(\sigma_m)\mathcal{B}_n(\sigma_m),$$

$$\eta_n(\sigma_m, \sigma_{m'}) = \mathcal{A}_{n-1}(\sigma_m)\mathcal{B}_n(\sigma_{m'})G\left(w_n; \phi_{\sigma_{m'}}\right)\pi_{\sigma_m \to \sigma_{m'}}.$$

– Use $\eta_n(\sigma_m, \sigma_{m'})$ to compute

$$\xi_{\sigma_m}(\sigma_{m'}) = \sum_{n=2}^{N} \eta_n(\sigma_m, \sigma_{m'}).$$

- M-step:

– Update ρ by replacing with

$$\left(\frac{\zeta_1(\sigma_1)}{\sum_{\sigma_m} \zeta_1(\sigma_m)}, \dots, \frac{\zeta_1(\sigma_M)}{\sum_{\sigma_m} \zeta_1(\sigma_m)} \right).$$

– Update $\mathbf{\Pi}$ by replacing each $\boldsymbol{\pi}_{\sigma_m}$ with

$$\left(\frac{\xi_{\sigma_m}(\sigma_1)}{\sum_{\sigma_{m'}} \xi_{\sigma_m}(\sigma_{m'})}, \dots, \frac{\xi_{\sigma_m}(\sigma_M)}{\sum_{\sigma_{m'}} \xi_{\sigma_m}(\sigma_{m'})} \right).$$

– Update $\boldsymbol{\phi}$ by replacing each ϕ_{σ_m} with the maximizer of

$$\sum_{n=1}^{N} \zeta_n(\sigma_m) \log G_{\phi_{\sigma_m}}(w_n).$$

The iterations are terminated either after a fixed number of repetitions or when the improvement between successive approximations of $\boldsymbol{\rho}, \mathbf{\Pi}, \boldsymbol{\phi}$ falls below a predetermined threshold.

Like any EM method, convergence of Algorithm 8.4 to the *global* optimizer $\boldsymbol{\rho}^*, \mathbf{\Pi}^*, \boldsymbol{\phi}^*$ is *not* guaranteed. In practice, we need to try multiple initial guesses spanning a wide region of parameter space and, at the end, select the best optimizer found according to, say, the numerical value for the likelihood, $p(w_{1:N}|\boldsymbol{\rho}^*, \mathbf{\Pi}^*, \boldsymbol{\phi}^*)$. As Algorithm 8.4 does not provide the value of $p(w_{1:N}|\boldsymbol{\rho}^*, \mathbf{\Pi}^*, \boldsymbol{\phi}^*)$ to compare the resulting maximizers, we need to compute it separately through, for example, Eq. (8.7).

Expectation Step*

In the E-step, we start from an initial approximation $\boldsymbol{\rho}^{\text{old}}, \mathbf{\Pi}^{\text{old}}, \boldsymbol{\phi}^{\text{old}}$ of the maximizer $\boldsymbol{\rho}^*, \mathbf{\Pi}^*, \boldsymbol{\phi}^*$ we wish to obtain and compute the expectation function that we will later maximize in the M-step. Specifically, we compute the expectation of

$$\log p(s_{1:N}, w_{1:N}|\boldsymbol{\rho}, \mathbf{\Pi}, \boldsymbol{\phi}) = \log \rho_{s_1} + \sum_{n=2}^{N} \log \pi_{s_{n-1} \to s_n} + \sum_{n=1}^{N} \log G_{\phi_{s_n}}(w_n) \tag{*8.17*}$$

with respect to the probability distribution of $s_{1:N}|w_{1:N}, \boldsymbol{\rho}^{\text{old}}, \mathbf{\Pi}^{\text{old}}, \boldsymbol{\phi}^{\text{old}}$. Since this expectation is a function of $\boldsymbol{\rho}, \mathbf{\Pi}, \boldsymbol{\phi}$ and also depends on $\boldsymbol{\rho}^{\text{old}}, \mathbf{\Pi}^{\text{old}}, \boldsymbol{\phi}^{\text{old}}$, we denote it by $Q_{\boldsymbol{\rho}^{\text{old}}, \mathbf{\Pi}^{\text{old}}, \boldsymbol{\phi}^{\text{old}}}(\boldsymbol{\rho}, \mathbf{\Pi}, \boldsymbol{\phi})$. In particular, this expectation is given by

* This is an advanced topic and could be skipped on a first reading.

$$
\begin{aligned}
&\mathcal{Q}_{\rho^{\text{old}},\boldsymbol{\Pi}^{\text{old}},\boldsymbol{\phi}^{\text{old}}}(\boldsymbol{\rho},\boldsymbol{\Pi},\boldsymbol{\phi}) \\
&= \sum_{s_1} p\left(s_1|w_{1:N},\boldsymbol{\rho}^{\text{old}},\boldsymbol{\Pi}^{\text{old}},\boldsymbol{\phi}^{\text{old}}\right)\log \rho_{s_1} \\
&+ \sum_{s_{n-1}}\sum_{n=2}^{N}\sum_{s_n} p\left(s_{n-1},s_n|w_{1:N},\boldsymbol{\rho}^{\text{old}},\boldsymbol{\Pi}^{\text{old}},\boldsymbol{\phi}^{\text{old}}\right)\log \pi_{s_{n-1}\to s_n} \\
&+ \sum_{s_n}\sum_{n=1}^{N} p\left(s_n|w_{1:N},\boldsymbol{\rho}^{\text{old}},\boldsymbol{\Pi}^{\text{old}},\boldsymbol{\phi}^{\text{old}}\right)\log G_{\phi_{s_n}}(w_n). \qquad (*8.18*)
\end{aligned}
$$

The distributions over $s_n|w_{1:N},\boldsymbol{\rho}^{\text{old}},\boldsymbol{\Pi}^{\text{old}},\boldsymbol{\phi}^{\text{old}}$ and $s_{n-1},s_n|w_{1:N},\boldsymbol{\rho}^{\text{old}},\boldsymbol{\Pi}^{\text{old}},\boldsymbol{\phi}^{\text{old}}$ can be computed in terms of the forward and backward terms of Eqs. (8.8) and (8.13). As these are computed using only $\boldsymbol{\rho}^{\text{old}},\boldsymbol{\Pi}^{\text{old}},\boldsymbol{\phi}^{\text{old}}$, we use the associated $\mathcal{A}_n^{\text{old}}(s_n)$ and $\mathcal{B}_n^{\text{old}}(s_n)$. These distributions are

$$
p\left(s_n|w_{1:N},\boldsymbol{\rho}^{\text{old}},\boldsymbol{\Pi}^{\text{old}},\boldsymbol{\phi}^{\text{old}}\right) = \frac{\mathcal{A}_n^{\text{old}}(s_n)\mathcal{B}_n^{\text{old}}(s_n)}{p\left(w_{1:N}|\boldsymbol{\rho}^{\text{old}},\boldsymbol{\Pi}^{\text{old}},\boldsymbol{\phi}^{\text{old}}\right)}, \qquad (*8.19*)
$$

$$
p\left(s_{n-1},s_n|w_{1:N},\boldsymbol{\rho}^{\text{old}},\boldsymbol{\Pi}^{\text{old}},\boldsymbol{\phi}^{\text{old}}\right) = \frac{\mathcal{A}_{n-1}^{\text{old}}(s_{n-1})\mathcal{B}_n^{\text{old}}(s_n)G_{\phi_{s_n}^{\text{old}}}(w_n)\,\pi_{s_{n-1}\to s_n}^{\text{old}}}{p\left(w_{1:N}|\boldsymbol{\rho}^{\text{old}},\boldsymbol{\Pi}^{\text{old}},\boldsymbol{\phi}^{\text{old}}\right)}. \qquad (*8.20*)
$$

Finally, because $p\left(w_{1:N}|\boldsymbol{\rho}^{\text{old}},\boldsymbol{\Pi}^{\text{old}},\boldsymbol{\phi}^{\text{old}}\right)$ does not depend upon $\boldsymbol{\rho},\boldsymbol{\Pi},\boldsymbol{\phi}$, this term does not affect the maximization of $\mathcal{Q}_{\rho^{\text{old}},\boldsymbol{\Pi}^{\text{old}},\boldsymbol{\phi}^{\text{old}}}(\boldsymbol{\rho},\boldsymbol{\Pi},\boldsymbol{\phi})$ with respect to $\boldsymbol{\rho},\boldsymbol{\Pi},\boldsymbol{\phi}$. As such, we can safely drop it to obtain

$$
\begin{aligned}
\mathcal{Q}_{\rho^{\text{old}},\boldsymbol{\Pi}^{\text{old}},\boldsymbol{\phi}^{\text{old}}}(\boldsymbol{\rho},\boldsymbol{\Pi},\boldsymbol{\phi}) \propto &\sum_{m=1}^{M}\zeta_1^{\text{old}}(\sigma_m)\log \rho_{\sigma_m} \\
&+ \sum_{m=1}^{M}\sum_{n=2}^{N}\sum_{m'=1}^{M}\eta_n^{\text{old}}(\sigma_m,\sigma_{m'})\log \pi_{\sigma_m\to\sigma_{m'}} \\
&+ \sum_{m=1}^{M}\sum_{n=1}^{N}\zeta_n^{\text{old}}(\sigma_m)\log G_{\phi_{\sigma_m}}(w_n), \qquad (8.21)
\end{aligned}
$$

where

$$
\zeta_n^{\text{old}}(\sigma_m) = \mathcal{A}_n^{\text{old}}(\sigma_m)\mathcal{B}_n^{\text{old}}(\sigma_m), \qquad n=1:N, \qquad (8.22)
$$

$$
\eta_n^{\text{old}}(\sigma_m,\sigma_{m'}) = \mathcal{A}_{n-1}^{\text{old}}(\sigma_m)\mathcal{B}_n^{\text{old}}(\sigma_{m'})G_{\phi_{\sigma_{m'}}^{\text{old}}}(w_n)\,\pi_{\sigma_m\to\sigma_{m'}}^{\text{old}}, \qquad n=2:N. \qquad (8.23)
$$

Maximization Step*

In the M-step, we obtain an improved approximation $\boldsymbol{\rho}^{\text{new}}, \boldsymbol{\Pi}^{\text{new}}, \boldsymbol{\phi}^{\text{new}}$ of the maximizer sought by maximizing the expectation function obtained in the E-step. Specifically, we maximize $Q_{\boldsymbol{\rho}^{\text{old}},\boldsymbol{\Pi}^{\text{old}},\boldsymbol{\phi}^{\text{old}}}(\boldsymbol{\rho}, \boldsymbol{\Pi}, \boldsymbol{\phi})$ under the constraints

$$\sum_{m=1}^{M} \rho_{\sigma_m} = 1, \qquad \sum_{m'=1}^{M} \pi_{\sigma_m \to \sigma'_m} = 1,$$

needed to ensure that $\boldsymbol{\rho}^{\text{new}}, \boldsymbol{\Pi}^{\text{new}}$ consists of valid probability vectors. As our objective function in Eq. (8.21) is separable, the computation of the new maximizer can be broken down into separate maximizations:

$$\begin{aligned}
\boldsymbol{\rho}^{\text{new}} &= \underset{\boldsymbol{\rho}}{\operatorname{argmax}} \sum_{m=1}^{M} \zeta_1^{\text{old}}(\sigma_m) \log \rho_{\sigma_m}, \\
\boldsymbol{\pi}_{\sigma_m}^{\text{new}} &= \underset{\boldsymbol{\pi}_{\sigma_m}}{\operatorname{argmax}} \sum_{n=2}^{N} \sum_{m'=1}^{M} \eta_n^{\text{old}}(\sigma_m, \sigma_{m'}) \log \pi_{\sigma_m \to \sigma_{m'}}, \\
\phi_{\sigma_m}^{\text{new}} &= \underset{\phi_{\sigma_m}}{\operatorname{argmax}} \sum_{n=1}^{N} \zeta_n^{\text{old}}(\sigma_m) \log G_{\phi_{\sigma_m}}(w_n).
\end{aligned}$$

Maximization for Initial Probabilities

The first optimization entails one constraint, which we can solve by using a single Lagrange multiplier λ under the Lagrangian

$$\mathbb{L}\left(\lambda, \rho_{\sigma_1}, \ldots, \rho_{\sigma_M}\right) = \left(1 - \sum_{m=1}^{M} \rho_{\sigma_m}\right)\lambda + \sum_{m=1}^{M} \zeta_1^{\text{old}}(\sigma_m) \log \rho_{\sigma_m}.$$

Accordingly, the optimizer solves

$$\frac{\partial \mathbb{L}\left(\lambda, \rho_{\sigma_1}, \ldots, \rho_{\sigma_M}\right)}{\partial \lambda} = 0, \qquad \frac{\partial \mathbb{L}\left(\lambda, \rho_{\sigma_1}, \ldots, \rho_{\sigma_M}\right)}{\partial \rho_{\sigma_m}} = 0.$$

This system can be solved analytically. The solution, which provides the improved value $\boldsymbol{\rho}^{\text{new}}$ of the optimizer $\boldsymbol{\rho}^*$, is

$$\boldsymbol{\rho}^{\text{new}} = \left(\frac{\zeta_1^{\text{old}}(\sigma_1)}{\sum_{m=1}^{M} \zeta_1^{\text{old}}(\sigma_m)}, \ldots, \frac{\zeta_1^{\text{old}}(\sigma_M)}{\sum_{m=1}^{M} \zeta_1^{\text{old}}(\sigma_m)} \right). \tag{*8.24*}$$

Maximization for Transition Probabilities

For each m, the second optimization also entails one constraint, which we can solve using a single Lagrange multiplier κ_m under the Langrangian

* This is an advanced topic and could be skipped on a first reading.

$$\mathbb{K}_m\left(\kappa_m, \pi_{\sigma_m\to\sigma_1}, \dots, \pi_{\sigma_m\to\sigma_M}\right) = \left(1 - \sum_{m'=1}^{M} \pi_{\sigma_m\to\sigma_{m'}}\right)\kappa_m + \sum_{n=2}^{N}\sum_{m'=1}^{M} \eta_n^{\text{old}}(\sigma_m, \sigma_{m'}) \log \pi_{\sigma_m\to\sigma_{m'}}.$$

Accordingly, the optimizer solves

$$\frac{\partial \mathbb{K}_m\left(\kappa_m, \pi_{\sigma_m\to\sigma_1}, \dots, \pi_{\sigma_m\to\sigma_M}\right)}{\partial \kappa_m} = 0,$$

$$\frac{\partial \mathbb{K}_m\left(\kappa_m, \pi_{\sigma_m\to\sigma_1}, \dots, \pi_{\sigma_m\to\sigma_M}\right)}{\partial \pi_{\sigma_m\to\sigma_{m'}}} = 0.$$

Again, this system can be solved analytically. The solution, which provides the improved value $\pi_{\sigma_m}^{\text{new}}$ of the optimizer $\pi_{\sigma_m}^*$, is

$$\pi_{\sigma_m}^{\text{new}} = \left(\frac{\xi_{\sigma_m}^{\text{old}}(\sigma_1)}{\sum_{m'=1}^{M} \xi_{\sigma_m}^{\text{old}}(\sigma_{m'})}, \dots, \frac{\xi_{\sigma_m}^{\text{old}}(\sigma_M)}{\sum_{m'=1}^{M} \xi_{\sigma_m}^{\text{old}}(\sigma_{m'})}\right), \tag{*8.25*}$$

where

$$\xi_{\sigma_m}^{\text{old}}(\sigma_{m'}) = \sum_{n=2}^{N} \eta_n^{\text{old}}(\sigma_m, \sigma_{m'}).$$

Maximization for Emission Parameters

Unlike the first two optimizations, the third one generally cannot be solved analytically. Instead, depending on the functional form of the density $G_\phi(w)$, numerical techniques are needed to compute improved values $\phi_{\sigma_m}^{\text{new}}$ of the optimizers $\phi_{\sigma_m}^*$. In Example 8.1, we illustrate a simpler case where numerical optimization is unnecessary.

Example 8.1 Estimation in a HMM with normal observations

We consider a HMM with state-space $\sigma_{1:M}$ and normal emissions,

$$G_{\mu_{\sigma_m}, v_{\sigma_m}}(w) = \text{Normal}\left(w; \mu_{\sigma_m}, v_{\sigma_m}\right),$$

where the state parameters are $\phi_{\sigma_m} = (\mu_{\sigma_m}, v_{\sigma_m})$. Further, we suppose that an approximation $\rho^{\text{old}}, \boldsymbol{\Pi}^{\text{old}}, \phi^{\text{old}}$ of the maximizer $\rho^*, \boldsymbol{\Pi}^*, \phi^*$ has already been computed and we seek an improved one $\rho^{\text{new}}, \boldsymbol{\Pi}^{\text{new}}, \phi^{\text{new}}$ using the Baum–Welch algorithm.

Due to the exponential form of $G_{\mu_{\sigma_m}, v_{\sigma_m}}(w)$, we can derive a maximization procedure for the emission parameters analytically. That is, for each σ_m, the improved emission parameters $\mu_{\sigma_m}^{\text{new}}, v_{\sigma_m}^{\text{new}}$ maximize

$$\sum_{n=1}^{N} \zeta_n^{\text{old}}(\sigma_m) \log G_{\mu_{\sigma_m}, v_{\sigma_m}}(w_n) \propto \sum_{n=1}^{N} \zeta_n^{\text{old}}(\sigma_m)\left(-\log v_{\sigma_m} - \frac{\left(w_n - \mu_{\sigma_m}\right)^2}{v_{\sigma_m}}\right).$$

Since the maximizers are obtained by maximizing the above, they are found by solving

$$\frac{\partial}{\partial \mu_{\sigma_m}} \sum_{n=1}^{N} \zeta_n^{\text{old}}(\sigma_m) \left(-\log v_{\sigma_m} - \frac{\left(w_n - \mu_{\sigma_m}\right)^2}{v_{\sigma_m}} \right) = 0,$$

$$\frac{\partial}{\partial v_{\sigma_m}} \sum_{n=1}^{N} \zeta_n^{\text{old}}(\sigma_m) \left(-\log v_{\sigma_m} - \frac{\left(w_n - \mu_{\sigma_m}\right)^2}{v_{\sigma_m}} \right) = 0.$$

The solution is

$$\mu_{\sigma_m}^{\text{new}} = \frac{\sum_{n=1}^{N} \zeta_n^{\text{old}}(\sigma_m) w_n}{\sum_{n=1}^{N} \zeta_n^{\text{old}}(\sigma_m)}, \qquad v_{\sigma_m}^{\text{new}} = \frac{\sum_{n=1}^{N} \zeta_n^{\text{old}}(\sigma_m) \left(w_n - \mu_{\sigma_m}^{\text{new}}\right)^2}{\sum_{n=1}^{N} \zeta_n^{\text{old}}(\sigma_m)}.$$

8.3.4 Some Computational Considerations

The forward $\mathcal{A}_n(s_n)$ and backward $\mathcal{B}_n(s_n)$ variables are central to nearly every algorithm we have encountered so far and their accurate evaluation is essential in an HMM. Unfortunately, the computations in Algorithms 8.1 and 8.2, which rely on the recursions of Eqs. (*8.9*) and (*8.14*), involve a large number of multiplications between small numbers. Consequently, these algorithms are of limited practical value as most often they lead to rapid *underflow* and erroneous results.

Underflow is prevented if we consider *normalized* forward and backward terms,

$$\hat{\mathcal{A}}_n(s_n) = \mathcal{A}_n(s_n) \frac{1}{p(w_{1:n}|\boldsymbol{\rho}, \boldsymbol{\Pi}, \boldsymbol{\phi})}, \tag{8.26}$$

$$\check{\mathcal{B}}_n(s_n) = \mathcal{B}_n(s_n) \frac{1}{p\left(w_{n+1:N}|w_{1:n}, \boldsymbol{\rho}, \boldsymbol{\Pi}, \boldsymbol{\phi}\right)}, \tag{8.27}$$

and perform the recursions for $\hat{\mathcal{A}}_n(s_n)$ and $\check{\mathcal{B}}_n(s_n)$ instead of $\mathcal{A}_n(s_n)$ and $\mathcal{B}_n(s_n)$. In these cases, the recursions needed rely on

$$\hat{\mathcal{A}}_1(s_1) = \frac{1}{\hat{\mathcal{C}}_1} G_{\phi_{s_1}}(w_1)\, \rho_{s_1},$$

$$\hat{\mathcal{A}}_n(s_n) = \frac{1}{\hat{\mathcal{C}}_n} G_{\phi_{s_n}}(w_n) \sum_{s_{n-1}} \pi_{s_{n-1} \to s_n} \hat{\mathcal{A}}_{n-1}(s_{n-1}), \qquad n = 2 : N, \quad (*8.28*)$$

$$\check{\mathcal{B}}_n(s_n) = \frac{1}{\hat{\mathcal{C}}_{n+1}} \sum_{s_{n+1}} \check{\mathcal{B}}_{n+1}(s_{n+1}) G_{\phi_{s_{n+1}}}(w_{n+1})\, \pi_{s_n \to s_{n+1}}, \qquad n = N-1 : 1, \quad (*8.29*)$$

$$\check{\mathcal{B}}_N(s_N) = 1,$$

with the constants $\hat{\mathcal{C}}_n$ given by

$$\hat{\mathcal{C}}_1 = p(w_1|\boldsymbol{\rho}, \boldsymbol{\phi}),$$

$$\hat{\mathcal{C}}_n = p(w_n|w_{1:n-1}, \boldsymbol{\rho}, \boldsymbol{\Pi}, \boldsymbol{\phi}), \qquad n = 2 : N.$$

As the normalized terms $\hat{\mathcal{A}}_n(\sigma_1), \ldots, \hat{\mathcal{A}}_n(\sigma_M)$ are valid probabilities themselves, they are already scaled self-consistently and underflow is avoided. Further, because $\sum_{m=1}^{M} \hat{\mathcal{A}}_n(\sigma_m) = 1$, the constants $\hat{C}_n$, can be easily computed during the forward recursion. The steps involved in both recursions are summarized in Algorithms 8.5 and 8.6.

Algorithm 8.5 Forward recursion for HMM (stable version)

Given observations $w_{1:N}$ and parameters $\boldsymbol{\rho}, \boldsymbol{\Pi}, \boldsymbol{\phi}$, the forward terms $\hat{\mathcal{A}}_n(\sigma_m)$ for each time level n and each state σ_m are computed as follows.

- At $n = 1$ initialize by

$$\hat{\mathcal{A}}'_1(\sigma_m) = G_{\phi_{\sigma_m}}(w_1)\rho_{\sigma_m},$$

$$\hat{C}_1 = \sum_{m=1}^{M} \hat{\mathcal{A}}'_n(\sigma_m),$$

$$\hat{\mathcal{A}}_n(\sigma_m) = \frac{1}{\hat{C}_1}\hat{\mathcal{A}}'_n(\sigma_m).$$

- For $n = 2 : N$, compute recursively

$$\hat{\mathcal{A}}'_n(\sigma_m) = G_{\phi_{\sigma_m}}(w_n) \sum_{m'=1}^{M} \pi_{\sigma_{m'} \to \sigma_m} \hat{\mathcal{A}}_{n-1}(\sigma_{m'}),$$

$$\hat{C}_n = \sum_{m=1}^{M} \hat{\mathcal{A}}'_n(\sigma_m),$$

$$\hat{\mathcal{A}}_n(\sigma_m) = \frac{1}{\hat{C}_n}\hat{\mathcal{A}}'_n(\sigma_m).$$

Upon completion, the algorithm provides every $\hat{\mathcal{A}}_n(\sigma_m)$ and $\hat{C}_n$, which may be tabulated as

$$\begin{array}{c} \\ t_1 \\ t_2 \\ \vdots \\ t_n \\ \vdots \\ t_N \end{array} \begin{array}{c} \begin{array}{cccc} \sigma_1 & \sigma_2 & \cdots & \sigma_M \end{array} \\ \begin{bmatrix} \hat{\mathcal{A}}_1(\sigma_1) & \hat{\mathcal{A}}_1(\sigma_2) & \cdots & \hat{\mathcal{A}}_1(\sigma_M) \\ \hat{\mathcal{A}}_2(\sigma_1) & \hat{\mathcal{A}}_2(\sigma_2) & \cdots & \hat{\mathcal{A}}_2(\sigma_M) \\ \vdots & \vdots & \ddots & \vdots \\ \hat{\mathcal{A}}_n(\sigma_1) & \hat{\mathcal{A}}_n(\sigma_2) & \cdots & \hat{\mathcal{A}}_n(\sigma_M) \\ \vdots & \vdots & \ddots & \vdots \\ \hat{\mathcal{A}}_N(\sigma_1) & \hat{\mathcal{A}}_N(\sigma_2) & \cdots & \hat{\mathcal{A}}_N(\sigma_M) \end{bmatrix} \end{array} \begin{array}{c} \\ \begin{bmatrix} \hat{C}_1 \\ \hat{C}_2 \\ \vdots \\ \hat{C}_n \\ \vdots \\ \hat{C}_N \end{bmatrix} \end{array}.$$

Algorithm 8.6 Backward recursion for HMM (stable version)

Given observations $w_{1:N}$, parameters $\boldsymbol{\Pi}, \boldsymbol{\phi}$, and $\hat{C}_{2:N}$, the backward terms $\mathcal{B}_n(\sigma_m)$ for each time level n and each state σ_m are computed as follows

- At $n = N$, initialize by

$$\check{\mathcal{B}}_N(\sigma_m) = 1.$$

• For $n = N - 1, \ldots, 1$, compute recursively

$$\check{\mathcal{B}}_n(\sigma_m) = \frac{1}{\hat{\mathcal{C}}_{n+1}} \sum_{m'=1}^{M} \check{\mathcal{B}}_{n+1}(\sigma_{m'}) G_{\phi_{\sigma_{m'}}}(w_{n+1})\, \pi_{\sigma_m \to \sigma_{m'}}.$$

Upon completion, the algorithm provides every $\check{\mathcal{B}}_n(\sigma_m)$, which may be tabulated as follows

$$\begin{array}{c} t_1 \\ t_2 \\ \vdots \\ t_n \\ \vdots \\ t_N \end{array} \overset{\begin{array}{cccc} \sigma_1 & \sigma_2 & \cdots & \sigma_M \end{array}}{\begin{bmatrix} \check{\mathcal{B}}_1(\sigma_1) & \check{\mathcal{B}}_1(\sigma_2) & \cdots & \check{\mathcal{B}}_1(\sigma_M) \\ \check{\mathcal{B}}_2(\sigma_1) & \check{\mathcal{B}}_2(\sigma_2) & \cdots & \check{\mathcal{B}}_2(\sigma_M) \\ \vdots & \vdots & \ddots & \vdots \\ \check{\mathcal{B}}_n(\sigma_1) & \check{\mathcal{B}}_n(\sigma_2) & \cdots & \check{\mathcal{B}}_n(\sigma_M) \\ \vdots & \vdots & \ddots & \vdots \\ \check{\mathcal{B}}_N(\sigma_1) & \check{\mathcal{B}}_N(\sigma_2) & \cdots & \check{\mathcal{B}}_N(\sigma_M) \end{bmatrix}}.$$

As we can see, Algorithms 8.5 and 8.6 involve more computations than Algorithms 8.1 and 8.2. Nevertheless, this difference is almost negligible as the most expensive operation in both versions is a matrix-vector multiplication, appearing in Eq. (*8.28*) and Eq. (*8.29*), which scales with M^2N. In any case, although less efficient than $\mathcal{A}_n(s_n)$ and $\mathcal{B}_n(s_n)$, computing $\hat{\mathcal{A}}_n(s_n)$ and $\check{\mathcal{B}}_n(s_n)$ avoids underflow, which is indispensable.

Furthermore, on account of Eqs. (8.26) and (8.27), the normalized terms $\hat{\mathcal{A}}_n(s_n)$ and $\check{\mathcal{B}}_n(s_n)$ can be used almost anywhere that both $\mathcal{A}_n(s_n)$ and $\mathcal{B}_n(s_n)$ are required. For example, both ways of decoding an HMM, *e.g.*, Eqs. (*8.11*) and (*8.12*) or Eqs. (8.15) and (*8.16*), are unaffected by the normalization. Similarly, the maximization of Eq. (8.21) for estimating an HMM, *e.g.*, Eqs. (8.22) and (8.23), remains similarly unaffected.

However, an important exception occurs when evaluating an HMM. In particular, because Eq. (8.7) depends explicitly upon $\mathcal{A}_N(s_N)$, the normalization *does* have an effect and the marginal likelihood $p(w_{1:N}|\boldsymbol{\rho}, \boldsymbol{\Pi}, \boldsymbol{\phi})$ needs to be evaluated differently. The most convenient way is through the factorization

$$p(w_{1:N}|\boldsymbol{\rho}, \boldsymbol{\Pi}, \boldsymbol{\phi}) = \prod_{n=1}^{N} \hat{\mathcal{C}}_n, \tag{*8.30*}$$

with the constants $\hat{\mathcal{C}}_n$ computed most efficiently using the forward recursion of Algorithm 8.5 and stored in logarithmic form.

8.3.5 State-Space Labeling and Likelihood*

The algorithms presented so far are routinely used to answer questions pertaining to an HMM. These algorithms exhibit maximum efficiency for

* This is an advanced topic and could be skipped on a first reading.

their tasks. However, they are limited to yielding point estimates *only*. That is, at best these algorithms provide a single choice for the values of the variables of interest, for example $s_n^*, s_{1:N}^\#$ or $\boldsymbol{\rho}^*, \boldsymbol{\Pi}^*, \boldsymbol{\phi}^*$. Unfortunately, they fail to quantify the uncertainty associated with each estimator, which is a serious limitation by itself.

Error bars around the estimators may be obtained with generic likelihood-based strategies, for example through Fisher information or bootstrapping techniques that we do not dwell upon here. Indeed, such approaches are possible, at least in theory, under Monte Carlo sampling or greedy computations where a portion of all possible sequences $s_{1:N}$ are computed. However, even with greedy computations, there is a fundamental degeneracy in likelihoods constructed from Eqs. (8.4) to (8.6), prohibiting the uniqueness of any computed estimator.

Namely, similar to what we saw in Section 7.1.3, HMM likelihoods, $p(w_{1:N}|\boldsymbol{\rho}, \boldsymbol{\Pi}, \boldsymbol{\phi})$ or $p(w_{1:N}, s_{1:N}|\boldsymbol{\rho}, \boldsymbol{\Pi}, \boldsymbol{\phi})$, are *invariant to permutations* of the constitutive state labels. That is, relabeling of the constitutive states results in the same value of the likelihood. Consequently, irrespective of how an estimator is obtained, there are always additional $M!-1$ equally optimal ones leading to $M!$-fold degeneracy.

Example 8.2 State relabeling

To illustrate the likelihood's degeneracy, we consider a simplified HMM containing $N = 3$ time levels and $M = 2$ constitutive states. Further, for clarity we adopt pedantic notation and let $s_{1:3}, \boldsymbol{\rho}, \boldsymbol{\Pi}, \boldsymbol{\phi}$ stand for the corresponding random variables. Considering realized values for these random variables, invariance of the (marginal) likelihood reads

$$p\left(w_{1:3}\middle|\boldsymbol{\rho}=(\rho_\alpha,\rho_\beta), \boldsymbol{\Pi}=\begin{pmatrix}\pi_{\alpha\to\alpha} & \pi_{\alpha\to\beta}\\ \pi_{\beta\to\alpha} & \pi_{\beta\to\beta}\end{pmatrix}, \boldsymbol{\phi}=(\phi_\alpha,\phi_\beta)\right)$$

$$= p\left(w_{1:3}\middle|\boldsymbol{\rho}=(\rho_\beta,\rho_\alpha), \boldsymbol{\Pi}=\begin{pmatrix}\pi_{\beta\to\beta} & \pi_{\beta\to\alpha}\\ \pi_{\alpha\to\beta} & \pi_{\alpha\to\alpha}\end{pmatrix}, \boldsymbol{\phi}=(\phi_\beta,\phi_\alpha)\right).$$

Similarly, invariance of the (joint) likelihood reads

$$p\left(w_{1:3}, s_{1:3}=(\alpha,\beta,\beta)\middle|\boldsymbol{\rho}=(\rho_\alpha,\rho_\beta), \boldsymbol{\Pi}=\begin{pmatrix}\pi_{\alpha\to\alpha} & \pi_{\alpha\to\beta}\\ \pi_{\beta\to\alpha} & \pi_{\beta\to\beta}\end{pmatrix}, \boldsymbol{\phi}=(\phi_\alpha,\phi_\beta)\right)$$

$$= p\left(w_{1:3}, s_{1:3}=(\beta,\alpha,\alpha)\middle|\boldsymbol{\rho}=(\rho_\beta,\rho_\alpha), \boldsymbol{\Pi}=\begin{pmatrix}\pi_{\beta\to\beta} & \pi_{\beta\to\alpha}\\ \pi_{\alpha\to\beta} & \pi_{\alpha\to\alpha}\end{pmatrix}, \boldsymbol{\phi}=(\phi_\beta,\phi_\alpha)\right).$$

Both cases are produced by considering every possible permutation of the constitutive states, which, for this simple example, are

$$\begin{matrix}\sigma_1 & \sigma_2\end{matrix}$$
$$\begin{pmatrix}\alpha & \beta\\ \beta & \alpha\end{pmatrix}.$$

8.4 The Hidden Markov Model in the Bayesian Paradigm

A Bayesian formulation provides more modeling flexibility than its frequentist counterpart. Such flexibility is quite useful when modeling dynamical systems. For example, it provides a recipe by which to rigorously back-propagate measurement error into uncertainty over the parameters we seek, or even characterize state-spaces in themselves as we will discover later in the context of HMMs within the Bayesian nonparametric paradigm.

In the Bayesian setting, we sample posteriors. Therefore, every question about a system formulated with a Bayesian HMM is answered through the posterior $p(\boldsymbol{\rho}, \boldsymbol{\Pi}, \boldsymbol{\phi}|w_{1:N})$ or the completed posterior $p(s_{1:N}, \boldsymbol{\rho}, \boldsymbol{\Pi}, \boldsymbol{\phi}|w_{1:N})$. The HMM of Eqs. (8.4) to (8.6) provides probability distributions only for the passing states $p(s_{1:N}|\boldsymbol{\rho}, \boldsymbol{\Pi})$ and the measurements $p(w_{1:N}|s_{1:N}, \boldsymbol{\phi})$, which do not suffice in fully specifying our posterior. For this reason, a Bayesian HMM requires the specification of additional distributions that supply statistics to the parameters $\boldsymbol{\rho}, \boldsymbol{\Pi}, \boldsymbol{\phi}$. These distributions are our priors and, as anticipated, several reasonable choices can be devised to accommodate the system at hand. Below, we describe suitable choices and subsequently appropriate sampling techniques for a generic Bayesian HMM. We present more specialized versions, tailored to specific cases, in subsequent sections.

8.4.1 Priors for the HMM

The simplest choice for the initial ρ_{σ_m} and transition $\pi_{\sigma_m \to \sigma_{m'}}$ probabilities are offered by independent priors on $\boldsymbol{\rho}$ and each $\boldsymbol{\pi}_{\sigma_m}$. For instance, draws from Dirichlet distributions,

$$\boldsymbol{\rho} \sim \text{Dirichlet}_{\sigma_{1:M}}(\eta \boldsymbol{\zeta}),$$
$$\boldsymbol{\pi}_{\sigma_m} \sim \text{Dirichlet}_{\sigma_{1:M}}\left(\alpha_{\sigma_m} \boldsymbol{\beta}_{\sigma_m}\right),$$

ensure valid probability arrays. In these priors, η and α_{σ_m} are positive scalar constants, while, $\boldsymbol{\zeta} = [\zeta_{\sigma_1}, \ldots, \zeta_{\sigma_M}]$ and $\boldsymbol{\beta}_{\sigma_m} = [\beta_{\sigma_m \to \sigma_1}, \ldots, \beta_{\sigma_m \to \sigma_1}]$ are probability arrays. Due to the conjugacy between the categorical and Dirichlet distributions, as we will see shortly, such prior choices are also computationally favored.

Despite the generality in the priors over the dynamical parameters, a choice for the emission parameters ϕ_{σ_m} depends heavily on the distribution $\mathbb{G}_{\phi}$, which, in turn, varies widely between systems. Computational tractability is facilitated when we consider iid priors,

$$\phi_{\sigma_m} \sim \mathbb{H},$$

under a common, system-specific probability distribution $\mathbb{H}$. Additionally, we see below that the computations involved are greatly simplified if $\mathbb{H}$ is conjugate to $\mathbb{G}_{\phi}$.

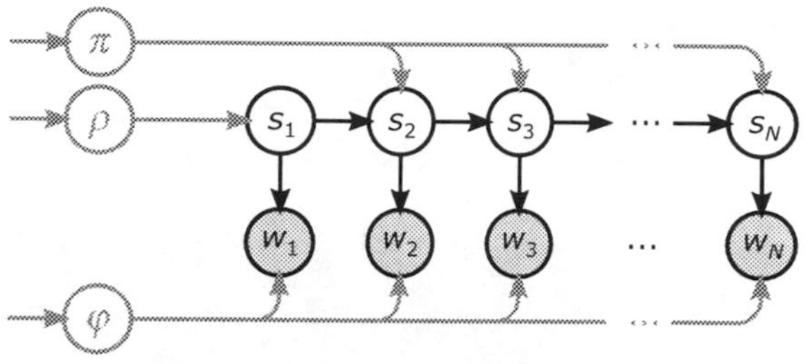

Fig. 8.2 Graphical representation of a Bayesian HMM. Here, the parameters ρ, $\mathbf{\Pi}$, and $\boldsymbol{\phi}$ are assumed unknown.

8.4.2 MCMC Inference in the Bayesian HMM

With the choices described above, an entire Bayesian HMM forward model is summarized as

$$\rho \sim \text{Dirichlet}_{\sigma_{1:M}}(\eta\zeta), \tag{8.31}$$
$$\pi_{\sigma_m} \sim \text{Dirichlet}_{\sigma_{1:M}}\left(\alpha_{\sigma_m}\beta_{\sigma_m}\right), \tag{8.32}$$
$$\phi_{\sigma_m} \sim \mathbb{H}, \tag{8.33}$$
$$s_1|\rho \sim \text{Categorical}_{\sigma_{1:M}}(\rho), \tag{8.34}$$
$$s_n|s_{n-1}, \mathbf{\Pi} \sim \text{Categorical}_{\sigma_{1:M}}(\pi_{\sigma_m}), \qquad n = 2:N, \tag{8.35}$$
$$w_n|s_n, \boldsymbol{\phi} \sim \mathbb{G}_{\phi_{s_n}}, \qquad n = 1:N, \tag{8.36}$$

and illustrated in Fig. 8.2. Inference on this HMM is more complicated than its non-Bayesian counterpart. Below, we describe two complementary MCMC sampling schemes. One is based on the Gibbs sampler, appropriate for routine applications, and the other is based on the Metropolis–Hastings sampler, appropriate for demanding applications where mixing of the Gibbs sampler becomes inefficient.

Gibbs Sampling

A Gibbs sampling scheme is most efficient when it generates MCMC samples from the HMM's completed posterior $p\left(s_{1:N}, \rho, \mathbf{\Pi}, \boldsymbol{\phi}|w_{1:N}\right)$. In a basic implementation, we iterate between successive updates of $s_{1:N}|\rho, \mathbf{\Pi}, \boldsymbol{\phi}, w_{1:N}$ and $\rho, \mathbf{\Pi}, \boldsymbol{\phi}|s_{1:N}, w_{1:N}$. Due to the formulation of the HMM, the latter reduces to independent updates for each parameter. Specifically, once $s_{1:N}|\rho, \mathbf{\Pi}, \boldsymbol{\phi}, w_{1:N}$ is sampled, we can update the parameters by sampling separately $\rho|s_1$ and $\pi_{\sigma_m}|s_{1:N}$ and $\phi_{\sigma_m}|s_{1:N}, w_{1:N}$ for each σ_m. The entire scheme is summarized in Algorithm 8.7.

Algorithm 8.7 Gibbs sampling for Bayesian HMM

Given an initial sample $\rho^{\text{old}}, \mathbf{\Pi}^{\text{old}}, \boldsymbol{\phi}^{\text{old}}$, which may be generated from the corresponding priors, MCMC updates are carried out by iterating the following steps.

- Update the state sequence by sampling $s_{1:N}^{\text{new}}$ with *forward filtering backward sampling* based on $\boldsymbol{\rho}^{\text{old}}, \boldsymbol{\Pi}^{\text{old}}, \boldsymbol{\phi}^{\text{old}}$.
- Compute state indices $\mathcal{N}_{\sigma_m}^{\text{new}}$ and counts $\boldsymbol{d}^{\text{new}}, \boldsymbol{C}^{\text{new}}$ based on $s_{1:N}^{\text{new}}$ as described shortly.
- Update the dynamic parameters by sampling from

$$\boldsymbol{\rho}^{\text{new}} \sim \text{Dirichlet}_M\left(\eta\boldsymbol{\zeta} + \boldsymbol{d}^{\text{new}}\right),$$

$$\boldsymbol{\pi}_{\sigma_m}^{\text{new}} \sim \text{Dirichlet}_M\left(\alpha_{\sigma_m}\boldsymbol{\beta}_{\sigma_m} + \boldsymbol{c}_{\sigma_m}^{\text{new}}\right).$$

- Update the emission parameters by sampling $\phi_{\sigma_m}^{\text{new}}$ for each σ_m based on $\mathcal{N}_{\sigma_m}^{\text{new}}$.

Next, we examine the steps involved in this Gibbs scheme in more detail. For clarity, we designate by $\boldsymbol{\rho}^{\text{old}}, \boldsymbol{\Pi}^{\text{old}}, \boldsymbol{\phi}^{\text{old}}$ a sample in the MCMC chain and by $s_{1:N}^{\text{new}}, \boldsymbol{\rho}^{\text{new}}, \boldsymbol{\Pi}^{\text{new}}, \boldsymbol{\phi}^{\text{new}}$ the very next sample.

Updates of the State Sequence

In the Gibbs sampler, the state sequence is updated by sampling from $p(s_{1:N}|\boldsymbol{\rho}^{\text{old}}, \boldsymbol{\Pi}^{\text{old}}, \boldsymbol{\phi}^{\text{old}}, w_{1:N})$. This distribution is factorized as

$$p(s_{1:N}|\boldsymbol{\rho}^{\text{old}}, \boldsymbol{\Pi}^{\text{old}}, \boldsymbol{\phi}^{\text{old}}, w_{1:N})$$
$$= p(s_N|\boldsymbol{\rho}^{\text{old}}, \boldsymbol{\Pi}^{\text{old}}, \boldsymbol{\phi}^{\text{old}}, w_{1:N}) \prod_{n=1}^{N-1} p(s_n|s_{n+1:N}, \boldsymbol{\rho}^{\text{old}}, \boldsymbol{\Pi}^{\text{old}}, \boldsymbol{\phi}^{\text{old}}, w_{1:N}),$$

which allows s_N^{new} to be sampled first and, subsequently, each s_n^{new} to be sampled recursively backwards. We can perform such sampling using the forward terms $\hat{\mathcal{A}}_n(\sigma_m)$, which need to be precomputed through filtering, for example, via Algorithm 8.5. As these terms need to be computed under $\boldsymbol{\rho}^{\text{old}}, \boldsymbol{\Pi}^{\text{old}}, \boldsymbol{\phi}^{\text{old}}$, we designate them using $\hat{\mathcal{A}}_n^{\text{old}}(\sigma_m)$. Once every $\hat{\mathcal{A}}_n^{\text{old}}(\sigma_m)$ is computed with a forward recursion, sampling begins with

$$s_N^{\text{new}} \sim \text{Categorical}_{\sigma_{1:M}}\left(\hat{\mathcal{A}}_N^{\text{old}}(\sigma_1), \ldots, \hat{\mathcal{A}}_N^{\text{old}}(\sigma_M)\right) \tag{8.37}$$

and recurses backward based on

$$s_n^{\text{new}} \sim \text{Categorical}_{\sigma_{1:M}}$$
$$\times\left(\frac{\hat{\mathcal{A}}_n^{\text{old}}(\sigma_1)\pi^{\text{old}}_{\sigma_1\to s_{n+1}^{\text{new}}}}{\sum_{m=1}^M \hat{\mathcal{A}}_n^{\text{old}}(\sigma_m)\pi^{\text{old}}_{\sigma_m\to s_{n+1}^{\text{new}}}}, \ldots, \frac{\hat{\mathcal{A}}_n^{\text{old}}(\sigma_M)\pi^{\text{old}}_{\sigma_M\to s_{n+1}^{\text{new}}}}{\sum_{m=1}^M \hat{\mathcal{A}}_n^{\text{old}}(\sigma_m)\pi^{\text{old}}_{\sigma_m\to s_{n+1}^{\text{new}}}}\right),$$
$$n = N-1:1. \tag{*8.38*}$$

The entire processes is termed *forward filtering backward sampling* and the steps involved are summarized in Algorithm 8.8.

Algorithm 8.8 Forward filtering backward sampling

Given observations $w_{1:N}$ and parameters $\boldsymbol{\rho}, \boldsymbol{\Pi}, \boldsymbol{\phi}$, a state sequence $s_{1:N}$ is sampled as follows.

- Use Algorithm 8.5 and $\boldsymbol{\rho}, \boldsymbol{\Pi}, \boldsymbol{\phi}$ to compute the forward terms $\hat{\mathcal{A}}_n(\sigma_m)$.
- At $n = N$, generate

$$s_N \sim \text{Categorical}_{\sigma_{1:M}}\left(\hat{\mathcal{A}}_N(\sigma_1), \ldots, \hat{\mathcal{A}}_N(\sigma_M)\right).$$

- For $n = N - 1 : 1$, generate recursively

$$s_n \sim$$
$$\text{Categorical}_{\sigma_{1:M}}\left(\frac{\hat{\mathcal{A}}_n(\sigma_1)\pi_{\sigma_1\to s_{n+1}}}{\sum_{m=1}^M \hat{\mathcal{A}}_n(\sigma_m)\pi_{\sigma_m\to s_{n+1}}}, \ldots, \frac{\hat{\mathcal{A}}_n(\sigma_M)\pi_{\sigma_M\to s_{n+1}}}{\sum_{m=1}^M \hat{\mathcal{A}}_n(\sigma_m)\pi_{\sigma_m\to s_{n+1}}}\right).$$

Updates of the Dynamic Parameters

In the Gibbs sampler, the initial probabilities are updated by sampling from $p(\boldsymbol{\rho}|s_1^{\text{new}})$. Due to conjugacy, this sampling reduces to

$$\boldsymbol{\rho}^{\text{new}} \sim \text{Dirichlet}_{\sigma_{1:M}}(\eta\boldsymbol{\zeta} + \boldsymbol{d}^{\text{new}}),$$

where $\boldsymbol{d}^{\text{new}} = [d_{\sigma_1}^{\text{new}}, \ldots, d_{\sigma_M}^{\text{new}}]$ is an array of zeroes and ones whose σ_m entry indicates whether $s_1^{\text{new}} = \sigma_m$.

Similarly, the transition probabilities out of each σ_m are updated by sampling from $p(\boldsymbol{\pi}_{\sigma_m}|s_{1:N}^{\text{new}})$. Again, due to conjugacy, this sampling reduces to

$$\boldsymbol{\pi}_{\sigma_m}^{\text{new}} \sim \text{Dirichlet}_{\sigma_{1:M}}\left(\alpha_{\sigma_m}\boldsymbol{\beta}_{\sigma_m} + \boldsymbol{c}_{\sigma_m}^{\text{new}}\right),$$

where $\boldsymbol{c}_{\sigma_m}^{\text{new}} = [c_{\sigma_m\to\sigma_1}^{\text{new}}, \ldots, c_{\sigma_m\to\sigma_M}^{\text{new}}]$ is a vector whose $\sigma_m \to \sigma_{m'}$ entry counts how many times the transition $\sigma_m \to \sigma_{m'}$ occurs in $s_{1:N}^{\text{new}}$.

Note 8.7 Transition count matrix

Bookkeeping in Algorithm 8.7 is simpler if we tabulate the count arrays $\boldsymbol{c}_{\sigma_m}$ into

$$\begin{array}{c} \\ \sigma_1 \\ \sigma_2 \\ \vdots \\ \sigma_M \end{array} \begin{array}{c} \begin{array}{cccc} \sigma_1 & \sigma_2 & \cdots & \sigma_M \end{array} \\ \begin{bmatrix} c_{\sigma_1\to\sigma_1} & c_{\sigma_1\to\sigma_2} & \cdots & c_{\sigma_1\to\sigma_M} \\ c_{\sigma_2\to\sigma_1} & c_{\sigma_2\to\sigma_2} & \cdots & c_{\sigma_2\to\sigma_M} \\ \vdots & \vdots & \ddots & \vdots \\ c_{\sigma_M\to\sigma_1} & c_{\sigma_M\to\sigma_2} & \cdots & c_{\sigma_M\to\sigma_M} \end{bmatrix} \end{array} = \begin{bmatrix} \boldsymbol{c}_{\sigma_1} \\ \boldsymbol{c}_{\sigma_2} \\ \vdots \\ \boldsymbol{c}_{\sigma_M} \end{bmatrix} = \boldsymbol{C},$$

similar to the tabulation of $\boldsymbol{\Pi}$.

Updates of the Observation Parameters

In the Gibbs sampler, the emission parameters of σ_m are updated by sampling from $p(\phi_{\sigma_m}|s_{1:N}^{\text{new}}, w_{1:N})$. Using Bayes' rule, this distribution factorizes as

$$p(\phi_{\sigma_m}|s_{1:N}^{\text{new}}, w_{1:N}) \propto H(\phi_{\sigma_m}) \prod_{n \in \mathcal{N}_{\sigma_m}^{\text{new}}} G_{\phi_{\sigma_m}}(w_n), \tag{8.39}$$

where $\mathcal{N}_{\sigma_m}^{\text{new}}$ gathers the indices n of the time levels when $s_n^{\text{new}} = \sigma_m$. For arbitrary $\mathbb{H}$ and $\mathbb{G}_\phi$, sampling of $p(\phi_{\sigma_m}|s_{1:N}^{\text{new}}, w_{1:N})$ cannot be performed directly and a *within Gibbs* scheme is required. Nevertheless, as we show in Example 8.3, distributions $\mathbb{G}_\phi$ with conjugate priors $\mathbb{H}$ are sampled directly.

Example 8.3 Bayesian HMM with normal observations

We consider a HMM with state-space $\sigma_{1:M}$ and normal emission densities

$$G_{\mu,\tau}(w) = \text{Normal}\left(w; \mu, \frac{1}{\tau}\right).$$

Gibbs sampling in this HMM is greatly simplified if we apply the conditionally conjugate prior

$$\mu \sim \text{Normal}\left(\xi, \frac{1}{\psi}\right), \qquad \tau \sim \text{Gamma}\left(\alpha, \beta\right).$$

With this choice of prior, a full update of the parameters $\mu_{\sigma_m}, \tau_{\sigma_m}$ is achieved by successively sampling $\mu_{\sigma_m}|\tau_{\sigma_m}^{\text{old}}, s_{1:N}^{\text{new}}, w_{1:N}$ and $\tau_{\sigma_m}^{\text{new}}|\mu_{\sigma_m}^{\text{new}}, s_{1:N}^{\text{new}}, w_{1:N}$. With these choices, the factorization of Eq. (8.39) leads to

$$\mu_{\sigma_m}^{\text{new}} \sim \text{Normal}\left(\frac{\psi\xi + \tau^{\text{old}} \sum_{n \in \mathcal{N}_{\sigma_m}^{\text{new}}} w_n}{\psi + \tau^{\text{old}}|\mathcal{N}_{\sigma_m}^{\text{new}}|}, \frac{1}{\psi + \tau^{\text{old}}|\mathcal{N}_{\sigma_m}^{\text{new}}|}\right),$$

$$\tau_{\sigma_m}^{\text{new}} \sim \text{Gamma}\left(\alpha + \frac{1}{2}|\mathcal{N}_{\sigma_m}^{\text{new}}|, \frac{1}{\frac{1}{\beta} + \frac{1}{2}\sum_{n \in \mathcal{N}_{\sigma_m}^{\text{new}}}\left(w_n - \mu_{\sigma_m}^{\text{new}}\right)^2}\right).$$

Updating each component, μ_{σ_m} and τ_{σ_m}, *once* per Gibbs iteration yields a valid sampler. However, mixing is better if these samplings are repeated *several* times per iteration. As inner iterations typically require considerably fewer computations than forward filtering backward sampling, for most HMM applications they add little to the sampler's overall computational cost while greatly improving its mixing.

Metropolis–Hastings Sampling*

The Gibbs sampler described so far is most often sufficient for HMM applications whenever the total number of time levels with measurements, N, is low or the emission distributions, $\mathbb{G}_{\phi_{\sigma_m}}$, appreciably overlap. However, for long sequences and/or well-separated emission distributions, mixing of the Gibbs sampler may become poor. For such cases, an alternative sampler, implemented in Algorithm 8.9, that updates $\boldsymbol{\rho}, \boldsymbol{\Pi}, \boldsymbol{\phi}$ while keeping the state sequence $s_{1:N}$ marginalized, is preferable.

* This is an advanced topic and could be skipped on a first reading.

Algorithm 8.9 Metropolis–Hastings sampling for Bayesian HMM

Given an initial sample $\rho^{\text{old}}, \mathbf{\Pi}^{\text{old}}, \phi^{\text{old}}$, which may be generated from the corresponding priors, MCMC updates are carried out as follows until convergence.

- First, compute $\hat{C}^{\text{old}}_{1:N}$ based on $\rho^{\text{old}}, \mathbf{\Pi}^{\text{old}}, \phi^{\text{old}}$ and set $\mathcal{L}^{\text{old}} = \sum_{n=1}^{N} \log \hat{C}^{\text{old}}_n$.
- Then, iterate the steps:
  - Generate proposals $\rho^{\text{prop}}, \mathbf{\Pi}^{\text{prop}}, \phi^{\text{prop}}$ based on $\rho^{\text{old}}, \mathbf{\Pi}^{\text{old}}, \phi^{\text{old}}$.
  - Compute $\hat{C}^{\text{prop}}_{1:N}$ based on $\rho^{\text{prop}}, \mathbf{\Pi}^{\text{prop}}, \phi^{\text{prop}}$ and set $\mathcal{L}^{\text{prop}} = \sum_{n=1}^{N} \log \hat{C}^{\text{prop}}_n$.
  - Perform the Metropolis–Hastings acceptance test based on $\{\mathcal{L}^{\text{old}}, \rho^{\text{old}}, \mathbf{\Pi}^{\text{old}}, \phi^{\text{old}}\}$ and $\{\mathcal{L}^{\text{prop}}, \rho^{\text{prop}}, \mathbf{\Pi}^{\text{prop}}, \phi^{\text{prop}}\}$.
    * If the acceptance test *succeeds*, set $\rho^{\text{new}} = \rho^{\text{prop}}$, $\mathbf{\Pi}^{\text{new}} = \mathbf{\Pi}^{\text{prop}}$, $\phi^{\text{new}} = \phi^{\text{prop}}$, and $\mathcal{L}^{\text{new}} = \mathcal{L}^{\text{prop}}$.
    * If the acceptance test *fails*, set $\rho^{\text{new}} = \rho^{\text{old}}$, $\mathbf{\Pi}^{\text{new}} = \mathbf{\Pi}^{\text{old}}$, $\phi^{\text{new}} = \phi^{\text{old}}$, and $\mathcal{L}^{\text{new}} = \mathcal{L}^{\text{old}}$.

Such a sampler may be developed based upon the same principles as the generic Metropolis–Hastings sampler of Section 5.2.1. In particular, to sample from an HMM's posterior $p(\rho, \mathbf{\Pi}, \phi | w_{1:N})$, a Metropolis–Hastings sampler requires the selection of a suitable proposal $Q_{\rho^{\text{old}}, \mathbf{\Pi}^{\text{old}}, \phi^{\text{old}}}(\rho^{\text{prop}}, \mathbf{\Pi}^{\text{prop}}, \phi^{\text{prop}})$. Although such a proposal may attempt to update all parameters at once, in general, it is more practical to update one or at most few parameters at a time. This may be achieved by a mixture proposal, for example, of the form

$$
\begin{aligned}
&Q_{\rho^{\text{old}}, \mathbf{\Pi}^{\text{old}}, \phi^{\text{old}}}(\rho^{\text{prop}}, \mathbf{\Pi}^{\text{prop}}, \phi^{\text{prop}}) \\
&\quad = \omega\, Q_{\rho^{\text{old}}, \mathbf{\Pi}^{\text{old}}}(\rho^{\text{prop}}, \mathbf{\Pi}^{\text{prop}})\, \delta_{\phi^{\text{old}}}(\phi^{\text{prop}}) \\
&\qquad + (1-\omega)\, Q_{\phi^{\text{old}}}(\phi^{\text{prop}})\, \delta_{\rho^{\text{old}}, \mathbf{\Pi}^{\text{old}}}(\rho^{\text{prop}}, \mathbf{\Pi}^{\text{prop}}),
\end{aligned}
$$

where, with probability ω, only proposals for the dynamical parameters are made; while, with probability $1 - \omega$, only proposals for the observational parameters are made. In turn, each of the partial proposals $Q_{\rho^{\text{old}}, \mathbf{\Pi}^{\text{old}}}(\rho^{\text{prop}}, \mathbf{\Pi}^{\text{prop}})$ and $Q_{\phi^{\text{old}}}(\phi^{\text{prop}})$ may consist of further mixtures themselves that propose $\rho^{\text{prop}}, \pi^{\text{prop}}_{\sigma_m}, \phi^{\text{prop}}_{\sigma_m}$ separately.

Note 8.8 Choice of proposals

Generally, $Q_{\phi^{\text{old}}}(\phi^{\text{prop}})$ is problem specific; however, for $Q_{\rho^{\text{old}}, \mathbf{\Pi}^{\text{old}}}(\rho^{\text{prop}}, \mathbf{\Pi}^{\text{prop}})$ we may construct generic proposals by considering products of Dirichlet distributions. For example,

$$
\begin{aligned}
&Q_{\rho^{\text{old}}, \mathbf{\Pi}^{\text{old}}}(\rho^{\text{prop}}, \mathbf{\Pi}^{\text{prop}}) \\
&\quad = \text{Dirichlet}_{\sigma_{1:M}}\left(\rho^{\text{prop}}; \kappa \rho^{\text{old}}\right) \prod_{m=1}^{M} \text{Dirichlet}_{\sigma_{1:M}}\left(\pi^{\text{prop}}_{\sigma_m}; \lambda \pi^{\text{old}}_{\sigma_m}\right).
\end{aligned}
$$

This choice ensures that the proposed $\rho^{\text{prop}}, \mathbf{\Pi}^{\text{old}}$ consist of valid probability arrays and also allows for tuning of the resulting acceptance rate through the values of κ and λ.

Finally, once a proposal $\rho^{\text{prop}}, \mathbf{\Pi}^{\text{prop}}, \phi^{\text{prop}}$ is made, either through $Q_{\rho^{\text{old}},\mathbf{\Pi}^{\text{old}}}(\rho^{\text{prop}}, \mathbf{\Pi}^{\text{prop}})$ or $Q_{\phi^{\text{old}}}(\phi^{\text{prop}})$, an acceptance ratio,

$$A_{\rho^{\text{old}},\mathbf{\Pi}^{\text{old}},\phi^{\text{old}}}(\rho^{\text{prop}}, \mathbf{\Pi}^{\text{prop}}, \phi^{\text{prop}}) = \frac{p(\rho^{\text{prop}}, \mathbf{\Pi}^{\text{prop}}, \phi^{\text{prop}}|w_{1:N})}{p\left(\rho^{\text{old}}, \mathbf{\Pi}^{\text{old}}, \phi^{\text{old}}|w_{1:N}\right)} \times \frac{Q_{\rho^{\text{prop}},\mathbf{\Pi}^{\text{prop}},\phi^{\text{prop}}}\left(\rho^{\text{old}}, \mathbf{\Pi}^{\text{old}}, \phi^{\text{old}}\right)}{Q_{\rho^{\text{old}},\mathbf{\Pi}^{\text{old}},\phi^{\text{old}}}(\rho^{\text{prop}}, \mathbf{\Pi}^{\text{prop}}, \phi^{\text{prop}})},$$

is computed to complete the Metropolis–Hastings acceptance test. The second ratio depends on the specific choices for the proposals made and can be easily computed. The first ratio arises from the product of the ratio of priors as well as marginal likelihoods,

$$\frac{p(\rho^{\text{prop}}, \mathbf{\Pi}^{\text{prop}}, \phi^{\text{prop}}|w_{1:N})}{p\left(\rho^{\text{old}}, \mathbf{\Pi}^{\text{old}}, \phi^{\text{old}}|w_{1:N}\right)} = \underbrace{\frac{p(w_{1:N}|\rho^{\text{prop}}, \mathbf{\Pi}^{\text{prop}}, \phi^{\text{prop}})}{p\left(w_{1:N}|\rho^{\text{old}}, \mathbf{\Pi}^{\text{old}}, \phi^{\text{old}}\right)}}_{\text{marginal likelihoods}} \underbrace{\frac{p(\rho^{\text{prop}}, \mathbf{\Pi}^{\text{prop}}, \phi^{\text{prop}})}{p\left(\rho^{\text{old}}, \mathbf{\Pi}^{\text{old}}, \phi^{\text{old}}\right)}}_{\text{priors}}.$$

The last ratio depends exclusively on the priors and can also be easily computed. The other ratio is formed by the marginal likelihoods, which we need to evaluate through Eq. (*8.30*). In particular, filtering such as Algorithm 8.5 needs to be invoked twice: once for $p\left(w_{1:N}|\rho^{\text{old}}, \mathbf{\Pi}^{\text{old}}, \phi^{\text{old}}\right)$ and once for $p(w_{1:N}|\rho^{\text{prop}}, \mathbf{\Pi}^{\text{prop}}, \phi^{\text{prop}})$. As filtering makes up the most computationally intensive part, both likelihoods can be retained and updated upon acceptance. By doing so, at the next iteration we may avoid recomputing $p\left(w_{1:N}|\rho^{\text{old}}, \mathbf{\Pi}^{\text{old}}, \phi^{\text{old}}\right)$, thereby reducing the computational load from two to only *one* filtering operation per iteration, which renders this scheme competitive with the earlier Gibbs scheme.

Note 8.9 Sampling the state sequence

If needed, Algorithm 8.9 can also sample a state sequence to generate samples from the completed posterior $p(s_{1:N}, \rho, \mathbf{\Pi}, \phi|w_{1:N})$ at little additional cost. For instance, following each filtering operation, if instead of only computing marginal likelihoods we also maintain and update every forward term $\hat{\mathcal{A}}_{1:N}(\sigma_m)$, a new state sequence $s_{1:N}^{\text{new}}$ can be obtained at the end of each Metropolis–Hastings iteration by executing only the backward sampling stage of Algorithm 8.8.

8.4.3 Interpretation and Label Switching*

Similar to the HMM's likelihoods we saw in Section 8.3.5, the posteriors (both marginal or completed) of the Bayesian HMM in Eqs. (8.31) to (8.36) are invariant to label permutations. In this case, the invariance can be seen in the factorization

$$p(\rho, \mathbf{\Pi}, \phi | w_{1:N}) \propto p(w_{1:N} | \rho, \mathbf{\Pi}, \phi) \times \text{Dirichlet}_{\sigma_{1:M}}(\rho; \eta\zeta) \\ \times \prod_{m=1}^{M} \text{Dirichlet}_{\sigma_{1:M}}\left(\pi_{\sigma_m}; \alpha_{\sigma_m}\beta_{\sigma_m}\right) \prod_{m=1}^{M} H(\phi_{\sigma_m})$$

and arises from the invariance of the likelihood as well as of the priors of Section 8.4.1 with respect to label permutation.

Note 8.10 Breaking the posterior's invariance

Unlike the frequentist HMM, in a Bayesian HMM we may avoid the posterior's invariance if we assign *label-specific* priors on the parameters. For instance, an alternative Bayesian HMM may be constructed with the following prior choices:

$$\rho \sim \text{Dirichlet}_{\sigma_{1:M}}(\eta(\zeta_1, \ldots, \zeta_M)),$$
$$\pi_{\sigma_m} \sim \text{Dirichlet}_{\sigma_{1:M}}\left(\alpha_m(\beta_{m\to 1}, \ldots, \beta_{m\to M})\right),$$
$$\phi_{\sigma_m} \sim \mathbb{H}_m.$$

In this version, priors are label specific and, as a result, relabeling of the state-space leads to different posterior values to each one of the $M!$ samples produced by every label permutation. This way, we need not invoke post hoc heuristics to resolve identifiability problems.

However, it is better if label-specific priors are avoided. This is because priors, informed by state labels, may hinder the mixing of the MCMC samplers applied. For the best computational efficiency, it is preferable to use priors that are state, but not label, specific.

The posterior's invariance to label permutations leads to multimodal posteriors. For example, for any MAP estimate $\rho^*, \mathbf{\Pi}^*, \phi^*$, there are $M!$ total maximizers produced by the label permutations. Each one of these $M!$ maximizers, under vague priors, is a local mode of the posterior and is associated with a unique labeling. As Eqs. (8.31) to (8.36) do not exhibit preference for a particular labeling of the state-space, in general, MCMC samplers produce samples that may use any of them. In fact, a sampler that performs well samples the entire posterior and thus, in the long run, switches between state labels producing samples from all $M!$ posterior modes.

As long as we are interested in deriving estimates that depend only on the constitutive states and not on the particular labeling chosen, the MCMC chains generated are sufficient. For example, if all we care about is

* This is an advanced topic and could be skipped on a first reading.

quantifying the emission parameters attained at a particular level, we may focus on $p(\phi_{s_n}|w_{1:N})$ alone, which, in itself, has only one permutation.

To derive label-specific estimates, and therefore to allow for full interpretation of our estimates, we can impose post hoc identifiability constraints in terms of the state labels similar to the frequentist HMM of Section 8.3.5. For instance, because all $M!$ modes are equally probable, for each MCMC sample computed we can consider all other $M!-1$ permutations by forming every possible permutation and selecting the one that satisfies our constraints. Below we explain the steps involved in more detail.

Suppose that an MCMC sampler has already been employed and that, for clarity, we denote by $\boldsymbol{\theta}_k^{(j)} = \left\{\boldsymbol{s}_k^{(j)}, \boldsymbol{\rho}_k^{(j)}, \boldsymbol{\Pi}_k^{(j)}, \boldsymbol{\phi}_k^{(j)}\right\}$ and $k = 1$ the values sampled at the jth iteration. As we have $M!$ total permutations, we have $K = M!$ total posterior samples, which we index with $k = 1 : K$. We use $k > 1$ to denote every other sample value that can be formed by $\boldsymbol{\theta}_1^{(j)}$ through permutations of the state labels. As we mentioned above, due to label invariance all such samples are equiprobable,

$$p\left(\boldsymbol{\theta}_1^{(j)} | w_{1:N}\right) = p\left(\boldsymbol{\theta}_k^{(j)} | w_{1:N}\right).$$

To restore identifiability, it is sufficient to select a single $\boldsymbol{\theta}_k^{(j)}$ out of $\boldsymbol{\theta}_{1:K}^{(j)}$ satisfying our constraints. Since the permutation satisfying the constraints generally may differ from iteration to iteration, we designate it using $k^{(j)}$.

Perhaps the simplest way to impose identifiability relies on an ordering of the emission parameters, if one exists. For instance, provided ϕ_{σ_m} are real scalars, a labeling of the state-space $\sigma'_{1:M}$ may be selected such that it leads to a unique arrangement $\phi_{\sigma'_1} < \cdots < \phi_{\sigma'_M}$ (such as increasing mean signal level). In this simple case, $k^{(j)}$ may be easily identified and $\boldsymbol{\theta}_{k^{(j)}}^{(j)}$ readily found. This strategy, of course, is problem specific and very sensitive to the parameterization of the mother distribution $\mathbb{G}_\phi$ as well as to the imposed arrangement of the emission parameters ϕ. Further, it is unable to handle multivariate emission parameters or parameters that cannot be arranged in a sensible way. Below, we describe an alternative strategy with higher computational cost, which although heuristic in nature, is less reliant on parametrizations.

For this strategy, we first need to select a reference point $\hat{\boldsymbol{\theta}}$ against which we can compare $\boldsymbol{\theta}_k^{(j)}$. Subsequently, for each j, from $\boldsymbol{\theta}_{1:K}^{(j)}$ we select $\boldsymbol{\theta}_{k^{(j)}}^{(j)}$ that yields the best match. The reference $\hat{\boldsymbol{\theta}}$ can be either an ad hoc chosen point in the space of $\boldsymbol{s}, \boldsymbol{\rho}, \boldsymbol{\Pi}, \boldsymbol{\phi}$ or the MCMC sample with the highest posterior value. The latter can be readily found post hoc among the computed MCMC values $\boldsymbol{\theta}_1^{(j)}$.

Once an appropriate reference $\hat{\boldsymbol{\theta}}$ is selected, the comparison can be based on a dissimilarity function $\mathcal{D}(\boldsymbol{\theta}, \boldsymbol{\theta}')$ that we also need to choose. For example, if $\mathcal{D}(\boldsymbol{\theta}, \boldsymbol{\theta}')$ is based on the Euclidean distance, then selection from $\boldsymbol{\theta}_{1:K}^{(j)}$ results in finding the k belonging to the same semi-orthant with $\hat{\boldsymbol{\theta}}$. Of course, such a k is unique.

Note 8.11 Dissimilarity function

A *dissimilarity function* $\mathcal{D}(\boldsymbol{\theta}, \boldsymbol{\theta}')$ associated to every pair $\boldsymbol{\theta}$ and $\boldsymbol{\theta}'$ yields a positive real scalar quantifying the dissimilarity between $\boldsymbol{\theta}$ and $\boldsymbol{\theta}'$. For example, for two identical samples $\boldsymbol{\theta} = \boldsymbol{\theta}'$, the dissimilarity must be zero; while, for different samples $\boldsymbol{\theta} \neq \boldsymbol{\theta}'$ the dissimilarity must be strictly positive. Solely for restoring identifiability, $\mathcal{D}(\boldsymbol{\theta}, \boldsymbol{\theta}')$ need not be symmetric. For instance, $\mathcal{D}(\boldsymbol{\theta}, \boldsymbol{\theta}')$ and $\mathcal{D}(\boldsymbol{\theta}', \boldsymbol{\theta})$ could attain different values.

A computationally convenient family of $\mathcal{D}(\boldsymbol{\theta}, \boldsymbol{\theta}')$ is offered by those additive over the dissimilarities of the individual state labels,

$$\mathcal{D}(\boldsymbol{\theta}, \boldsymbol{\theta}') = \sum_{m=1}^{M} \mathcal{E}_m(\boldsymbol{\theta}, \boldsymbol{\theta}'),$$

where $\mathcal{E}_m(\boldsymbol{\theta}, \boldsymbol{\theta}')$ is a dissimilarity function that compares *only* σ_m of $\boldsymbol{\theta}$ with σ'_m of $\boldsymbol{\theta}'$. In this case, finding the best $\boldsymbol{\theta}_k^{(j)}$ out of $\boldsymbol{\theta}_{1:K}^{(j)}$ reduces to a linear assignment problem, namely to finding the best association between the labeling $\sigma_{1:M}$ employed in $\boldsymbol{\theta}$ and the labeling $\sigma'_{1:M}$ employed in $\boldsymbol{\theta}'$. As such, it can be solved efficiently through the *Hungarian algorithm* without explicitly forming each one of the K samples $\boldsymbol{\theta}_{1:K}^{(j)}$.

8.5 Dynamical Variants of the Bayesian HMM*

As we mentioned earlier, the Bayesian HMM affords a level of flexibility otherwise unavailable within the frequentist paradigm. For example, we may consider hierarchical formulations with hyperpriors on $\boldsymbol{\beta}_{\sigma_m}$ and, as we see in the next section, develop a HMM whose state-space $\sigma_{1:M}$ may grow arbitrarily in size. In doing so, such a formulation avoids the pitfalls of having to specify a particular size M to begin with, which is often a serious limitation when studying uncharacterized dynamical systems.

Before we turn to the study of uncharacterized systems, however, we focus on systems for which M is assumed known. For several such systems, properly tuning the prior on $\boldsymbol{\rho}$ and $\boldsymbol{\Pi}$ is sufficient in introducing flexibility in modeling dynamics. While the scenarios we may consider are endless, we restrict ourselves to only a few key cases.

8.5.1 Modeling Time Scales

Earlier, in Section 8.4.1, we spoke of priors on transition probabilities. As we now show, these priors directly impact the induced prior on the escape time which, for some applications, constitutes a more natural quantity with which to work. Here, we discuss how Bayesian methods provide us with the ability to directly place priors on escape times and thus model time scales.

* This is an advanced topic and could be skipped on a first reading.

In particular, on account of the Markov assumption built into the dynamics of the state sequence $s_{1:N}$, once a system modeled by an HMM visits a constitutive state σ_m, it remains for a random number of *additional* steps D_{σ_m} before escaping and selecting another constitutive state. Specifically, in Section 2.4.3 we derived the distribution

$$D_{\sigma_m}|\pi_{\sigma_m\to\sigma_m} \sim \text{Geometric}(1-\pi_{\sigma_m\to\sigma_m}),$$

which exclusively depends upon the self-transition probability $\pi_{\sigma_m\to\sigma_m}$. Under the prior of Section 8.4.1, we immediately obtain $\pi_{\sigma_m\to\sigma_m} \sim \text{Beta}\left(\alpha_{\sigma_m}\beta_{\sigma_m\to\sigma_m}, \alpha_{\sigma_m}(1-\beta_{\sigma_m\to\sigma_m})\right)$, which we may use to derive the induced prior on D_{σ_m}, namely,

$$D_{\sigma_m} \sim \text{BetaNegBinomial}\left(1, \alpha_{\sigma_m}\beta_{\sigma_m\to\sigma_m}, \alpha_{\sigma_m}(1-\beta_{\sigma_m\to\sigma_m})\right).$$

This illustrates how the priors applied on an HMM's transition probabilities, in essence, also act as priors on induced time scales. For instance, since the mean of D_{σ_m} is

$$\langle D_{\sigma_m}\rangle = \frac{\alpha_{\sigma_m}\beta_{\sigma_m\to\sigma_m}}{\alpha_{\sigma_m}(1-\beta_{\sigma_m\to\sigma_m})-1},$$

we can tune the hyperparameters α_{σ_m} and $\beta_{\sigma_m\to\sigma_m}$ to influence priors on dwell periods selecting an a priori desired duration. For example, if a duration $\langle D_{\sigma_m}\rangle$ is specified, setting

$$\alpha_{\sigma_m} = \frac{\langle D_{\sigma_m}\rangle}{(1-\beta_{\sigma_m\to\sigma_m})\langle D_{\sigma_m}\rangle - \beta_{\sigma_m\to\sigma_m}}$$

provides a recipe for adjusting the values of α_{σ_m} that allows for state specific time scales.

Note 8.12 The sticky HMM

One way of influencing the *same* timescale across all constitutive states in a Bayesian HMM proceeds via setting every $\alpha_{\sigma_m} = \alpha$ equal and reparametrizating $\boldsymbol{\beta}_{\sigma_m}$ as

$$\boldsymbol{\beta}_{\sigma_m} = (1-c)\boldsymbol{B} + c\boldsymbol{D}_{\sigma_m},$$

where c is a scalar selected between 0 and 1; $\boldsymbol{B} = [B_{\sigma_1}, \ldots, B_{\sigma_M}]$ is a probability array; and $\boldsymbol{D}_{\sigma_m} = [D_{\sigma_m\to\sigma_1}, \ldots, D_{\sigma_m\to\sigma_M}]$ is a probability array specific to each constitutive state σ_m. The latter can be used to separate self-transitions by setting $D_{\sigma_m\to\sigma_m} = 1$ and $D_{\sigma_m\to\sigma_{m'}} = 0$. The resulting prior on $\boldsymbol{\Pi}$ now takes the form

$$\pi_{\sigma_m} \sim \text{Dirichlet}_{\sigma_{1:M}}\left(\alpha(1-c)\boldsymbol{B} + \alpha c\boldsymbol{D}_{\sigma_m}\right).$$

With this prior, self-transitions over the entire state-space are reinforced with only a limited number of hyperparameters $\alpha, c, \boldsymbol{B}$. Due to its ability to reinforce self-transitions and long dwells, this prior is termed *sticky*. Under the sticky prior, the induced dwell durations are

$$\langle D_{\sigma_m} \rangle = \frac{c + (1-c)B_{\sigma_m}}{(1-c)\,(1-B_{\sigma_m}) - \frac{1}{\alpha}},$$

which become uniform over $\sigma_{1:M}$ by setting $B_{\sigma_m} = 1/M$.

8.5.2 Modeling Equilibrium

Provided every $\beta_{\sigma_m \to \sigma_{m'}}$ is nonzero, the prior on $\mathbf{\Pi}$ ensures that the transition probabilities $\pi_{\sigma_m \to \sigma_{m'}}$ in a Bayesian HMM are strictly positive. This, in turn, ensures that transitions between any pair of constitutive states are possible in all resulting $s_{1:N}$. Therefore, a system modeled by such an HMM is ergodic, *i.e.*, may explore the entire state-space. Such systems, if allowed to evolve for sufficiently long time, may reach equilibrium.

For a dynamical system at equilibrium, initialization and kinetics are interrelated. In particular, ρ_{σ_m} and $\pi_{\sigma_m \to \sigma_{m'}}$ satisfy the balance condition

$$\rho_{\sigma_m} = \sum_{m'=1}^{M} \rho_{\sigma_{m'}} \pi_{\sigma_{m'} \to \sigma_m}.$$

For a dynamical system at equilibrium, we can use this condition to express ρ in terms of $\mathbf{\Pi}$, suggesting that at equilibrium ρ is a dependent parameter. Accordingly, to model a system at equilibrium, we need to place priors only on $\mathbf{\Pi}$. For example,

$$\begin{aligned}
\pi_{\sigma_m} &\sim \text{Dirichlet}_{\sigma_{1:M}}\left(\alpha_{\sigma_m}\boldsymbol{\beta}_{\sigma_m}\right), \\
\phi_{\sigma_m} &\sim \mathbb{H}, \\
s_1|\mathbf{\Pi} &\sim \text{Categorical}_{\sigma_{1:M}}\left(\boldsymbol{\rho}_{\mathbf{\Pi}}\right), \\
s_n|s_{n-1}, \mathbf{\Pi} &\sim \text{Categorical}_{\sigma_{1:M}}\left(\boldsymbol{\pi}_{\sigma_m}\right), && n = 2:N, \\
w_n|s_n, \boldsymbol{\phi} &\sim \mathbb{G}_{\phi_{s_n}}, && n = 1:N,
\end{aligned}$$

in which the initial probability array $\boldsymbol{\rho}_{\mathbf{\Pi}}$ is now dictated by the balance condition.

Although the prior on $\mathbf{\Pi}$ is the same as seen before, due to its implicit effect on $\boldsymbol{\rho}_{\mathbf{\Pi}}$ it is no longer conjugate to $s_{1:N}|\mathbf{\Pi}$. Consequently, we cannot use the Gibbs sampler of Algorithm 8.7 to obtain MCMC samples $\pi_{\sigma_m}|s_{1:N}$ and inference is only possible by means of a Metropolis–Hastings sampler such as an appropriately adjusted Algorithm 8.9.

8.5.3 Modeling Reversible Systems

The prior on $\mathbf{\Pi}$ we just discussed enforces equilibrium on the HMM; this is somewhat stronger than simply ensuring reversibility of the kinetics irrespective of equilibrium being reached by the time of the first measurement. To model a reversible dynamical system, which may not necessarily have reached equilibrium before the measurement's onset, we need to consider independent priors on ρ and $\mathbf{\Pi}$. In such cases, $\rho \sim \text{Dirichlet}_{\sigma_{1:M}}(\eta\zeta)$ remains an appropriate choice; however, ensuring reversible kinetics requires fundamentally different choices for $\mathbf{\Pi}$.

Note 8.13 A reversible HMM

One way to ensure a reversible $\mathbf{\Pi}$ is to reparametrize the transition probabilities as

$$\pi_{\sigma_m \to \sigma_{m'}} = \frac{\lambda_{\sigma_m \leftrightarrow \sigma_{m'}}}{\sum_{m''=1}^{M} \lambda_{\sigma_m \leftrightarrow \sigma_{m''}}}.$$

Reversibility is ensured by requiring that the new parameters be pairwise symmetric,

$$\lambda_{\sigma_m \leftrightarrow \sigma_{m'}} = \lambda_{\sigma_{m'} \leftrightarrow \sigma_m}.$$

On account of symmetry, in the new parametrization we need only $M(M+1)/2$ priors that we may select independently. For instance,

$$\lambda_{\sigma_m \leftrightarrow \sigma_{m'}} \sim \text{Gamma}\left(f E_{\sigma_m} E_{\sigma_{m'}}, 1\right),$$

where f and $E_{\sigma_1}, \ldots, E_{\sigma_M}$ are hyperparameters controlling how tightly each constitutive state couples to the others.

The kinetic scheme induced by the symmetric prior of $\lambda_{\sigma_m \leftrightarrow \sigma_{m'}}$ leads to the reparametrized equilibrium distribution

$$\rho_* = \left[\frac{\sum_{m=1}^{M} \lambda_{\sigma_1 \leftrightarrow \sigma_m}}{\sum_{m=1}^{M} \sum_{m'=1}^{M} \lambda_{\sigma_{m'} \leftrightarrow \sigma_m}}, \ldots, \frac{\sum_{m=1}^{M} \lambda_{\sigma_M \leftrightarrow \sigma_m}}{\sum_{m=1}^{M} \sum_{m'=1}^{M} \lambda_{\sigma_{m'} \leftrightarrow \sigma_m}} \right].$$

As a result, whenever equilibrium needs to be imposed as a stronger condition to reversibility, we may proceed by setting the initial probabilities ρ equal to ρ_*.

8.5.4 Modeling Kinetic Schemes

Unlike the ergodic HMM where the system may evolve to and from any constitutive state, some physical scenarios require that some transitions be prohibited. For example, modeling irreversible chemical reactions such as *photo-bleaching* where molecules undergo a chemical change rendering them unable to fluoresce at a designated wavelength.

From the modeling perspective, we can take advantage of the flexibility allowed by the hyperparameters $\beta_{\sigma_m \to \sigma_{m'}}$ to model kinetic schemes. Under the prior of Section 8.4.1, a transition probability $\pi_{\sigma_m \to \sigma_{m'}}$ is zero only when the corresponding $\beta_{\sigma_m \to \sigma_{m'}}$ is zero. Essentially, to ensure that the system modeled cannot undergo some transitions, or undergoes other transitions in a certain order, we need to properly set the sparsity pattern of

$$\begin{array}{c} \\ \sigma_1 \\ \sigma_2 \\ \vdots \\ \sigma_M \end{array} \begin{array}{c} \begin{array}{cccc} \sigma_1 & \sigma_2 & \cdots & \sigma_M \end{array} \\ \begin{bmatrix} \beta_{\sigma_1 \to \sigma_1} & \beta_{\sigma_1 \to \sigma_2} & \cdots & \beta_{\sigma_1 \to \sigma_M} \\ \beta_{\sigma_2 \to \sigma_1} & \beta_{\sigma_2 \to \sigma_2} & \cdots & \beta_{\sigma_2 \to \sigma_M} \\ \vdots & \vdots & \ddots & \vdots \\ \beta_{\sigma_M \to \sigma_1} & \beta_{\sigma_M \to \sigma_2} & \cdots & \beta_{\sigma_M \to \sigma_M} \end{bmatrix} \end{array} = \begin{bmatrix} \boldsymbol{\beta}_{\sigma_1} \\ \boldsymbol{\beta}_{\sigma_2} \\ \vdots \\ \boldsymbol{\beta}_{\sigma_M} \end{bmatrix}.$$

Example 8.4 A left-to-right HMM

To model a system, such as an idealized molecular motor with no reverse stepping, where returning to previous constitutive states is *prohibited*, we may use a left-to-right structure of the form

$$\begin{array}{c} \\ \sigma_1 \\ \sigma_2 \\ \sigma_3 \\ \sigma_4 \\ \sigma_5 \end{array} \begin{array}{c} \begin{array}{ccccc} \sigma_1 & \sigma_2 & \sigma_3 & \sigma_4 & \sigma_5 \end{array} \\ \begin{bmatrix} 1/5 & 1/5 & 1/5 & 1/5 & 1/5 \\ 0 & 1/4 & 1/4 & 1/4 & 1/4 \\ 0 & 0 & 1/3 & 1/3 & 1/3 \\ 0 & 0 & 0 & 1/2 & 1/2 \\ 0 & 0 & 0 & 0 & 1 \end{bmatrix} \end{array} = \begin{bmatrix} \boldsymbol{\beta}_{\sigma_1} \\ \boldsymbol{\beta}_{\sigma_2} \\ \boldsymbol{\beta}_{\sigma_3} \\ \boldsymbol{\beta}_{\sigma_4} \\ \boldsymbol{\beta}_{\sigma_5} \end{bmatrix},$$

where, for simplicity, we have chosen a state-space of size $M = 5$. Observed for a sufficiently long period, $N \gg 1$, a system modeled as such eventually reaches σ_5. Since the prior imposed on π_{σ_5} is deterministic, allowing only for $\pi_{\sigma_5} = [0, 0, 0, 0, 1]$, this model is then equivalent to modeling absorbing dynamics at the boundary.

8.5.5 Modeling Factorial Dynamics

Occasionally, the underlying system of interest consists of multiple components that evolve *independently*. Although in such systems each component follows its own dynamics, it may be possible that the entire system is assessed only through a common observation. That is, all components may give rise to a single collective observation.

Example 8.5 Photo-blinking

Imagine that we observe a specimen consisting of K fluorescent probes undergoing photo-blinking. That is, each probe switches between a bright and a dark state, which we may model by σ_1 and σ_2, respectively. Under idealized conditions, *i.e.*, when probes are far away from each other, we assume that each probe switches between σ_1 and σ_2 independently. We can readily model such a scenario with

$$s_n^k | s_{n-1}^k, \boldsymbol{\Pi} \sim \text{Categorical}_{\sigma_{1:2}}(\boldsymbol{\pi}_{s_n^k}).$$

Here, s_n^k is the state, termed *photo-state*, of the kth probe at the time of the nth assessment and $\boldsymbol{\Pi}$ gathers the transition probabilities $\boldsymbol{\pi}_{\sigma_1}$ and $\boldsymbol{\pi}_{\sigma_2}$.

When a probe is bright, it emits photons with a rate $\mu_{\sigma_1} > 0$. However, when the probe is dark, it emits no photons, which we model with a rate $\mu_{\sigma_2} = 0$. Since photon emissions from all probes are additive, in total our specimen emits photons with a rate that combines contributions from all

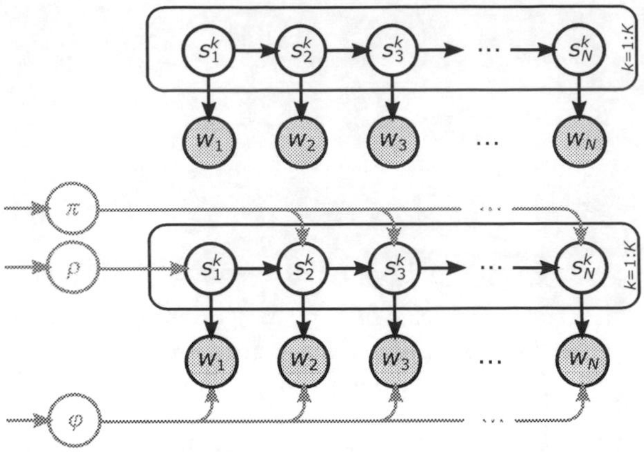

Fig. 8.3 Graphical representation of a factorial HMM. (Top) The basic structure; (Bottom) Also indicates dependencies on the parameters.

probes. As such, the photon emission rate driving the nth assessment is $\sum_{k=1}^{K} \mu_{s_n^k}$. Considering a detector with exposure time $\tau_{\exp}$, this leads to an emission distribution

$$w_n | s_n^{1:K} \sim \text{Poisson}\left(\tau_{\exp} \sum_{k=1}^{K} \mu_{s_n^k}\right),$$

where w_n denotes the net amount of photon detections at the nth time level.

Naturally, such a system may be formulated by a generalization of the HMM as follows:

$$\begin{aligned} s_1^k | \boldsymbol{\rho} &\sim \text{Categorical}_{\sigma_{1:M}}(\boldsymbol{\rho}), & & k = 1:K, \\ s_n^k | s_{n-1}^k, \boldsymbol{\Pi} &\sim \text{Categorical}_{\sigma_{1:M}}(\boldsymbol{\pi}_{s_n^k}), & n = 2:N, \quad & k = 1:K, \\ w_n | s_n^{1:K}, \boldsymbol{\phi} &\sim \mathbb{G}_{\sum_{k=1}^{K} \phi_{s_n^k}}, & n = 1:N. \end{aligned}$$

In this model, each component is modeled by its own state s_n^k. However, at each time level observations are coupled by a common emission distribution $\mathbb{G}_{\sum_{k=1}^{K} \phi_{s_n^k}}$. This formulation is termed *factorial hidden Markov model* and it is depicted graphically in Fig. 8.3. Inference in this system relies on the posterior $p\left(s_{1:N}^{1:K}, \boldsymbol{\rho}, \boldsymbol{\Pi}, \boldsymbol{\phi} | w_{1:N}\right)$ and follows the same filtering and smoothing algorithms seen earlier.

8.6 The Infinite Hidden Markov Model*

In the previous section, we saw how a Bayesian HMM constructed around a fixed state-space $\sigma_{1:M}$ can identify the characteristics of each constitutive state σ_m, for example dynamical and observational parameters represented by $\boldsymbol{\rho}, \boldsymbol{\Pi}$, and $\boldsymbol{\phi}$, respectively. These characteristics are captured in the

* This is an advanced topic and could be skipped on a first reading.

posterior $p(\rho, \mathbf{\Pi}, \boldsymbol{\phi}|w_{1:N})$ that, for the models presented so far, inevitably depends upon the size, M, of the state-space employed.

In practice, we often need to study dynamical systems whose state-space is uncharacterized. In this case, our knowledge of the system at hand does not allow us to specify a unique M. Indeed, despite the generality and elegance of our formulations, the dependence of our posteriors upon M is a limiting factor. Luckily, extensions of the Bayesian formulation are possible, resulting in posterior distributions independent of M, which may remain unspecified or arbitrarily large.

In particular, by building upon the Bayesian HMM and using appropriate hyperpriors, described shortly, we may develop an HMM version whose state-space is infinite. Such a formulation remains valid and may be applied even when our primary goal is to identify the characteristics of the constitutive states visited by the system while the total number of available states is unknown.

Note 8.14 Dynamics on infinite state-spaces

With an infinite state-space, our system has access to infinite constitutive states. Specifically, each time the system departs from a passing state s_n it may escape to infinitely many σ_m. Provided that the system has already visited only a finite number of them, this means that at every transition the systems can always explore new states that will be visited for the first time. In principle, such a system may be allowed to visit an unvisited state every time it transitions. Although such scenarios may arise, for example the birth processes of Example 2.4, most often we are interested in studying systems that frequently or sporadically *revisit* states. For the latter systems, the number of constitutive states visited during the time course of our measurements, which is finite, is drastically lower than the total number of observations.

As we mentioned, the posterior $p(\rho, \mathbf{\Pi}, \boldsymbol{\phi}|w_{1:N})$ of the model in Eqs. (8.31) to (8.36) depends upon M. Such dependency signifies that with a different number of constitutive states available, different choices of kinetic $\rho, \mathbf{\Pi}$, and emission $\boldsymbol{\phi}$ parameters are assigned under the measurements $w_{1:N}$. To eliminate such dependence on M, we need to be able to reinforce state revisiting in an infinite state model. This can be achieved by properly selecting the priors on the initial and transition probabilities, ρ and $\mathbf{\Pi}$.

One way to do this is to consider placing a common prior among all constitutive states. In this case, setting η and all α_{σ_m} equal. For simplicity, we denote the latter by α. Also, we may set all elements of the arrays ζ and $\boldsymbol{\beta}_{\sigma_m}$ equal and denote them by $\boldsymbol{\beta} = [\beta_{\sigma_1}, \ldots, \beta_{\sigma_M}]$. Under this common prior, constitutive states with high β_{σ_m} generally receive more transitions into them than constitutive states with low β_{σ_m}.

Of course, for an uncharacterized system, we cannot identify beforehand how often the constitutive states are visited or even which of them are visited more often than others. Thus, in principle, the prior $\boldsymbol{\beta}$ is unknown too and we need to estimate it in parallel with other quantities of interest. For this reason, we place a hyperprior on $\boldsymbol{\beta}$ and, as $\boldsymbol{\beta}$ is a probability array, the most natural choice for it is also a Dirichlet distribution. This leads to the hierarchical Dirichlet formulation

$$\boldsymbol{\beta} \sim \text{Dirichlet}_{\sigma_{1:M}}(\gamma\boldsymbol{\xi}), \tag{8.40}$$

$$\boldsymbol{\rho}|\boldsymbol{\beta} \sim \text{Dirichlet}_{\sigma_{1:M}}(\alpha\boldsymbol{\beta}), \tag{8.41}$$

$$\boldsymbol{\pi}_{\sigma_m}|\boldsymbol{\beta} \sim \text{Dirichlet}_{\sigma_{1:M}}(\alpha\boldsymbol{\beta}), \tag{8.42}$$

where γ is a positive scalar and $\boldsymbol{\xi}$ a probability array. As our system is uncharacterized, at this stage, as we cannot distinguish between the constitutive states, we need to ensure symmetry of $\boldsymbol{\beta}$, which we may achieve through

$$\boldsymbol{\xi} = \left[\frac{1}{M}, \dots, \frac{1}{M}\right].$$

As anticipated, the hierarchical prior of Eqs. (8.40) and (8.41), when combined with the HMM's kinetics and emissions,

$$\begin{aligned}
\phi_{\sigma_m} &\sim \mathbb{H}, \\
s_1|\boldsymbol{\rho} &\sim \text{Categorical}_{\sigma_{1:M}}(\boldsymbol{\rho}), \\
s_n|s_{n-1}, \boldsymbol{\Pi} &\sim \text{Categorical}_{\sigma_{1:M}}(\boldsymbol{\pi}_{\sigma_m}), && n = 2:N, \\
w_n|s_n, \boldsymbol{\phi} &\sim \mathbb{G}_{\phi_{s_n}}, && n = 1:N,
\end{aligned}$$

results in a posterior $p(\boldsymbol{\rho}, \boldsymbol{\Pi}, \boldsymbol{\phi}|w_{1:N})$ that converges in the limit $M \to \infty$. Consequently, so long as M is sufficiently large, the HMM above provides estimates independent of the particular M values chosen.

Computational inference on this model can be based on appropriate modifications of the Gibbs or Metropolis–Hastings samplers of Algorithms 8.7 and 8.9. The modifications for the latter are straightforward and, for this reason, here we focus only on a presentation of the Gibbs sampler that targets the completed posterior $p(s_{1:N}, \boldsymbol{\beta}, \boldsymbol{\rho}, \boldsymbol{\Pi}, \boldsymbol{\phi}|w_{1:N})$. For this target, only an additional step to update $\boldsymbol{\beta}$ is required of Algorithm 8.7. This update needs to sample $\boldsymbol{\beta}$ from its full conditional $p(\boldsymbol{\beta}|s_{1:N}, \boldsymbol{\rho}, \boldsymbol{\Pi}, \boldsymbol{\phi}, w_{1:N})$, which reduces to $p(\boldsymbol{\beta}|\boldsymbol{\rho}, \boldsymbol{\Pi})$. However, since Eq. (8.40) is not conjugate with Eqs. (8.41) and (8.42), a Metropolis–Hastings step is necessary.

Note 8.15 iHMM

The description and the associated computational schemes we presented in this section rely on a finite approximation of the *infinite hidden Markov model* (iHMM). Formally, the latter is the model achieved in the limiting case $M = \infty$ and entails a truly infinite state-space $\sigma_{1:\infty}$. In this limit, a detailed description of the corresponding generative model involves the Dirichlet and hierarchical Dirichlet processes. It is also possible to carry out our computational inference on the exact iHMM instead of relying on finite approximations. For example, it is possible to carry out MCMC sampling involving an infinite state-space by completing the posterior,

$$p\left(s_{1:N}, \boldsymbol{\beta}, \boldsymbol{\rho}, \boldsymbol{\Pi} | w_{1:N}\right) = \int du_{1:N}\, p\left(u_{1:N}, s_{1:N}, \boldsymbol{\beta}, \boldsymbol{\rho}, \boldsymbol{\Pi} | w_{1:N}\right),$$

with auxiliary slice variables $u_{1:N}$ as we developed in Example 5.14. The resulting sampler gives rise to *beam sampling* schemes.

8.7 A Case Study in Fluorescence Spectroscopy*

Favoring simplicity, so far we have focused on problems where observations depend directly on the underlying hidden states or, as we might call them, on first-order HMM. To help illustrate why the methods presented here are more general than first appears, we describe a case study involving dynamics in continuous time that necessarily leads to a second-order HMM as observations occur precisely at jump times. In this case study, we introduce an auxiliary variable method, inspired by Section 5.4.3, in order reduce the second-order HMM to a first-order HMM for which the algorithms provided in this chapter hold. We also demonstrate how to discretize time in order to incorporate continuous time observations. This treatment here is necessary for observations occurring at jump times. More complex models with continuous dynamics and observations at arbitrary times are dealt with in Chapter 10.

8.7.1 Time Resolved Spectroscopy

We start by considering an important class of experiments that does not probe the state of the dynamical system of interest but rather jumps in the system's trajectory. For instance, *time-resolved* spectroscopic experiments collect individual photons and report on their detection time. Since the detected photons stem from the probed physical system, they are emitted precisely when the system (an atom or, more typically, molecule) jumps across energy levels. Since the time a photon needs to reach the detector in such experiments is insignificant, the recorded photon detection times report upon transitions between, rather than instantaneous, states of the system.

In this case study, we consider a fluorescent molecule, *i.e.*, a *fluorophore*, with three energy states labeled G, S, and T. Respectively, these are: the fluorophore's *ground state* (state with the lowest energy); the first excited *singlet state* (state with the highest energy); and the first excited *triplet state* (state with intermediate energy). These are typically depicted schematically, in increasing energy order, using a *Jablonski diagram* as in Fig. 8.4.

* This is an advanced topic and could be skipped on a first reading.

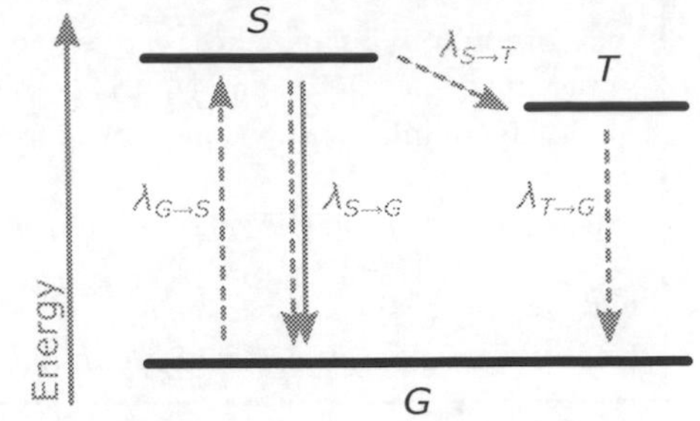

Fig. 8.4 Jablonski diagram of a fluorophore possessing three energy states: G, S, and T. Arrows indicate Markovian transitions at the rates shown. Solid and dashed arrows distinguish between detectable and nondetectable transitions, respectively.

During an experiment, while residing in G a fluorophore absorbs energy at a random time and undergoes a transition $G \rightarrow S$. Subsequently, after residing for a short period in S, the fluorophore undergoes either an $S \rightarrow G$ or an $S \rightarrow T$ transition. If in T, the fluorophore may only undergo a $T \rightarrow G$ transition. All such transitions are denoted by arrows in Fig. 8.4. Terminating at G, the fluorophore is re-excited and the same cycle repeats until the conclusion of the experiment. Physical chemistry often models dwells in each one of the three states as memoryless. This leads to a kinetic scheme fully determined by the transition rates $\lambda_{G\rightarrow S}, \lambda_{S\rightarrow G}, \lambda_{S\rightarrow T}, \lambda_{T\rightarrow G}$ also shown in Fig. 8.4.

Of interest is often the mean dwell time in the excited state S, *i.e.*, the so-called fluorescence lifetime, which helps in characterizing the fluorophore. This is because lifetime is often unique to each molecule or alternative chemical forms of a molecule (assuming that, within error, lifetimes are sufficiently well separated so that they can be distinguished).

On the theoretical front, what makes this set-up challenging to analyze is the fact that photons are emitted and detected only whenever the fluorophore undergoes the transition $S \rightarrow G$; while, in a typical experiment, all other transitions are either *nonradiative* or emit photons not otherwise detected. The situation is even more complicated due to the fact that, even when the fluorophore undergoes $S \rightarrow G$ transitions, photons may not always be emitted or may not always be detected. Here, we formulate this system and show how the general framework for HMMs can be used to estimate transition rates and, eventually, through them, the fluorescence lifetime.

8.7.2 Discretization of Time

For clarity, we consider an experiment that starts at time $T_{\min}$ and concludes at time $T_{\max}$. Further, we use T_k, with indices $k = 1 : K$, to denote the reported photon detection times, which we arrange in ascending order, *i.e.*, $T_{k-1} < T_k$.

To operate within the HMM framework, we must first discretize time. For this, we break the experiment's time course into a total of N hypothetical windows separated by the time levels

$$t_n = T_{\min} + \frac{n}{N}(T_{\max} - T_{\min}), \qquad n = 0 : N.$$

These time levels define N windows that we successively index by $n = 1 : N$. Specifically, our nth window spans the time interval between t_{n-1} and t_n.

8.7.3 Formulation of the Dynamics

Following notation first introduced in Section 2.3, we denote $\mathcal{S}(t)$ to be the passing state at time t of our fluorophore. Due to memorylessness, the trajectory $\mathcal{S}(\cdot)$ is a Markov jump process with state-space G, S, and T, and its transition rate matrix is given by

$$\mathbf{\Lambda} = \begin{bmatrix} 0 & \lambda_{G\to S} & \lambda_{G\to T} \\ \lambda_{S\to G} & 0 & \lambda_{S\to T} \\ \lambda_{T\to G} & \lambda_{T\to S} & 0 \end{bmatrix}.$$

Our end goal is to estimate the unknown entries of $\mathbf{\Lambda}$. To do so, we do not need the full trajectory $\mathcal{S}(\cdot)$. Instead, we focus on the passing states only at the time levels t_n that are already sufficient to link $\mathbf{\Lambda}$ with our measurements. Accordingly, for each time level we consider the corresponding passing state,

$$s_n = \mathcal{S}(t_n), \qquad n = 0 : N.$$

As the underlying trajectory is a Markov jump process, we can easily deduce the transition rules of our dynamical model,

$$s_n | s_{n-1} \sim \text{Categorical}_{G,S,T}\left(\boldsymbol{\pi}_{s_{n-1}}\right), \qquad n = 1 : N.$$

According to Eq. (2.15), the transition probabilities stem from the rows of the propagator

$$\mathbf{\Pi} = \begin{bmatrix} \boldsymbol{\pi}_G \\ \boldsymbol{\pi}_S \\ \boldsymbol{\pi}_T \end{bmatrix} = \begin{bmatrix} \pi_{G\to G} & \pi_{G\to S} & \pi_{G\to T} \\ \pi_{S\to G} & \pi_{S\to S} & \pi_{S\to T} \\ \pi_{T\to G} & \pi_{T\to S} & \pi_{T\to T} \end{bmatrix} = \boldsymbol{Q}^{t_{n-1}\to t_n} = \exp\left(\frac{T_{\max} - T_{\min}}{N}\boldsymbol{G}\right) \tag{8.43}$$

that corresponds to the generator $\boldsymbol{G}$ of the rate matrix $\mathbf{\Lambda}$. As with every dynamical system seen so far, the kinetic model does not specify the initial conditions. Consequently, we need to model the initialization rule separately,

$$s_0 \sim \text{Categorical}_{G,S,T}(\boldsymbol{\rho}),$$

with appropriate initial probabilities $\boldsymbol{\rho} = [\rho_G, \rho_S, \rho_T]$ that may or may not be related to $\mathbf{\Lambda}$ depending upon the specifics of the experiment.

8.7.4 Formulation of the Measurements

The most convenient way to model the photon detection times is to consider a set of observation variables $w_{1:N}$, where each one of our windows is associated with its own w_n. We encode the photon detection times $T_{1:K}$ by setting $w_n = 1$ when at least one photon is detected and setting $w_n = 0$ when no photon is detected during our nth window.

> **Note 8.16** Observations
>
> If we use N_k to denote the window that encodes the kth photon detection time, T_k, we see that
>
> $$N_k = \left\lceil N \frac{T_k - T_{\min}}{T_{\max} - T_{\min}} \right\rceil, \qquad k = 1 : K,$$
>
> where $\lceil x \rceil$ is the ceiling function, *i.e.*, the smallest index that is larger than x.
>
> When we attempt to model our photon detections with a low N, our windows may be large and, misleadingly, some of them may absorb more than one photon detection. However, as N grows large, and our windows correspondingly shrink, the photon detections times $T_{1:K}$ are encoded in different, well-separated, windows. For *sufficiently* large N, our observation variables $w_{1:N}$ follow the pattern
>
> $$\underbrace{0,\ldots,0}_{\substack{\text{windows}\\1:N_1-1}}, \overbrace{1}^{T_1}, \underbrace{0,\ldots,0}_{\substack{\text{windows}\\N_1+1:N_2-1}}, \overbrace{1}^{T_2}, \underbrace{0,\ldots,0}_{\substack{\text{windows}\\N_2+1:N_3-1}}, \overbrace{1}^{T_3}, 0\ldots$$
>
> $$\ldots 0, \overbrace{1}^{T_K}, \underbrace{0,\ldots,0}_{\substack{\text{windows}\\N_K+1:N}}.$$
>
> On account of this pattern, our observation sequence $w_{1:N}$ contains *no successive* windows with $w_n = 1$. By contrast, it contains multiple successive windows with $w_n = 0$.

Under the variables $w_{1:N}$, it is straightforward to model our assessment rules by

$$w_n | s_{n-1}, s_n \sim \text{Bernoulli}\left(\beta_{s_{n-1} \to s_n}\right), \qquad n = 1 : N,$$

and, as we have nine possible pairs $s_{n-1} \to s_n$, we need to specify nine different Bernoulli weights. To a good approximation, these are given by

$$\begin{aligned}
&\beta_{G \to G} \approx 0, && \beta_{G \to S} \approx 0, && \beta_{G \to T} \approx 0,\\
&\beta_{S \to G} \approx \eta, && \beta_{S \to S} \approx 0, && \beta_{S \to T} \approx 0,\\
&\beta_{T \to G} \approx 0, && \beta_{T \to S} \approx 0, && \beta_{T \to T} \approx 0,
\end{aligned}$$

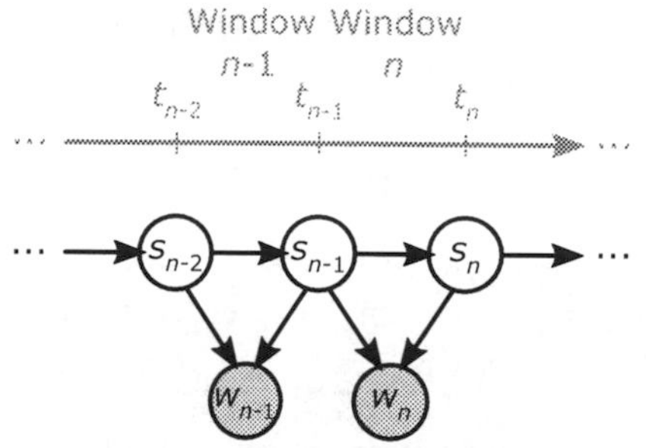

Fig. 8.5 An HMM for fluorescence spectroscopy representing time resolved measurements, $T_{1:K}$, by observation variables w_n that are linked to the passing states s_n of the underlying fluorophore. By contrast to the observation variables $w_{1:N}$ measured in an experiment, the passing states $s_{0:N}$ remain hidden.

where η is the fraction of detectable transitions $S \to G$ to total transitions $S \to G$. To be clear, all approximately zero terms above become strictly zero in the limit that N tends to infinity.

Our approximations on $\beta_{s_{n-1} \to s_n}$ improve and eventually become exact as $N \to \infty$, for which our hypothetical windows become so narrow that they accommodate no more than one transition. For this reason, our end goal is to devise a training method for our model supporting this limit. Put differently, our strategy is to derive a set of training equations on which we can formally reach the $N \to \infty$ limit.

8.7.5 Modeling Overview

In summary, the model of time-resolved fluorescence spectroscopy developed so far reads

$$
\begin{aligned}
s_0 &\sim \text{Categorical}_{G,S,T}(\rho), & \\
s_n | s_{n-1} &\sim \text{Categorical}_{G,S,T}\left(\pi_{s_{n-1}}\right), & n = 1:N, \\
w_n | s_{n-1}, s_n &\sim \text{Bernoulli}\left(\beta_{s_{n-1} \to s_n}\right), & n = 1:N,
\end{aligned}
$$

and is depicted graphically in Fig. 8.5. An immediate challenge that we face is that each observation variable $w_n | s_{n-1}, s_n$ depends on *two*, rather than one, hidden states. On account of this almost imperceptible difference, with dramatic theoretical consequences, few of the algorithms developed in Section 8.3 apply.

8.7.6 Reformulation

To continue, we must reformulate our model in such a way that it becomes similar to the HMM devised earlier. Namely, we need to transform it such that each observation is associated with *only one* hidden state.

One way to achieve a transformation is to consider a positive time period τ that is otherwise sufficiently small, $\tau < (T_{\max} - T_{\min})/N$. With the aid of τ, we introduce two auxiliary variables. That is, we consider two additional passing states per time level,

$$u_n = \mathcal{S}\left(t_{n-1} + \frac{\tau}{2}\right), \qquad v_n = \mathcal{S}\left(t_n - \frac{\tau}{2}\right), \qquad n = 1:N.$$

From these new states, u_n occurs near the very beginning and v_n occurs near the very end of their respective window. Due to memorylessness, we can exactly represent the dynamics of our system,

$$\begin{aligned} u_n|s_{n-1} &\sim \text{Categorical}_{G,S,T}\left(\psi'_{s_{n-1}}\right), & n &= 1:N, \\ v_n|u_n &\sim \text{Categorical}_{G,S,T}\left(\pi'_{u_n}\right), & n &= 1:N, \\ s_n|v_n &\sim \text{Categorical}_{G,S,T}\left(\psi'_{v_n}\right), & n &= 1:N. \end{aligned}$$

The new transition probabilities are obtained through the rows of the propagators

$$\begin{aligned} \mathbf{\Psi}' &= \begin{bmatrix} \psi'_G \\ \psi'_S \\ \psi'_T \end{bmatrix} = \begin{bmatrix} \psi'_{G\to G} & \psi'_{G\to S} & \psi'_{G\to T} \\ \psi'_{S\to G} & \psi'_{S\to S} & \psi'_{S\to T} \\ \psi'_{T\to G} & \psi'_{T\to S} & \psi'_{T\to T} \end{bmatrix} \\ &= \boldsymbol{Q}^{t_{n-1}\to t_{n-1}+\frac{\tau}{2}} = \boldsymbol{Q}^{t_n-\frac{\tau}{2}\to t_n} = \exp\left(\frac{\tau}{2}\boldsymbol{G}\right), \\ \mathbf{\Pi}' &= \begin{bmatrix} \pi'_G \\ \pi'_S \\ \pi'_T \end{bmatrix} = \begin{bmatrix} \pi'_{G\to G} & \pi'_{G\to S} & \pi'_{G\to T} \\ \pi'_{S\to G} & \pi'_{S\to S} & \pi'_{S\to T} \\ \pi'_{T\to G} & \pi'_{T\to S} & \pi'_{T\to T} \end{bmatrix} \\ &= \boldsymbol{Q}^{t_{n-1}+\frac{\tau}{2}\to t_n-\frac{\tau}{2}} = \exp\left(\left(\frac{T_{\max} - T_{\min}}{N} - \tau\right)\boldsymbol{G}\right). \end{aligned}$$

Taking advantage of the new states, and provided τ is sufficiently small, we can introduce another approximation to the observations,

$$\beta_{s_{n-1}\to s_n} \approx \beta_{u_n\to v_n}, \qquad n = 1:N.$$

This approximation becomes exact as $\tau \to 0^+$, at which u_n and v_n essentially merge with s_{n-1} and s_n, respectively. Of course, since $\tau < (T_{\max} - T_{\min})/N$, this limiting condition does not introduce further restrictions in our formulation since it is already fulfilled under $N \to \infty$.

Gathering everything together, our reformulated Markov model that leverages auxiliary variables reads

$$\begin{aligned} s_0 &\sim \text{Categorical}_{G,S,T}(\rho), \\ u_n|s_{n-1} &\sim \text{Categorical}_{G,S,T}\left(\psi'_{s_{n-1}}\right), & n &= 1:N, \\ v_n|u_n &\sim \text{Categorical}_{G,S,T}\left(\pi'_{u_n}\right), & n &= 1:N, \\ s_n|v_n &\sim \text{Categorical}_{G,S,T}\left(\psi'_{v_n}\right), & n &= 1:N, \\ w_n|u_n, v_n &\sim \text{Bernoulli}\left(\beta_{u_n\to v_n}\right), & n &= 1:N, \end{aligned}$$

and it is depicted graphically in Fig. 8.6. Now, because the states $s_{0:N}$ are no longer directly associated with observations, we can afford to discard

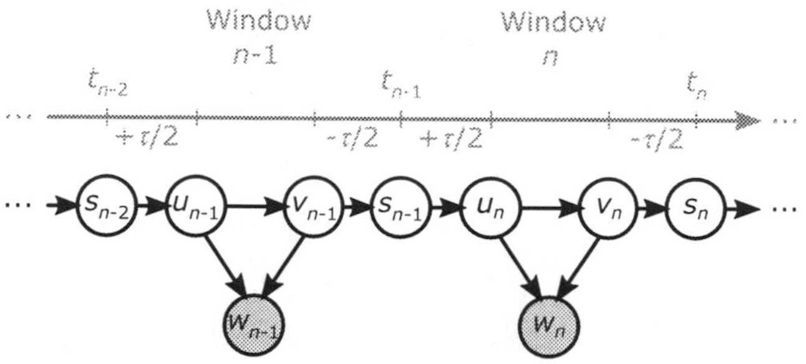

Fig. 8.6 An HMM, augmented with additional hidden states $u_{1:N}$, $v_{1:N}$, is used to decouple successive passing states $s_{0:N}$ from their respective observations $w_{1:N}$.

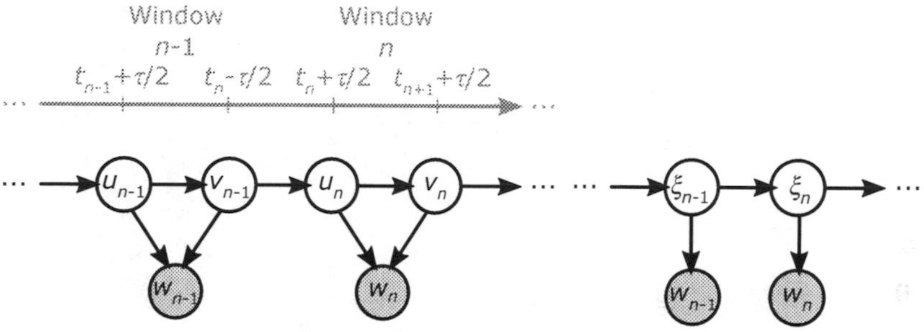

Fig. 8.7 (Left) A modified HMM with two decoupled passing states per observation. (Right) An equivalent HMM with one passing state per observation.

them through marginalization, which leads to an equivalent, but somewhat simpler model,

$$
\begin{aligned}
u_1 &\sim \text{Categorical}_{G,S,T}(\boldsymbol{\rho}'), \\
v_1|u_1 &\sim \text{Categorical}_{G,S,T}\left(\boldsymbol{\pi}'_{u_1}\right), \\
u_n|v_{n-1} &\sim \text{Categorical}_{G,S,T}\left(\boldsymbol{\psi}''_{v_{n-1}}\right), && n = 2:N, \\
v_n|u_n &\sim \text{Categorical}_{G,S,T}\left(\boldsymbol{\pi}'_{u_n}\right), && n = 2:N, \\
w_n|u_n, v_n &\sim \text{Bernoulli}\left(\beta_{u_n \to v_n}\right), && n = 1:N,
\end{aligned}
$$

which we depict graphically in the left panel of Fig. 8.7. Marginalization implies that, in this model, the initial probabilities are given by

$$
\boldsymbol{\rho}' = \begin{bmatrix} \rho'_G & \rho'_S & \rho'_T \end{bmatrix} = \boldsymbol{\rho}\boldsymbol{\Phi}' = \boldsymbol{\rho}\exp\left(\frac{\tau}{2}\boldsymbol{G}\right)
$$

and the transition probabilities by the rows of

$$
\boldsymbol{\Psi}'' = \begin{bmatrix} \boldsymbol{\psi}''_G \\ \boldsymbol{\psi}''_S \\ \boldsymbol{\psi}''_T \end{bmatrix} = \begin{bmatrix} \psi''_{G\to G} & \psi''_{G\to S} & \psi''_{G\to T} \\ \psi''_{S\to G} & \psi''_{S\to S} & \psi''_{S\to T} \\ \psi''_{T\to G} & \psi''_{T\to S} & \psi''_{T\to T} \end{bmatrix} = \boldsymbol{\Psi}'\boldsymbol{\Psi}' = \exp\left(\tau\boldsymbol{G}\right).
$$

Note 8.17 HMM order reduction

The last version of our model represents a conventional HMM as introduced in Section 8.2. To make the correspondence clearer, we consider superstates $\xi_n = (u_n, v_n)$, depicted graphically on the right panel of Fig. 8.7, and rewrite the model in the equivalent form,

$$\xi_1 \sim \text{Categorical}_{\chi_{1:9}}(\boldsymbol{r}),$$
$$\xi_n | \xi_{n-1} \sim \text{Categorical}_{\chi_{1:9}}\left(\boldsymbol{P}_{\xi_{n-1}}\right), \qquad n = 2:N,$$
$$w_n | \xi_n \sim \text{Bernoulli}\left(\beta_{\xi_n}\right), \qquad n = 1:N.$$

The initial, $\boldsymbol{r}$, and transition, $\boldsymbol{P}_\xi$, probabilities are determined according to $\rho', \boldsymbol{\Pi}', \boldsymbol{\Psi}''$. In particular, these are

$$\begin{aligned} r_{\xi_1} &= p(\xi_1) = p(u_1, v_1) \\ &= p(v_1|u_1)p(u_1) = \pi'_{u_1 \to v_1} \rho'_{u_1}, \\ P_{\xi_{n-1} \to \xi_n} &= p(\xi_n|\xi_{n-1}) = p(u_n, v_n|u_{n-1}, v_{n-1}) \\ &= p(v_n|u_n, u_{n-1}, v_{n-1})p(u_n|u_{n-1}, v_{n-1}) \\ &= p(v_n|u_n)p(u_n|v_{n-1}) = \pi'_{u_n \to v_n} \psi''_{v_{n-1} \to u_n}. \end{aligned}$$

As each superstate is formed by a pair of G, S, and T, our new state-space consists of

$$\chi_1 = GG, \qquad \chi_2 = GS, \qquad \chi_3 = GT,$$
$$\chi_4 = SG, \qquad \chi_5 = SS, \qquad \chi_6 = ST,$$
$$\chi_7 = TG, \qquad \chi_8 = TS, \qquad \chi_9 = TT,$$

and, because each constitutive superstate is *derived* from G, S, and T, similar to Example 2.8, we follow the common convention and order $\chi_{1:9}$ lexicographically.

8.7.7 Computational Training

Via auxiliary variables, we have reformulated our second-order Markov model problem in order to make it amenable to a similar training strategy as the conventional HMM of Sections 8.3 and 8.4. In its final version, the unknown parameters are still those of the initial problem, namely the entries of $\boldsymbol{\Lambda}$ and potentially ρ, η. The likelihood of our model, formally given by $p(w_{1:N}|\boldsymbol{\Lambda}, \rho, \eta)$, can be computed according to Eq. (8.7) by completion with the terminal states

$$p\left(w_{1:N}|\boldsymbol{\Lambda}, \rho, \eta\right) = \sum_{u_N, v_N} p\left(w_{1:N}, u_N, v_N|\boldsymbol{\Lambda}, \rho, \eta\right) = \sum_{u_N, v_N} \mathcal{A}_N(u_N, v_N).$$

The terms of the filter $\mathcal{A}_N(u_N, v_N)$ can, in turn, be computed by forward filtering. Nevertheless, because N needs to be large, such that our approximate observation representation holds, naive filtering with Algorithm 8.1 is impractical. Additionally, even if we were able to perform the filtering recursion in Algorithm 8.1 for excessively large N, directly training our model suffers from the approximations induced by having a nonzero τ and finite N. Now we show how to eliminate such approximations altogether and derive a tractable version of the filtering algorithm that carries over onto the limit $N \to \infty$.

Limit $\tau \to 0^+$.

As all of our propagators depend continuously on τ, we can formally apply the $\tau \to 0^+$ limit. Specifically, Note 2.18 implies that

$$\exp\left(\frac{\tau}{2}\boldsymbol{G}\right) \to \mathbb{1}, \quad \exp(\tau\boldsymbol{G}) \to \mathbb{1}, \quad \exp\left(\left(\frac{T_{\max} - T_{\min}}{N} - \tau\right)\boldsymbol{G}\right) \to \boldsymbol{\Pi}.$$

Here, $\mathbb{1}$ is the identity matrix of size three. In this limit, we can safely replace our model with the limiting one,

$$\begin{aligned} u_1 &\sim \text{Categorical}_{G,S,T}(\rho), \\ v_1|u_1 &\sim \text{Categorical}_{G,S,T}\left(\pi_{u_1}\right), \\ u_n|v_{n-1} &\sim \text{Categorical}_{G,S,T}\left(\mathbb{1}_{v_{n-1}}\right), && n = 2:N, \\ v_n|u_n &\sim \text{Categorical}_{G,S,T}\left(\pi_{u_n}\right), && n = 2:N, \\ w_n|u_n, v_n &\sim \text{Bernoulli}\left(\beta_{u_n \to v_n}\right), && n = 1:N, \end{aligned}$$

thereby relaxing any approximation mediated by τ.

Marginal Likelihood

Having relaxed the dependency of the model on τ, we now show how to apply forward filtering, *i.e.*, Algorithm 8.1. To make our calculations more transparent, we adopt the superstate formalism over superstate $\xi_n = (u_n, v_n)$ of Note 8.17 and show how to recursively compute the forward terms of the filter, which, in this case, read $\mathcal{A}_n(u_n, v_n) = \mathcal{A}_n(\xi_n)$. Further, to maintain the notation to a minimum, we follow Note 8.5 and gather our forward terms in row arrays,

$$\begin{aligned} \mathbb{A}_n = [&\mathcal{A}_n(\chi_1)\ \mathcal{A}_n(\chi_2)\ \mathcal{A}_n(\chi_3)\ \mathcal{A}_n(\chi_4)\ \mathcal{A}_n(\chi_5) \\ &\mathcal{A}_n(\chi_6)\ \mathcal{A}_n(\chi_7)\ \mathcal{A}_n(\chi_8)\ \mathcal{A}_n(\chi_9)]\,. \end{aligned}$$

With this convention, the computation of the (marginal) likelihood, L, reduces to

$$L = \mathbb{A}_N \boldsymbol{\Sigma}, \qquad \boldsymbol{\Sigma} = \sigma \otimes \sigma, \qquad \sigma = \begin{bmatrix} 1 \\ 1 \\ 1 \end{bmatrix},$$

where, to be clear, $\boldsymbol{\Sigma}$ is simply a row vector populated by ones.

Note 8.18 Vectorization of $\boldsymbol{r}$ and $\boldsymbol{P}$

According to Note 8.17, the model in Section 8.7.7, leads to the tabulations

$$\boldsymbol{r} = \left[\rho_G\pi_{G\to G}\ \rho_G\pi_{G\to S}\ \rho_G\pi_{G\to T}\ \rho_S\pi_{S\to G}\ \rho_S\pi_{S\to S}\ \rho_S\pi_{S\to T}\ \rho_T\pi_{T\to G}\ \rho_T\pi_{T\to S}\ \rho_T\pi_{T\to T}\right],$$

$$\boldsymbol{P} = \begin{bmatrix} \pi_{G\to G} & \pi_{G\to S} & \pi_{G\to T} & 0 & 0 & 0 & 0 & 0 & 0 \\ 0 & 0 & 0 & \pi_{S\to G} & \pi_{S\to S} & \pi_{S\to T} & 0 & 0 & 0 \\ 0 & 0 & 0 & 0 & 0 & 0 & \pi_{T\to G} & \pi_{T\to S} & \pi_{T\to T} \\ \pi_{G\to G} & \pi_{G\to S} & \pi_{G\to T} & 0 & 0 & 0 & 0 & 0 & 0 \\ 0 & 0 & 0 & \pi_{S\to G} & \pi_{S\to S} & \pi_{S\to T} & 0 & 0 & 0 \\ 0 & 0 & 0 & 0 & 0 & 0 & \pi_{T\to G} & \pi_{T\to S} & \pi_{T\to T} \\ \pi_{G\to G} & \pi_{G\to S} & \pi_{G\to T} & 0 & 0 & 0 & 0 & 0 & 0 \\ 0 & 0 & 0 & \pi_{S\to G} & \pi_{S\to S} & \pi_{S\to T} & 0 & 0 & 0 \\ 0 & 0 & 0 & 0 & 0 & 0 & \pi_{T\to G} & \pi_{T\to S} & \pi_{T\to T} \end{bmatrix}.$$

Adopting array operations, both are vectorized as

$$\boldsymbol{r} = \left(\boldsymbol{\rho} \otimes \boldsymbol{\sigma}^t\right) \odot \left(\boldsymbol{a}_G^t \boldsymbol{\Pi} \boldsymbol{B}_G + \boldsymbol{a}_S^t \boldsymbol{\Pi} \boldsymbol{B}_S + \boldsymbol{a}_T^t \boldsymbol{\Pi} \boldsymbol{B}_T\right),$$

$$\boldsymbol{P} = \left(\boldsymbol{\sigma} \otimes \boldsymbol{I} \otimes \boldsymbol{\sigma}^t\right) \odot \left(\boldsymbol{A}_G^t \boldsymbol{\Pi} \boldsymbol{B}_G + \boldsymbol{A}_S^t \boldsymbol{\Pi} \boldsymbol{B}_S + \boldsymbol{A}_T^t \boldsymbol{\Pi} \boldsymbol{B}_T\right),$$

where $\otimes$, $\odot$ denote the Kronecker and Hadamard product, respectively, and the auxiliary arrays are

$$\boldsymbol{a}_G = \begin{bmatrix} 1 \\ 0 \\ 0 \end{bmatrix}, \qquad \boldsymbol{A}_G = \begin{bmatrix} 1 & 1 & 1 & 1 & 1 & 1 & 1 & 1 & 1 \\ 0 & 0 & 0 & 0 & 0 & 0 & 0 & 0 & 0 \\ 0 & 0 & 0 & 0 & 0 & 0 & 0 & 0 & 0 \end{bmatrix},$$

$$\boldsymbol{B}_G = \begin{bmatrix} 1 & 0 & 0 & 0 & 0 & 0 & 0 & 0 & 0 \\ 0 & 1 & 0 & 0 & 0 & 0 & 0 & 0 & 0 \\ 0 & 0 & 1 & 0 & 0 & 0 & 0 & 0 & 0 \end{bmatrix},$$

$$\boldsymbol{a}_S = \begin{bmatrix} 0 \\ 1 \\ 0 \end{bmatrix}, \qquad \boldsymbol{A}_S = \begin{bmatrix} 0 & 0 & 0 & 0 & 0 & 0 & 0 & 0 & 0 \\ 1 & 1 & 1 & 1 & 1 & 1 & 1 & 1 & 1 \\ 0 & 0 & 0 & 0 & 0 & 0 & 0 & 0 & 0 \end{bmatrix},$$

$$\boldsymbol{B}_S = \begin{bmatrix} 0 & 0 & 0 & 1 & 0 & 0 & 0 & 0 & 0 \\ 0 & 0 & 0 & 0 & 1 & 0 & 0 & 0 & 0 \\ 0 & 0 & 0 & 0 & 0 & 1 & 0 & 0 & 0 \end{bmatrix},$$

$$\boldsymbol{a}_T = \begin{bmatrix} 0 \\ 0 \\ 1 \end{bmatrix}, \qquad \boldsymbol{A}_T = \begin{bmatrix} 0 & 0 & 0 & 0 & 0 & 0 & 0 & 0 & 0 \\ 0 & 0 & 0 & 0 & 0 & 0 & 0 & 0 & 0 \\ 1 & 1 & 1 & 1 & 1 & 1 & 1 & 1 & 1 \end{bmatrix},$$

$$\boldsymbol{B}_T = \begin{bmatrix} 0 & 0 & 0 & 0 & 0 & 0 & 1 & 0 & 0 \\ 0 & 0 & 0 & 0 & 0 & 0 & 0 & 1 & 0 \\ 0 & 0 & 0 & 0 & 0 & 0 & 0 & 0 & 1 \end{bmatrix}.$$

From Eq. (*8.9*), we see that the filtering updates follow the recursion

$$\mathcal{A}_n(\xi_n) = \sum_{\xi_{n-1}} \text{Bernoulli}(w_n; \beta_{\xi_n}) P_{\xi_{n-1}\to\xi_n} \mathcal{A}_{n-1}(\xi_{n-1}), \qquad n = 2:N,$$

which we can vectorize as

$$\mathbb{A}_n = \mathbb{A}_{n-1} \boldsymbol{P}_{w_n}, \qquad n = 2:N.$$

Note 8.19 Vectorization of $\boldsymbol{P}_0$ and $\boldsymbol{P}_1$

The matrices $\boldsymbol{P}_0$ and $\boldsymbol{P}_1$, required in the filtering updates, are tabulated as

$$\boldsymbol{P}_0 = \begin{bmatrix} \pi_{G\to G} & \pi_{G\to S} & \pi_{G\to T} & 0 & 0 & 0 & 0 & 0 & 0 \\ 0 & 0 & 0 & \zeta_0\pi_{S\to G} & \pi_{S\to S} & \pi_{S\to T} & 0 & 0 & 0 \\ 0 & 0 & 0 & 0 & 0 & 0 & \pi_{T\to G} & \pi_{T\to S} & \pi_{T\to T} \\ \pi_{G\to G} & \pi_{G\to S} & \pi_{G\to T} & 0 & 0 & 0 & 0 & 0 & 0 \\ 0 & 0 & 0 & \zeta_0\pi_{S\to G} & \pi_{S\to S} & \pi_{S\to T} & 0 & 0 & 0 \\ 0 & 0 & 0 & 0 & 0 & 0 & \pi_{T\to G} & \pi_{T\to S} & \pi_{T\to T} \\ \pi_{G\to G} & \pi_{G\to S} & \pi_{G\to T} & 0 & 0 & 0 & 0 & 0 & 0 \\ 0 & 0 & 0 & \zeta_0\pi_{S\to G} & \pi_{S\to S} & \pi_{S\to T} & 0 & 0 & 0 \\ 0 & 0 & 0 & 0 & 0 & 0 & \pi_{T\to G} & \pi_{T\to S} & \pi_{T\to T} \end{bmatrix},$$

$$\boldsymbol{P}_1 = \begin{bmatrix} 0 & 0 & 0 & 0 & 0 & 0 & 0 & 0 & 0 \\ 0 & 0 & 0 & \zeta_1\pi_{S\to G} & 0 & 0 & 0 & 0 & 0 \\ 0 & 0 & 0 & 0 & 0 & 0 & 0 & 0 & 0 \\ 0 & 0 & 0 & 0 & 0 & 0 & 0 & 0 & 0 \\ 0 & 0 & 0 & \zeta_1\pi_{S\to G} & 0 & 0 & 0 & 0 & 0 \\ 0 & 0 & 0 & 0 & 0 & 0 & 0 & 0 & 0 \\ 0 & 0 & 0 & 0 & 0 & 0 & 0 & 0 & 0 \\ 0 & 0 & 0 & \zeta_1\pi_{S\to G} & 0 & 0 & 0 & 0 & 0 \\ 0 & 0 & 0 & 0 & 0 & 0 & 0 & 0 & 0 \end{bmatrix},$$

with $\zeta_0 = 1 - \eta$ and $\zeta_1 = \eta$. Similar to $\boldsymbol{P}$, these are vectorized by

$$\boldsymbol{P}_0 = \left(\sigma \otimes \boldsymbol{I} \otimes \sigma^t\right) \odot \left(\boldsymbol{A}^t_G \boldsymbol{\Pi}_0 \boldsymbol{B}_G + \boldsymbol{A}^t_S \boldsymbol{\Pi}_0 \boldsymbol{B}_S + \boldsymbol{A}^t_T \boldsymbol{\Pi}_0 \boldsymbol{B}_T\right),$$

$$\boldsymbol{P}_1 = \left(\sigma \otimes \boldsymbol{I} \otimes \sigma^t\right) \odot \left(\boldsymbol{A}^t_G \boldsymbol{\Pi}_1 \boldsymbol{B}_G + \boldsymbol{A}^t_S \boldsymbol{\Pi}_1 \boldsymbol{B}_S + \boldsymbol{A}^t_T \boldsymbol{\Pi}_1 \boldsymbol{B}_T\right).$$

In $\boldsymbol{P}_0$ and $\boldsymbol{P}_1$, we use $\boldsymbol{\Pi}_0$ and $\boldsymbol{\Pi}_1$ to discriminate between detection-less and detection-full pseudo-propagators

$$\boldsymbol{\Pi}_0 = \boldsymbol{Z}_0 \odot \boldsymbol{\Pi}, \qquad \boldsymbol{\Pi}_1 = \boldsymbol{Z}_1 \odot \boldsymbol{\Pi},$$

where with $\boldsymbol{Z}_0$ and $\boldsymbol{Z}_1$, termed "masks," we encode detection-less and detection-full transitions

$$\boldsymbol{Z}_0 = \begin{bmatrix} 1 & 1 & 1 \\ \zeta_0 & 1 & 1 \\ 1 & 1 & 1 \end{bmatrix}, \qquad \boldsymbol{Z}_1 = \begin{bmatrix} 0 & 0 & 0 \\ \zeta_1 & 0 & 0 \\ 0 & 0 & 0 \end{bmatrix}.$$

As $\boldsymbol{Z}_0$ and $\boldsymbol{Z}_1$ encode our observation rules, *i.e.*, encode time windows with either 0 or 1 detections, the pseudo-propagators are related by $\boldsymbol{\Pi} = \boldsymbol{\Pi}_0 + \boldsymbol{\Pi}_1$. Additionally, because photons are emitted only when our system *jumps* across constitutive states, the diagonal entries in $\boldsymbol{Z}_0$ are all one. By contrast, the diagonal entries in $\boldsymbol{Z}_1$ are all zero.

Finally, according to Eq. (*8.10*), the filter is initialized with $\mathcal{A}_1(\xi_1) = \text{Bernoulli}(w_1; \beta_{\xi_1}) r_{\xi_1}$, which in vectorized form reads

$$\mathbb{A}_1 = \left(\rho \otimes \sigma^t\right) \odot \left(\boldsymbol{a}^t_G \boldsymbol{\Pi}_{w_1} \boldsymbol{B}_G + \boldsymbol{a}^t_S \boldsymbol{\Pi}_{w_1} \boldsymbol{B}_S + \boldsymbol{a}^t_T \boldsymbol{\Pi}_{w_1} \boldsymbol{B}_T\right).$$

Note 8.20 Vectorization

With the aid of two operators,

$$\mathbb{L}(\boldsymbol{C}) = (\rho \boldsymbol{C}) \otimes \sigma^t, \qquad \mathbb{D}(\boldsymbol{C}) = \boldsymbol{a}^t_G \boldsymbol{C} \boldsymbol{B}_G + \boldsymbol{a}^t_S \boldsymbol{C} \boldsymbol{B}_S + \boldsymbol{a}^t_T \boldsymbol{C} \boldsymbol{B}_T,$$

defined over the 3×3 matrices $\boldsymbol{C}$, the initial forward term takes a much simpler form:

$$\mathbb{A}_1 = \mathbb{L}\left(\boldsymbol{I}\right) \odot \mathbb{D}\left(\boldsymbol{\Pi}_{w_1}\right).$$

By induction, we can now show that the forward variables, $\mathbb{A}_{1:N}$, satisfy an important relationship,

$$\mathbb{A}_n = \mathbb{L}\left(\boldsymbol{I}\boldsymbol{\Pi}_{w_1} \cdots \boldsymbol{\Pi}_{w_{n-1}}\right) \odot \mathbb{D}\left(\boldsymbol{\Pi}_{w_n}\right), \qquad n = 1 : N.$$

Accordingly, the (marginal) likelihood is given by

$$\begin{aligned} L &= \left[\mathbb{L}\left(\boldsymbol{\Pi}_{w_1} \cdots \boldsymbol{\Pi}_{w_{N-1}}\right) \odot \mathbb{D}\left(\boldsymbol{\Pi}_{w_N}\right)\right] \boldsymbol{\Sigma} \\ &= \mathbb{L}\left(\boldsymbol{\Pi}_{w_1} \cdots \boldsymbol{\Pi}_{w_{N-1}}\right)\left[\mathbb{D}\left(\boldsymbol{\Pi}_{w_N}\right)\right]^t = \boldsymbol{\rho}\, \boldsymbol{\Pi}_{w_1} \cdots \boldsymbol{\Pi}_{w_N} \boldsymbol{\sigma}. \end{aligned}$$

Limit $N \to \infty$.

According to Note 8.16, the product of the pseudo-propagators in our likelihood takes the form

$$\begin{aligned} \boldsymbol{\Pi}_{w_1} \cdots \boldsymbol{\Pi}_{w_N} &= \overbrace{\boldsymbol{\Pi}_0 \cdots \boldsymbol{\Pi}_0}^{\substack{\text{windows} \\ 1:N_1-1}} \boldsymbol{\Pi}_1 \overbrace{\boldsymbol{\Pi}_0 \cdots \boldsymbol{\Pi}_0}^{\substack{\text{windows} \\ N_1+1:N_2-1}} \boldsymbol{\Pi}_1 \boldsymbol{\Pi}_0 \cdots \qquad \cdots \boldsymbol{\Pi}_0 \boldsymbol{\Pi}_1 \overbrace{\boldsymbol{\Pi}_0 \cdots \boldsymbol{\Pi}_0}^{\substack{\text{windows} \\ N_K+1:N}} \\ &= \boldsymbol{\Pi}_0^{N_1-1} \boldsymbol{\Pi}_1 \boldsymbol{\Pi}_0^{N_2-N_1-1} \boldsymbol{\Pi}_1 \cdots \qquad \cdots \boldsymbol{\Pi}_0^{N_K-N_{K-1}-1} \boldsymbol{\Pi}_1 \boldsymbol{\Pi}_0^{N-N_K}. \end{aligned}$$

Note 8.21 Asymptotics

Considering the limit $N \to \infty$, from Eq. (8.43), we see that

$$\boldsymbol{\Pi} = \boldsymbol{I} + \frac{T_{\max} - T_{\min}}{N} \boldsymbol{G} + \mathcal{O}\left(\frac{1}{N^2}\right),$$

which we may also use to approximate the pseudo-propagators. Specifically, as $\boldsymbol{\Pi}_0 = \boldsymbol{\Pi} \odot \boldsymbol{Z}_0$ and $\boldsymbol{\Pi}_1 = \boldsymbol{\Pi} \odot \boldsymbol{Z}_1$, we readily derive

$$\begin{aligned} \boldsymbol{\Pi}_0 &= \boldsymbol{I} + \frac{T_{\max} - T_{\min}}{N} \boldsymbol{G}_0 + \mathcal{O}\left(\frac{1}{N^2}\right) = \exp\left(\frac{T_{\max} - T_{\min}}{N} \boldsymbol{G}_0\right) + \mathcal{O}\left(\frac{1}{N^2}\right), \\ \boldsymbol{\Pi}_1 &= \frac{T_{\max} - T_{\min}}{N} \boldsymbol{G}_1 + \mathcal{O}\left(\frac{1}{N^2}\right), \end{aligned}$$

where $\boldsymbol{G}_0 = \boldsymbol{G} \odot \boldsymbol{Z}_0$ and $\boldsymbol{G}_1 = \boldsymbol{G} \odot \boldsymbol{Z}_1$.

Additionally, according to the definition of N_k in Note 8.16, we have

$$N_k \frac{T_{\max} - T_{\min}}{N} = \frac{T_{\max} - T_{\min}}{N} + \mathcal{O}\left(\frac{1}{N^2}\right), \qquad k = 1 : K.$$

Putting everything together, we obtain an asymptotic expression of our likelihood,

$$L = \left(\frac{T_{\max} - T_{\min}}{N}\right)^K \ell + \mathcal{O}\left(\frac{1}{N^{K+1}}\right), \tag{8.44}$$

where ℓ is *independent* of N. Specifically, ℓ is given by

$$\ell = \rho \exp(g_0 \boldsymbol{G}_0)\boldsymbol{G}_1 \exp(g_1 \boldsymbol{G}_0)\, \boldsymbol{G}_1 \cdots \boldsymbol{G}_1 \exp(g_{K-1} \boldsymbol{G}_0)\, \boldsymbol{G}_1 \exp(g_K \boldsymbol{G}_0)\, \sigma. \tag{8.45}$$

As we can see, ℓ depends only on $\boldsymbol{\Lambda}, \rho, \eta$, and the successive time lags

$$g_0 = T_1 - T_{\min}, \qquad g_1 = T_2 - T_1, \qquad \dots$$
$$g_{K-1} = T_K - T_{K-1}, \qquad g_K = T_{\max} - T_K.$$

8.7.8 Bayesian Considerations

From Eq. (8.44), it becomes clear that the unknown parameters in our formulation enter the model's likelihood in a complicated way, rendering it pointless to seek training through the Baum–Welch method of Section 8.3.3 simply because closed-form expressions do not follow from the derivatives in the M-step. Similarly, as conjugate priors are unavailable, Bayesian training as in Section 8.4.1 is also not possible.

A viable training strategy, however, is through a Metropolis–Hastings MCMC scheme where, under non-conjugate prior assignments, proposals are drawn and subsequently accepted or rejected according to the (marginal) posterior. As this strategy is quite general, here we consider a wider problem, where the unknown parameters may include not only entries of the transition rate matrix $\boldsymbol{\Lambda}$, but also initial probabilities ρ and observation parameter η.

For clarity, we gather the unknown parameters in $\boldsymbol{\theta}$ and, to stress their dependence, we denote by $\ell(\boldsymbol{\theta})$ the product in Eq. (8.45). With this formalism, our priors, which need to be specified, are encoded in $p(\boldsymbol{\theta})$ and our likelihood is given, only asymptotically, by

$$p(w_{1:N}|\boldsymbol{\theta}) = \left(\frac{T_{\max} - T_{\min}}{N}\right)^K \ell(\boldsymbol{\theta}) + \mathcal{O}\left(\frac{1}{N^{K+1}}\right).$$

As in Section 5.2.1, using an appropriate Metropolis–Hastings proposal $q(\boldsymbol{\theta}^{\text{prop}}|\boldsymbol{\theta}^{\text{old}})$, we arrive at the acceptance ratio, Eq. (5.8), of the form

$$A_N\left(\boldsymbol{\theta}^{\text{prop}}|\boldsymbol{\theta}^{\text{old}}\right) = \frac{p\left(w_{1:N}|\boldsymbol{\theta}^{\text{prop}}\right)}{p\left(w_{1:N}|\boldsymbol{\theta}^{\text{old}}\right)} \frac{p\left(\boldsymbol{\theta}^{\text{prop}}\right)}{p\left(\boldsymbol{\theta}^{\text{old}}\right)} \frac{Q\left(\boldsymbol{\theta}^{\text{old}}|\boldsymbol{\theta}^{\text{prop}}\right)}{Q\left(\boldsymbol{\theta}^{\text{prop}}|\boldsymbol{\theta}^{\text{old}}\right)}.$$

For any finite choice of N, this ratio is intractable. However, the limiting case $N \to \infty$ leads to

$$A_\infty\left(\boldsymbol{\theta}^{\text{prop}}|\boldsymbol{\theta}^{\text{old}}\right) = \frac{\ell\left(\boldsymbol{\theta}^{\text{prop}}\right)}{\ell\left(\boldsymbol{\theta}^{\text{old}}\right)} \frac{p\left(\boldsymbol{\theta}^{\text{prop}}\right)}{p\left(\boldsymbol{\theta}^{\text{old}}\right)} \frac{Q\left(\boldsymbol{\theta}^{\text{old}}|\boldsymbol{\theta}^{\text{prop}}\right)}{Q\left(\boldsymbol{\theta}^{\text{prop}}|\boldsymbol{\theta}^{\text{old}}\right)},$$

which we can readily evaluate numerically.

Example 8.6 Bayesian fluorescence spectroscopy

In the most general case, the unknowns in a typical problem of interest in fluorescence spectroscopy may include: all transition rates $\lambda_{G\to S}, \lambda_{S\to G}, \lambda_{S\to T}, \lambda_{T\to G}$, all initial probabilities ρ_G, ρ_S, ρ_T, and η. Convenient prior choices then include

$$\lambda_{G\to S} \sim \text{Gamma}\left(2, \frac{\lambda_{\text{ref}}}{2}\right), \qquad \lambda_{S\to G} \sim \text{Gamma}\left(2, \frac{\lambda_{\text{ref}}}{2}\right),$$

$$\lambda_{S\to T} \sim \text{Gamma}\left(2, \frac{\lambda_{\text{ref}}}{2}\right), \qquad \lambda_{T\to G} \sim \text{Gamma}\left(2, \frac{\lambda_{\text{ref}}}{2}\right),$$

$$\rho \sim \text{Dirichlet}_3\left(\frac{1}{3}, \frac{1}{3}, \frac{1}{3}\right), \qquad \eta \sim \text{Beta}\,(1,1).$$

In these priors, the hyperparameters may be adjusted to incorporate prior confidence on certain values and λ_{ref} can be used to set a priori appropriate timescales.

For numerical stability, it is preferable to use unitless priors

$$\tilde{\lambda}_{G\to S} \sim \text{Gamma}\left(2, \frac{1}{2}\right), \qquad \tilde{\lambda}_{S\to G} \sim \text{Gamma}\left(2, \frac{1}{2}\right),$$

$$\tilde{\lambda}_{S\to T} \sim \text{Gamma}\left(2, \frac{1}{2}\right), \qquad \tilde{\lambda}_{T\to G} \sim \text{Gamma}\left(2, \frac{1}{2}\right),$$

and implement the timescale through Eq. (8.45) cast in the form

$$\ell = \rho \exp\left(\tilde{g}_0 \tilde{\boldsymbol{G}}_0\right) \tilde{\boldsymbol{G}}_1 \exp\left(\tilde{g}_1 \tilde{\boldsymbol{G}}_0\right) \tilde{\boldsymbol{G}}_1 \cdots \tilde{\boldsymbol{G}}_1 \exp\left(\tilde{g}_{K-1} \tilde{\boldsymbol{G}}_0\right) \tilde{\boldsymbol{G}}_1 \exp\left(\tilde{g}_K \tilde{\boldsymbol{G}}_0\right) \sigma,$$

with $\tilde{g}_k = g_k \lambda_{\text{ref}}$ and where $\tilde{\boldsymbol{G}}$ is the coinciding $\boldsymbol{G}$ constructed from $\tilde{\boldsymbol{\Lambda}}$. Numerical stability can be further increased if the fastest timescale is separated and evaluated analytically. In particular, if $\tilde{\lambda}_{\text{fast}}$ denotes the fastest rate, $\tilde{\boldsymbol{G}}_0$ can be replaced by $\tilde{\boldsymbol{\Gamma}}_0 - \tilde{\lambda}_{\text{fast}} \mathbb{1}$. This way, ℓ results in

$$\ell = e^{-\tilde{\lambda}_{\text{fast}} \lambda_{\text{ref}} (T_{\max} - T_{\min})} \rho \exp\left(\tilde{g}_0 \tilde{\boldsymbol{\Gamma}}_0\right) \tilde{\boldsymbol{G}}_1 \exp\left(\tilde{g}_1 \tilde{\boldsymbol{\Gamma}}_0\right) \tilde{\boldsymbol{G}}_1 \cdots \tilde{\boldsymbol{G}}_1 \exp\left(\tilde{g}_{K-1} \tilde{\boldsymbol{\Gamma}}_0\right) \tilde{\boldsymbol{G}}_1 \exp\left(\tilde{g}_K \tilde{\boldsymbol{\Gamma}}_0\right) \sigma.$$

8.8 Exercise Problems

Exercise 8.1 EM for Poisson HMM

Adapt the Baum–Welch algorithm to train an HMM with Poisson emissions. For concreteness, consider the model

$$s_1 | \rho \sim \text{Categorical}_{\sigma_{1:M}}(\rho),$$
$$s_n | s_{n-1}, \boldsymbol{\Pi} \sim \text{Categorical}_{\sigma_{1:M}}(\pi_{\sigma_m}), \qquad n = 2:N,$$
$$w_n | s_n, \phi \sim \text{Poisson}\left(\phi_{s_n}\right), \qquad n = 1:N.$$

Compare your parameters $\rho, \boldsymbol{\Pi}$, and ϕ estimated using Baum–Welch with the ground truth you used to generate your synthetic data.

Exercise 8.2 Implementing Viterbi

Generate observations $w_{1:N}$ using ancestral sampling for a simple HMM with two states and a normal emission model. Assume known kinetic and emission parameters. Implement the Viterbi algorithm, Algorithm 8.3, to find the sequence $s^{\#}_{1:N}$. Compare your $s^{\#}_{1:N}$ to the ground truth.

Exercise 8.3 Bayesian model for Poisson HMM

Consider the same model as in Exercise 8.1 and provide a Bayesian formulation that estimates all unknown model parameters. Make your own choices for the priors and briefly justify your choices. Histogram your MCMC samples and indicate the ground truth.

Exercise 8.4 HMMs with common parameters

Consider a total of Q independent HMMs whose dynamics and observations are influenced by the same $\rho, \mathbf{\Pi}, \phi$. This scenario is typical of experiments where we try to estimate $\rho, \mathbf{\Pi}, \phi$ from a number of short traces. Here, we need to consider a joint likelihood over all traces and apply a common prior over the parameters. For each trace, we have

$$\begin{aligned} s_1^q|\rho &\sim \text{Categorical}_{\sigma_{1:M}}(\rho), & & q = 1:Q, \\ s_n^q|s_{n-1}^q, \mathbf{\Pi} &\sim \text{Categorical}_{\sigma_{1:M}}\left(\pi_{s_{n-1}^q}\right), & n = 2:N, \quad & q = 1:Q, \\ w_n^q|s_n^q, \phi &\sim \mathbb{G}_{\phi_{s_n^q}}, & n = 1:N, \quad & q = 1:Q. \end{aligned}$$

1. Adapt the Baum–Welch algorithm to train the resulting model assuming normal emission.
2. Implement a Bayesian model that estimates all model parameters and represent your model graphically. Histogram your MCMC samples and indicate the ground truth in your histogram.

Exercise 8.5 A sticky HMM

Here, we provide a Bayesian model that estimates all dynamic parameters for a HMM with two states and normal emissions. We assume that the emission parameters are known and our only goal is to estimate dynamical parameters.

Start by generating synthetic data and assume that your escape probabilities coincide with escape rates of the same order of magnitude for each state.

Next, perform inference using a sticky HMM. As you implement the sticky HMM, consider three cases: one where the hyperparameters of Note 8.12 are tuned to approximately match the dwell time of each state in your synthetic data; and two more cases where the hyperparameters under- and overestimate the dwell by an order of magnitude.

For all three cases, histogram your MCMC samples for your kinetic parameters and indicate the ground truth in your histogram.

Exercise 8.6 The iHMM

Here we consider the iHMM of Section 8.6.

1. Generate synthetic data with three states using the usual ancestral sampling scheme of an HMM model. Assume a normal emission model with, for simplicity, the same known variance in each state.
2. Implement the Gibbs sampler proposed in Section 8.6 to sample kinetic parameters and mean levels of the emission distributions assuming $M = 10$ and $M = 50$.
3. Using your MCMC samples, histogram the fraction of time spent in each state. Compare your results to the ground truth and the mean expected time derived from your prior.

Project 8.1 De-drifting a trace in HMM analysis

In experimental techniques, such as *force spectroscopy*, the apparatus collecting data drifts over time, giving rise to an apparently low frequency undulation added on top of the signal. In force spectroscopy, the slow drift of an optical trap holding a micron-sized bead corrupts our assessment of its position used as a microscopic measure of force impinged upon the bead. Often, this force can be imparted by a molecule undergoing transitions in a discrete state-space through a dual optical trap setup (*e.g.*, M. J. Comstock, T. Ha, Y. R. Chemla. Ultrahigh-resolution optical trap with single-fluorophore sensitivity. *Nat. Meth.* 8:335, 2011).

To learn the properties of a system free from the corruption introduced by drift, we consider an HMM with two states in the presence of drift, $d(\cdot)$, captured by the following generative model:

$$\begin{aligned} d(\cdot) &\sim \text{GaussianP}(\mu_{\text{drift}}(\cdot), C_{\text{drift}}(\cdot,\cdot)), \\ s_1|\boldsymbol{\rho} &\sim \text{Categorical}_{\sigma_{1:2}}(\boldsymbol{\rho}), \\ s_{n+1}|s_n, \boldsymbol{\Pi} &\sim \text{Categorical}_{\sigma_{1:2}}(\boldsymbol{\pi}_{\sigma_m}), \\ w_n|s_n, d(\cdot) &\sim \text{Normal}\left(\mu_{s_n} + d(t_n), v\right). \end{aligned}$$

1. Simulate about a 10^3 point trajectory using the familiar squared exponential $C_{\text{drift}}(\cdot,\cdot)$ with prefactor equal to 2 and length scale equal to 500 times the time step size. Set $M = 2$, $\mu_{\text{drift}} = 0$, $\pi_{\sigma_1\to\sigma_1} = \pi_{\sigma_2\to\sigma_2} = 0.9$, $\pi_{\sigma_1\to\sigma_2} = \pi_{\sigma_2\to\sigma_1} = 0.1$, $\mu_{\sigma_1} = -5$, $\mu_{\sigma_2} = 5$, $v = 1$ (in rescaled "unitless" units). That is, drift should occur on a slow time scale as compared to other time scales of the problem.
2. Place appropriate priors and implement an MCMC sampling scheme to estimate $d(\cdot)$, $\boldsymbol{\Pi}$ under known v. As a prior on $d(\cdot)$ use a GaussianP with a squared exponential $C_{\text{drift}}(\cdot,\cdot)$ whose parameters are close to that used to generate the data.
3. Plot various samples of your $d(\cdot)$ and compare to the ground truth. Also, histogram your values for $\boldsymbol{\Pi}$ and compare with the ground truth.

Project 8.2 A Bayesian HMM for raw FRET measurements

In fluorescence experiments relying on *Förster resonance energy transfer* (FRET) measurements, we typically obtain two scalar measurements, w_n^D and w_n^A, at each time level t_n. These are the number of photons emitted by a fluorescent label (called a fluorophore) designated as *donor* and the number of photons emitted by a second fluorophore designated as *acceptor*, respectively.

The donor and acceptor may be located on opposing ends of a molecule. When the donor and acceptor move close to one another, as a molecule collapses on itself or folds, energy can be transferred from a donor (typically directly excited by a laser light) to an acceptor. As such, the origin of the photons (whether higher energy photons from the donor or lower energy photons from the acceptor) report back on the conformational state of a molecule.

As individual photons are emitted by the fluorophores independently, the raw measurements are described by

$$w_n^D | s_n \sim \text{Poisson}\left(\mu_{s_n}^D\right), \qquad w_n^A | s_n \sim \text{Poisson}\left(\mu_{s_n}^A\right), \qquad n = 1 : N,$$

where s_n is the conformational state of the molecule attached to the two fluorophores and the state-dependent parameters $\mu_{\sigma_1}^D, \ldots, \mu_{\sigma_M}^D$ and $\mu_{\sigma_1}^A, \ldots, \mu_{\sigma_M}^A$ are the corresponding average photon emissions per unit time.

1. Set up a Bayesian HMM for the analysis of measurements $w_{1:N}^D$ and $w_{1:N}^A$, generated from synthetic data, from the donor and acceptor channels. Typical values are $N = 1000$, $M = 3$, and $\mu_{\sigma_m}^D, \mu_{\sigma_m}^A$ in the range 100–1,000 photons/s.
2. Describe an MCMC sampling scheme for the model posterior in part 1.
3. Implement the MCMC sampling scheme of step 2.
4. Verify, using synthetic data, that your implementation of step 3 generates samples with the correct statistics.

   In FRET experiments, a common issue is the *crossover* of photons into the wrong photon detector due to spectral overlap. Crossover is generally given as a matrix of probabilities,

$$C = \begin{bmatrix} c_{D\to D} & c_{D\to A} \\ c_{A\to D} & c_{A\to A} \end{bmatrix},$$

   where, for example, $c_{D\to A}$ is the probability of a donor photon detected in the acceptor channel. Due to conservation, these probabilities satisfy $c_{D\to D} + c_{D\to A} = 1$ and $c_{A\to D} + c_{A\to A} = 1$. Typical values are $c_{D\to A}, c_{A\to D}$ in the range 5–15%.
5. Show that with crossover, the measurements are described by

$$w_n^D | s_n \sim \text{Poisson}\left(c_{D\to D}\mu_{s_n}^D + c_{A\to D}\mu_{s_n}^A\right),$$
$$w_n^A | s_n \sim \text{Poisson}\left(c_{D\to A}\mu_{s_n}^D + c_{A\to A}\mu_{s_n}^A\right).$$

6. Modify the Bayesian model of step 1 to incorporate crossover, assuming known crossover probabilities, and implement and verify your MCMC.

Project 8.3 A Bayesian HMM for FRET efficiency measurements

In a FRET experiment like that in Project 8.2, most often w_n^D and w_n^A are combined into a single scalar quantity,

$$\epsilon_n = \frac{w_n^A}{w_n^A + w_n^D},$$

which is termed the (apparent) *FRET efficiency*. In this case, the observation model takes a simpler form,

$$\epsilon_n | s_n \sim \mathbb{G}_{\phi_{\sigma_m}},$$

where $\phi_{\sigma_m} = (\mu^D_{\sigma_m}, \mu^A_{\sigma_m})$. In general, the probability density $G_\phi(\epsilon)$ is analytically intractable. However, provided *all* emission levels are high enough, we can safely use the approximations

$$\text{Poisson}\left(w^D; \mu^D\right) \approx \text{Gamma}\left(w^D; \mu^D, 1\right),$$
$$\text{Poisson}\left(w^A; \mu^A\right) \approx \text{Gamma}\left(w^A; \mu^A, 1\right).$$

1. Considering these approximations, derive an analytic formula for the resulting emission density $G_\phi(\epsilon)$.
2. Set up a Bayesian HMM for the analysis of apparent FRET efficiencies $\epsilon_{1:N}$.
3. Describe an MCMC sampling scheme for the posterior of the model in step 2.
4. Implement the MCMC sampling scheme of step 3.
5. Verify, using synthetic data, that your implementation of step 4 generates samples with the correct statistics. As in Project 8.2, typical values are $N = 1{,}000$, $M = 3$, and $\mu^D_{\sigma_m}, \mu^A_{\sigma_m}$ in the range 100–1,000 photons/s.

Additional Reading

C. Bishop. *Pattern recognition and machine learning*. Springer, 2006.

O. Cappé, E. Moulines, T. Rydén. *Inference in hidden Markov models*. Springer, 2005.

S. Särkkä. *Bayesian filtering and smoothing*. Cambridge University Press, 2013.

L. R. Rabiner, B. Juang. An introduction to hidden Markov models. *IEEE ASSP Magazine*, 3:4, 1986.

L. R. Rabiner. A tutorial on hidden Markov models and selected applications in speech recognition. *Proc. IEEE*, 77:257, 1989.

M. Beal, Z. Ghahramani, C. Rasmussen. The infinite hidden Markov model. *NIPS*, 2001.

Z. Ghahramani, M. Jordan. Factorial hidden Markov models. *NIPS*, 1995.

J. van Gael, Y. Teh, Z. Ghahramani. The infinite factorial hidden Markov model. *NIPS*, 2008.

E. B. Fox, E. B. Sudderth, M. Jordan, A. S. Willsky. A sticky HDP-HMM with application to speaker diarization. *Ann. Appl. Stat.*, 5:1020, 2011.

J. van Gael, Y. Saatci, T. W. Teh, Z. Ghahramani. Beam sampling for the infinite hidden Markov model. *Proc. 25th Intl. Conf. Mach. Learn.*, 1088, 2008.

A. Saurabh, M. Safar, I. Sgouralis, M. Fazel, S. Pressé. Single photon smFRET. I. Theory and conceptual basis. *Biophys. Rep.*, 3:100089, 2023.

I. V. Gopich, A. Szabo. Theory of the statistics of kinetic transitions with application to single-molecule enzyme catalysis. *J. Chem. Phys.*, 124:154712, 2006.

I. V. Gopich, A. Szabo. Theory of the energy transfer efficiency and fluorescence lifetime distribution in single-molecule FRET. *Proc. Natl. Acad. Sc.*, 109:7747, 2012.

I. V. Gopich, A. Szabo. Theory of photon statistics in single-molecule Förster resonance energy transfer. *J. Chem. Phys.*, 122:014707, 2005.

S. A. McKinney, C. Joo, T. Ha. Analysis of single-molecule FRET trajectories using hidden Markov modeling. *Biophys. J.*, 91:1941, 2006.

H. Mazal, G. Haran. Single-molecule FRET methods to study the dynamics of proteins at work. *Current Op. Biomed. Eng.*, 12:8, 2019.

B. Schuler, and H. Hofmann. Single-molecule spectroscopy of protein folding dynamics–expanding scope and timescales. *Current Op. Struct. Bio.*, 23:36, 2013.

Z. Kilic, I. Sgouralis, S. Pressé. Residence time analysis of RNA polymerase transcription dynamics: A Bayesian sticky HMM approach. *Biophys. J.*, 120:1665, 2021.

I. Sgouralis, S. Madaan, F. Djutanta, R. Kha, R. Hariadi, S. Pressé. A Bayesian nonparametric approach to single molecule Förster resonance energy transfer. *J. Phys. Chem. B.*, 123:675, 2019.

I. Sgouralis, S. Pressé. ICON: An adaptation of infinite HMMs for time traces with drift. *Biophys. J.*, 112:2117, 2017.

I. Sgouralis, S. Pressé. An introduction to infinite HMMs for single molecule data analysis. *Biophys. J.*, 112:2021, 2017.

9 State-Space Models

By the end of this chapter, we will have presented

- *Generalized ways of modeling dynamical systems*
- *Fundamentals of continuous state-space models*
- *Kalman filters and their foundation*

In this chapter we continue in our study of *time-dependent measurements*. Unlike in the previous chapter, where we focused on discrete space and discrete time systems, here we focus on continuous space systems evolving in discrete time. Due to computational limitations, we consider linear Gaussian systems and present a detailed description of the theory associated with *Kalman filters*, the continuous space generalization of the filters visited in Chapter 8.

9.1 State-Space Models

In direct analogy to HMMs with discrete state-spaces, in many cases of practical interest the dynamics underlying the behavior of physical systems in real, continuous space are fully or partially *unobserved*. When experiments report back on corrupted realizations of dynamical random variables, we must apply specialized models assimilating empirical data to learn about the underlying physical system while remaining faithful to its mathematical structure. Indeed, we saw in Chapter 8 how HMMs can be used in conjunction with measurements for this task.

Here, we see how to do so for continuous state-spaces, such as those involving positions of moving objects, or magnitudes of recorded waves, and forces, for which dynamical variables may attain a *continuum* of values. In such applications, *continuous state-space models*, often simply abbreviated to *state-space models* that we describe in this chapter provide flexible extensions preserving the statistical structure of the HMM but otherwise accommodating continuous dynamical variables.

In particular, state-space models allow us to model a sequence of observations $w_{1:N}$, obtained at times $t_{1:N}$, as being driven by a sequence of state variables $r_{1:N}$. Each state r_n in this sequence is, in turn, linked across time to past $r_{1:n-1}$ and future $r_{n+1:N}$ states by stochastic events. Similar to an HMM, in our formulation the sequence of state variables remains unobserved but, as we will see, can be estimated.

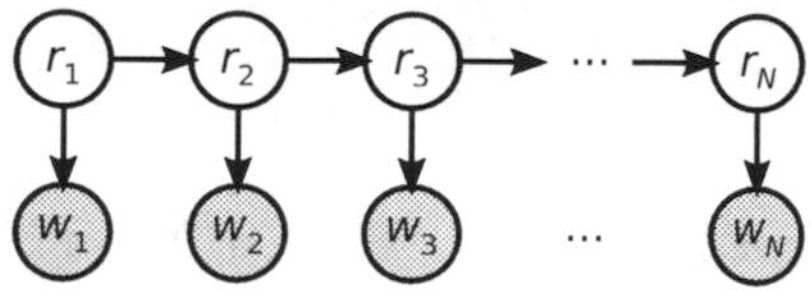

Fig. 9.1 Graphical representation of a state-space model.

A *state-space model*, in fairly general form, can be summarized as

$$r_1 \sim \mathbb{R}, \tag{9.1}$$

$$r_n | r_{n-1} \sim \mathbb{Q}^n_{r_{n-1}}, \qquad n = 2 : N, \tag{9.2}$$

$$w_n | r_n \sim \mathbb{F}^n_{r_n}, \qquad n = 1 : N, \tag{9.3}$$

and depicted graphically in Fig. 9.1. In this modeling framework, $\mathbb{R}, \mathbb{Q}^n_{r'}, \mathbb{F}^n_r$ are probability distributions providing statistics for the dynamical variables r and $r|r'$ and measurement variables $w|r$, respectively.

As we saw in Chapter 2, Eq. (9.1) is the system's initialization rule, Eq. (9.2) is the transition rule, and Eq. (9.3) is the assessment rule. Equation (9.2) formulates the dynamics of our system's unobserved state r_n, influenced only by its predecessor r_{n-1}, and Eq. (9.3) describes the generation of the measurements w_n influenced only by r_n.

The distributions $\mathbb{R}, \mathbb{Q}^n_r$ describe a causal stochastic Markovian relation between successive states and may account for effects internal to the evolving system, while $\mathbb{F}^n_r$ describes a stochastic relation between measurements and occupying states that may account for external effects such as observation noise. Both dynamical and observational relations may change over time irrespective of the changes induced by the state variables allowing for the modeling of stationary or nonstationary systems.

Example 9.1 Tracking of a single diffusive particle

At times $t_{1:N}$, we have noisy measurements $w_{1:N}$ of a particle's positions $r_{1:N}$. In this setting, our hidden states are $r_{1:N}$ and our observations are $w_{1:N}$.

- Our state-space consists of all possible positions attained by the particle and can be represented in three-dimensional Cartesian space.
- Our initialization rule is represented by the probability distribution over the first position r_1.
- Our transition rule is represented by the probability distributions of successive positions $r_2|r_1, r_3|r_2, \ldots$.
- Our observation rule is represented by the probability distributions of $w_1|r_1, w_2|r_2, \ldots$.

To decode the particles's position, we may use a state-space model of the form

$$r_1 \sim \text{Normal}_3\left((0,0,0), v\begin{pmatrix}1 & 0 & 0\\ 0 & 1 & 0\\ 0 & 0 & 1\end{pmatrix}\right),$$

$$r_n|r_{n-1} \sim \text{Normal}_3\left(r_{n-1}, 2D(t_n - t_{n-1})\begin{pmatrix}1 & 0 & 0\\0 & 1 & 0\\0 & 0 & 1\end{pmatrix}\right), \qquad n = 2:N,$$

$$w_n|r_n \sim \text{Normal}_2\left(r_n\begin{pmatrix}1 & 0\\0 & 1\\0 & 0\end{pmatrix}, \sigma^2\begin{pmatrix}1 & 0\\0 & 1\end{pmatrix}\right), \qquad n = 1:N,$$

where initialization, transition, and observation rules are all represented by multivariate normal distributions similar to those explored in Chapter 3.

This model assumes that the particle is initially located near the origin and subsequently undergoes free Brownian motion characterized by a diffusion coefficient D. Further, this model assumes that our measurements probe only the first two Cartesian coordinates of the particle while leaving the third unobserved. This is consistent, say, with observations under a microscope using one imaging plane. In addition, it assumes that our positional assessments are contaminated with zero-mean additive noise characterized by the variance σ^2.

In the general form provided above, the formulation of a state-space model places no restrictions on the distributions $\mathbb{R}, \mathbb{Q}^n_{r'}, \mathbb{F}^n_r$ besides that these describe *continuous* random variables and, as such, are characterized by probability *densities*. For most practical applications, however, it is sufficient or even desirable to consider more specialized model classes.

9.2 Gaussian State-Space Models

To facilitate computational tractability, the standard approach is to replace $\mathbb{R}, \mathbb{Q}^n_r$, and $\mathbb{F}^n_r$ with normal distributions, possibly multivariate, and cast our initialization, transition, and observation rules as

$$\mathbb{R} = \text{Normal}_K\left(\mu, \upsilon\right), \qquad \mathbb{Q}^n_r = \text{Normal}_K\left(q_n(r), V_n\right),$$
$$\mathbb{F}^n_r = \text{Normal}_L\left(f_n(r), U_n\right),$$

for some *functions* $q_n(r)$ and $f_n(r)$ that may also vary in time. In this class of state-space models, V_n and U_n are the covariances of the dynamics and measurements, respectively, that together with $q_n(r)$ and $f_n(r)$ assume problem-specific forms.

This modeling framework allows for scalar or vector-valued variables r_n and w_n. For clarity, in this chapter we assume that the state variables are K-dimensional vectors while observation variables are L-dimensional. To represent vectors, in the following, we exclusively adopt *row* vector convention with scalars recovered when $K = 1$ or $L = 1$.

In this setting, our generative model becomes a *Gaussian* state-space model,

$$r_1 \sim \text{Normal}_K(\mu, \upsilon), \tag{9.4}$$

$$r_n|r_{n-1} \sim \text{Normal}_K\left(q_n(r_{n-1}), V_n\right), \qquad n = 2:N, \tag{9.5}$$

$$w_n|r_n \sim \text{Normal}_L\left(f_n(r_n), U_n\right), \qquad n = 1:N. \tag{9.6}$$

As this model is already in generative form, we can easily characterize its joint distribution $p(w_{1:N}, r_{1:N})$ via MC sampling. That is, because of the factorization

$$p(w_{1:N}, r_{1:N}) = p(r_1)p(w_1|r_1)\left(\prod_{n=2}^{N} p(r_n|r_{n-1})p(w_n|r_n)\right), \qquad (*9.7*)$$

this can be achieved via ancestral sampling as in Algorithm 9.1.

Algorithm 9.1 Ancestral sampling for Gaussian state-space models

To simulate states and observations from the model of Eqs. (9.4) to (9.6), iterate as follows.

- Generate $r_1 \sim \text{Normal}_K(\mu, \upsilon)$.
- Generate $w_1 \sim \text{Normal}_L(f_1(r_1), U_1)$.
- For $n = 2 : N$,
  - Generate $r_n \sim \text{Normal}_K(q_n(r_{n-1}), V_n)$.
  - Generate $w_n \sim \text{Normal}_L(f_n(r_n), U_n)$.

With a linear Gaussian state-space model, we can address four main questions.

1. Given parameters and measurements, what are the properties of the *previous* states?
2. Given parameters and measurements, what are the properties of the *subsequent* states?
3. Given parameters and measurements, what are the properties of *all* states?
4. Given measurements, what are the properties of the states and parameters?

These questions involve *smoothing, forecasting, simulation*, and *estimation*, respectively.

As we will see in the next section, smoothing, forecasting, and simulation can be answered through the distributions $p(r_n|w_{1:N}), p(r_{N+d}|w_{1:N})$, and $p(r_{1:N}|w_{1:N})$, respectively. The computation of these distributions requires specialized schemes. Since these schemes heavily rely on analytic formulas, the three problems can be solved without approximations only in the very specialized class of linear Gaussian state-space models that we investigate in Section 9.3. Finally, estimation requires the characterization of $p(r_{1:N}, \theta|w_{1:N})$ where θ gathers any parameter that may have unknown value. This requires a Bayesian formulation and, generally, MCMC sampling. We investigate this case in Section 9.4.

* Reminder: The asterisks by some equations indicates that the detailed derivation can be found in Appendix F.

9.3 Linear Gaussian State-Space Models

For a number of systems, it is sufficient to model the transition $q_n(r)$ and observation $f_n(t)$ functions of a Gaussian state-space model as *linear polynomials*. For instance,

$$q_n(r) = a_n + rA_n, \qquad f_n(r) = b_n + rB_n,$$

for some vectors a_n, b_n and matrices A_n, B_n of appropriate dimensions consistent with the covariances V_n, U_n and state and observation variables. With this convention, our statistical model becomes a *linear* Gaussian state-space model,

$$r_1 \sim \text{Normal}_K(\mu, \upsilon), \tag{9.8}$$

$$r_n | r_{n-1} \sim \text{Normal}_K\left(a_n + r_{n-1}A_n, V_n\right), \qquad n = 2 : N, \tag{9.9}$$

$$w_n | r_n \sim \text{Normal}_L\left(b_n + r_n B_n, U_n\right), \qquad n = 1 : N. \tag{9.10}$$

The great advantage of mapping a problem onto a linear Gaussian state-space model is that, as we demonstrate below, we can compute almost every probability distribution of interest analytically. For instance, smoothing and forecasting are answered via $p(r_n|w_{1:N})$ and $p(r_{N+d}|w_{1:N})$, both of which are obtained in closed form. As the underlying theory parallels the theory of the HMM to a large degree, below we only provide a brief derivation of key results.

Example 9.2 Identification of a homogenous linear Gaussian state-space model

In its *homogenous* form, a linear Gaussian state-space model only admits parameters that remain constant in time. Such a model is summarized in

$$r_1 \sim \text{Normal}_K\left(\mu, \upsilon\right),$$

$$r_n | r_{n-1} \sim \text{Normal}_K\left(a + r_{n-1}A, V\right), \qquad n = 2 : N,$$

$$w_n | r_n \sim \text{Normal}_L\left(b + r_n B, U\right), \qquad n = 1 : N.$$

The ingredients of this model are:

- r_n, a random vector of size K.
- w_n, a random vector of size L.
- μ, a vector of size K.
- a, a vector of size K.
- b, a vector of size L.
- A, a square matrix of size K.
- B, a rectangular matrix of size $L \times K$.
- υ, a square matrix of size K.
- V, a square matrix of size K.
- U, a square matrix of size L.

9.3.1 Filtering

As with HMMs, it is convenient to start our study of linear Gaussian state-space models with the computation of *filters*. These are defined by

$$\hat{A}_n(r_n) = p\left(r_n | w_{1:n}\right)$$

and capture statistics of hidden states under measurements *only* up to each state's time level.

Starting with the first time level and moving forward, each filter's density can be computed sequentially using Bayes' rule,

$$\hat{\mathcal{A}}_n(r_n) \propto p\,(w_n|r_n)\,p\,(r_n|w_{1:n-1}). \tag{*9.11*}$$

The evaluation of this formula can be carried out in two stages: a *prediction stage* where the density $p\,(r_n|w_{1:n-1})$ is computed first and a *correction stage* where the product $p\,(w_n|r_n)\,p\,(r_n|w_{1:n-1})$ is formed next. Once both stages are completed, the missing proportionally constant is easily recovered by normalization.

In particular, at $n = 1$ the prediction relies on the initialization rule

$$p\,(r_1) = \text{Normal}_K\left(r_1;\, \bar{\mu}_1, \bar{\upsilon}_1\right),$$

where $\bar{\mu}_1 = \mu$ and $\bar{\upsilon}_1 = \upsilon$. Subsequently, the correction relies on

$$p(w_1|r_1)p(r_1) \propto \text{Normal}_K(r_1;\, \hat{\mu}_1, \hat{\upsilon}_1), \tag{*9.12*}$$

where $\hat{\mu}_1, \hat{\upsilon}_1$ are functions of $\bar{\mu}_1, \bar{\upsilon}_1$. Similarly, at $n=2$ the prediction relies on

$$p\,(r_2|w_1) = \text{Normal}_K\left(r_2;\, \bar{\mu}_2, \bar{\upsilon}_2\right), \tag{*9.13*}$$

where $\bar{\mu}_2, \bar{\upsilon}_2$ are functions of $\hat{\mu}_1, \hat{\upsilon}_1$. Subsequently, as with Eq. (*9.12*), the correction relies on

$$p(w_2|r_2)p(r_2|w_1) \propto \text{Normal}_K(r_2;\, \hat{\mu}_2, \hat{\upsilon}_2),$$

where $\hat{\mu}_2, \hat{\upsilon}_2$, are functions of $\bar{\mu}_2, \bar{\upsilon}_2$, and so on. Prediction and correction formulas at subsequent time levels immediately follow from Eqs. (*9.12*) and (*9.13*).

As will become apparent, due to the specific structure of a linear Gaussian state-space model, its filter densities attain a unified form,

$$\hat{\mathcal{A}}_n(r_n) = \text{Normal}_K\left(r_n;\, \hat{\mu}_n, \hat{\upsilon}_n\right).$$

These are termed *Kalman filters*, or Kálmán filters, and their parameters, often termed statistics, are given by the formulas

$$\begin{aligned}
&\bar{\mu}_1 = \mu, && \bar{\mu}_n = a_n + \hat{\mu}_{n-1}A_n, && n = 2:N,\\
&\bar{\upsilon}_1 = \upsilon, && \bar{\upsilon}_n = V_n + A_n^T\hat{\upsilon}_{n-1}A_n, && n = 2:N,\\
& && G_n = \left(U_n + B_n^T\bar{\upsilon}_nB_n\right)^{-1}B_n\bar{\upsilon}_n, && n = 1:N,\\
& && \hat{\mu}_n = \bar{\mu}_n + \left(w_n - b_n - \bar{\mu}_nB_n\right)G_n, && n = 1:N,\\
& && \hat{\upsilon}_n = \bar{\upsilon}_n\,(\mathbb{1} - B_nG_n), && n = 1:N,
\end{aligned}$$

which can be computed in the forward recursion described in Algorithm 9.2. The correcting factors G_n are termed *Kalman gains* and are best evaluated, both accuracy- and time-wise, by solving the equivalent linear systems of equations $\left(U_n + B_n^T\bar{\upsilon}_nB_n\right)G_n = B_n\bar{\upsilon}_n$.

Algorithm 9.2 Forward recursion for Kalman filters

Given measurements $w_{1:N}$ and parameter values, the Kalman filters are computed as follows.

- At $n = 1$, *initialize* with:

$$\begin{aligned}
\bar{\mu}_1 &= \mu, \\
\bar{\upsilon}_1 &= \upsilon, \\
G_1 &= \left(U_1 + B_1^T \bar{\upsilon}_1 B_1\right)^{-1} B_1 \bar{\upsilon}_1, \\
\hat{\mu}_1 &= \bar{\mu}_1 + \left(w_1 - b_1 - \bar{\mu}_n B_1\right) G_1, \\
\hat{\upsilon}_1 &= \bar{\upsilon}_1 (\mathbb{1} - B_1 G_1).
\end{aligned}$$

- At $n = 2 : N$, *march forward* with:

$$\begin{aligned}
\bar{\mu}_n &= a_n + \hat{\mu}_{n-1} A_n, \\
\bar{\upsilon}_n &= V_n + A_n^T \hat{\upsilon}_{n-1} A_n, \\
G_n &= \left(U_n + B_n^T \bar{\upsilon}_n B_n\right)^{-1} B_n \bar{\upsilon}_n, \\
\hat{\mu}_n &= \bar{\mu}_n + \left(w_n - b_n - \bar{\mu}_n B_n\right) G_n, \\
\hat{\upsilon}_n &= \bar{\upsilon}_n (\mathbb{1} - B_n G_n).
\end{aligned}$$

Example 9.3 A toy case of forward filtering

Perhaps the simplest form of a linear Gaussian state-space model is described by the equations

$$\begin{aligned}
r_1 &\sim \text{Normal}\left(\mu, \upsilon\right), \\
r_n | r_{n-1} &\sim \text{Normal}\left(r_{n-1}, V\right), && n = 2 : N, \\
w_n | r_n &\sim \text{Normal}\left(r_n, U\right), && n = 1 : N.
\end{aligned}$$

In this case, we have only scalar parameters and the filters reduce to univariate distributions. In particular, the formulas read

$$\begin{aligned}
\hat{\mathcal{A}}_1(r_1) &= \text{Normal}\left(r_1; \hat{\mu}_1, \hat{\upsilon}_1\right) \\
&= \text{Normal}\left(r_1; \frac{\frac{\mu}{\upsilon} + \frac{w_1}{U}}{\frac{1}{\upsilon} + \frac{1}{U}}, \frac{1}{\frac{1}{\upsilon} + \frac{1}{U}}\right), && G_1 = \frac{\upsilon}{U + \upsilon}. \\
\hat{\mathcal{A}}_2(r_2) &= \text{Normal}\left(r_2; \hat{\mu}_2, \hat{\upsilon}_2\right) \\
&= \text{Normal}\left(r_2; \frac{\frac{\hat{\mu}_1}{V + \hat{\upsilon}_1} + \frac{w_2}{U}}{\frac{1}{V + \hat{\upsilon}_1} + \frac{1}{U}}, \frac{1}{\frac{1}{V + \hat{\upsilon}_1} + \frac{1}{U}}\right), && G_2 = \frac{V + \hat{\upsilon}_1}{U + V + \hat{\upsilon}_1}. \\
\hat{\mathcal{A}}_3(r_3) &= \text{Normal}\left(r_3; \hat{\mu}_3, \hat{\upsilon}_3\right) \\
&= \text{Normal}\left(r_3; \frac{\frac{\hat{\mu}_2}{V + \hat{\upsilon}_2} + \frac{w_3}{U}}{\frac{1}{V + \hat{\upsilon}_2} + \frac{1}{U}}, \frac{1}{\frac{1}{V + \hat{\upsilon}_2} + \frac{1}{U}}\right), && G_3 = \frac{V + \hat{\upsilon}_2}{U + V + \hat{\upsilon}_2}, \\
&\;\;\vdots && \;\;\vdots
\end{aligned}$$

9.3.2 Smoothing

Given a sequence of observations $w_{1:N}$, we can apply the terminal filter $\hat{\mathcal{A}}_N(r_N)$, based on $r_N|w_{1:N}$, to estimate the state at the very end of our observation sequence. However, to estimate states at any time level *before* the terminal one, we need a different set of distributions. This is because state estimates based on $r_n|w_{1:N}$ are more accurate than those provided by the corresponding filters, which are based on $r_n|w_{1:n}$, as the former incorporate every available observation.

Similar in spirit to the HMM, estimates here rely on *smoothers* defined as

$$\check{\mathcal{D}}_n(r_n) = p(r_n|w_{1:N}).$$

However, unlike with the HMM where these distributions are computed through a factorization in terms of forward and backward variables, *e.g.* Eq. (*8.12*), now it is more convenient to retain the unfactorized $\check{\mathcal{D}}_n(r_n)$. For instance, at $n = N$ we readily obtain

$$\check{\mathcal{D}}_N(r_N) = \text{Normal}_K\left(r_N; \check{\mu}_N, \check{\upsilon}_N\right), \tag{*9.14*}$$

where $\check{\mu}_N = \hat{\mu}_N$ and $\check{\upsilon}_N = \hat{\upsilon}_N$. Subsequently, at $n = N - 1$, we obtain

$$\check{\mathcal{D}}_{N-1}(r_{N-1}) = \text{Normal}_K\left(r_{N-1}; \check{\mu}_{N-1}, \check{\upsilon}_{N-1}\right), \tag{*9.15*}$$

where $\check{\mu}_{N-1}, \check{\upsilon}_{N-1}$ are functions of $\check{\mu}_N, \check{\upsilon}_N$. Marching backward, we can obtain earlier smoothers via formulas akin to Eq. (*9.15*). As will become apparent, the smoother densities of a linear Gaussian state-space model also attain a unified form,

$$\check{\mathcal{D}}_n(r_n) = \text{Normal}_K\left(r_n; \check{\mu}_n, \check{\upsilon}_n\right).$$

These smoothers are termed *Rauch–Tung–Striebel smoothers* and their statistics are given by the formulas

$$\begin{aligned}
&\check{\mu}_N = \hat{\mu}_N, \qquad && \check{\mu}_n = \hat{\mu}_n + \left(\check{\mu}_{n+1} - \bar{\mu}_{n+1}\right) J_n, \qquad && n = 1 : N-1,\\
&\check{\upsilon}_N = \hat{\upsilon}_N, && \check{\upsilon}_n = \hat{\upsilon}_n + J_n^T \left(\check{\upsilon}_{n+1} - \bar{\upsilon}_{n+1}\right) J_n, && n = 1 : N-1,\\
& && J_n = \bar{\upsilon}_{n+1}^{-1} A_{n+1} \hat{\upsilon}_n, && n = 1 : N-1,
\end{aligned}$$

which can be computed efficiently in the backward recursion described in Algorithm 9.3. The correcting factors J_n are termed *Rauch–Tung–Striebel gains* and, similar to the Kalman gains, are best evaluated by solving the equivalent linear systems of equations $\bar{\upsilon}_{n+1} J_n = A_{n+1}\hat{\upsilon}_n$.

Algorithm 9.3 Backward recursion for Rauch–Tung–Striebel smoothers

Given measurements $w_{1:N}$ and parameter values, the Rauch–Tung–Striebel smoothers are computed as follows.

- Use Algorithm 9.2 to obtain $\hat{\mu}_{1:N}, \hat{\upsilon}_{1:N}$, and $\bar{\mu}_{1:N}, \bar{\upsilon}_{1:N}$.

- At $n = N$, *initialize* with:

$$\check{\mu}_N = \hat{\mu}_N,$$
$$\check{v}_N = \hat{v}_N.$$

- At $n = N - 1 : 1$, *march backward* with:

$$J_n = \bar{v}_{n+1}^{-1} A_{n+1} \hat{v}_n,$$
$$\check{\mu}_n = \hat{\mu}_n + \left(\check{\mu}_{n+1} - \bar{\mu}_{n+1}\right) J_n,$$
$$\check{v}_n = \hat{v}_n + J_n^T \left(\check{v}_{n+1} - \bar{v}_{n+1}\right) J_n.$$

9.3.3 Forecasting

Given a sequence of observations $w_{1:N}$, we can estimate future states one or multiple time levels ahead. In fact, forecasting d time levels ahead, *i.e.*, characterizing the state statistics at time level $n = N + d$, is almost identical to the prediction stage we have already seen.

Note 9.1 Augmented state-space model

Forecasting requires an *augmented* model that contains states at time levels *beyond* those with available measurements. For instance, Eqs. (9.8) to (9.10) must be replaced by

$$r_1 \sim \text{Normal}_K(\mu, v),$$
$$r_n | r_{n-1} \sim \text{Normal}_K\left(a_n + r_{n-1} A_n, V_n\right), \qquad n = 2 : N + D,$$
$$w_n | r_n \sim \text{Normal}_L\left(b_n + r_n B_n, U_n\right), \qquad n = 1 : N,$$

where now the transition rules incorporate D *additional* steps that do not have a corresponding observation, as shown in Fig. 9.2. Such a model can support forecasting up to D steps ahead.

Although setting up such a general augmented model appears pedantic, it makes one aspect clear. Namely, that forecasting requires the specification of parameters $a_{N+1:N+D}$, $A_{N+1:N+D}$, $V_{N+1:N+D}$ not required in either filtering or smoothing.

Similar to filters and smoothers, we can also consider a *forecaster*,

$$\bar{\bar{\mathcal{E}}}_{N+d} = p(r_{N+d} | w_{1:N}).$$

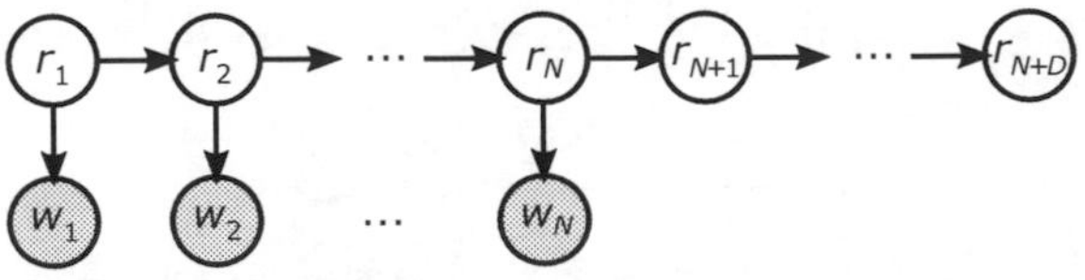

Fig. 9.2 Augmented state-space model that supports forecasting up to D time levels ahead.

Forecasting one step ahead is based on

$$\bar{\bar{\mathcal{E}}}_{N+1}(r_{N+1}) = \text{Normal}_K\left(r_{N+1}; \bar{\bar{\mu}}_{N+1}, \bar{\bar{\upsilon}}_{N+1}\right), \quad (*9.16*)$$

where $\bar{\bar{\mu}}_{N+1}, \bar{\bar{\upsilon}}_{N+1}$ depend upon $\hat{\mu}_N, \hat{\upsilon}_N$. Similarly, forecasting two steps ahead is based on

$$\bar{\bar{\mathcal{E}}}_{N+2}(r_{N+2}) = \text{Normal}_K\left(r_{N+2}; \bar{\bar{\mu}}_{N+2}, \bar{\bar{\upsilon}}_{N+2}\right), \quad (*9.17*)$$

where $\bar{\bar{\mu}}_{N+2}, \bar{\bar{\upsilon}}_{N+2}$ depend upon $\bar{\bar{\mu}}_{N+1}, \bar{\bar{\upsilon}}_{N+1}$. Subsequent steps are similar to Eq. (*9.17*). It becomes apparent that, as with the filter and smoother densities, our forecaster densities also attain a unified form,

$$\bar{\bar{\mathcal{E}}}_{N+d}(r_{N+d}) = \text{Normal}_K\left(r_{N+d}; \bar{\bar{\mu}}_{N+d}, \bar{\bar{\upsilon}}_{N+d}\right),$$

and the statistics are now given by

$$\bar{\bar{\mu}}_{N+1} = a_{N+1} + \hat{\mu}_N A_{N+1}, \qquad \bar{\bar{\upsilon}}_{N+1} = V_{N+1} + A_{N+1}^T \hat{\upsilon}_N A_{N+1},$$

$$\bar{\bar{\mu}}_{N+d} = a_{N+d} + \bar{\bar{\mu}}_{N+d-1} A_{N+d}, \quad \bar{\bar{\upsilon}}_{N+d} = V_{N+d} + A_{N+d}^T \bar{\bar{\upsilon}}_{N+d-1} A_{N+d}, \quad d = 2 : D.$$

9.3.4 Simulation

Given a sequence of observations $w_{1:N}$, the filters and smoothers characterize the properties of each state $r_n|w_{1:n}$ and $r_n|w_{1:N}$ irrespective of the others. However, to characterize the *joint* properties of the entire trajectory $r_{1:N}|w_{1:N}$, we need to perform *backward sampling* in a procedure akin to the HMM. That is, to generate trajectories $r_{1:N}|w_{1:N}$ we need to sample from

$$p(r_{1:N}|w_{1:N}) = p(r_1|r_2, w_1)p(r_2|r_3, w_{1:2}) \cdots p(r_{N-1}|r_N, w_{1:N-1})\hat{\mathcal{A}}_N(r_N). \quad (*9.18*)$$

Due to this factorization, we can first generate r_N directly from the filter $\hat{\mathcal{A}}_N(r_N)$ and subsequently generate each earlier state from the individual factors,

$$p(r_n|r_{n+1}, w_{1:n}) = \text{Normal}_K\left(r_n; \hat{\mu}_n + (r_{n+1} - \bar{\mu}_{n+1})J_n, \hat{\upsilon}_n - J_n^T \bar{\upsilon}_{n+1} J_n\right), \quad n = 1 : N-1. \quad (*9.19*)$$

The entire scheme, also termed *forward filtering backward sampling* (FFBS), is summarized in Algorithm 9.4.

Algorithm 9.4 FFBS in linear Gaussian state-space models

Given measurements $w_{1:N}$ and parameter values (in Eqs. (9.8)–(9.10)), trajectories $r_{1:N}|w_{1:N}$ are generated as follows.

- Using Algorithm 9.2 to obtain $\hat{\mu}_{1:N}, \hat{\upsilon}_{1:N}$ and $\bar{\mu}_{1:N}, \bar{\upsilon}_{1:N}, J_{1:N}$.
- At $n = N$, generate r_N by sampling from:

$$r_N \sim \text{Normal}_K\left(\hat{\mu}_N, \hat{\upsilon}_N\right).$$

- At $n = N - 1 : 1$, generate r_n by sampling from:

$$r_n \sim \text{Normal}_K \left(r_n; \hat{\mu}_n + (r_{n+1} - \bar{\mu}_{n+1}) J_n, \hat{\upsilon}_n - J_n^T \bar{\upsilon}_{n+1} J_n \right).$$

9.4 Bayesian State-Space Models and Estimation

Commonly, a state-space model contains uncharacterized parameters. For instance, in the model of Example 9.1 the unknown parameters may include the noise variance σ^2 or the diffusion coefficient D. For convenience here, we gather all unknown parameters of our model under a single variable θ.

The parameter estimation problem is best dealt with within the Bayesian framework. In particular, to be able to model a system with unknown parameters we need appropriate equations to form a joint distribution $p(\theta, r_{1:N}, w_{1:N})$. A straightforward extension of Eqs. (9.1) to (9.3) reads

$$\begin{aligned} \theta &\sim \mathbb{T}, \\ r_1|\theta &\sim \mathbb{R}_\theta, \\ r_n|r_{n-1}, \theta &\sim \mathbb{Q}^n_{\theta, r_{n-1}}, & n &= 2 : N, \\ w_n|r_n, \theta &\sim \mathbb{F}^n_{\theta, r_n}, & n &= 1 : N, \end{aligned}$$

where $\mathbb{T}$ is our prior that must also be specified. Under this *Bayesian* state-space model, our main objective is to characterize its completed posterior $p(r_{1:N}, \theta|w_{1:N})$.

To this end, we may develop a Gibbs sampler, as described in Algorithm 9.5, producing MCMC samples $r_{1:N}^{(j)}, \theta^{(j)}$ from the model's posterior. At a minimum, this sampler must iterate between two stages: one updating the trajectory, $r_{1:N}$, and one updating the parameters, θ. In general, both are computationally demanding updates and their implementation depends heavily upon the specifics of the model at hand.

Algorithm 9.5 Gibbs sampler for state-space models

A Gibbs sampler generates MCMC samples $r_{1:N}, \theta$ from the posterior $p(r_{1:N}, \theta|w_{1:N})$ by iterating these steps.

- Generate *states* $r_{1:N}^{\text{new}}$ by sampling from

$$r_{1:N}^{\text{new}} \sim p\left(r_{1:N}|\theta^{\text{old}}, w_{1:N}\right).$$

- Generate *parameters* θ^{new} by sampling from

$$\theta^{\text{new}} \sim p\left(\theta|r_{1:N}^{\text{new}}, w_{1:N}\right).$$

In the most general case, both stages of Algorithm 9.5 need to be implemented via Metropolis–Hastings updates which often results in poor MCMC performance, especially for long measurement sequences, $w_{1:N}$. Avoiding such *within Gibbs* schemes can be achieved in two special circumstances allowing for direct sampling. In particular, linear Gaussian state-space models allow for direct sampling of $r_{1:N}$ with *forward filtering backward sampling* as in Algorithm 9.4. Further, certain models allow for *conditionally conjugate* priors on the model parameters greatly facilitating the updates for θ. Example 9.4 takes advantage of both features.

Example 9.4 Bayesian tracking of a single diffusive particle

In this example, we consider the same setting as in Example 9.1. That is, we wish to decode successive positions of a single particle undergoing free Brownian motion assessed through noisy observations. Further, we assume that neither the diffusion coefficient D nor the noise variance σ^2 are known.

Within a Bayesian formulation both D and σ^2 must be inferred. To help with inference, we first *reparametrize* our model using

$$\eta = \frac{1}{D}, \qquad \tau = \frac{1}{\sigma^2},$$

in terms of precision parameters η, τ. On these we apply independent Gamma priors. Overall, our *Bayesian linear Gaussian state-space model* takes the form

$$\begin{aligned}
\eta &\sim \text{Gamma}\left(\phi, \psi\right),\\
\tau &\sim \text{Gamma}\left(\Phi, \Psi\right),\\
r_1 &\sim \text{Normal}_3\left((0,0,0), v\begin{pmatrix}1 & 0 & 0\\ 0 & 1 & 0\\ 0 & 0 & 1\end{pmatrix}\right),\\
r_n|r_{n-1}, \eta &\sim \text{Normal}_3\left(r_{n-1}, 2\frac{t_n - t_{n-1}}{\eta}\begin{pmatrix}1 & 0 & 0\\ 0 & 1 & 0\\ 0 & 0 & 1\end{pmatrix}\right), \qquad n = 2:N,\\
w_n|r_n, \tau &\sim \text{Normal}_2\left(r_n\begin{pmatrix}1 & 0\\ 0 & 1\\ 0 & 0\end{pmatrix}, \frac{1}{\tau}\begin{pmatrix}1 & 0\\ 0 & 1\end{pmatrix}\right), \qquad n = 1:N.
\end{aligned}$$

The posterior we seek to characterize is $p(r_{1:N}, \eta, \tau|w_{1:N})$. Similar to Algorithm 9.5, MCMC sampling of this posterior requires a Gibbs sampler with the following updates:

- Generate $r_{1:N}$ by sampling from $p(r_{1:N}|\eta, \tau, w_{1:N})$.
- Generate η, τ by sampling from $p(\eta, \tau|r_{1:N}, w_{1:N})$.

As the underlying dynamical model is linear and Gaussian, the first update is performed via forward filtering backward sampling as in Algorithm 9.4. Next, to update η, τ we develop an inner Gibbs sampler with the following two updates:

- Generate η by sampling from $p(\eta|\tau, r_{1:N}, w_{1:N})$.
- Generate τ by sampling from $p(\tau|\eta, r_{1:N}, w_{1:N})$.

Due to the structure of our model, the last two distributions reduce to $p(\eta|r_{1:N})$ and $p(\tau|r_{1:N}, w_{1:N})$, respectively. Taking advantage of *conditional conjugacy*, these are derived analytically and sampled directly,

$$p(\eta|r_{1:N}) = \text{Gamma}\left(\eta; \phi + \frac{3}{2}(N-1), \frac{1}{\frac{1}{\psi} + \frac{1}{2}\sum_{n=2}^{N}(r_n - r_{n-1})(r_n - r_{n-1})^T}\right), \tag{*9.20*}$$

$$p(\tau|r_{1:N}, w_{1:N}) = \text{Gamma}\left(\tau; \Phi + N, \frac{1}{\frac{1}{\Psi} + \frac{1}{2}\sum_{n=1}^{N}(w_n - r_n B)(w_n - r_n B)^T}\right). \tag{*9.21*}$$

9.5 Exercise Problems

Exercise 9.1 Gaussian linear state-space model

1. Select parameters for a two-dimensional Gaussian linear state-space model with 100 time levels and implement Algorithm 9.1 to simulate synthetic measurements.
2. Implement the forward recursion of Algorithm 9.2 and compute filters based on your synthetic measurements.
3. Implement the backward recursion of Algorithm 9.3 and compute smoothers based on your synthetic measurements.
4. Augment your model with 25 additional time points and compute forecasters based on your synthetic measurements.

Exercise 9.2 State-space models with missing observations

Consider a linear Gaussian state-space model where every other observation is missing.

1. Derive its graphical representation.
2. Derive its filter densities.
3. Derive its smoother densities.
4. Derive its backward sampling densities.

For concreteness, consider an even number of time levels and observations missing from every odd time level.

Project 9.1 EM for linear Gaussian state-space models

In Chapter 8 we saw two general training strategies for an HMM: one frequentist and one Bayesian. Both strategies also carry over linear Gaussian state-space models and in Section 9.4 we saw the latter. Here, we develop the former.

1. Consider a one-dimensional linear Gaussian state-space model of the form

$$
\begin{aligned}
r_1 &\sim \text{Normal}\left(\mu, v\right), \\
r_n | r_{n-1} &\sim \text{Normal}\left(r_{n-1}, V_n\right), & n = 2:N, \\
w_n | r_n &\sim \text{Normal}\left(\beta r_n, U\right), & n = 1:N.
\end{aligned}
$$

   Following Section 8.3.3, develop an EM method for the estimation of B and U that parallels Algorithm 8.4.
2. Extend your EM method for cases with multi-dimensional r_n and w_n.

Project 9.2 Estimation under Brownian motion

In this project we expand upon the setting of Examples 9.1 and 9.4. Namely, we consider the study of *free Brownian motion* via empirical data. Unlike the earlier examples where we modeled a single particle assessed under idealized conditions, here we consider modeling a realistic experiment that typically provides measurements of the positions of multiple particles corrupted by noise and other artifacts. Due to out-of-focus movement (or photo-blinking), experiments typically assess positions for a total number of time points that differs from particle to particle. Under these conditions, a realistic model reads

$$
\begin{aligned}
r_1^m &\sim \text{Normal}_3\left((0,0,0), v \begin{pmatrix} 1 & 0 & 0 \\ 0 & 1 & 0 \\ 0 & 0 & 1 \end{pmatrix}\right), \\
r_n^m | r_{n-1}^m &\sim \text{Normal}_3\left(r_{n-1}^m, 2D^m(t_n^m - t_{n-1}^m) \begin{pmatrix} 1 & 0 & 0 \\ 0 & 1 & 0 \\ 0 & 0 & 1 \end{pmatrix}\right), & n = 2:N^m, \\
w_n^m | r_n^m &\sim \text{Normal}_2\left((g_x, g_y)t_n^m + \begin{pmatrix} 1 & 0 \\ 0 & 1 \\ 0 & 0 \end{pmatrix} r_n^m, \sigma^2 \begin{pmatrix} 1 & 0 \\ 0 & 1 \end{pmatrix}\right), & n = 1:N^m.
\end{aligned}
$$

This model captures a total of M individual particles indexed by superscripts $m = 1:M$. The mth particle's position is assessed N^m times. Its corresponding positions are r_n^m, its measurements are w_n^m, and the time points are t_n^m.

In our formulation, all particles share the same initialization rule, though each one is characterized by transition rules with different diffusion coefficients that we model using $D^m = D_{\text{ref}} S^m$. Here, D_{ref} is a diffusion coefficient common to all particles and S^m is a unitless constant that allows for differences among individual particles.

The particles share the same assessment rules, which incorporate additive noise and drift. Noise is characterized by a variance σ^2 and drift is characterized by a velocity (g_x, g_y).

Use this formulation to:

1. Identify the dimensions and units of all quantities of interest.
2. Develop a Bayesian model that can estimate v, D_{ref}, $S^{1:M}$ and g_x, g_y, σ^2. Apply your own priors and briefly provide reasons for your choices.

3. Provide a graphical representation of your model.
4. Describe a Gibbs sampler that can characterize your model's posterior probability distribution.
5. Implement your Gibbs sampler.
6. Use synthetic data to validate your implementation of the Gibbs sampler.

Typical values are $M = 50$, $\upsilon = 1\ \mu\text{m}^2$, $D_{\text{ref}} = 0.1\ \mu\text{m}^2/s$, $\sigma^2 = 500\ \text{nm}^2$, $g_x = g_y = 1$ nm/ms. The values of N^m range between 5 and 25, the values of S^m range between 0.5 and 2.0, and successive times t_n^m are separated by 30 ms.

Additional Reading

C. Bishop. *Pattern recognition and machine learning*. Springer, 2006.

S. Särkkä. *Bayesian filtering and smoothing*. Cambridge University Press, 2013.

K. Law, A. Stuart, K. Zygalakis. *Data assimilation: A mathematical introduction*. Springer, 2015.

O. Cappé, E. Moulines, T. Rydén. *Inference in hidden Markov models*. Springer, 2005.

A. C. Harvey. *Forecasting, structural time series models and the Kalman filter*. Cambridge University Press, 1990.

R. Shumway, D. Stoffer. *Time series analysis and its applications with R examples*. 4th ed. Springer, 2017.

J. Bechhoefer. *Control theory for physicists*. Cambridge University Press, 2021.

S. Jazani, I. Sgouralis, O. M. Shafraz, M. Levitus, S. Sivasankar, S. Pressé. An alternative framework for fluorescence correlation spectroscopy. *Nat. Comm.*, 10:1, 2019.

M. Tavakoli, S. Jazani, I. Sgouralis, O. M. Shafraz, B. Donaphon, S. Sivasankar, M. Levitus, S. Pressé. Pitching single focus confocal data analysis one photon at a time with Bayesian nonparametrics. *Phys. Rev. X.*, 10:011021, 2020.

10 Continuous Time Models*

By the end of this chapter, we will have presented

- *Inference on Markov jump processes*
- *Uniformization and virtual jumps*
- *Filtering on irregular grids*

10.1 Modeling in Continuous Time

In Chapters 8 and 9, we modeled systems evolving in discrete time where the system's passing states were assumed to change precisely at predefined time levels. Until now, we had assumed that these coincide with the time levels of our measurements. However, in physical systems whose transitions may invariably occur anytime, these assumptions amount to:

- Rounding off the transition time to the nearest time level. The error introduced may be small when the system's transitions occur on time scales much slower than the periods between successive measurements. In general, however, these errors may be considerable.
- Relinquishing our ability to resolve continuous time trajectories. Continuous time trajectories may contain one or multiple jump times occurring between successive measurements.
- Relinquishing our ability to resolve kinetic rates. Transition probabilities do not uniquely map to rates, *i.e.*, only the converse is true as knowing rates provides a unique transition probability.

From a more philosophical standpoint, assuming slow kinetics also highlights a potential internal inconsistency of applications to physical systems as we often assume a priori that transition probabilities are unknown. It is appealing, therefore, to explore an analysis paradigm allowing us to model and study continuous time trajectories and transition rates.

On account of their mathematical structure, our previous models (HMMs and state-space models) render these quantities beyond our reach. Taken together, these considerations motivate the use of *hidden Markov jump processes* (hidden MJPs) for the estimation of continuous time trajectories and rates without artificially restricting rate values or introducing artifactual round-off errors. This treatment therefore generalizes our earlier modeling efforts.

* This is an advanced topic and could be skipped on a first reading.

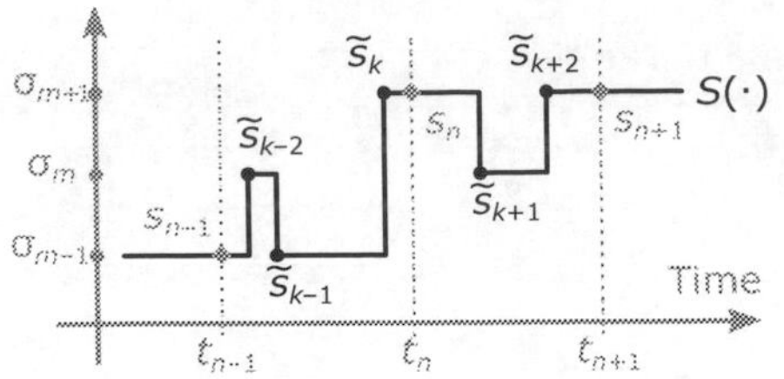

Fig. 10.1 An MJP trajectory. A comparison of the triplets $\{N, t_{1:N}, s_{1:N}\}$ and $\{K, \tilde{t}_{1:K}, \tilde{s}_{1:K}\}$ reveals that only $\{K, \tilde{t}_{1:K}, \tilde{s}_{1:K}\}$ allow for reconstruction of the trajectory $\mathcal{S}(\cdot)$. By contrast, a trajectory $\mathcal{S}(\cdot)$ allows for reconstruction of both $\{N, t_{1:N}, s_{1:N}\}$ and $\{K, \tilde{t}_{1:K}, \tilde{s}_{1:K}\}$.

Hidden MJPs are necessarily more complex and, from the onset, demand a generalization of the formalism we used in Chapters 8 and 9. To help introduce hidden MJPs, we begin by considering measurements performed at times t_n, whose passing states coincide with $s_n = \mathcal{S}(t_n)$. Borrowing notation from Section 2.3.2, we write

$$\mathcal{S}(\cdot) \sim \mathrm{MJP}_{\sigma_{1:M}}(\boldsymbol{\rho}, \boldsymbol{\Lambda}),$$

$$w_n|\mathcal{S}(\cdot) \sim \mathbb{G}_{\phi_{\mathcal{S}(t_n)}},$$

where $\boldsymbol{\rho}$ gathers the initial state occupation probabilities, $\boldsymbol{\Lambda}$ is the reaction rate matrix, and ϕ_{σ_m} are the emission parameters for each constitutive state in our system. Since our system evolves in continuous time, we label the system's jump times as $\tilde{t}_k$ and the states occupied just preceding these jumps as $\tilde{s}_k = \mathcal{S}(\tilde{t}_k)$. Our conventions are shown in Fig. 10.1.

Following established convention, we let the total number of acquired data points be N and the total number of jumps be K. These K jumps may occur anywhere between the trajectory's start, t_0, and end, t_{end}. Indeed, unlike HMMs, data points and jump times are differently indexed as K may be larger than, the same as, or smaller than N. Now that we have introduced our indexing convention, for convenience, we repeat in Note 10.1 the definition of a MJP trajectory.

Note 10.1 Trajectory definition

Following the convention first introduced in Section 2.3.2, we define the trajectory $\mathcal{S}(\cdot)$ over the time interval from t_0 to t_{end} as

$$\mathcal{S}(\cdot) = \begin{cases} \tilde{s}_1 & \text{if} \quad t_0 \le t < t_0 + \tilde{h}_1, \\ \tilde{s}_2 & \text{if} \quad t_0 + \tilde{h}_1 \le t < t_0 + \tilde{h}_1 + \tilde{h}_2, \\ & \vdots \\ \tilde{s}_K & \text{if} \quad t_0 + \tilde{h}_1 + \cdots + \tilde{h}_{K-1} \le t < t_{\mathrm{end}}, \end{cases}$$

where the $\tilde{h}_k$'s denote holding periods, $\tilde{h}_k = \tilde{t}_k - \tilde{t}_{k-1}$. The definition of $\tilde{s}(\cdot)$ is thus encoded in the triplet $\{K, \tilde{h}_{1:K-1}, \tilde{s}_{1:K}\}$ or, equivalently, $\{K, \tilde{t}_{1:K}, \tilde{s}_{1:K}\}$.

The posterior of relevance to this chapter is $p(\mathcal{S}(\cdot), \rho, \mathbf{\Lambda}, \boldsymbol{\phi}|w_{1:N})$. Before embarking on sampling the full posterior $p(\mathcal{S}(\cdot), \rho, \mathbf{\Lambda}, \boldsymbol{\phi}|w_{1:N})$, it is useful to establish how to sample $p(\mathcal{S}(\cdot)|\rho, \mathbf{\Lambda}, \boldsymbol{\phi}, w_{1:N})$ in the first place. Sampling continuous time trajectories requires new tools and this immediately leads us to the topic of the next section.

10.2 MJP Uniformization and Virtual Jumps

So far, we have introduced the triplet $\{K, \tilde{t}_{1:K}, \tilde{s}_{1:K}\}$ coinciding with trajectories of the physical system, and the triplet $\{N, t_{1:N}, s_{1:N}\}$ coinciding with the measurement. As we will shortly be describing an algorithm to sample trajectories, *i.e.*, to sample $\{K, \tilde{t}_{1:K}, \tilde{s}_{1:K}\}$, we introduce yet another time grid $\bar{t}_{1:L}$ whose triplet coincides with the proposed jump times ($\bar{t}_{1:L}$), and their associated passing states ($\bar{s}_\ell = \mathcal{S}(\bar{s}_\ell)$). This new triplet reads $\{L, \bar{t}_{1:L}, \bar{s}_{1:L}\}$. The jumps at $\bar{t}_{1:L}$ are often termed *virtual jumps*; see Fig. 10.2.

Note 10.2 States across time grids

As jumps denote state *changes*, by definition, successive states $\tilde{s}_k$ and $\tilde{s}_{k+1}$ must attain different values. By contrast, successive states s_n and s_{n+1} need not necessarily attain different values as our system may or may not remain in the *same* constitutive state over multiple measurements. By necessity, successive states $\bar{s}_\ell$ and $\bar{s}_{\ell+1}$ may or may not remain in the same state as proposed jump times may turn out to coincide with *self-transitions*.

Put differently, one $\{N, t_{1:N}, s_{1:N}\}$ is consistent with multiple $\{K, \tilde{t}_{1:K}, \tilde{s}_{1:K}\}$ and $\{L, \bar{t}_{1:L}, \bar{s}_{1:L}\}$. Similarly, one $\{L, \bar{t}_{1:L}, \bar{s}_{1:L}\}$ is consistent with multiple $\{K, \tilde{t}_{1:K}, \tilde{s}_{1:K}\}$. By contrast, one $\{K, \tilde{t}_{1:K}, \tilde{s}_{1:K}\}$ leads to unique $\{N, t_{1:N}, s_{1:N}\}$ and $\{L, \bar{t}_{1:L}, \bar{s}_{1:L}\}$.

10.2.1 Uniformization Sampler

In Section 2.3.3, we saw how to sample a trajectory $\mathcal{S}(\cdot)$ given $\rho, \mathbf{\Lambda}$ by realizing a triplet $\{K, \tilde{t}_{1:K}, \tilde{s}_{1:K}\}$ by means of a Gillespie simulation (special-

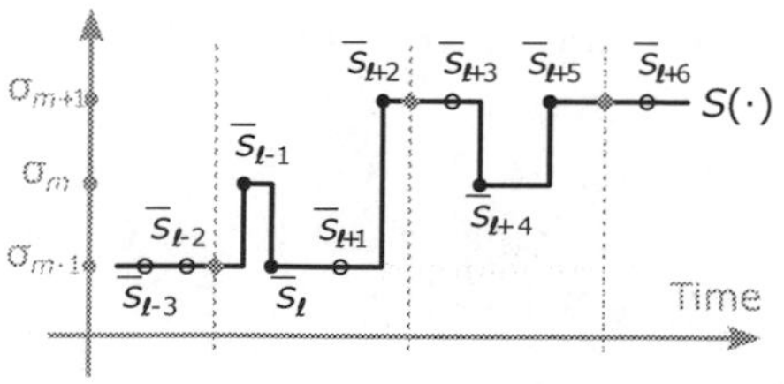

Fig. 10.2 Encoding of an MJP trajectory with the triplet $\{L, \bar{t}_{1:L}, \bar{s}_{1:L}\}$. Although the triplet $\{L, \bar{t}_{1:L}, \bar{s}_{1:L}\}$ allows for reconstruction of the trajectory $\mathcal{S}(\cdot)$, a trajectory $\mathcal{S}(\cdot)$ accommodates multiple triplets $\{L, \bar{t}_{1:L}, \bar{s}_{1:L}\}$.

ized to the case of one particle undergoing transitions between different constitutive states). Put differently, in the language of this chapter, the Gillespie simulation provides a strategy by which to draw samples from the *prior* distribution $p(\mathcal{S}(\cdot)|\boldsymbol{\rho}, \boldsymbol{\Lambda})$. Yet the distribution of relevance to this chapter is the *posterior* $p(\mathcal{S}(\cdot)|\boldsymbol{\rho}, \boldsymbol{\Lambda}, \boldsymbol{\phi}, w_{1:N})$. The latter differs critically from the former in that it is also conditioned on data $w_{1:N}$. As we do not know where the jump times may be located from times t_0 to t_{end}, we propose $(\bar{t}_\ell)$ in order to sample from $p(\mathcal{S}(\cdot)|\boldsymbol{\rho}, \boldsymbol{\Lambda}, \boldsymbol{\phi}, w_{1:N})$ though these proposed times may not coincide with jump times $\tilde{t}_k$. As such, the Gillespie simulation for trajectories is inappropriate. For this reason, we turn to an alternative, termed *uniformization*, as described in Algorithm 10.1.

Algorithm 10.1 MJP uniformization

Given t_0, t_{end}, $\boldsymbol{\rho}$, $\boldsymbol{\Lambda}$, and $\boldsymbol{\phi}$, we use $\boldsymbol{\Lambda}$ to compute any rate, Ω, large enough to satisfy $\Omega > \max_m \lambda_{\sigma_m}$. Here, we recall that λ_{σ_m} is the resulting escape rate from state σ_m.

Given $\boldsymbol{\Lambda}$, and defining a transition probability matrix $\bar{\boldsymbol{\Pi}} = \mathbb{1} + \boldsymbol{G}/\Omega$, where we recall that $\boldsymbol{G}$ is the generator matrix constructed from $\boldsymbol{\Lambda}$ according to Note 2.17, uniformization proceeds as follows:

$$\begin{aligned} L &\sim \text{Poisson}\left((t_{\text{end}} - t_0)\Omega\right), \\ \bar{s}_0 &\sim \text{Categorical}_{\sigma_{1:M}}(\boldsymbol{\rho}), \\ \bar{s}_\ell | \bar{s}_{\ell-1} &\sim \text{Categorical}_{\sigma_{1:M}}\left(\bar{\boldsymbol{\Pi}}_{\bar{s}_{\ell-1}}\right), && \ell = 1:L, \\ \bar{\bar{t}}_\ell &\sim \text{Uniform}_{[t_0, t_{\text{end}}]}, && \ell = 1:L. \end{aligned}$$

As a final step, times $\bar{t}_{1:L}$ are obtained by re-ordering $\bar{\bar{t}}_{1:L}$ in increasing order.

Algorithm 10.1 yields a triplet $\{L, \bar{t}_{1:L}, \bar{s}_{1:L}\}$ that encodes a trajectory $\mathcal{S}(\cdot)$ with the same statistics as the trajectory obtained by sampling a triplet $\{K, \tilde{t}_{1:K}, \tilde{s}_{1:K}\}$ using a Gillespie simulation.

10.2.2 Why Does Uniformization Work?

Our goal is to justify why we can use matrix elements of $\bar{\boldsymbol{\Pi}}$ to compute transition probabilities for $\bar{s}_\ell|\bar{s}_{\ell-1}$, *i.e.*, $p(\bar{s}_\ell = \sigma_m|\bar{s}_{\ell-1} = \sigma_{m'}, \boldsymbol{\Lambda}) = \bar{\pi}_{\sigma_{m'} \to \sigma_m}$, in Algorithm 10.1. To start, following convention from Eq. (2.22), we use $\boldsymbol{G}$ to define the transition probability over an interval of duration τ,

$$\boldsymbol{\Pi} = \exp(\tau \boldsymbol{G}). \tag{10.1}$$

Next, rearranging $\bar{\boldsymbol{\Pi}} = \mathbb{1} + \boldsymbol{G}/\Omega$ to obtain $\boldsymbol{G} = \Omega\left(\bar{\boldsymbol{\Pi}} - \mathbb{1}\right)$, we insert this expression for $\boldsymbol{G}$ into Eq. (10.1), which yields

$$\boldsymbol{\Pi} = e^{-\tau\Omega} \exp\left(\tau\Omega\bar{\boldsymbol{\Pi}}\right) = \sum_{\ell=0}^{\infty} \left(e^{-\tau\Omega} \frac{(\tau\Omega)^\ell}{\ell!}\right) \bar{\boldsymbol{\Pi}}^\ell.$$

As can be seen, the resulting transition matrix $\boldsymbol{\Pi}$, is a weighted sum over ℓ events in interval τ of another transition matrix $\bar{\boldsymbol{\Pi}}^\ell$. Thus, propagating state occupancies over the interval τ using $\mathbf{\Pi}$ can be achieved by sampling the number of events ℓ from a Poisson distribution with rate Ω over a period τ and with marginal distribution over states after ℓ steps of $\bar{\mathbf{\Pi}}^{\ell}$. This marginal distribution is precisely the one we use in uniformization and thus coincides with the same Markov jump process.

10.3 Hidden MJP Sampling with Uniformization and Filtering

Now that we have described uniformization, we are ready to develop a method to characterize $p(\mathcal{S}(\cdot)|\boldsymbol{\rho}, \mathbf{\Lambda}, \boldsymbol{\phi}, w_{1:N})$. Our goal is to sample this distribution and to accomplish this we will develop an appropriate MCMC sampler. Specifically, starting from some $s^{\text{old}}(\cdot)$ we consider three major steps resulting in $s^{\text{new}}(\cdot)$.

- From $\{K^{\text{old}}, \bar{t}^{\text{old}}_{1:K^{\text{old}}}, \bar{s}^{\text{old}}_{1:K^{\text{old}}}\}$, sample $\{L^{\text{old}}, \bar{t}^{\text{old}}_{1:L^{\text{old}}}, \bar{s}^{\text{old}}_{1:L^{\text{old}}}\}$.
- Use forward filtering backward sampling to pass from $\{L^{\text{old}}, \bar{t}^{\text{old}}_{1:L^{\text{old}}}, \bar{s}^{\text{old}}_{1:L^{\text{old}}}\}$ to $\{L^{\text{old}}, \bar{t}^{\text{old}}_{1:L^{\text{old}}}, \bar{s}^{\text{new}}_{1:L^{\text{old}}}\}$.
- Drop self-transitions to pass from $\{L^{\text{old}}, \bar{t}^{\text{old}}_{1:L^{\text{old}}}, \bar{s}^{\text{new}}_{1:L^{\text{old}}}\}$ to $\{K^{\text{new}}, \bar{t}^{\text{new}}_{1:K^{\text{new}}}, \bar{s}^{\text{new}}_{1:K^{\text{new}}}\}$.

In Section 10.3.1 we provide details for the first step, and in Section 10.3.2 we provide details for the second and third steps. For now, we provide the full algorithm containing these three steps.

Algorithm 10.2 Hidden MJP sampling

Given t_0, t_{end}, $\boldsymbol{\rho}$, $\mathbf{\Lambda}$, $\boldsymbol{\phi}$, and $w_{1:N}$, we use $\mathbf{\Lambda}$ to compute Ω and define a transition probability matrix $\bar{\mathbf{\Pi}} = \mathbb{1} + \mathbf{G}/\Omega$. Next, initialize $\{K^{\text{old}}, \bar{t}^{\text{old}}_{1:K^{\text{old}}}, \bar{s}^{\text{old}}_{1:K^{\text{old}}}\}$ and iterate as follows:

- For each k from $1 : K^{\text{old}}$, sample jump times in the interval from $\bar{t}^{\text{old}}_{k-1}$ to $\bar{t}^{\text{old}}_{k}$ according to Exponential$\left(\Omega - \lambda_{\bar{s}^{\text{old}}_{k-1}}\right)$.
- Collect the set of all jump times, both $\bar{t}^{\text{old}}_{1:K^{\text{old}}}$ and newly introduced in the last step, under the set $\bar{t}_{1:L}$. We now have $\left\{L^{\text{old}}, \bar{t}^{\text{old}}_{1:L^{\text{old}}}, \bar{s}^{\text{old}}_{1:L^{\text{old}}}\right\}$.
- Discard all state labels and construct $\bar{w}_{1:L}$ as per Section 10.3.2.
- Identify the transition probability $p(\bar{s}_\ell = \sigma_m|\bar{s}_{\ell-1} = \sigma_{m'}, \mathbf{\Lambda})$, appearing in the filter of Eq. (*10.2*), as the (m', m) element of $\bar{\mathbf{\Pi}}$, and perform a forward filter backward sampling pass following Section 10.3.2. This yields $\{L^{\text{old}}, \bar{t}^{\text{old}}_{1:L^{\text{old}}}, \bar{s}^{\text{new}}_{1:L^{\text{old}}}\}$.
- Discard self-transitions and collect the remaining jump times and passing states in $\{K^{\text{new}}, \bar{t}^{\text{new}}_{1:K^{\text{new}}}, \bar{s}^{\text{new}}_{1:K^{\text{new}}}\}$.

In the second step, $\bar{s}^{\text{old}}_{1:L^{\text{old}}}$ can be left unspecified because state labels are immediately discarded in the third step.

10.3.1 Embedding Uniformization into the MJP Trajectory Sampler

This section deals with passing from $\left\{K^{\text{old}}, \tilde{t}^{\text{old}}_{1:K^{\text{old}}}, \tilde{s}^{\text{old}}_{1:K^{\text{old}}}\right\}$ to $\left\{L^{\text{old}}, \bar{t}^{\text{old}}_{1:L^{\text{old}}}, \bar{s}^{\text{old}}_{1:L^{\text{old}}}\right\}$. To do so, we propose new jump times at each interval from $\tilde{t}^{\text{old}}_{k-1}$ to $\tilde{t}^{\text{old}}_{k}$ according to Exponential $\left(\Omega - \lambda_{\tilde{s}^{\text{old}}_{k-1}}\right)$. We terminate before the latest jump time exceeds $\tilde{t}_k$. This procedure, in general, adds new jumps between time intervals. The set of new and old jumps define the set of *virtual jumps*, $\bar{t}^{\text{old}}_{1:L^{\text{old}}}$. Since Ω exceeds the fastest escape rate, it is quite possible for many of the virtual jumps to coincide with self-transitions.

10.3.2 Irregular Grid Filtering in Hidden MJP Sampling

This section first deals with using forward filtering backward sampling to pass from $\left\{L^{\text{old}}, \bar{t}^{\text{old}}_{1:L^{\text{old}}}, \bar{s}^{\text{old}}_{1:L^{\text{old}}}\right\}$ to $\left\{L^{\text{old}}, \bar{t}^{\text{old}}_{1:L^{\text{old}}}, \bar{s}^{\text{new}}_{1:L^{\text{old}}}\right\}$. To be able to write down the required filter, we begin by defining a new array of data, $\bar{w}_{1:L}$, developed from $w_{1:N}$. Here, $\bar{w}_\ell$ coincides with all data points immediately following the time of the $(\ell - 1)$th virtual jump, $\bar{t}^{\text{old}}_{\ell-1}$, up to but excluding jump ℓ. Thus, the wider the spacing between jumps, the more data points are contained between jumps and contained in $\bar{w}_\ell$. Put differently, if multiple jumps occur between measured data, then $\bar{w}_\ell$ may be empty.

Now that we have defined an array of data appropriate for each virtual jump time, we introduce the forward filter at each jump time,

$$\mathcal{A}_\ell(\bar{s}_\ell) = p(\bar{w}_{1:\ell}, \bar{s}_\ell | \boldsymbol{\rho}, \boldsymbol{\Lambda}, \boldsymbol{\phi}),$$

entirely analogous to the filter introduced in Chapter 8 on a regular temporal grid coinciding with the times of the measurements. With this proviso, and closely following Eq. (*8.9*), the forward filter recursion then reads

$$\mathcal{A}_\ell(\bar{s}_\ell) = p(\bar{w}_\ell | \bar{s}_\ell, \boldsymbol{\phi}) \sum_{\bar{s}_{\ell-1}} p(\bar{s}_\ell | \bar{s}_{\ell-1}, \boldsymbol{\Lambda}) \mathcal{A}_{\ell-1}(\bar{s}_{\ell-1}), \tag{*10.2*}$$

where, from Eq. (2.22), $p(\bar{s}_\ell = \sigma_m | \bar{s}_{\ell-1} = \sigma_{m'}, \boldsymbol{\Lambda})$ is understood as the (m', m) element of $\exp\left((\bar{t}_\ell - \bar{t}_{\ell-1})\boldsymbol{G}\right)$, where $\boldsymbol{G}$ is, as before, the generator matrix constructed from $\boldsymbol{\Lambda}$. Similarly, following Eq. (*8.10*), the initial filter reads

$$\mathcal{A}_1(\bar{s}_1) = p(\bar{w}_1 | \bar{s}_1, \boldsymbol{\phi}) p(\bar{s}_1 | \boldsymbol{\rho}). \tag{*10.3*}$$

Equations (*10.2*) and (*10.3*) can be used to set up the forward recursion and expressions analogous to the normalized filters and the stable recursion immediately follow in direct analogy to Eq. (*8.28*) and Algorithm 8.5. Following the evaluation of the filters, backward sampling is performed similarly to Eqs. (8.37) and (*8.38*) and Algorithm 8.8.

Note 10.3 Emission likelihoods on the irregular grid

In our filters, defined on the irregular grid $\bar{t}_{1:L}$, appears the new term $p(\bar{w}_\ell|\bar{s}_\ell, \phi)$, which differs from the emission distribution, $p(w_n|s_n, \phi)$, appearing in the filter on the regular grid $t_{1:N}$. To evaluate $p(\bar{w}_\ell|\bar{s}_\ell, \phi)$, we consider two possibilities: either no data point is contained between $\bar{t}_{\ell-1}$ and $\bar{t}_\ell$; or one or more data points are contained between $\bar{t}_{\ell-1}$ and $\bar{t}_\ell$. In the former case, $\bar{w}_\ell$ is the empty set and $p(\bar{w}_\ell|\bar{s}_\ell, \phi)$ is independent of observations and thus proportional to a constant. In the latter case, $p(\bar{w}_\ell|\bar{s}_\ell, \phi)$ is simply the iid product of each independent observation occurring between $\bar{t}_{\ell-1}$ and $\bar{t}_\ell$.

10.4 Sampling Trajectories and Model Parameters

Given Algorithm 10.2, it is possible to develop a global Gibbs sampling scheme where we iterate between parameter, $\rho, \mathbf{\Lambda}, \phi|\mathcal{S}(\cdot), w_{1:N}$, and trajectory $\mathcal{S}(\cdot)|\rho, \mathbf{\Lambda}, \phi, w_{1:N}$, sampling. For this, following the discussion in Section 2.3.2, we present two sets of generative models, one expressed in terms of transition rates and the other in terms of transition probabilities and escape rates.

10.4.1 Hidden MJP Formulation with Transition Rates

From Eq. (2.5) and Eq. (2.6), a generative model expressed solely in terms of rates,

$$\tilde{s}_{k+1}|\tilde{s}_k \sim \text{Categorical}_{\sigma_{1:M}}\left(\frac{\lambda_{\tilde{s}_k\to\sigma_1}}{\lambda_{\tilde{s}_k}}, \ldots, \frac{\lambda_{\tilde{s}_k\to\sigma_M}}{\lambda_{\tilde{s}_k}}\right), \qquad k = 2:K,$$
$$\tilde{h}_k|\tilde{s}_k \sim \text{Exponential}\left(\lambda_{\tilde{s}_k}\right), \qquad k = 2:K-1,$$

lends itself to a physically intuitive formulation. That is,

$$\begin{aligned}
\lambda_{\sigma_m\to\sigma_{m'}} &\sim \text{Gamma}\left(fE_{\sigma_m}E_{\sigma_{m'}}, \lambda_{\text{ref}}\right),\\
\rho &\sim \text{Dirichlet}_{\sigma_{1:M}}(\eta\zeta),\\
\phi_{\sigma_m} &\sim \mathbb{H},\\
\tilde{s}_0|\rho &\sim \text{Categorical}_{\sigma_{1:M}}(\rho),\\
\tilde{s}_k|\tilde{s}_{k-1}, \mathbf{\Lambda} &\sim \text{Categorical}_{\sigma_{1:M}}\left(\frac{\lambda_{\tilde{s}_{k-1}\to\sigma_1}}{\sum_{m=1}^M \lambda_{\tilde{s}_{k-1}\to\sigma_m}}, \ldots, \frac{\lambda_{\tilde{s}_{k-1}\to\sigma_M}}{\sum_{m=1}^M \lambda_{\tilde{s}_{k-1}\to\sigma_m}}\right),\\
\tilde{h}_k|\tilde{s}_k &\sim \text{Exponential}\left(\lambda_{\tilde{s}_k}\right),\\
w_n|\mathcal{S}(\cdot) &\sim \mathbb{G}_{\phi_{\mathcal{S}(t_n)}}.
\end{aligned}$$

In this model, K is implicitly defined by the first holding period that ends after t_{end}. As seen in Note 8.13, f and $E_{\sigma_{1:M}}$ are hyperparameters controlling how tightly each constitutive state couples to the others. In Algorithm 10.3 we sketch a basic Metropolis–Hastings within Gibbs sampling scheme that implements this model.

Algorithm 10.3 Metropolis–Hastings within Gibbs for Markov jump processes

Given t_0, t_{end}, and $w_{1:N}$, we initialize ρ^{old}, $\mathbf{\Lambda}^{\text{old}}$, ϕ^{old}, and $\left\{K^{\text{old}}, \tilde{t}^{\text{old}}_{1:K^{\text{old}}}, \tilde{s}^{\text{old}}_{1:K^{\text{old}}}\right\}$. We then iterate the following:

- Sample a trajectory $\{K^{\text{new}}, \tilde{t}^{\text{new}}_{1:K^{\text{new}}}, \tilde{s}^{\text{new}}_{1:K^{\text{new}}}\}$ following Algorithm 10.2.
- Given $s^{\text{new}}_{1:K^{\text{new}}}$, sample the parameters ρ, $\mathbf{\Lambda}$, ϕ using a Metropolis–Hastings scheme.

10.4.2 Hidden MJP Formulation with Transition Probabilities

In a separate formulation, appearing earlier in Section 2.3.2, we recall that we may write

$$\begin{aligned} \tilde{s}_{k+1}|\tilde{s}_k &\sim \text{Categorical}_{\sigma_{1:M}}\left(\pi_{\tilde{s}_k}\right), & k &= 2:K, \\ \tilde{h}_k|\tilde{s}_k &\sim \text{Exponential}\left(\lambda_{\tilde{s}_k}\right), & k &= 2:K-1, \end{aligned}$$

with the understanding that π_{σ_m} has self-transitions set to zero.

In contrast to the previous representation expressed solely in terms of rates, a representation expressed both in terms of rates and transition probabilities, while less physically intuitive, allows us to place Dirichlet priors on the transition probabilities.

Following the ideas of Section 8.4.2, we write

$$\begin{aligned} \lambda_{\sigma_m} &\sim \text{Gamma}\left(E_{\sigma_m}, \lambda_{\text{ref}}\right), \\ \rho &\sim \text{Dirichlet}_{\sigma_{1:M}}\left(\eta\zeta\right), \\ \pi_{\sigma_m} &\sim \text{Dirichlet}_{\sigma_{1:M}}\left(\alpha_{\sigma_m}\beta_{\sigma_m}\right), \\ \phi_{\sigma_m} &\sim \mathbb{H}, \\ \tilde{s}_0|\rho &\sim \text{Categorical}_{\sigma_{1:M}}\left(\rho\right), \\ \tilde{s}_k|\tilde{s}_{k-1}, \Pi &\sim \text{Categorical}_{\sigma_{1:M}}\left(\pi_{\sigma_m}\right), & k &= 2:K, \\ \tilde{h}_k|\tilde{s}_k &\sim \text{Exponential}\left(\lambda_{\tilde{s}_k}\right), & k &= 1:K-1, \\ w_n|s_n &\sim \mathbb{G}_{\phi_{s_n}}. \end{aligned}$$

Again, K is implicitly defined by the first holding period that ends after t_{end}. As we discussed in Note 7.7, the element of the base distribution, β_{σ_m}, along the diagonal must be set to zero so that π_{σ_m} has zero weight associated to self-transitions. As earlier, it is now possible to develop an algorithm akin to Algorithm 10.3 to sample both trajectories and other parameters.

10.4.3 Trajectory Marginalization

As we now understand, different model features and parameters dictate MCMC mixing. These include, but are not limited to: the size of the state-space M, scale separation between rates, presumptive differences in values for the emission parameters, and measurement noise models. In general, improving mixing often qualifies as an active field of research worthy of independent investigation. In fact, the diversity of solutions to this problem are only outnumbered by the potential failure points we may encounter along the way for our applications. At any rate, naturally, there comes the question as to what exactly we wish to estimate with our model and whether it is worth the computational cost. In light of this argument, for the hidden MJP we may opt to marginalize over trajectories and simply estimate rates or, equivalently, transition probabilities.

For example, we may marginalize over passing states by summing the terminal forward filter, $\mathcal{A}_L(\bar{s}_L) = p(\bar{w}_{1:L}, \bar{s}_L|\rho, \Lambda, \phi)$, to compute the marginal likelihood $p(\bar{w}_{1:L}|\rho, \Lambda, \phi)$. Armed with this likelihood, we may then estimate parameters by placing priors on these and subsequently using a modification of Algorithm 10.3 such as shown below in Algorithm 10.4.

Algorithm 10.4 **Hidden MJP sampling with trajectory marginalization**

Given t_0, t_{end}, $w_{1:N}$, $\rho^{\text{old}}, \Lambda^{\text{old}}, \phi^{\text{old}}$, and $\{K^{\text{old}}, \bar{t}^{\text{old}}_{1:K^{\text{old}}}, \bar{s}^{\text{old}}_{1:K^{\text{old}}}\}$, iterate the following.

- Sample a trajectory, $\{L^{\text{old}}, \bar{t}^{\text{old}}_{1:L^{\text{old}}}, \bar{s}^{\text{old}}_{1:L^{\text{old}}}\}$, following Algorithm 10.2.
- Compute the forward filter and obtain the marginal likelihood $p(w_{1:N}|\rho^{\text{old}}, \Lambda^{\text{old}}, \phi^{\text{old}}, L^{\text{old}}, \bar{t}^{\text{old}}_{1:L^{\text{old}}})$.
- Propose values for $\rho^{\text{prop}}, \Lambda^{\text{prop}}, \phi^{\text{prop}}$ with a Metropolis–Hastings proposal.
- Compute the forward filter and obtain the marginal likelihood $p(w_{1:N}|\rho^{\text{prop}}, \Lambda^{\text{prop}}, \phi^{\text{prop}}, L^{\text{old}}, \bar{t}^{\text{old}}_{1:L^{\text{old}}})$.
- Carry out the Metropolis–Hastings acceptance test.

As compared to Algorithm 10.2, this algorithm exhibits better mixing as we have eliminated states $\bar{s}_{1:L}$.

In Algorithm 10.4, we see that we still need to sample the trajectory even if only to marginalize over the passing states to compute the likelihood at the next iteration. This then raises the question as to why we need to propose L and $\bar{t}_{1:L}$ in the first place and motivates filtering in continuous time.

Note 10.4 Filtering in continuous time

We end with a note on continuous time filtering with which we may sample states at specific, preselected times. We inherit, unchanged, the notion of the triplet $\{N, t_{1:N}, s_{1:N}\}$ coinciding with our measurements. We eliminate the triplet $\{K, \tilde{t}_{1:K}, \tilde{s}_{1:K}\}$ as we no longer care to estimate trajectories, thereby also eliminating the notion of trajectory proposals, $\{L, \bar{t}_{1:L}, \bar{s}_{1:L}\}$.

Instead, we define a new triplet $\{Q, \bar{\bar{t}}_{1:Q}, \bar{\bar{s}}_{1:Q}\}$ coinciding with the times and states associated with the points at which we choose to evaluate our filter. We may define forward filters $\mathcal{A}_q(\bar{\bar{s}}_q)$ and develop forward recursions. These are similar to Eq. (*10.2*) except for a trivial change,

$$\mathcal{A}_q(\bar{\bar{s}}_q) = p(\bar{\bar{w}}_q|\bar{\bar{s}}_q, \boldsymbol{\phi}) \sum_{\bar{\bar{s}}_{q-1}} p(\bar{\bar{s}}_q|\bar{\bar{s}}_{q-1}, \boldsymbol{\Lambda})\mathcal{A}_{q-1}(\bar{\bar{s}}_{q-1}),$$

in the definition of $\bar{\bar{w}}_q$. It follows that all we need to do is select Q and $\bar{\bar{t}}_{1:Q}$ and perform forward filtering backward sampling to sample $\bar{\bar{s}}_{1:Q}$. This procedure, in turn, can be embedded into a Gibbs sampling scheme while estimating $\rho, \boldsymbol{\Lambda}, \boldsymbol{\phi}$.

10.5 Exercise Problems

Exercise 10.1 Implementing uniformization

Specify a time interval t_0, t_{end} and parameters $\rho, \boldsymbol{\Lambda}$. Implement Algorithm 10.1 for a system transitioning between two discrete states. Repeat the simulation for the same $\rho, \boldsymbol{\Lambda}$ using the Gillespie simulation. Demonstrate that the following statistics of your trajectories are consistent using both methods: average time spent in each state; fraction of time spent in each state; and the number of transitions between different states.

Exercise 10.2 Sampling trajectories for a Markov jump process

Consider a system transitioning between two discrete states.

1. Generate synthetic data using the Gillespie simulation. Start by assuming that the measurement points occur on timescales on par with transition events. For simplicity assume that measurement points are equally spaced. Select measurement points and, to those, add normal noise.
2. Implement Algorithm 10.2 using a stable form of the forward recursion to sample trajectories assuming all other parameters are known.
3. Construct a 95% credible interval around your trajectories and compare with the ground truth trajectory.
4. Repeat the above twice more: once with events occurring on time scales much slower than data acquisition and once with events occurring on time scales faster than data acquisition.
5. Discuss differences in the credible interval for all three cases.

Exercise 10.3 Performance of Viterbi versus hidden Markov jump processes

Once Exercise 10.2 is completed, apply the HMM paradigm and compare your credible intervals to the Viterbi trajectory of Algorithm 8.3. Compare your results in all three cases considered in Exercise 10.2, *i.e.*, where transition events are faster than, on par with, or slower than the time between measurements.

Exercise 10.4 Selecting a Poisson rate in uniformization

Generate data as in Exercise 10.2 with transition events occurring on time scales on par with data acquisition. Develop a uniformization procedure where Ω is barely larger than, twice as large as, or exceeds by tenfold the largest escape rate. Plot your posterior versus MCMC iteration and discuss differences in posterior convergence.

Exercise 10.5 Determining rates and trajectories

Generate data as in Exercise 10.2 with events occurring on time scales on par with the data acquisition. Develop a Gibbs sampling scheme, similar to Algorithm 10.3, to determine rates and trajectories from your synthetically generated data. Histogram the Gibbs samples of the rates after burn-in removal and compare with your ground truth.

Project 10.1 Determining trajectories for integrative detector models

A realistic measurement model often captures detector integration times. It does so by assuming that the detector reports on an average of the system states occupied over any given measurement period of duration Δt. That is, if the system spends time $\tilde{\tau}_1$ in state σ_1 with associated signal level μ_{σ_1} and time $\Delta t-\tilde{\tau}_1$ in state σ_2 with associated signal level μ_{σ_2}, then the measurement, w, over this period is sampled from $\text{Normal}\left(\left(\tilde{\tau}_1\mu_{\sigma_1} + (\Delta t - \tilde{\tau}_1)\mu_{\sigma_2}\right)/\Delta t, v\right)$. Of course, the states occupied and jump times within a measurement period are a priori unknown. Develop a sampling scheme to learn the trajectory, assuming known rates, under this new observation model.

Additional Reading

W. K. Grassmann. Transient solutions in Markovian queueing systems. *Comput. Operat. Res.*, 4:47, 1977.

W. K. Grassmann. Transient solutions in Markovian queues. *Eur. J. Operational Res.*, 1:396, 1977.

B. Zhang. Efficient path and parameter inference for Markov jump processes. PhD Thesis, Dept. of Statistics (Advisor: V. Rao), Purdue, 2019.

B. Zhang, V. Rao. Efficient parameter sampling for Markov jump processes. *J. Comp. Graph. Stat.*, 30:25, 2021.

V. Rao, Y. W. Teh. Fast MCMC sampling for Markov jump processes and extensions. *J. Mach. Learn. Res.*, 14:3295, 2013.

C. R. Shelton, G. Ciardo. Tutorial on structured continuous-time Markov processes. *J. AI Res.*, 51:725, 2014.

S. Ardavan, A. Bouchard-Côté. Priors over recurrent continuous time processes. *Adv. Neur. Inf. Proc. Sys.*, 24:2052, 2011.

Z. Kilic, I. Sgouralis, W. Heo, K. Ishii, T. Tahara, S. Pressé. Extraction of rapid kinetics from smFRET measurements using integrative detectors. *Cell Rep.: Phys. Sci.*, 2:100409, 2021.

Z. Kilic, I. Sgouralis, S. Pressé. Generalizing HMMs to continuous time for fast kinetics: Hidden Markov jump processes. *Biophys. J.*, 120:409 (2021).

PART III

APPENDICES

A Notation and Other Conventions

A.1 Time and Other Physical Quantities

Throughout this book, we adopt, to the degree possible, terminology that remains intuitive in the broader context of the natural sciences. For example, in the dynamical systems or machine learning literature it is customary to present "time" in a unitless integer-valued fashion. Unfortunately, such an approach might be confusing in our context and we discourage it.

We also pay special attention to maintaining physically correct units. So, in the examples and exercises, temporal variables are measured in units of time, spatial variables are measured in units of space, and so on.

Starting in Chapter 2, time becomes a critical notion and for this reason we reserve the most important indices for temporal quantities. Although some exceptions persist, we generally denote time using subscripts and reserve superscripts for other labeling conventions. As real-life experiments provide temporally arranged measurements, we use the same convention to index individual measurements even when our main interest is focused on time-independent problems.

A.2 Random Variables and Other Mathematical Notions

In Chapter 1, we distinguish between a random variable, most often by using capital letters, over the values that a random variable takes, most often by using lower case letters; for instance, R versus r. Additionally, we distinguish between probability distributions and probability densities; for instance, $\mathbb{P}$ and $p(r)$. However, starting in Chapter 2, this practice becomes cumbersome. For this reason, we generally relax these rules after Chapter 1.

Throughout the chapters, we sometimes denote vector or matrix quantities with bold faced letters; for instance $\boldsymbol{f}$ and $\boldsymbol{A}$. However, even so, we only do so sparingly when we wish to emphasize the vectorial or matricial properties of the quantity at hand. As we explain in detail below, we are careful to distinguish between lists, vectors, and arrays, and for all of these we adopt different notation.

A.3 Collections

Even in Chapter 1, we encounter models consisting of numerous variables and it is practical to arrange them into groups. Here, we gather our conventions concerning such grouping.

- When the ordering of the variables is *unimportant*, we denote groups using braces, for example $\{\alpha, \beta, \gamma\}$, and refer to them as *lists*. In these cases, it is valid to write down expressions like

$$\{\alpha, \beta, \gamma\} = \{\alpha, \gamma, \beta\}, \qquad \{1, 2, 3\} = \{1, 3, 2\}.$$

  In the particular case of lists of *indexed* variables, for example $w_1, w_2, \dots, w_N$, instead of $\{w_1, w_2, \dots, w_N\}$, often we use a compact notation with subscripts $w_{1:N}$ and it is valid to write

$$w_{1:N} = \{w_1, w_2, \dots, w_N\}.$$

- When the ordering of the variables is *important*, we denote groups using parenthesis, for example (x, y, z), and refer to them as *vectors*. In these cases, it is valid to write down expressions like

$$(x, y, z) \neq (x, z, y), \qquad (1, 2, 3) \neq (1, 3, 2).$$

- We also encounter cases where the ordering of indexed variables *depends on the indices*; for example, variables like $\pi_{\rho_1}, \pi_{\rho_2}, \dots, \pi_{\rho_M}$. We denote such groups using brackets with the ordering explicitly shown; for example, ${}_{\rho_1,\rho_2,\dots,\rho_M}[\pi_{\rho_1}, \pi_{\rho_2}, \dots, \pi_{\rho_M}]$, and refer to them as *arrays*. In these cases, it is valid to write down expressions like

$${}_{\sigma_1,\sigma_2,\sigma_3}[1, 2, 3] \neq {}_{\sigma_1,\sigma_2,\sigma_3}[1, 3, 2], \qquad {}_{\sigma_1,\sigma_2,\sigma_3}[1, 2, 3] = {}_{\sigma_1,\sigma_3,\sigma_2}[1, 3, 2].$$

  Of course, when the ordering adopted is obvious, generally, we avoid showing it and simply write $[\pi_{\sigma_1}, \pi_{\sigma_2}, \dots, \pi_{\sigma_M}]$ instead of the more elaborate ${}_{\rho_1,\rho_2,\dots,\rho_M}[\pi_{\rho_1}, \pi_{\rho_2}, \dots, \pi_{\rho_M}]$.

B Numerical Random Variables

Here, we list descriptions of some common *scalar* random variables. In our notation, $\mathbb{P}_\theta$ denotes a probability distribution that depends on θ, which, in general, might include one or more individual parameters. For a random variable $R \sim \mathbb{P}_\theta$ attaining values r, we use $p(r;\theta)$ to denote its probability density, which also depends on θ. As with all random variables whose distribution attains a density, probabilities are obtained by summing $dr\, p(r;\theta)$. This indicates that our densities $p(r;\theta)$ have *units* and these are equal to the *reciprocal units* of r.

To clarify the parametrizations of our distributions, for each random variable in the following sections, we also list its first and second moments. In particular, we use $\mathbb{E}(R)$ to denote the mean and $\mathbb{V}(R)$ to denote the variance of the random variable R. These are computed by the integrals

$$\mathbb{E}(R) = \int dr\, p(r;\theta) r, \qquad \mathbb{V}(R) = \int dr\, p(r;\theta)\,(r - \mathbb{E}(R))^2,$$

which are considered over the entire support of R.

The *support* of a random variable $R \sim \mathbb{P}_\theta$ gathers all values r that the random variable may realize. The associated density $p(r;\theta)$ is nonzero when r is in the support of $\mathbb{P}_\theta$ and zero otherwise.

Certain parameters in θ may characterize the spread of the values r over the support of a random variable $R \sim \mathbb{P}_\theta$. We refer to these parameters as *scales* when they have the same units as the mean of R and as *rates* when they have the reciprocal units of the mean of R.

B.1 Continuous Random Variables

B.1.1 With Bounded Support

Uniform Random Variable

Definition

A uniform random variable $R \sim \text{Uniform}_{[r_{\min}, r_{\max}]}$, takes real scalar values r, between $r_{\min}$ and $r_{\max}$, and has the probability density and moments

$$\text{Uniform}_{[r_{\min}, r_{\max}]}(r) = \frac{1}{r_{\max} - r_{\min}}, \qquad \mathbb{E}(R) = \frac{r_{\max} + r_{\min}}{2},$$

$$\mathbb{V}(R) = \frac{(r_{\max} + r_{\min})^2}{12}.$$

The associated density, $p(r)$, is plotted in Fig. B.1.

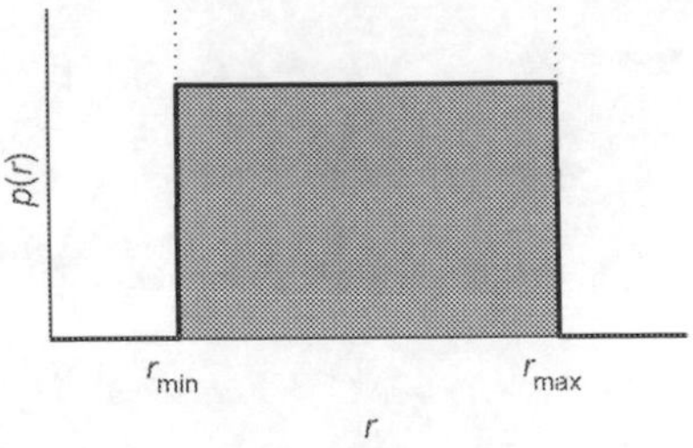

Fig. B.1 The uniform probability density.

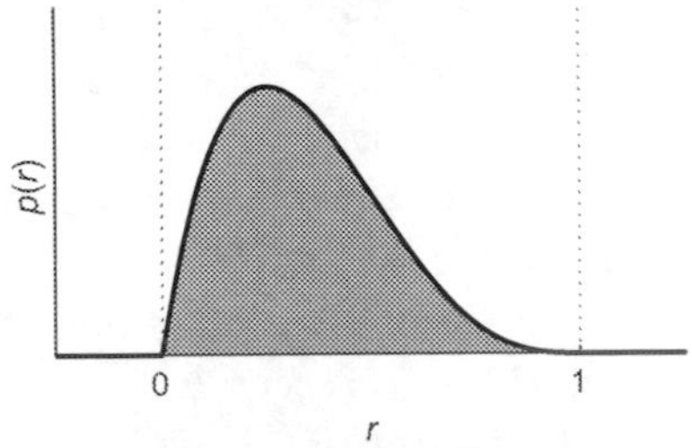

Fig. B.2 The beta probability density.

Parametrization

In the common parametrization, the *endpoints* $r_{\min}$ and $r_{\max}$ are real numbers with $r_{\min} < r_{\max}$.

Sampling Scheme

To simulate $R \sim \text{Uniform}_{[r_{\min}, r_{\max}]}$, first simulate $U \sim \text{Uniform}_{[0,1]}$, and then set $r = r_{\min} + (r_{\max} - r_{\min})u$.

Beta Random Variable

Definition

A beta random variable $R \sim \text{Beta}(\alpha, \beta)$ takes positive real scalar values r between 0 and 1 and has the probability density and moments

$$\text{Beta}(r; \alpha, \beta) = \frac{r^{\alpha-1}(1-r)^{\beta-1}}{B(\alpha, \beta)}, \quad \mathbb{E}(R) = \frac{\alpha}{\alpha+\beta},$$
$$\mathbb{V}(R) = \frac{\alpha\beta}{(\alpha+\beta)^2(\alpha+\beta+1)},$$

where $B(\cdot, \cdot)$ is the beta function. The associated density, $p(r)$, is plotted in Fig. B.2.

Parametrization

In the common parametrization, the *shapes* α and β are positive real numbers.

Related Distributions

The $\text{Uniform}_{[0,1]}$ distribution is a special of case of the Beta(1, 1) distribution.

Sampling Scheme

To simulate $R \sim \text{Beta}(\alpha, \beta)$, first simulate $U_1 \sim \text{Gamma}(\alpha, 1)$, $U_2 \sim \text{Gamma}(\beta, 1)$, and then set $r = u_1/(u_1 + u_2)$.

B.1.2 With Semibounded Support

Exponential Random Variable

Definition

An exponential random variable $R \sim \text{Exponential}(\lambda)$ takes positive real scalar values r and has the probability density and moments

$$\text{Exponential}(r; \lambda) = \lambda e^{-\lambda r}, \qquad \mathbb{E}(R) = \frac{1}{\lambda}, \qquad \mathbb{V}(R) = \frac{1}{\lambda^2}.$$

The associated density, $p(r)$, is plotted in Fig. B.3.

Parametrization

In the common parametrization, the *rate* λ is a positive real number.

Related Distributions

Alternative parametrizations include mean or scale $\mu = 1/\lambda$. According to this,

$$\text{Exponential}\left(r; \frac{1}{\mu}\right) = \frac{1}{\mu} e^{-\frac{r}{\mu}}, \qquad \mathbb{E}(R) = \mu, \qquad \mathbb{V}(R) = \mu^2.$$

Related Distributions

The Exponential(λ) distribution is a special of case of the Gamma(1, $1/\lambda$) distribution.

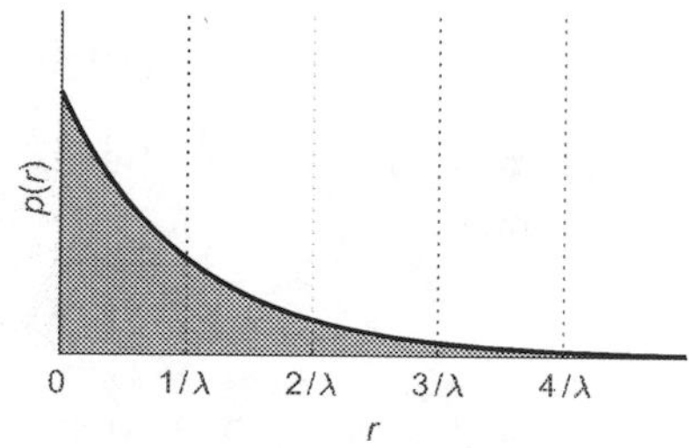

Fig. B.3 The exponential probability density.

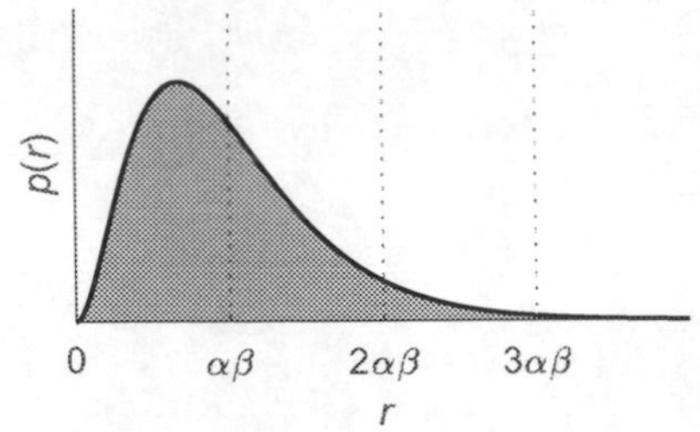

Fig. B.4 The gamma probability density.

Sampling Scheme

To simulate $R \sim \text{Exponential}(\lambda)$, first simulate $U \sim \text{Uniform}_{[0,1]}$, and set $r = -\frac{1}{\lambda}\log u$.

Gamma Random Variable

Definition

A gamma random variable $R \sim \text{Gamma}(\alpha, \beta)$ takes positive real scalar values r and has the probability density and moments

$$\text{Gamma}(r; \alpha, \beta) = \frac{1}{\beta\Gamma(\alpha)}\left(\frac{r}{\beta}\right)^{\alpha-1} e^{-\frac{r}{\beta}}, \quad \mathbb{E}(R) = \alpha\beta, \quad \mathbb{V}(R) = \alpha\beta^2,$$

where $\Gamma(\cdot)$ is the gamma function. The associated density, $p(r)$, is plotted in Fig. B.4.

Parametrization

In the common parametrization, the *shape* α and *scale* β are positive real numbers.

Alternative Parametrization

Alternative parametrizations include shape α and rate $\psi = 1/\beta$. According to this,

$$\text{Gamma}\left(r; \alpha, \frac{1}{\psi}\right) = \frac{\psi}{\Gamma(\alpha)}\left(\psi r\right)^{\alpha-1} e^{-\psi r}, \quad \mathbb{E}(R) = \frac{\alpha}{\psi}, \quad \mathbb{V}(R) = \frac{\alpha}{\psi^2}.$$

Related Distributions

The $\text{Exponential}(\lambda)$ distribution is a special of case of the $\text{Gamma}(1, 1/\lambda)$ distribution.

Sampling Scheme

To simulate $R \sim \text{Gamma}(\alpha, \beta)$, first simulate $U \sim \text{Gamma}(\alpha, 1)$, and then set $r = \beta u$.

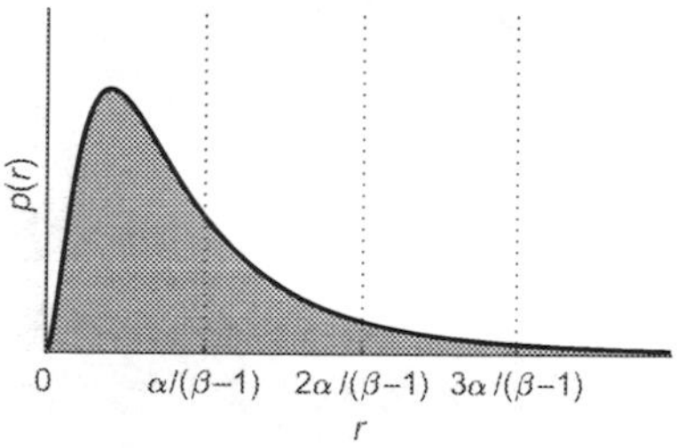

Fig. B.5 The beta prime probability density.

Beta Prime Random Variable

Definition

A betaprime random variable $R \sim \text{BetaPrime}(\alpha, \beta)$ takes positive real scalar values r and has the probability density and moments

$$\text{BetaPrime}(r; \alpha, \beta) = \frac{1}{B(\alpha, \beta)} \frac{r^{\alpha-1}}{(1+r)^{\alpha+\beta}}, \qquad \mathbb{E}(R) = \frac{\alpha}{\beta - 1},$$

$$\mathbb{V}(R) = \frac{\alpha(\alpha + \beta - 1)}{(\beta - 2)(\beta - 1)^2},$$

where $B(\cdot, \cdot)$ is the beta function. The associated density, $p(r)$, is plotted in Fig. B.5.

Parametrization

In the common parametrization, the *shapes* α and β are positive real numbers.

Sampling scheme

To simulate $R \sim \text{BetaPrime}(\alpha, \beta)$, first simulate $U_1 \sim \text{Gamma}(\alpha, 1)$, $U_2 \sim \text{Gamma}(\beta, 1)$, and then set $r = u_1/u_2$.

B.1.3 With Unbounded Support

Normal Random Variable

Definition

A normal random variable $R \sim \text{Normal}(\mu, \upsilon)$, also termed Gaussian, takes real scalar values r and has the probability density and moments

$$\text{Normal}(r; \mu, \upsilon) = \frac{1}{\sqrt{2\pi\upsilon}} \exp\left(-\frac{1}{2}\frac{(r-\mu)^2}{\upsilon}\right), \quad \mathbb{E}(R) = \mu, \quad \mathbb{V}(R) = \upsilon.$$

The associated density, $p(r)$, is plotted in Fig. B.6.

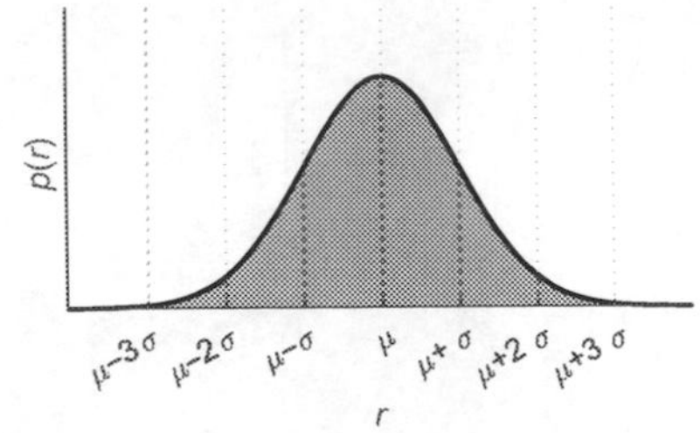

Fig. B.6 The normal probability density.

Parametrization

In the common parametrization, the *mean* μ and *variance* v are real numbers and v is positive.

Alternative Parametrization

Alternative parametrizations include mean μ and standard deviation $\sigma = \sqrt{v}$, or mean μ and precision $\tau = 1/v$. According to these,

$$\text{Normal}(r; \mu, \sigma^2) = \frac{1}{\sqrt{2\pi\sigma^2}} \exp\left(-\frac{1}{2}\frac{(r-\mu)^2}{\sigma^2}\right), \quad \mathbb{E}(R) = \mu, \quad \mathbb{V}(R) = \sigma^2,$$

$$\text{Normal}\left(r; \mu, \frac{1}{\tau}\right) = \sqrt{\frac{\tau}{2\pi}} \exp\left(-\frac{\tau}{2}(r-\mu)^2\right), \qquad \mathbb{E}(R) = \mu, \quad \mathbb{V}(R) = \frac{1}{\tau}.$$

Related Distributions

The Normal(μ, v) distribution is a special of case of the Normal$_M(\mu, v)$ distribution.

Sampling Scheme

To simulate $R \sim \text{Normal}(\mu, v)$, first simulate $U \sim \text{Normal}(0, 1)$, and then set $r = \mu + \sqrt{v}u$.

StudentT Random Variable

Definition

A StudentT random variable $R \sim \text{StudentT}_\nu(\mu, \sigma)$, takes real scalar values r and has the probability density and moments

$$\text{StudentT}_\nu(r; \mu, \sigma) = \frac{\Gamma\left(\frac{\nu+1}{2}\right)}{\Gamma\left(\frac{\nu}{2}\right)} \frac{1}{\sqrt{\pi\nu}} \frac{1}{\sigma} \left(1 + \frac{1}{\nu}\left(\frac{r-\mu}{\sigma}\right)^2\right)^{-\frac{\nu+1}{2}}, \mathbb{E}(R) = \mu,$$

$$\mathbb{V}(R) = \frac{\nu}{\nu - 2}\sigma^2,$$

where $\Gamma(\cdot)$ is the gamma function. The associated density, $p(r)$, is plotted in Fig. B.7.

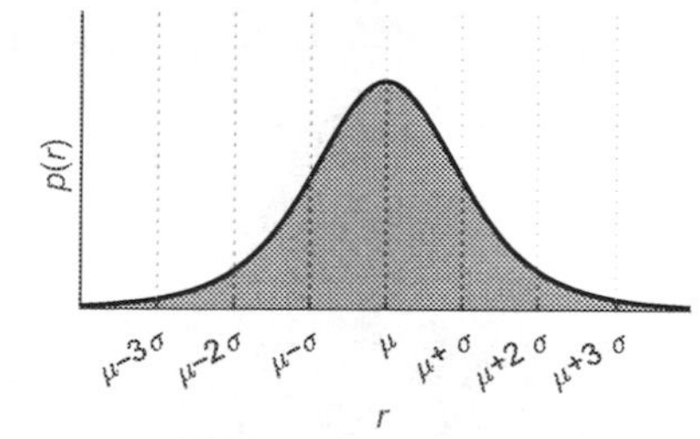

Fig. B.7 The StudentT probability density.

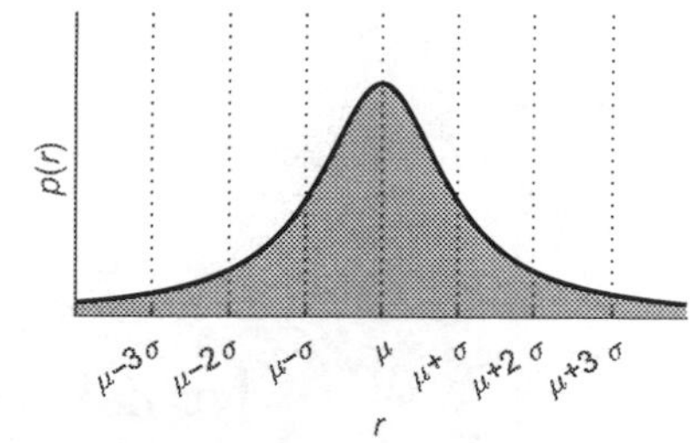

Fig. B.8 The Cauchy probability density.

Parametrization

In the common parametrization, the *mean* μ, *scale* σ, and degrees of freedom ν are real numbers and ν and σ are positive.

Related Distributions

The $\text{StudentT}_1(\mu, \sigma)$ distribution is a special of case of the $\text{Cauchy}(\mu, \sigma)$ distribution.

Sampling Scheme

To simulate $R \sim \text{StudentT}_\nu(\mu, \sigma)$, first simulate $U_1 \sim \text{Normal}(0, 1)$ and $U_2 \sim \text{Gamma}(\nu/2, 1)$, and set $r = \mu + \sigma\sqrt{\frac{\nu}{2u_2}}u_1$.

Cauchy Random Variable

Definition

A Cauchy random variable $R \sim \text{Cauchy}(\mu, \sigma)$, takes real scalar values r and has the probability density and moments

$$\text{Cauchy}(r; \mu, \sigma) = \frac{1}{\pi}\frac{\sigma}{\sigma^2 + \left(r - \mu\right)^2}, \qquad \mathbb{E}(R) = \infty, \qquad \mathbb{V}(R) = \infty.$$

The associated density, $p(r)$, is plotted in Fig. B.8.

Parametrization

In the common parametrization, the *mean* μ and *scale* σ are real numbers and σ is positive.

Related Distributions

The Cauchy(μ, σ) distribution is a special of case of the StudentT$_1(\mu, \sigma)$ distribution.

Sampling Scheme

To simulate $R \sim$ Cauchy(μ, σ), first simulate $U \sim$ Uniform$_{[0,1]}$, and then set $r = \mu + \sigma \tan\left(\pi\left(u - \frac{1}{2}\right)\right)$.

Laplace Random Variable

Definition

A Laplace random variable $R \sim$ Laplace(μ, β) takes real scalar values r and has the probability density and moments

$$\text{Laplace}(r; \mu, \beta) = \frac{1}{2\beta} \exp\left(-\frac{|r - \mu|}{\beta}\right), \qquad \mathbb{E}(R) = \mu, \qquad \mathbb{V}(R) = 2\beta^2.$$

The associated density, $p(r)$, is plotted in Fig. B.9.

Parametrization

In the common parametrization, the *mean* μ and *scale* β are real numbers and β is positive.

Sampling scheme

To simulate $R \sim$ Laplace(μ, β), first simulate $U_1 \sim$ Gamma$(1, 1)$, $U_2 \sim$ Gamma$(1, 1)$, and then set $r = \mu + \beta\,(u_1 - u_2)$.

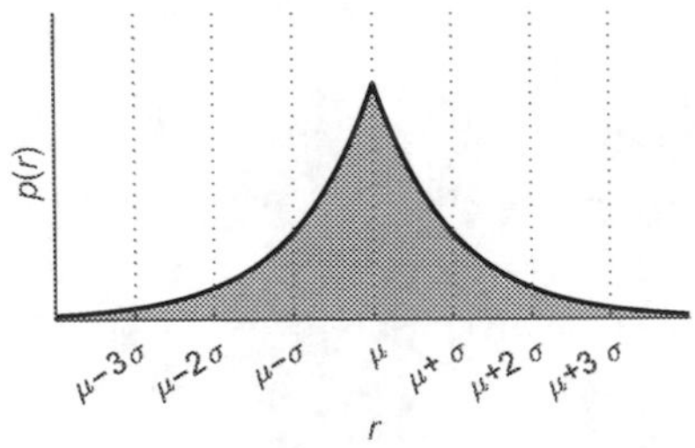

Fig. B.9 The Laplace probability density.

B.2 Discrete Random Variables

B.2.1 With Bounded Support

Bernoulli Random Variable

Definition

A Bernoulli random variable $R \sim \text{Bernoulli}(\pi)$, takes values 1 or 0 and has the probability density and moments

$$\text{Bernoulli}(r; \pi) = \pi\delta_1(r) + (1-\pi)\delta_0(r), \quad \mathbb{E}(R) = \pi, \quad \mathbb{V}(R) = \pi(1-\pi).$$

The associated density, $p(r)$, is plotted in Fig. B.10.

Parametrization

In the common parametrization, the *success probability* π is a real number between 0 and 1.

Binomial Random Variable

Definition

A binomial random variable $R \sim \text{Binomial}(J, \pi)$, takes nonnegative integer scalar values r between 0 and J and has the probability density and moments

$$\text{Binomial}(r; J, \pi) = \sum_{k=0}^{J} \frac{J!}{k!\,(J-k)!}\pi^k (1-\pi)^{J-k}\,\delta_k(r), \mathbb{E}(R) = J\pi,$$
$$\mathbb{V}(R) = J\pi(1-\pi).$$

The associated density, $p(r)$, is plotted in Fig. B.11.

Parametrization

In the common parametrization, the *number of trials* J is a positive integer and the *success probability* π is a real number between 0 and 1.

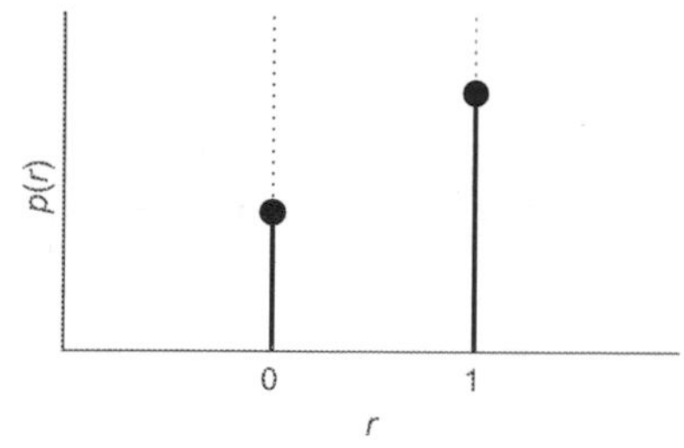

Fig. B.10 The Bernoulli probability density.

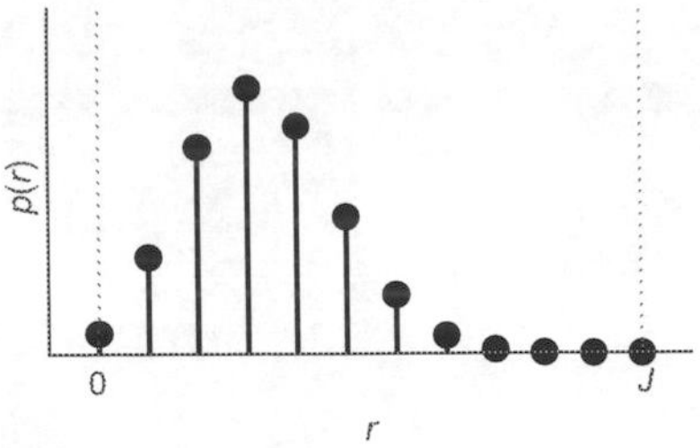

Fig. B.11 The binomial probability density.

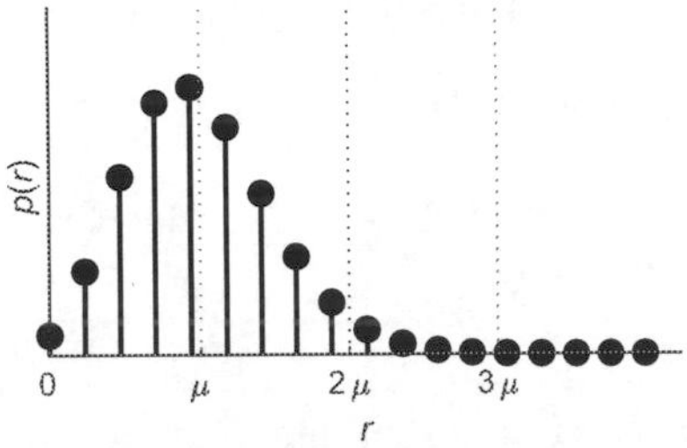

Fig. B.12 The Poisson probability density.

B.2.2 With Semibounded Support

Poisson Random Variable

Definition

A Poisson random variable $R \sim \text{Poisson}(\mu)$, takes nonnegative integer scalar values r and has the probability density and moments

$$\text{Poisson}(r; \mu) = \sum_{k=0}^{\infty} \frac{\mu^k}{k!} e^{-\mu}\, \delta_k(r), \qquad \mathbb{E}(R) = \mu, \qquad \mathbb{V}(R) = \mu.$$

The associated density, $p(r)$, is plotted in Fig. B.12.

Parametrization

In the common parametrization, the *mean* μ is a positive real number.

Geometric Random Variable

Definition

A geometric random variable $R \sim \text{Geometric}(\pi)$, takes nonnegative integer scalar values r and has the probability density and moments

$$\text{Geometric}(r; \pi) = \sum_{k=0}^{\infty} (1-\pi)^k \pi\, \delta_k(r), \quad \mathbb{E}(R) = \frac{1-\pi}{\pi}, \quad \mathbb{V}(R) = \frac{1-\pi}{\pi^2}.$$

The associated density, $p(r)$, is plotted in Fig. B.13.

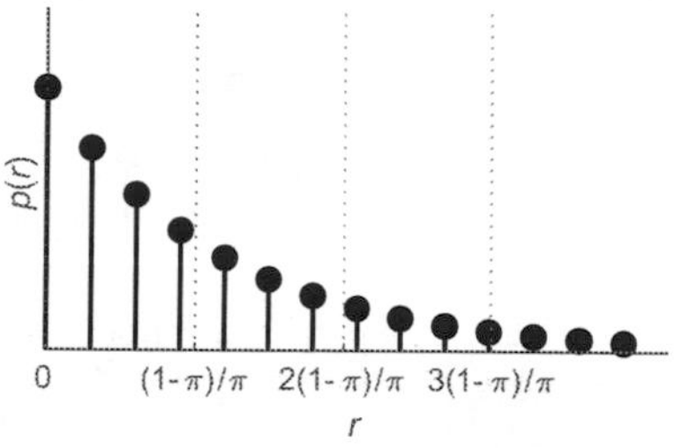

Fig. B.13 The geometric probability density.

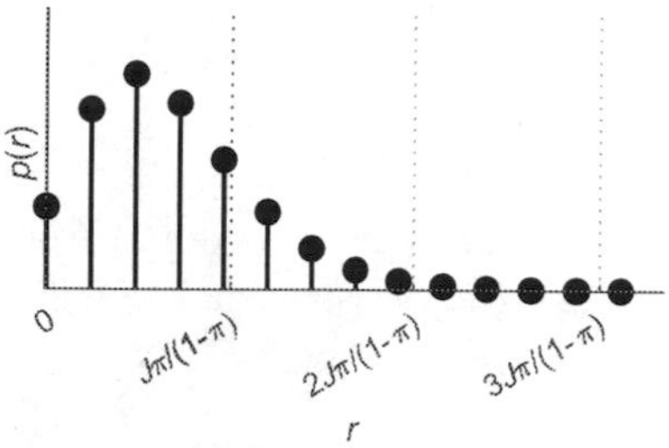

Fig. B.14 The NegBinomial probability density.

Parametrization

In the common parametrization, the *success probability* π is a real number between 0 and 1.

NegBinomial Random Variable

Definition

A NegBinomial random variable $R \sim \text{NegBinomial}(J, \pi)$, takes nonnegative integer scalar values r and has the probability density and moments

$$\text{NegBinomial}(r; J, \pi) = \sum_{k=0}^{\infty} \frac{\Gamma(J+k)}{\Gamma(k+1)\Gamma(J)}(1-\pi)^k \pi^k \, \delta_k(r),$$

$$\mathbb{E}(R) = J\frac{\pi}{1-\pi},$$

$$\mathbb{V}(R) = J\frac{\pi}{(1-\pi)^2},$$

where $\Gamma(\cdot)$ is the gamma function. The associated density, $p(r)$, is plotted in Fig. B.14.

Parametrization

In the common parametrization, the *number of trials* J is a positive integer and the *success probability* π is a real number between 0 and 1.

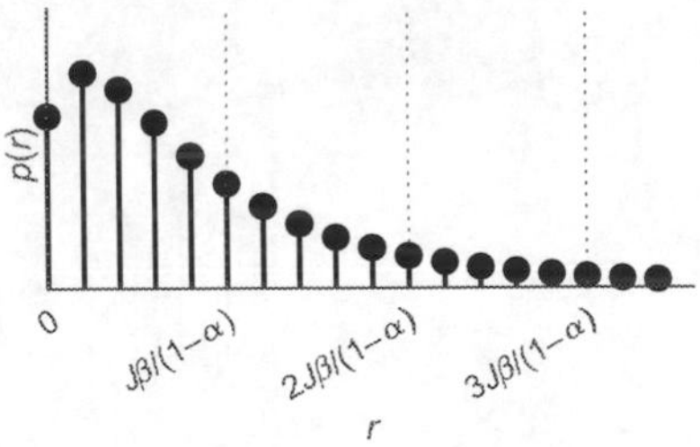

Fig. B.15 The BetaNegBinomial probability density.

BetaNegBinomial Random Variable

Definition

A BetaNegBinomial random variable $R \sim \text{BetaNegBinomial}(J, \alpha, \beta)$, takes nonnegative integer scalar values r and has the probability density and moments

$$
\begin{aligned}
&\text{BetaNegBinomial}(r; J, \alpha, \beta) \\
&\quad = \sum_{k=0}^{\infty} \frac{\Gamma(J+k)}{\Gamma(k+1)\Gamma(J)} \frac{\Gamma(\alpha+J)\Gamma(\beta+k)}{\Gamma(\alpha+J+\beta+k)} \frac{\Gamma(\alpha+\beta)}{\Gamma(\alpha)\Gamma(\beta)} \delta_k(r), \\
&\quad \mathbb{E}(R) = J\frac{\beta}{\alpha-1}, \qquad \mathbb{V}(R) = J\frac{(\alpha+J-1)\beta(\alpha+\beta-1)}{(\alpha-2)(\alpha-1)^2},
\end{aligned}
$$

where $\Gamma(\cdot)$ is the gamma function. The associated density, $p(r)$, is plotted in Fig. B.15.

Parametrization

In the common parametrization, the *number of trials* J is a positive integer and the *shapes* α and β are positive real numbers.

C The Kronecker and Dirac Deltas

C.1 Kronecker Δ

The Kronecker Δ, also called Kronecker delta, is denoted by $\Delta_\rho(r)$. This is an *indicator function* and is unitless. It takes on continuous or discrete arguments, r, and returns

$$\Delta_\rho(r) = \begin{cases} 0, & r \neq \rho, \\ 1, & r = \rho. \end{cases}$$

As suggested in Note 1.7, the Kronecker Δ is convenient when we need to derive explicit formulas for discrete densities such as Bernoulli or, more generally, categorical distributions.

C.2 Dirac δ

The Dirac δ, also called Dirac delta, is denoted by $\delta_\rho(r)$. As in Section 1.2, the Dirac δ is convenient when we describe probability densities over discrete or discretized random variables.

Our notation differs slightly from what is commonly used where $\delta_\rho(r)$ would normally be written as $\delta(r - \rho)$. For instance, in physics literature, $\delta(\rho - r)$ is used to represent how mass, charge, or other quantities are spread over time or space. In this setting, r and ρ typically denote points in space or time for which subtraction $r - \rho$ is meaningful. Here, however, $\delta_\rho(r)$ represents how probability is spread over the values of the random variables of interest. In the probabilistic context, r and ρ may stand for any quantity, even for such quantities where subtraction may be meaningless, such as the constitutive states in a system's state-space, as in Chapter 2. For this reason, to denote a Dirac δ centered at an arbitrary point ρ, we prefer the more general notation $\delta_\rho(r)$ instead of $\delta(r - \rho)$.

C.2.1 Definition

In one dimension, where subtraction of r and ρ is meaningful, a common description of the Dirac δ is through the limit of increasingly thinner Gaussians or normal densities

$$\delta_\rho(r) = \lim_{\sigma \to 0} \frac{1}{\sqrt{2\pi\sigma^2}} e^{-\frac{(r-\rho)^2}{2\sigma^2}} = \lim_{\sigma \to 0} \text{Normal}\left(r; \rho, \sigma^2\right), \qquad r \neq \rho. \qquad \text{(C.1)}$$

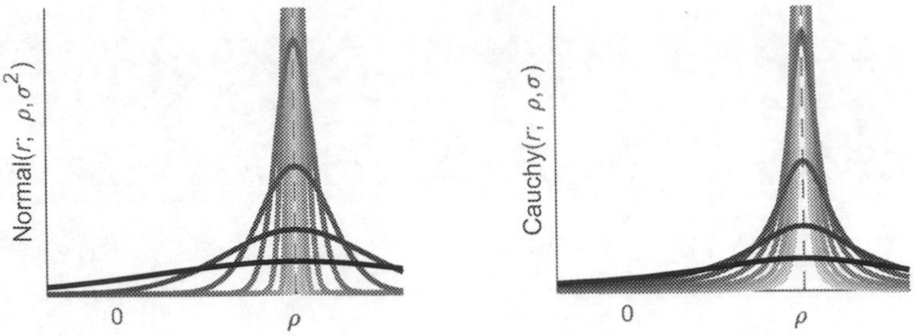

Fig. C.1 The Dirac $\delta_\rho(r)$ as a sequence of successively thinner probability densities.

Many descriptions equivalent to Eq. (C.1) also exist. For instance, the Gaussians are replaced by Lorentzians or Cauchy densities,

$$\delta_\rho(r) = \lim_{\sigma\to 0} \frac{1}{\pi} \frac{\sigma}{\sigma^2 + (r-\rho)^2} = \lim_{\sigma\to 0} \text{Cauchy}\left(r; \rho, \sigma\right), \qquad r \neq \rho. \tag{C.2}$$

Descriptions of this type are quite intuitive, for example see Fig. C.1, and are readily extended to more than one dimension by replacing normal or Cauchy densities with the appropriate multivariate extensions.

Integration of Eqs. (C.1) and (C.2) yields $\int dr\, \text{Normal}\left(r; \rho, \sigma^2\right) = 1$ and $\int dr\, \text{Cauchy}\left(r; \rho, \sigma\right) = 1$, with the resulting values remaining constant irrespective of σ, whose value becomes arbitrarily small. This property carries over to the $\sigma \to 0$ limit and also applies to $\delta_\rho(r)$. For instance,

$$\begin{aligned}\int dr\, \delta_\rho(r) &= \int dr \lim_{\sigma\to 0} \text{Normal}\left(r; \rho, \sigma^2\right) \\ &= \lim_{\sigma\to 0} \int dr\, \text{Normal}\left(r; \rho, \sigma^2\right) = \lim_{\sigma\to 0} 1 = 1.\end{aligned}$$

This last equality, combined with an explicit evaluation of the limits $\sigma \to 0$, namely,

$$\delta_\rho(r) = 0, \qquad r \neq \rho, \tag{C.3}$$

$$\int dr\, \delta_\rho(r) = 1, \tag{C.4}$$

are perhaps the most distinctive characteristics of the Dirac δ. Since Eqs. (C.3) and (C.4) are independent of the normal or Cauchy densities appearing in Eqs. (C.1) and (C.2), they offer a better starting point for the definition of Dirac δ.

As special cases, we invoke Eqs. (C.1) and (C.2) only in situations where subtractions like $r - \rho$ actually make sense. These include cases where r, ρ are scalars or vectors. In more general settings, however, the definition of $\delta_\rho(r)$ can be described only abstractly. In an abstract setting, it is sufficient to start directly from the conditions in Eqs. (C.3) and (C.4) without any reference to an underlying limiting scheme.

Note C.1 What is $\delta_\rho(\rho)$?

Motivated by Eqs. (C.1) and (C.2), which provide valid values for the limits even at $r = \rho$, *heuristic* definitions of $\delta_\rho(r)$ often assign a value of ∞ for $\delta_\rho(\rho)$. For instance, a typical heuristic is

$$\delta_\rho(r) = \begin{cases} 0, & r \neq \rho, \\ \infty, & r = \rho. \end{cases}$$

However, specifying a value for $\delta_\rho(r)$ at $r = \rho$ is *unnecessary.* In fact, assigning a value for $\delta_\rho(\rho)$ and, in particular, ∞ is misleading and may lead to inaccuracies. For example, since $2 \times 0 = 0$ and $2 \times \infty = \infty$, a reasoning stream may proceed as follows:

$$1 = \int dr\, \delta_\rho(r) = \int dr\, 2\delta_\rho(r) = 2\int dr\, \delta_\rho(r) = 2.$$

While Eqs. (C.3) and (C.4) are *necessary*, to avoid paradoxes like this, it is much safer to leave $\delta_\rho(\rho)$ defined abstractly. This way, $\delta_\rho(\rho)$ remains free to attain any value that is needed to make Eqs. (C.3) and (C.4) consistent.

Of course, values of $\delta_\rho(\rho)$ that make Eqs. (C.3) and (C.4) consistent with each other *cannot* be numeric, finite, or infinite. For this reason, the Dirac δ is designated as a *generalized* function as it attains *nonnumeric* values in a specific manner.

C.2.2 Properties

As we see, $\delta_\rho(r)$ evaluates to 0 when $r \neq \rho$. This means that, when integrating over a domain $\mathcal{R}$ that does *not* include ρ, it evaluates to zero. However, when integrating over a domain $\mathcal{R}$, which *does* include ρ, it evaluates to unity. That is,

$$\int_{\mathcal{R}} dr\, \delta_\rho(r) = \begin{cases} 0, & \rho \notin \mathcal{R}, \\ 1, & \rho \in \mathcal{R}. \end{cases}$$

The same property also carries over when $\delta_\rho(r)$ is multiplied by a *continuous* function,

$$\int_{\mathcal{R}} dr\, f(r)\delta_\rho(r) = \begin{cases} 0, & \rho \notin \mathcal{R}, \\ f(\rho), & \rho \in \mathcal{R}. \end{cases}$$

When r and ρ are multiplied by a scalar, integral rescaling immediately yields

$$\delta_{\lambda\rho}(\lambda r) = \frac{1}{|\lambda|^d}\delta_\rho(r),$$

where λ is a nonzero scalar and d is the dimension of r and ρ.

D Memoryless Distributions

In the context of discrete state-space systems evolving in continuous time of Section 2.3 and, in particular, of the Markov jump processes of Section 2.3.2, *memorylessness* implies that the probability of sampling a particular holding period h given that this period exceeds *any* threshold d, is the same as sampling a holding period equal to $h - d$ in the first place. In other words, memorylessness entails that the distribution of holding periods forgets the elapsed time represented by d.

Informally, the memorylessness requirement reads,

$$\begin{pmatrix}\text{Probability of}\\ H > h \text{ given } H > d\end{pmatrix} = \begin{pmatrix}\text{Probability of}\\ H > h - d\end{pmatrix}.$$

To express this requirement formally, we use $p(h)$ to denote the probability density of H. Under $p(h)$, we see that the right-hand side equals $\int_{h-d}^{\infty} d\eta\, p(\eta)$. Further, by the definition of conditional probability, the left-hand side is the same as the ratio

$$\frac{\begin{pmatrix}\text{Probability of}\\ H > h \text{ and } H > d\end{pmatrix}}{\begin{pmatrix}\text{Probability of}\\ H > d\end{pmatrix}} = \frac{\begin{pmatrix}\text{Probability of}\\ H > h\end{pmatrix}}{\begin{pmatrix}\text{Probability of}\\ H > d\end{pmatrix}},$$

which, under $p(h)$, is equal to $\frac{\int_h^\infty d\eta\, p(\eta)}{\int_d^\infty d\eta\, p(\eta)}$. Therefore, our requirement formally reads as

$$\frac{\int_h^\infty d\eta\, p(\eta)}{\int_d^\infty d\eta\, p(\eta)} = \int_{h-d}^{\infty} d\eta\, p(\eta),$$

which, after rearrangement of the terms, turns to

$$\int_h^\infty d\eta\, p(\eta) = \int_d^\infty d\eta\, p(\eta) \int_{h-d}^\infty d\eta\, p(\eta). \tag{D.1}$$

Differentiating Eq. (D.1) with respect to h, we obtain

$$p(h) = p(h-d) \int_d^\infty d\eta\, p(\eta). \tag{D.2}$$

Similarly, differentiating Eq. (D.1) with respect to d, we obtain

$$-p(d) \int_\infty^{h-d} d\eta\, p(\eta) = -p(h-d) \int_\infty^{d} d\eta\, p(\eta). \tag{D.3}$$

We notice that the righthand sides of Eq. (D.2) and Eq. (D.3) are identical and thus equate their left-hand sides,

$$p(h) = -p(d) \int_{\infty}^{h-d} d\eta\, p(\eta).$$

Finally, setting $d = 0$ and differentiating once more with respect to h, we obtain

$$p'(h) = -p(0)p(h).$$

The general solution of this differential equation is

$$p(h) = Ce^{-p(0)h}.$$

Normalization, $\int_0^\infty dh\, p(h) = 1$, gives $C = p(0)$. Therefore, a memoryless probability density attains the form

$$p(h) = p(0)e^{-p(0)h}.$$

Of course, this is the density of an *exponential* probability distribution with rate, which we commonly denote λ, equal to $p(0)$.

The converse is also true. Namely, an exponential random variable $H \sim \text{Exponential}(\lambda)$ is memoryless. Indeed, we can verify this directly:

$$\frac{\int_h^\infty d\eta\, \text{Exponential}(\eta; \lambda)}{\int_d^\infty d\eta\, \text{Exponential}(\eta; \lambda)} = \frac{e^{-\lambda h}}{e^{-\lambda d}} = e^{-\lambda(h-d)} = \int_{h-d}^{\infty} d\eta\, \text{Exponential}(\eta; \lambda).$$

Note D.1 Two properties of memoryless random variables

For a collection of memoryless random variables,

$$H_{\sigma_m} \sim \text{Exponential}(\lambda_{\sigma_m}), \qquad m = 1 : M,$$

the minimum value for this collection $H_* = \min_{m=1:M} H_{\sigma_m}$ is also a random variable. This is distributed according to

$$H_* \sim \text{Exponential}\,(\lambda_*), \qquad \lambda_* = \sum_{m=1}^{M} \lambda_{\sigma_m}.$$

Similarly, the identity of the minimum $M_* = \text{argmin}_{m=1:M} H_{\sigma_m}$ is also a random variable. This is distributed according to

$$M_* \sim \text{Categorical}_{\sigma_{1:M}}(\boldsymbol{\pi}_*), \qquad \boldsymbol{\pi}_* = \left(\frac{\lambda_{\sigma_1}}{\lambda_*}, \frac{\lambda_{\sigma_2}}{\lambda_*}, \ldots, \frac{\lambda_{\sigma_M}}{\lambda_*}\right).$$

We leave a proof of these as an exercise.

E Foundational Aspects of Probabilistic Modeling

In Chapter 1 we introduced random variables and probability distributions. Although we took a rather practical approach, below we give a self-contained presentation of the notions involved.

E.1 Outcomes and Events

To be able to talk about probabilities in a comprehensive setting, first, we need a notion of some kind of incident that generates outcomes at random. More concretely, we think of this incident as a *stochastic operation* that selects *a single* outcome *out of many*, finite or infinite, possible ones. As we will see shortly, a stochastic operation is only one ingredient needed to establish a stochastic experiment and an appropriate probability notion.

For most physical applications, it is most intuitive to think of the stochastic operation as the source of randomness in our formulation. Examples of such an operation include a coin toss, a cell reaching a critical time-point in its life cycle, the absorption of a photon from a molecule, the execution of a pseudo-random number generator, amongst other examples. Generally, for our purposes, an operation is an abstract model of some core phenomenon that is vaguely random. As we will see later on, this operation influences our observations very decisively but, as a notion, is distinct from the observations themselves.

The family of all possible outcomes ω that may be generated by our operation is commonly called the *sample space* and denoted as Ω. For example, when a coin is tossed once, the sample space consists of "heads" and "tails." Similarly, when the displacement of a moving object is assessed, the sample space consists of all possible values that can be measured, *i.e.*, all points in a three dimensional space.

Example 5.1 The photo-electric effect as a stochastic operation

As we mention on Example 1.1, when a photon falls into certain materials, sometimes a photo-electron is emitted and sometimes it is not. When two photons fall onto the same material, we may model the entire phenomenon as a stochastic operation with four outcomes:

- ω_1, where no photon causes an emission.
- ω_2, where the first photon causes an emission while the second photon causes no emission.

- ω_3, where the first photon causes no emission while the second photon causes an emission.
- ω_4, where both photons cause an emission.

In this setup, we may speak of a sample space $\Omega = \{\omega_1, \omega_2, \omega_3, \omega_4\}$.

In practice, we may not have access to the particular outcome, say ω_*, generated by the operation, instead we may have access only to quantities influenced by this ω_*. For example, when shuffling a deck of cards we may be keeping track only of particular face cards; or when measuring a displacement, we may be recording only measurements that fall within certain ranges set by our detectors; or when monitoring the times that certain events occur, we may be noting events that happen only during our observation periods, and so on.

When building a stochastic model, it is most useful to envision that, while the operation selects ω_* which initiates the generation of the appropriate observations, the value of ω_* remains unknown to us. In this setting, our objective from now is to try to identify ω_* within our sample space; that is, finding which particular ω in Ω might have caused our observations.

In practice, observations are typically consistent with multiple outcomes within the sample space and thus may not suffice to rule out all but a single one. For example, when measuring the distance between two points with noisy instruments, or when counting the number of photons emitted from a fluorescent molecule, it is impossible to distinguish a single distance or a single photon count below an error threshold. Consequently, almost always we are compelled to identify ω_* only within collections of multiple outcomes ω, rather than individual outcomes ω. To identify ω_* within a collection of ω in a concise manner we need to formulate an appropriate framework larger than just a sample space Ω.

Example 5.2 Photo-electron counting as a stochastic experiment

Consider a simplified situation where two photons fall into a photo-electric material. In a naive experiment we try to assess the sequence of successive photo-electron emissions by capturing and recording the total number of photo-electrons emitted.

As we have seen in Example 5.1, a reasonable sample space for our purpose consists of four outcomes $\Omega = \omega_{1:4}$. In this Ω, we try to identify ω_* by counting the resulting total number of photo-electrons which, for the outcomes listed in Example 5.1, are:

- 0 for ω_1.
- 1 for ω_2.
- 1 for ω_3.
- 2 for ω_4.

Obviously, outcomes ω_1 or ω_4 are individually identified by recordings of 0 or 2 photo-electrons, respectively. However, since outcomes ω_2 and ω_3 are both associated with recordings of 1 photo-electron, they can be only collectively identified. In other words, we can tell whether ω_* is equal to ω_1, or ω_4, or any of ω_2 and ω_3; but, we cannot distinguish whether ω_* is equal to ω_2 or ω_3.

In this larger framework, we can think of a *stochastic experiment* which combines the underlying stochastic operation, *i.e.*, a physical phenomenon, that selects ω_* out of Ω at random and which we cannot assess, as well as any combination of the possible outcomes ω that we may assess collectively either directly or indirectly. Commonly, assessable collections β of individual outcomes ω, in a stochastic experiment, are called *events* and the family $\mathcal{B}$ that gathers every event in an experiment is called the *event space*.

Example 5.3 Single-target localization as a stochastic experiment

Consider a situation where we try to localize a bright target, typically a fluorescent molecule, in conventional microscopy. In a real-life experiment we try to assess the target's position by recording its distance from certain points such as a set of fiducial markers in our field of view.

Since our goal is to estimate the position of the target, it is reasonable to consider a sample space Ω that consists of the entire three-dimensional space and ω_* as the target's position. In this Ω, even in the idealized case that our distance recordings are noiseless, we might run into multiple scenarios that depend upon how many markers we utilize. In particular:

- With only one marker at our disposal, we can identify the target's position only up to the surface of the spheres centered around the marker.
- With only two markers at our disposal, we can identify the target's position only up to circles that are inter-sections of the surfaces of the spheres centered around the markers.
- With three or more markers at our disposal, we can identify the target's position up to intersections of the surfaces of the spheres centered around the markers which, of course, except co-linear marker placements, span the entire space.

Depending on which of the three scenarios we run into, we need to adapt our events β accordingly as well as the entire event space $\mathcal{B}$ we choose to work with.

Note E.1 Set theory

To help with the formalism that follows, it might be useful to recall some notions from *set theory*.

A *set* is a collection of *elements* that we need to specify. To denote that a particular element belongs to a set, we write $\omega \in \Omega$. This reads "ω belongs to Ω." When there is no conflict, we may denote sets using capital letters and their elements using lowercase letters. For example, we have already met this convention with the sample space Ω, which has outcomes ω as its elements. Unfortunately, this might not always be possible; for example, we already encountered $\mathcal{B}$, which has elements that are also sets themselves, so we prefer to be explicit in our descriptions.

Sometimes we need collections of only a few of the elements in a set. In such cases, we may speak of a subset β and write $\beta \subset \Omega$. This reads "β is a subset of Ω" and means that each element of β is also an element of Ω. As we do not require the set Ω to contain elements not contained in β, the relationship $\beta \subset \Omega$ also includes cases where β and Ω are essentially the same set.

For any number, finite or infinite, of subsets $\beta_1, \beta_2, \ldots \subset \Omega$, we can consider *unions* $\beta_1 \cup \beta_2 \cup \cdots$ and *intersections* $\beta_1 \cap \beta_2 \cap \cdots$. These are also subsets of the set Ω. In particular, the union $\beta_1 \cup \beta_2 \cup \cdots$ contains only the elements ω contained in *at least one* of $\beta_1, \beta_2, \ldots$, while the intersection $\beta_1 \cap \beta_2 \cap \cdots$ contains only the elements contained in *all* $\beta_1, \beta_2, \ldots$.

If $\beta_1, \beta_2, \ldots$ have no element in common, then their intersection $\beta_1 \cap \beta_2 \cap \cdots$ contains no element and is called the *empty set*, which we denote using $\emptyset$. By convention, we consider $\emptyset$ to be subset of any set.

As an illustration, we consider a set consisting of five elements, $\Omega = \{\omega_1, \omega_2, \omega_3, \omega_4, \omega_5\}$, and we also consider two subsets, $\beta_1 = \{\omega_1, \omega_2, \omega_3\}$ and $\beta_2 = \{\omega_3, \omega_4\}$. In this case, the union and intersection are $\beta_1 \cup \beta_2 = \{\omega_1, \omega_2, \omega_3, \omega_4\}$ and $\beta_1 \cap \beta_2 = \{\omega_3\}$. If we further consider a third subset, $\beta_3 = \{\omega_4, \omega_5\}$, we see $\beta_1 \cap \beta_3 = \emptyset$ since there is no element in common in both β_1 and β_3.

Naturally, within this framework we encode our inquiries with *logical statements* about ω_* and the events β, or, in more formal terms, with statements of the form $\omega_* \in \beta$. For this reason, we require $\mathcal{B}$ to have certain properties that ensure the fundamental rules of logic hold. For instance, an event space:

1. Must include the entire sample space (in which case we refer to the event $\beta = \Omega$ as a *certain event*), the empty set (in which case we refer to the event $\beta = \emptyset$ as an *impossible event*), and possibly other subsets of the sample space.
2. For any event β, the event space must also contain an event β^c that, when combined with β, results in the entire sample space $\beta^c \cup \beta = \Omega$ (in which case we refer to β^c as the complement of β).
3. The event space must also include unions $\beta_1 \cup \beta_2 \cup \ldots$ of any collection of events $\beta_1, \beta_2, \ldots$.

In technical terms, those properties ensure that $\mathcal{B}$ has the mathematical structure of what is called a *σ-algebra*. In less technical terms, however, these requirements indicate that, besides accounting in our formulation for the events $\beta_1, \beta_2, \ldots$ that we may encounter directly in a single assessment of our experiment, we also need to account for additional events, such as $\beta_1 \cup \beta_2$ or $\beta_1 \cap \beta_2$, that may be deduced indirectly and only after considering assessments from multiple executions of our experiment.

These properties of $\mathcal{B}$ are reasonable requirements based on what we expect to hold in real-world experiments that probe real-world events. Nevertheless, as general as these properties may be, they sometimes allow the formulation of multiple event spaces consistent with the same sample space. For example, when tossing a coin twice, the sample space consists of $\Omega = \{HH, HT, TH, TT\}$, where H stands for "heads" and T stands for "tails." In this sample space, a $\mathcal{B}$ that contains every possible subset of Ω satisfies the properties listed above and so serves as a valid event space. However, on the same sample space, $\mathcal{B}' = \{\emptyset, \Omega\}$ also satisfies the properties listed above, and so it is another perfectly valid event space. Generally, depending on the specifics of the problem at hand, we could use these two different event spaces, $\mathcal{B}$ and $\mathcal{B}'$, to develop two different probabilistic frameworks, essentially two different notions of probability, both utilizing the same underlying sample space.

Of course, for most practical applications we need not strictly differentiate between sample space Ω and event space $\mathcal{B}$ as these can easily be deduced from the context. Nonetheless, conceptually Ω and $\mathcal{B}$ represent different notions and have different properties. As we will see later, probabilities and random variables are understood better on $\mathcal{B}$ than on Ω. Thus, in subtle cases it might be useful to recall this distinction, especially in modeling conceptually challenging problems.

Note E.2 How do we describe an event space over the real line?

Unlike Examples 5.1 and 5.2, where the underlying stochastic operations selects ω_* among ω's that are separated from each other and finitely many, often we model stochastic operations, as in Example 5.3, where the underlying ω's are infinitely many or form continua and cannot be separated discretely from each other. In fact, with real-life experiments, very often we need to formulate situations where the sample space Ω consists of the whole or a portion of the real line. For example, in an experiment that assesses the distance between two points in space, we can imagine individual outcomes as nonnegative real numbers and, as we will see later, to compute probabilities we will need to rely on an appropriate event space.

In such cases, forming $\mathcal{B}$ by explicitly enumerating every single event, as we could do in Example 5.1, is impossible. Instead, we have to invoke more elaborate descriptions and some theoretical arguments. One of these descriptions leads to the so-called *Borel σ-algebra*, which, because it is so common, we briefly outline below.

Such an event space $\mathcal{B}$ requires that Ω be the *real line*. On the real line, the events β consist of a collection of points, and to specify which collections of points need to be included in $\mathcal{B}$, we start with $(-\infty, y]$. These are continuous semi-infinite intervals that contain all numbers ranging from $-\infty$ up to and including some y. By definition, all such intervals are required to be in $\mathcal{B}$.

Once we account for events $\beta_y = (-\infty, y]$, to fulfill the event space requirements we also need to account for their complements $\beta_y^c = (y, +\infty)$, which are continuous semi-infinite intervals that contain all numbers ranging from y up to $+\infty$.

Furthermore, we also need to account for events that result from unions of such events. For example, we also need to include events $(y_{\min}, y_{\max}]$, which are finite intervals that contain all numbers ranging from $y_{\min}$ up to and including $y_{\max}$. These result from intersections $\beta_{y_{\max}} \cap \beta_{y_{\min}}^c$. Similarly, we need to include events that are noncontinuous intervals $(y^1_{\min}, y^1_{\max}] \cup (y^2_{\min}, y^2_{\max}]$ or $[y^1_{\min}, y^1_{\max}) \cup [y^2_{\min}, y^2_{\max})$, and so on.

In short, once we include β_y in $\mathcal{B}$, in order to fulfill the requirements mentioned above we also need to include any finite or infinite arrangement of any form of intervals conceived: finite, infinite, continuous, discontinuous, closed from one side, closed from both sides, and so on.

Instead of going through this constructive description every time, often we say that the *Borel σ-algebra* is the *smallest* $\mathcal{B}$ that meets the requirements and includes at least the intervals $(-\infty, y]$. Since the *Borel σ-algebra* is generated by the procedure just described, the intervals are termed *generating events*.

It might appear that $\mathcal{B}$ contains any perceivable collection of points; however, this is not the case. It can be shown, though not easily, that certain subsets of the real line cannot be obtained by the generating intervals exclusively based on unions and complements. As a result, the specification of $\mathcal{B}$ as the *smallest* one is necessary since it ensures that $\mathcal{B}$ contains no more than precisely those events that stem from the generating events.

E.2 The Measure of Probability

In stochastic models we comprehensively think of randomness as *induced* by stochastic operations and *assessed* in stochastic experiments. In this context, we quantify randomness by the probabilities of the experiments' events. For concreteness below we revisit the notion of probability and establish some of the fundamental tools involved in probabilistic reasoning.

Once we have established the basic framework by specifying a sample space Ω and an appropriate event space $\mathcal{B}$, we have met all prerequisites for establishing a *measure of probability* $Q(\cdot)$, which is nothing more than a function from $\mathcal{B}$ to the real numbers. More specifically, $Q(\cdot)$ maps an event β to the real number $Q(\beta)$ that we call "the probability of event β." Later on, we will use $Q(\cdot)$ to quantify the probabilities of our observations which, as it will turn out, are distinct from the events β.

For interpretational purposes, we think of $Q(\beta)$ as quantifying the probability of the particular outcome ω_*, selected by the underlying operation, to belong in the event β. Loosely, this is

$$Q(\beta) = \begin{pmatrix} \text{probability of} \\ \omega_* \in \beta \end{pmatrix}.$$

As we will see later, depending upon a frequentist or a Bayesian modeling approach taken, it is through the number $Q(\beta)$ that we measure consistently relative frequencies or plausibilities.

We may understand that the basic properties of probability can be justified as either consequences of relative frequencies or, as we now see, from Cox's theorem. From now on, however, these properties are our starting point which we have to take for granted. Irrespective of how we think about the meaning of probabilities, from now on we *require* from $Q(\cdot)$ that it obey some reasonable axioms:

1. The probability of any event must be non-negative, *e.g.*, $0 \leq Q(\beta)$.
2. The probability of a certain event must be 1, *e.g.*, $Q(\Omega) = 1$.
3. The probability of an impossible event must be 0, *e.g.*, $Q(\emptyset) = 0$.
4. The probability must be additive.

The last requirement means that the combination of *mutually exclusive* events must be equal to the sum of the probabilities of the individual events or, in formal terms, if $\beta \cap \alpha = \emptyset$ then $Q(\alpha \cup \beta) = Q(\alpha) + Q(\beta)$.

Example 5.4 Probabilities of photo-electric emissions

Our familiar sample space $\Omega = \omega_{1:4}$ of two incident photons in Example 5.1 can, due to its finite size, give rise to 16 events at most, *i.e.*, no more than the total number of different groups formed by its four outcomes. In total, these are

$$\beta_1 = \emptyset,$$
$$\beta_2 = \{\omega_1\}, \quad \beta_3 = \{\omega_2\}, \quad \beta_4 = \{\omega_3\}, \quad \beta_5 = \{\omega_4\},$$
$$\beta_6 = \{\omega_1, \omega_2\}, \quad \beta_7 = \{\omega_1, \omega_3\}, \quad \beta_8 = \{\omega_1, \omega_4\},$$
$$\beta_9 = \{\omega_2, \omega_3\}, \quad \beta_{10} = \{\omega_2, \omega_4\}, \quad \beta_{11} = \{\omega_3, \omega_4\},$$
$$\beta_{12} = \{\omega_1, \omega_2, \omega_3\}, \quad \beta_{13} = \{\omega_1, \omega_2, \omega_4\}, \quad \beta_{14} = \{\omega_1, \omega_3, \omega_4\},$$
$$\beta_{15} = \{\omega_2, \omega_3, \omega_4\}, \quad \beta_{16} = \{\omega_1, \omega_2, \omega_3, \omega_4\} = \Omega.$$

Considered together, these events offer an event space $\mathcal{B} = \beta_{1:16}$ that meets all requirements of a σ-algebra. As a result, we may seek to establish a probability measure $Q(\cdot)$ on this $\mathcal{B}$ that we might apply to targeted applications. To establish $Q(\cdot)$, we proceed as follows.

By definition, $Q(\beta_1) = 0$ and $Q(\beta_{16}) = 1$. From the remaining 14 events, due to additivity, we have to assign values only to $Q(\beta_2), Q(\beta_3), Q(\beta_4)$, and $Q(\beta_5)$. Further, because

$$Q(\beta_2) + Q(\beta_3) + Q(\beta_4) + Q(\beta_5) = Q(\beta_{16}) = 1, \tag{E.1}$$

we actually need only decide three values out of $Q(\beta_2), Q(\beta_3), Q(\beta_4)$, and $Q(\beta_5)$.

Assuming that individual emissions of photo-electrons exert no influence upon each other, then the two photons involved in the simplified photo-electric assessment put forward in Example 5.1 are equally likely (or unlikely) to induce the emission of a photo-electron. Consequently, because events β_3 and β_4 both involve the emission of exactly one photo-electron, we need to consider them at even odds, *i.e.*,

$$\frac{Q(\beta_3)}{Q(\beta_4)} = 1. \tag{E.2}$$

Additionally, because event β_5 involves the emission of two photo-electrons, while event β_1 involves the emission of no photo-electron, and these emissions are assumed independent, we need to consider them at the odds

$$\frac{Q(\beta_5)}{Q(\beta_3)} = \frac{Q(\beta_3)}{Q(\beta_2)}. \tag{E.3}$$

Searching for a common solution to Eqs. (E.1) to (E.3) that satisfies $Q(\beta_2) \geq 0$, $Q(\beta_3) \geq 0$, $Q(\beta_4) \geq 0$, $Q(\beta_5) \geq 0$, we reach the necessary conditions

$$Q(\beta_2) = \left(1 - \sqrt{Q(\beta_5)}\right)^2, \tag{E.4}$$

$$Q(\beta_3) = Q(\beta_4) = \left(1 - \sqrt{Q(\beta_5)}\right)\sqrt{Q(\beta_5)}, \tag{E.5}$$

which leave $Q(\beta_5)$ as a free parameter that lies between 0 and 1.

Essentially, $\phi = \sqrt{Q(\beta_5)}$, which must also lie between 0 and 1, represents the probability of either photon inducing a photo-electron. In the physics literature, ϕ is termed *quantum efficiency*. So, reparametrizing the conditions in Eqs. (E.4) and (E.5) in terms of the more intuitive ϕ, we obtain the probabilities we seek. Explicitly, these read:

$$\begin{aligned}
&Q(\beta_1) = 0, && Q(\beta_2) = (1-\phi)^2, && Q(\beta_3) = (1-\phi)\phi,\\
&Q(\beta_4) = (1-\phi)\phi, && Q(\beta_5) = \phi^2, && Q(\beta_6) = 1-\phi^2,\\
&Q(\beta_7) = 1-\phi^2, && Q(\beta_8) = 1-2\phi+2\phi^2, && Q(\beta_9) = 2(1-\phi)\phi,\\
&Q(\beta_{10}) = \phi, && Q(\beta_{11}) = \phi, && Q(\beta_{12}) = 1-\phi^2,\\
&Q(\beta_{13}) = 1-(1+\phi)\phi, && Q(\beta_{14}) = (2-\phi)\phi, && Q(\beta_{15}) = (2-\phi)\phi,\\
&Q(\beta_{16}) = 1.
\end{aligned}$$

Of course, the event space $\mathcal{B} = \beta_{1:16}$ contains far more events than those we can individually assess in the simplistic experiment described in Example 5.2. In this case, it suffices to consider an alternative event space $\mathcal{B}'$ consisting of only

$$\begin{gathered}
\beta_1' = \emptyset,\\
\beta_2' = \{\omega_1\}, \quad \beta_3' = \{\omega_2, \omega_3\}, \quad \beta_4' = \{\omega_4\},\\
\beta_5' = \{\omega_1, \omega_2, \omega_3\}, \quad \beta_6' = \{\omega_1, \omega_4\}, \quad \beta_7' = \{\omega_2, \omega_3, \omega_4\},\\
\beta_8' = \{\omega_1, \omega_2, \omega_3, \omega_4\}.
\end{gathered}$$

Similar arguments lead to an alternative probability measure $Q'(\cdot)$ on $\mathcal{B}' = \beta_{1:8}'$ with the values

$$\begin{aligned}
&Q'(\beta_1') = 0, && Q'(\beta_2') = (1-\phi)^2, && Q'(\beta_3') = 2(1-\phi)\phi,\\
&Q'(\beta_4') = \phi^2, && Q'(\beta_5') = 1-\phi^2, && Q'(\beta_6') = 1-2(1-\phi)\phi,\\
&Q'(\beta_7') = 1-(1-\phi)^2, && Q'(\beta_8') = 1.
\end{aligned}$$

Based on the four requirements above, we can derive all familiar properties of probability. For example, relations like $Q(\beta^c) = 1 - Q(\beta)$, or $Q(\alpha \cap \beta^c) = Q(\alpha) - Q(\alpha \cap \beta)$ and $Q(\alpha \cup \beta) = Q(\alpha) + Q(\beta) - Q(\alpha \cap \beta)$, follow directly. These straightforward relations allow us to compute accurately and self-consistently probabilities of events, such as $\alpha \cup \beta$ or $\alpha \cap \beta$, which arise when we combine two simultaneous assessments in our experiment and can also be extended, in the obvious way, to allow us to combine more than two simultaneous assessments. In particular, *independent* events in an experiment are those that do not exert any influence on each other and, thus, one's probability is unrelated to the other's. Accordingly, α and β are independent when $Q(\alpha \cap \beta) = Q(\alpha)Q(\beta)$. This is because $Q(\alpha \cap \beta) = Q(\alpha)Q(\beta)$ indicates that $\omega_* \in \alpha$ does not influence the probability of β and *vice versa* $\omega_* \in \beta$ does not influence the probability of α.

Example 5.5 Mutually exclusive and independent photo-electric events

In Example 5.4 we have established a probabilistic framework consisting of $\Omega = \omega_{1:4}$, $\mathcal{B} = \beta_{1:16}$, and $Q(\beta_{1:16})$. Here, the events $\beta_2 = \{\omega_1\}$ and $\beta_5 = \{\omega_4\}$ are mutually exclusive, since $\beta_2 \cap \beta_5 = \emptyset$. Based on the assigned probability values, $Q(\beta_2 \cap \beta_1) = Q(\beta_1) = 0$, while $Q(\beta_2)Q(\beta_5) = (1-\phi)^2\phi^2$. Excluding trivial cases with $\phi = 0$ or $\phi = 1$, these probabilities indicate that the events β_2 and β_4 depend on each other.

Note E.3 How do we compute probabilities over the real line?

As already mentioned, very often we have to compute probabilities of events over the real line. For example, in Section 1.2.1 we noted that the probability of sampling a real scalar value within a given interval is given by the expression in Eq. (1.4) and that this expression is mediated by an integral. Now that we have gathered the appropriate machinery, we can revisit this probability and discuss specifically how integrals arise in them in the first place.

First, to compute probabilities we need an event space and a probability measure. In Note E.2 we saw that when a sample space Ω is the real line, the event space $\mathcal{B}$ is the Borel σ-algebra. In this setting, events are semi-infinite intervals $(-\infty, y]$ or they can be decomposed into unions or complements of such intervals.

On such $\mathcal{B}$, to obtain $Q(\cdot)$ a standard process is first to decide how to compute $Q\left((-\infty, y]\right)$, and subsequently generalize to the other events. For instance, given a function $f(y)$ on the real line, one way is based on

$$Q\left((-\infty, y]\right) = \frac{\int_{-\infty}^{y} dy' f(y')}{\int_{-\infty}^{+\infty} dy' f(y')}. \tag{E.6}$$

This representation yields a valid probability measure provided the function $f(y)$ is real valued, non negative, and normalizable, *i.e.*, $\int_{-\infty}^{+\infty} dy' f(y') < \infty$. All these conditions are satisfied by *probability density functions*. Therefore, given a probability density we can readily compute probabilities over the intervals $(-\infty, y]$.

For concreteness, suppose we have a scalar random variable $Y \sim \mathbb{P}$ with probability density $p(y)$. In this case, $Q((-\infty, y])$ is simply than the probability of Y taking any value lower or equal to y and it is given by $\int_{-\infty}^{y} dy' p(y')$, which is just a special case of Eq. (E.6). In any case, once we know how to compute $Q((-\infty, y])$, we can compute $Q(\beta)$ for any other event β in the real line and below we list two characteristic examples.

- If β has the form $(y, +\infty)$, then β and $(-\infty, y]$ are mutually exclusive with union Ω. Accordingly,

$$Q((y, +\infty)) = Q(\Omega) - Q((-\infty, y])$$
$$= 1 - \frac{\int_{-\infty}^{y} dy' f(y')}{\int_{-\infty}^{+\infty} dy' f(y')} = \frac{\int_{y}^{+\infty} dy' f(y')}{\int_{-\infty}^{+\infty} dy' f(y')}.$$

- If β has the form $(y_{\min}, y_{\max}]$, then β and $(-\infty, y_{\min}]$ are mutually exclusive with union $(-\infty, y_{\max}]$. Accordingly,

$$Q((y_{\min}, y_{\max}]) = Q((-\infty, y_{\max}]) - Q((-\infty, y_{\min}])$$
$$= \frac{\int_{-\infty}^{y_{\max}} dy' f(y')}{\int_{-\infty}^{+\infty} dy' f(y')} - \frac{\int_{-\infty}^{y_{\min}} dy' f(y')}{\int_{-\infty}^{+\infty} dy' f(y')} = \frac{\int_{y_{\min}}^{y_{\max}} dy' f(y')}{\int_{-\infty}^{+\infty} dy' f(y')}.$$

In practice, we also encounter situations where we need to compute probabilities of events that arise over *successive* assessments. In one such situation, we need to quantify the probability of some event, provided that we have already assessed or assuming that we will assess a particular event β. For this reason, we use a related notion, namely that of *conditional probability measure* $Q(\cdot|\beta)$. Conditional probability is one of the most useful tools in statistical inference and we now provide a formal definition.

As with $Q(\cdot)$, the conditional probability is also a function that maps an event α to the real number $Q(\alpha|\beta)$ that is associated with "the probability of event α given the event β." This number is equal to the ratio

$$Q(\alpha|\beta) = \frac{Q(\alpha \cap \beta)}{Q(\beta)}.$$

Naturally, for an event α independent of β, because $Q(\alpha|\beta) = Q(\alpha)$, the conditional probability is of little use. However, for arbitrary events the conditional probability immediately leads to a broader string of equalities $Q(\alpha|\beta)Q(\beta) = Q(\alpha \cap \beta) = Q(\beta \cap \alpha) = Q(\beta|\alpha)Q(\alpha)$. In essence, this is *Bayes' rule*, which we most often write in the equivalent form

$$Q(\alpha|\beta) = \frac{Q(\beta|\alpha)Q(\alpha)}{Q(\beta)}.$$

Note E.4 The conditional probability measure can be a probability measure itself

Conditioned on β, which is an event in $\mathcal{B}$, which in turn is an event space on a sample space Ω, the conditional probability measure $Q(\cdot|\beta)$ can be thought of as a probability measure $Q'(\cdot)$ in its own right. In the latter case, the sample space is $\Omega' = \beta$ and the event space $\mathcal{B}'$ consists of the events $\beta' = \alpha \cap \beta$ formed by all α in $\mathcal{B}$.

E.3 Random Variables

Up to this point, we have laid down the foundations of our modeling framework which is commonly termed a *probability space*. This includes a sample space Ω, its event space $\mathcal{B}$, and a probability measure $Q(\cdot)$. As we have seen, Ω and $\mathcal{B}$ provide convenient representations of real-world phenomena characterized by stochasticity and, within this framework, the probability measure $Q(\cdot)$ is the bedrock on which we rely to quantify randomness logically and self-consistently.

In a real-world situation, while a phenomenon of interest takes place, most likely we make numerical observations, *i.e.*, we take measurements. For example, depending on the specifics of our setup, we might be recording one or multiple scalar values such as lengths, temperatures, intensities, or other quantities of interest. Since observation are distinct from the underlying phenomena *per se*, which constitute the event space, observations cannot be directly modeled as events in $\mathcal{B}$ and $Q(\cdot)$ cannot be readily used to quantify the probabilities of our measurements. Nonetheless, observed values are related to and can be represented by the events in $\mathcal{B}$. Since it is of great importance to be able to associate $Q(\cdot)$ meaningfully with our measurements in order to extract valid conclusions, below we provide a description of how this is achieved.

Note E.5 Why do we distinguish sample and observation spaces?

That measurements are distinct from the outcomes of the underlying stochastic operation and require an expanded framework becomes clearer if we consider that:

- *Outcomes model physical phenomena* that, typically, are independent of whether we observe them or not. For example, a cell grows and divides and carries out its normal functions irrespective of whether it is observed.
- *Observations model measurements* that, almost always, depend on the specific details of our measuring devices and our choice of a frame of reference and units that, obviously, have little in common with the underlying phenomena.

To incorporate observations in a probabilistic framework $\Omega, \mathcal{B}, Q(\cdot)$, we need to introduce into our formulation the notion of *random variable* Y, which, in essence, is nothing more than a function over Ω. Such a function maps outcomes ω from the sample space to values y in some target space S. In our context, we think of S as the *observation space*, for example the real line or the positive integers, or other appropriate space, and $y = Y(\omega)$ as the value in S that is observed under the outcome ω; see Fig. E.3.

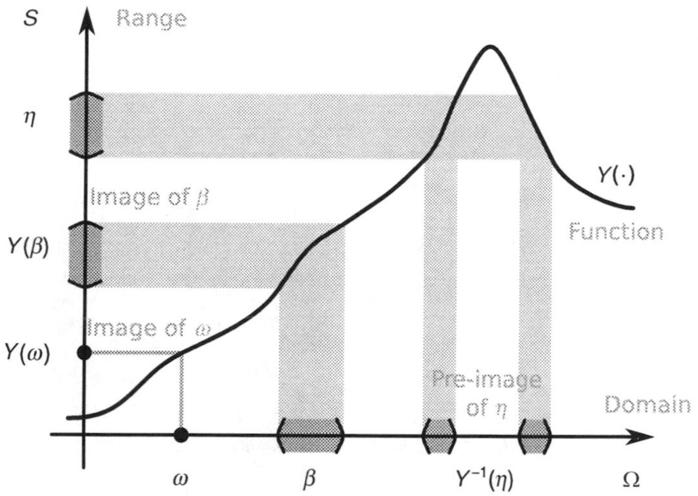

Fig. E.1 Illustration of a function $Y : \Omega \mapsto S$ and the notions of image and pre-image.

Note E.6 What is a function?

At first, it might appear somewhat controversial that a random variable is, in essence, a function because a function is neither variable nor random. So, at this stage it might be helpful to revisit some concepts.

As illustrated in Fig. E.1, a *function* Y is an *association* of each element of a set Ω with a *single* element of another set S. In this context, Ω is the *domain* and S is the *range* of Y and, to illustrate this relationship, we compactly write $Y : \Omega \mapsto S$. According to this association, for every $\omega \in \Omega$ there is associated a unique $y \in S$, which we nearly always denote $Y(\omega)$. However, the meaning of the notation $Y(\omega)$ is somewhat unclear. It may denote the particular value $y \in S$ that corresponds to ω or the association represented by the function Y. To distinguish between these two, we denote the latter simply Y or, when we need to emphasize the functional meaning, $Y(\cdot)$.

The value $Y(\omega)$ is sometimes called the *image* of ω and the same term applies also to subsets of Ω. In particular, for any $\beta \subset \Omega$, the image $Y(\beta)$ is the subset of S that gathers the images of each $\omega \in \beta$. We apply a similar convention to subsets of S. In particular, for any $\eta \subset S$, the *pre-image* $Y^{-1}(\eta)$ is the subset of Ω that gathers all ω with images in η. Although, by definition there is a single $y \in S$ assigned to each $\omega \in \Omega$, generally, there may be more than one $\omega \in \Omega$ assigned the same $y \in S$. As a result, the pre-image $Y^{-1}(\{y\})$ of a subset $\{y\} \subset S$ consisting of a single value $y \in S$ may, in general, contain more than one $\omega \in \Omega$.

Example 5.6 Dice rolls and random variables

Consider a dice roll. As we have seen, the sample space can be modeled as consisting of six outcomes $\Omega = \omega_{1:6}$, which we may identify with the six faces of the dice. In this setting, we can consider a random variable Y_1 that to each

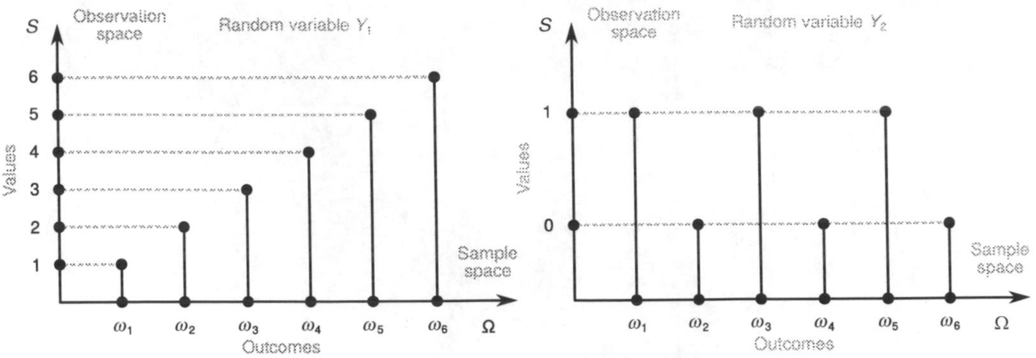

Fig. E.2 With a dice roll in mind, random variables are functions that assign a number to each one of the faces shown. Here, the sample space Ω consists of the outcomes $\omega_{1:6}$, each one associated with one of the faces, and the random variables Y_1 and Y_2 take values $1:6$ and $0, 1$, respectively.

outcome assigns the number shown. Explicitly, $Y_1 \colon \Omega \mapsto \{1, 2, 3, 4, 5, 6\}$ and the precise formula is

$$Y_1(\omega_m) = m, \qquad\qquad m = 1:6.$$

Alternatively, we can consider a random variable Y_2 that assigns 1 to the faces with an even number and 0 to the faces with an odd number. Explicitly, $Y_2 \colon \Omega \mapsto \{0, 1\}$ and the precise formula is

$$Y_2(\omega_m) = \begin{cases} 1, & m = 1, 3, 5, \\ 0, & m = 2, 4, 6. \end{cases}$$

These random variables are shown in Fig. E.2.

Following a roll, suppose we observe $y_1 = Y_1(\omega_*)$. In this case, we can use y_1 to uniquely identity ω_*. By contrast, suppose we observe $y_2 = Y_2(\omega_*)$. In this case, we can use y_2; however, we cannot identify ω_* uniquely, though we can winnow down possibilities.

E.4 The Measurables

Very often, random variables Y that model observations in real-world experiments associate the same value y with multiple different outcomes ω, like in Examples 5.2 and 5.6. Due to such inability to distinguish among some or all ω, the particular outcome ω_* selected by the stochastic operation remains generally unknown even when $y_* = Y(\omega_*)$ is revealed; though, y_* brings about important insight on ω_* as it can significantly narrow down the possibilities.

From a modeling perspective, the unique advantage of adopting a random variable is that, through the association $y = Y(\omega)$, randomness in the observation space S is consistently linked to the randomness in the underlying sample space Ω. As a result, we place both spaces in a unified setting on which we can now apply the same notions to express and use any insight we gain from y_* to identify ω_*. Next, we describe how we can quantify rigorously such insight through a probability notion $P(\cdot)$ that, unlike $Q(\cdot)$, applies on S, rather than on Ω.

Of course, since in a stochastic experiment we only assess events β and not individual outcomes ω, it makes sense to consider quantifying probabilities only of collections η of individual observations y. Such collections of observations are commonly termed *measurables* and the family $\mathcal{H}$ that gathers every measurable in our framework is called *measurable space*. Assigning probabilities even to measurables η is a subtle task that sometimes leads to conceptual paradoxes if Ω and $\mathcal{B}$ are inconsistently associated with S and $\mathcal{H}$, respectively.

To avoid paradoxes, the measurable space $\mathcal{H}$, similar to the event space $\mathcal{B}$, must also have the mathematical structure of a σ-algebra. That is, a measurable space:

1. Must include the entire observation space (in which case we refer to the measurable $\eta = S$ as a *certain measurement*), the empty set (in which case we refer to the measurable $\eta = \emptyset$ as an *impossible measurement*), and possibly other subsets of the observation space.
2. For any measurable η, the measurable space must contain also its complement η^c.
3. The measurable space must also include unions $\eta_1 \cup \eta_2 \cup \cdots$ of any collection of measurables $\eta_1, \eta_2, \ldots$.

Additionally, the association $y = Y(\omega)$ must ensure that, while there may be outcomes ω in some events β that result in observations y not contained in any measurable η, there are no observations y in any of the measurables η that do not stem from an outcome ω in some event β. In other words, Y must ensure that it sends on each measurable η observations from outcomes necessarily inside one or more events.

Under these rather technical, nevertheless critical, requirements it is ensured that each measurable η is compatible with the events β. Also, it is ensured that the probabilities of the events $Q(\beta)$ from $\mathcal{B}$ propagate consistently, through $y = Y(\omega)$ to $\mathcal{H}$, *i.e.*, we reach the same conclusions either by considering probabilities over events or probabilities over measurables. Accordingly, we can safely use $Q(\cdot)$ to derive a probability measure $P(\cdot)$ on $\mathcal{H}$ void of logical inconsistencies.

Since the probability $P(\eta)$ of a measurable η has to reflect the probability of the events associated with η, it is natural to consider $P(\eta) = Q\left(Y^{-1}(\eta)\right)$. Descriptively, this reads

$$P(\eta) = \begin{pmatrix} \text{probability of} \\ \omega_* \in Y^{-1}(\eta) \end{pmatrix}.$$

In plain terms, the probability, denoted $P(\eta)$, of observing a particular value in η is the same as the probability, denoted $Q(\beta)$, of the event $\beta = Y^{-1}(\eta)$ that gathers all outcomes leading to η.

Note E.7 Probability distributions of random variable

Revisiting Section 1.2, the distribution of a random variable $Y \sim \mathbb{P}$ is adequately described when we can compute the probabilities $P(\eta)$ of all of its measurables η.

We summarize the terminology and the notions that couple sample and observation spaces in Table E.1 and Fig. E.3. So far, due to clarity, we have been strict to differentiate between them and this is the reason we adopted different notation for the probability measures $Q(\cdot)$ and $P(\cdot)$. However, because the entire framework is set around this coupling, most often we use the same notation to indicate either $Q(\beta)$ or $P(\eta)$ and pay little attention to differentiate between events β and measurables η. In fact, most of the time, it is easy to conclude from the context whether we quantify the probability of a measurable η of an event β. A rigorous exposition might be necessary,

Table E.1. Summary of the main notions on an extended probabilistic framework that incorporates random variables.

Phenomena			Observations
Outcome	ω	y	Measurement
Sample space	Ω	S	Observation space
Event space	$\mathcal{B}$	$\mathcal{H}$	Measurable space
Event	β	η	Measurable
Probability measure	$Q(\cdot)$	$P(\cdot)$	Probability measure

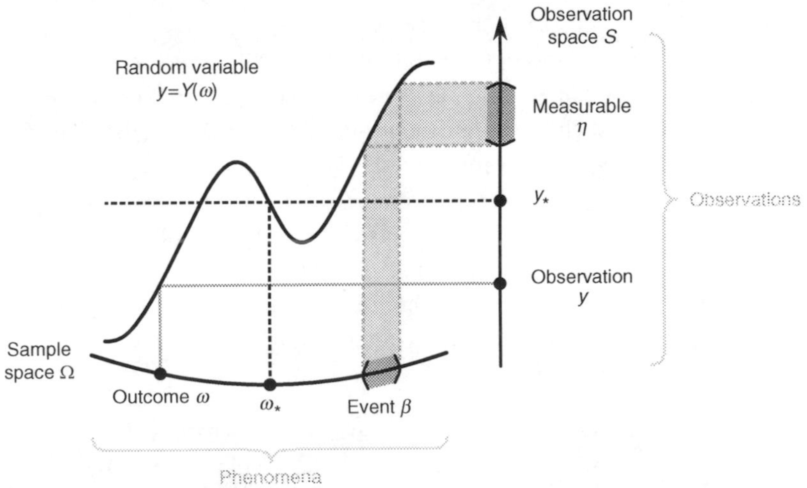

Fig. E.3 Schematic representation of the relationship between an observation space S and the underlying sample space Ω. Briefly, a phenomenon of interest takes place in Ω while it is observed in S. Individual observations y are related to individual outcomes ω by means of the random variable $y = Y(\omega)$. Events β and measurables η model our assessments in the two spaces.

however, when doubts about the validity or consistency of our framework are raised.

E.5 A Comprehensive Modeling Overview

In practice, very rarely do we need to think of random variables as functions, or even to consider the fine differences between events, measurables, and their associated spaces. This is because, typically, the development of a probabilistic framework that adheres to our prototype experiment takes the reverse route and, like in many other scientific activities, intuition combined with experience are often sufficient to ensure the validity of the resulting framework.

For instance, we saw that before a model takes concrete shape, we only have a description of the observation space S, which, as it can often be experimentally controlled, we tend to model faithfully. Although we can work with arbitrary observation spaces, most often the observation spaces encountered are finite or the real line, a portion of the real line, or multidimensional generalizations. On such observation spaces, it is easy to identify appropriate measurables η and lay a measurable space $\mathcal{H}$; for example, either by exhaustively enumerating each measurable as in Example 5.4, or by invoking some Borel description based on generating intervals along the lines of Note E.2.

Additionally, we have a good sense of the random variables $Y_n \sim \mathbb{P}_n$ and the distributions $\mathbb{P}_n$ associated with our measurements; these mostly depend on the measuring devices and often attain one of very few choices like normal, Poisson, categorical. Given the densities $p(y_n)$ of these distributions, we can easily develop a probability measure $P(\cdot)$ for our measurables; for example, based on integral representations along the lines of Note E.3. Accordingly, there are few modeling choices to be made about Y_n and $S, \mathcal{H}, P(\cdot)$. Since they are limited to more of less standard choices, a rigorous development of the observation side of our framework tends to be less critical.

In turn, once Y_n and $S, \mathcal{H}, P(\cdot)$ are adequately specified, extra assumptions are typically needed to specify a compatible sample space Ω and a probability measure $Q(\cdot)$. Such assumptions can be based on intuition or physical considerations on the phenomena at hand; for example, whether we model the motion of a mechanical system influenced by random perturbations, or the thermodynamics of an assembly of interacting components.

Finally, once we have determined $Y_n, S, \mathcal{H}, P(\cdot)$ and $\Omega, Q(\cdot)$, the only part left is to come up with an appropriate event space $\mathcal{B}$ that satisfies all requirements set by either the sample or observation sides. Again, provided standard choices are made and the random variables represent functions from Ω to S that are *at least* continuous and invertible, such an event space can be established again by invoking a Borel description, as in Note E.2, where the generating events this time are the pre-images of the measurables.

As such, so long as our model adheres to standard choices and the random variables in it meet some general conditions, we need not worry about the conclusions we draw either quantifying events or measurables.

Additional Reading

A. Papoulis. *Probability, random variables and stochastic processes*. 4th ed. McGraw-Hill, 2002.

R. Schilling. *Measure, integral, probability & processes: a concise introduction to probability and random processes. Probab(ilistical)ly the theoretical minimum*. Independently published, 2021.

J. A. Rice. *Mathematical statistics and data analysis*. 3rd ed. Duxbury, 2007.

G. Folland. *Real analysis: modern techniques and their applications*. 2nd ed. Wiley, 2007.

F Derivation of Key Relations

F.1 Relations in Chapter 2

Derivation of Eq. (*2.16*)

$$
\begin{aligned}
\frac{d}{dt}\sum_m P_{\sigma_m}(t) &= \sum_m \frac{d}{dt}P_{\sigma_m}(t) = \sum_m \left(-\lambda_{\sigma_m} P_{\sigma_m}(t) + \sum_{m'\neq m} \lambda_{\sigma_{m'}\to\sigma_m} P_{\sigma_{m'}}(t)\right) \\
&= -\sum_m \lambda_{\sigma_m} P_{\sigma_m}(t) + \sum_m \left(\sum_{m'\neq m} \lambda_{\sigma_{m'}\to\sigma_m} P_{\sigma_{m'}}(t)\right) \\
&= -\sum_m \lambda_{\sigma_m} P_{\sigma_m}(t) + \sum_{m'} \left(\sum_{m\neq m'} \lambda_{\sigma_{m'}\to\sigma_m} P_{\sigma_{m'}}(t)\right) \\
&= -\sum_m \lambda_{\sigma_m} P_{\sigma_m}(t) + \sum_{m'} \lambda_{\sigma'_m} P_{\sigma_{m'}}(t) \\
&= 0.
\end{aligned}
$$

Derivation of Eq. (*2.19*)

$$
\begin{aligned}
\lambda_b \sum_{c\mathcal{A}=0}^{\infty} c\mathcal{A}\,\tilde{P}_{c\mathcal{A}_{-1}}(t) &= \lambda_b \sum_{\ell=-1}^{\infty} (\ell+1) P_\ell(t) \\
&= \lambda_b + \lambda_b \sum_{\ell=-1}^{\infty} \ell P_\ell(t) \\
&= \lambda_b + \lambda_b \sum_{\ell=0}^{\infty} \ell P_\ell(t) \\
&= \lambda_b + \lambda_b \left\langle c\mathcal{A}(t)\right\rangle,
\end{aligned}
$$

where in the second line, we took $P_{-1}(t) = 0$ and in the last line we resubstituted the dummy index ℓ for $c\mathcal{A}$.

Derivation of Eq. (*2.69*)

We start with Eq. (2.68) reproduced below,

$$\frac{\partial}{\partial t}p(\mathbf{r},t)+\nabla_q\cdot\left(\mu p(\mathbf{r},t)\right)=-\nabla_q\cdot\left(\sigma\dot{W}_t\left(p(\mathbf{r}_0)\,e^{-t\nabla_q\cdot\mu}\right.\right.$$
$$-\int_0^t dse^{-(t-s)\nabla_q\cdot\mu}\mu\cdot\nabla_q p(\mathbf{r},s)$$
$$\left.\left.-\int_0^t dse^{-(t-s)\nabla_q\cdot\mu}\nabla_q\cdot\left(\sigma\dot{W}_s p(\mathbf{r},s)\right)\right)\right).$$

The last term, whose expectation value we take, reads

$$\left\langle\nabla_q\cdot\left(\sigma\dot{W}_t\left(\int_0^t dse^{-(t-s)\nabla_q\cdot\mu}\nabla_q\cdot\left(\sigma\dot{W}_s p(\mathbf{r},s)\right)\right)\right)\right\rangle$$
$$=\int_0^t dse^{-(t-s)\nabla_q\cdot\mu}\left\langle\nabla_q\cdot\left(\sigma\dot{W}_t\left(\nabla_q\cdot\left(\sigma\dot{W}_s p(\mathbf{r},s)\right)\right)\right)\right\rangle$$
$$=2\int_0^t dse^{-(t-s)\nabla_q\cdot\mu}\left\langle\nabla_q\cdot\sigma\left(\nabla_q\cdot\left(\sigma p(\mathbf{r},s)\,\delta_t(s)\right)\right)\right\rangle.$$

We recall that when we integrate a Dirac δ whose argument lies at the boundary, we recover half the area under the curve. This yields

$$=\nabla_q\cdot\sigma\left(\nabla_q\cdot\left(\sigma p(\mathbf{r},s)\right)\right).$$

For constant σ, this immediately simplifies to

$$=\sigma^2\nabla_q^2 p(\mathbf{r},s).$$

F.2 Relations in Chapter 3

Derivation of Eq. (*3.10*)

$$f_{\boldsymbol{\theta}^{\text{old}}}(s_{1:N})=p\left(s_{1:N}|w_{1:N},\boldsymbol{\theta}^{\text{old}}\right)$$
$$=\frac{p\left(s_{1:N},w_{1:N}|\boldsymbol{\theta}^{\text{old}}\right)}{p\left(w_{1:N}|\boldsymbol{\theta}^{\text{old}}\right)}$$
$$=\frac{p\left(w_{1:N}|s_{1:N},\boldsymbol{\theta}^{\text{old}}\right)}{p\left(w_{1:N}|\boldsymbol{\theta}^{\text{old}}\right)}p\left(s_{1:N}|\boldsymbol{\theta}^{\text{old}}\right)$$
$$=\prod_{n=1}^{N}\left[\frac{p\left(w_n|s_n,\boldsymbol{\theta}^{\text{old}}\right)}{p\left(w_n|\boldsymbol{\theta}^{\text{old}}\right)}p\left(s_n|\boldsymbol{\theta}^{\text{old}}\right)\right]$$

$$
\begin{aligned}
&= \prod_{n=1}^{N} \left[\frac{\mathrm{Normal}\left(w_n; \mu_{s_n}^{\mathrm{old}}, v_{s_n}^{\mathrm{old}}\right) (\pi_1^{\mathrm{old}})^{\Delta_1(s_n)} (\pi_2^{\mathrm{old}})^{\Delta_2(s_n)}}{\pi_1^{\mathrm{old}}\mathrm{Normal}\left(w_n; \mu_1^{\mathrm{old}}, v_1^{\mathrm{old}}\right) + \pi_2^{\mathrm{old}}\mathrm{Normal}\left(w_n; \mu_2^{\mathrm{old}}, v_2^{\mathrm{old}}\right)} \right] \\
&= \prod_{n=1}^{N} \left[\frac{\left(\pi_1^{\mathrm{old}}\mathrm{Normal}\left(w_n; \mu_1^{\mathrm{old}}, v_1^{\mathrm{old}}\right)\right)^{\Delta_1(s_n)} \left(\pi_2^{\mathrm{old}}\mathrm{Normal}\left(w_n; \mu_2^{\mathrm{old}}, v_2^{\mathrm{old}}\right)\right)^{\Delta_2(s_n)}}{\pi_1^{\mathrm{old}}\mathrm{Normal}\left(w_n; \mu_1^{\mathrm{old}}, v_1^{\mathrm{old}}\right) + \pi_2^{\mathrm{old}}\mathrm{Normal}\left(w_n; \mu_2^{\mathrm{old}}, v_2^{\mathrm{old}}\right)} \right] \\
&= \prod_{n=1}^{N} \left[\left(\frac{\pi_1^{\mathrm{old}}\mathrm{Normal}\left(w_n; \mu_1^{\mathrm{old}}, v_1^{\mathrm{old}}\right)}{\pi_1^{\mathrm{old}}\mathrm{Normal}\left(w_n; \mu_1^{\mathrm{old}}, v_1^{\mathrm{old}}\right) + \pi_2^{\mathrm{old}}\mathrm{Normal}\left(w_n; \mu_2^{\mathrm{old}}, v_2^{\mathrm{old}}\right)} \right)^{\Delta_1(s_n)} \right. \\
&\qquad \left. \times \left(\frac{\pi_2^{\mathrm{old}}\mathrm{Normal}\left(w_n; \mu_2^{\mathrm{old}}, v_2^{\mathrm{old}}\right)}{\pi_1^{\mathrm{old}}\mathrm{Normal}\left(w_n; \mu_1^{\mathrm{old}}, v_1^{\mathrm{old}}\right) + \pi_2^{\mathrm{old}}\mathrm{Normal}\left(w_n; \mu_2^{\mathrm{old}}, v_2^{\mathrm{old}}\right)} \right)^{\Delta_2(s_n)} \right] \\
&= \prod_{n=1}^{N} (\gamma_{1n}^{\mathrm{old}})^{\Delta_1(s_n)} (\gamma_{2n}^{\mathrm{old}})^{\Delta_2(s_n)},
\end{aligned}
$$

where

$$
\begin{aligned}
\gamma_{1n}^{\mathrm{old}} &= \frac{\pi_1^{\mathrm{old}}\mathrm{Normal}\left(w_n; \mu_1^{\mathrm{old}}, v_1^{\mathrm{old}}\right)}{\pi_1^{\mathrm{old}}\mathrm{Normal}\left(w_n; \mu_1^{\mathrm{old}}, v_1^{\mathrm{old}}\right) + \pi_2^{\mathrm{old}}\mathrm{Normal}\left(w_n; \mu_2^{\mathrm{old}}, v_2^{\mathrm{old}}\right)}, \\
\gamma_{2n}^{\mathrm{old}} &= \frac{\pi_2^{\mathrm{old}}\mathrm{Normal}\left(w_n; \mu_2^{\mathrm{old}}, v_2^{\mathrm{old}}\right)}{\pi_1^{\mathrm{old}}\mathrm{Normal}\left(w_n; \mu_1^{\mathrm{old}}, v_1^{\mathrm{old}}\right) + \pi_2^{\mathrm{old}}\mathrm{Normal}\left(w_n; \mu_2^{\mathrm{old}}, v_2^{\mathrm{old}}\right)}.
\end{aligned}
$$

Derivation of Eq. (*3.11*)

$$
\begin{aligned}
Q_{\boldsymbol{\theta}^{\mathrm{old}}}(\boldsymbol{\theta}) &= \sum_{s_{1:N}} \log p(s_{1:N}, w_{1:N}|\boldsymbol{\theta}) f_{\boldsymbol{\theta}^{\mathrm{old}}}(s_{1:N}) \\
&= \sum_{s_{1:N}} \left[\log p(s_{1:N}, w_{1:N}|\boldsymbol{\theta}) \prod_{n=1}^{N} (\gamma_{1n}^{\mathrm{old}})^{\Delta_1(s_n)} (\gamma_{2n}^{\mathrm{old}})^{\Delta_2(s_n)} \right] \\
&= \sum_{s_{1:N}} \left[\sum_{n'=1}^{N} [\log p(s_{n'}, w_{n'}|\boldsymbol{\theta})] \prod_{n=1}^{N} \left[(\gamma_{1n}^{\mathrm{old}})^{\Delta_1(s_n)} (\gamma_{2n}^{\mathrm{old}})^{\Delta_2(s_n)} \right] \right] \\
&= \sum_{n=1}^{N} \left[\sum_{s_n=1}^{2} (\gamma_{1n}^{\mathrm{old}})^{\Delta_1(s_n)} (\gamma_{2n}^{\mathrm{old}})^{\Delta_2(s_n)} \log p(s_n, w_n|\boldsymbol{\theta}) \right] \\
&= \sum_{n=1}^{N} \left[\sum_{s_n=1}^{2} (\gamma_{1n}^{\mathrm{old}})^{\Delta_1(s_n)} (\gamma_{2n}^{\mathrm{old}})^{\Delta_2(s_n)} [\log p(w_n|s_n, \boldsymbol{\theta}) p(s_n|\boldsymbol{\theta})] \right] \\
&= \sum_{n=1}^{N} \left[\sum_{s_n=1}^{2} (\gamma_{1n}^{\mathrm{old}} [\log \mathrm{Normal}(w_n; \mu_1, v_1)\pi_1])^{\Delta_1(s_n)} \right. \\
&\qquad \left. (\gamma_{2n}^{\mathrm{old}} [\log \mathrm{Normal}(w_n; \mu_2, v_2)\pi_2])^{\Delta_2(s_n)} \right]
\end{aligned}
$$

$$
\begin{aligned}
&= \sum_{n=1}^{N} \Big[\gamma_{1n}^{\text{old}}[\log \text{Normal}(w_n;\, \mu_1, v_1)\pi_1] \\
&\quad + \gamma_{2n}^{\text{old}}[\log \text{Normal}(w_n;\, \mu_2, v_2)\pi_2]\Big] \\
&= \sum_{n=1}^{N} \left[\gamma_{1n}^{\text{old}}\left(\log \pi_1 - \frac{\log(2\pi v_1)}{2} - \frac{(w_n-\mu_1)^2}{2v_1}\right)\right] \\
&\quad + \sum_{n=1}^{N} \left[\gamma_{2n}^{\text{old}}\left(\log \pi_2 - \frac{\log(2\pi v_2)}{2} - \frac{(w_n-\mu_2)^2}{2v_2}\right)\right] \\
&= \sum_{n=1}^{N} \left[\gamma_{1n}^{\text{old}}\left(\log \pi_1 - \frac{\log v_1}{2} - \frac{(w_n-\mu_1)^2}{2v_1}\right)\right] \\
&\quad + \sum_{n=1}^{N} \left[\gamma_{2n}^{\text{old}}\left(\log \pi_2 - \frac{\log v_2}{2} - \frac{(w_n-\mu_2)^2}{2v_2}\right)\right] + \text{constants} \\
&= \left(\log \pi_1 - \frac{\log v_1}{2}\right)\left(\sum_{n=1}^{N} \gamma_{1n}^{\text{old}}\right) + \left(\log \pi_2 - \frac{\log v_2}{2}\right)\left(\sum_{n=1}^{N} \gamma_{2n}^{\text{old}}\right) \\
&\quad - \frac{1}{2}\sum_{n=1}^{N}\left[\gamma_{1n}^{\text{old}}\frac{(w_n-\mu_1)^2}{v_1} + \gamma_{2n}^{\text{old}}\frac{(w_n-\mu_2)^2}{v_2}\right] + \text{constants.}
\end{aligned}
$$

Derivation of Eq. (*3.12*)

$$
\begin{aligned}
\log p\left(w_{1:N}, r_{1:N}|\boldsymbol{\theta}\right) &= \log p\left(r_{1:N}|w_{1:N}, \boldsymbol{\theta}\right) + \log p\left(w_{1:N}|\boldsymbol{\theta}\right) \\
\log p\left(w_{1:N}|\boldsymbol{\theta}\right) &= \log p\left(w_{1:N}, r_{1:N}|\boldsymbol{\theta}\right) - \log p\left(r_{1:N}|w_{1:N}, \boldsymbol{\theta}\right) \\
[\log p\left(w_{1:N}|\boldsymbol{\theta}\right)] & \\
\times p\left(r_{1:N}|w_{1:N}, \boldsymbol{\theta}^{\text{old}}\right) &= [\log p\left(w_{1:N}, r_{1:N}|\boldsymbol{\theta}\right) - \log p\left(r_{1:N}|w_{1:N}, \boldsymbol{\theta}\right)] \\
&\quad \times p\left(r_{1:N}|w_{1:N}, \boldsymbol{\theta}^{\text{old}}\right) \\
\sum_{r_{1:N}} [\log p\left(w_{1:N}|\boldsymbol{\theta}\right)] & \\
\times p\left(r_{1:N}|w_{1:N}, \boldsymbol{\theta}^{\text{old}}\right) &= \sum_{r_{1:N}} [\log p\left(w_{1:N}, r_{1:N}|\boldsymbol{\theta}\right) - \log p\left(r_{1:N}|w_{1:N}, \boldsymbol{\theta}\right)] \\
&\quad \times p\left(r_{1:N}|w_{1:N}, \boldsymbol{\theta}^{\text{old}}\right) \\
\log p\left(w_{1:N}|\boldsymbol{\theta}\right) &= \underbrace{\sum_{r_{1:N}} [\log p\left(w_{1:N}, r_{1:N}|\boldsymbol{\theta}\right)]\, p\left(r_{1:N}|w_{1:N}, \boldsymbol{\theta}^{\text{old}}\right)}_{Q_{\boldsymbol{\theta}^{\text{old}}}(\boldsymbol{\theta})} \\
&\quad - \sum_{r_{1:N}} [\log p\left(r_{1:N}|w_{1:N}, \boldsymbol{\theta}\right)]\, p\left(r_{1:N}|w_{1:N}, \boldsymbol{\theta}^{\text{old}}\right).
\end{aligned}
$$

F.3 Relations in Chapter 5

Derivation of Eq. (*5.4*)

$$
\begin{aligned}
p\left(\tau|w_{1:N}\right) &\propto p(w_{1:N}|\tau)p(\tau) \\
&= \left(\int d\mu\, p(w_{1:N},\mu|\tau)\right)p(\tau) \\
&= \left(\int d\mu\, p(w_{1:N}|\mu,\tau)p(\mu|\tau)\right)p(\tau) \\
&= \left(\int d\mu \left(\prod_{n=1}^{N} p(w_n|\mu,\tau)\right)p(\mu|\tau)\right)p(\tau) \\
&= \left(\int d\mu \left(\prod_{n=1}^{N} \text{Normal}\left(w_n;\mu,\frac{1}{\tau}\right)\right)\text{Normal}\left(\mu;\xi,\frac{1}{\psi_0\tau}\right)\right) \\
&\quad \times \text{Gamma}\left(\tau;\alpha,\beta\right) \\
&= \left(\prod_{n=1}^{N} \text{Normal}\left(w_n;\xi,\frac{\psi_0+1}{\psi_0\tau}\right)\right)\text{Gamma}\left(\tau;\alpha,\beta\right) \\
&\propto \text{Gamma}\left(\tau;\alpha+\frac{N}{2},\frac{1}{\frac{1}{\beta}+\frac{1}{2}\frac{\psi_0}{\psi_0+1}\sum_{n=1}^{N}(w_n-\xi)^2}\right).
\end{aligned}
$$

Derivation of Eq. (*5.5*)

$$
\begin{aligned}
p\left(\mu|\tau,w_{1:N}\right) &\propto p(w_{1:N}|\mu,\tau)p(\mu|\tau) \\
&= \left(\prod_{n=1}^{N} \text{Normal}\left(w_n;\mu,\frac{1}{\tau}\right)\right)\text{Normal}\left(\mu;\xi,\frac{1}{\psi_0\tau}\right) \\
&= \text{Normal}\left(\mu;\frac{\psi_0\xi+\sum_{n=1}^{N}w_n}{\psi_0+N},\frac{1}{\psi_0+N}\frac{1}{\tau}\right).
\end{aligned}
$$

Derivation of Eq. (*5.10*)

$$
p\left(v|h,f,w_{1:N},\bar{w}_{1:\bar{N}}\right) = p\left(v|h,f,w_{1:N}\right) = \text{Binomial}\left(v;W,\frac{h}{h+f}\right).
$$

Derivation of Eq. (*5.11*)

$$
\begin{aligned}
p\left(h|v,f,w_{1:N},\bar{w}_{1:\bar{N}}\right) &\propto p\left(v,h,f,w_{1:N},\bar{w}_{1:\bar{N}}\right)\\
&= p\left(v|h,f,w_{1:N},\bar{w}_{1:\bar{N}}\right) p\left(h,f,w_{1:N},\bar{w}_{1:\bar{N}}\right)\\
&= p\left(v|h,f,w_{1:N},\bar{w}_{1:\bar{N}}\right) p\left(w_{1:N}|h,f,\bar{w}_{1:\bar{N}}\right)\\
&\quad\times\ p\left(\bar{w}_{1:\bar{N}}|h,f\right) p\left(h,f\right)\\
&= p\left(v|h,f,w_{1:N}\right) p\left(w_{1:N}|h,f\right) p\left(\bar{w}_{1:\bar{N}}|f\right) p\left(h\right) p\left(f\right)\\
&\propto p\left(v|h,f,w_{1:N}\right) p\left(w_{1:N}|h,f\right) p\left(h\right)\\
&= \text{Binomial}\left(v;\ W, \frac{h}{h+f}\right)\left[\prod_{n=1}^{N}\text{Poisson}\left(w_n;\ \tau\beta(h+f)\right)\right]\\
&\quad\times\ \text{Gamma}\left(h;\ H, \frac{h_{\text{ref}}}{H}\right)\\
&\propto h^{H+v-1}e^{-\left(\frac{H}{h_{\text{ref}}}+N\tau\beta\right)h}\\
&\propto \text{Gamma}\left(h;\ H+v, \frac{1}{\frac{H}{h_{\text{ref}}}+N\tau\beta}\right).
\end{aligned}
$$

Derivation of Eq. (*5.12*)

$$
\begin{aligned}
p\left(f|v,h,w_{1:N},\bar{w}_{1:\bar{N}}\right) &\propto p\left(v,h,f,w_{1:N},\bar{w}_{1:\bar{N}}\right)\\
&= p\left(v|h,f,w_{1:N},\bar{w}_{1:\bar{N}}\right) p\left(h,f,w_{1:N},\bar{w}_{1:\bar{N}}\right)\\
&= p\left(v|h,f,w_{1:N},\bar{w}_{1:\bar{N}}\right) p\left(w_{1:N}|h,f,\bar{w}_{1:\bar{N}}\right)\\
&\quad\times\ p\left(\bar{w}_{1:\bar{N}}|h,f\right) p\left(h,f\right)\\
&= p\left(v|h,f,w_{1:N}\right) p\left(w_{1:N}|h,f\right) p\left(\bar{w}_{1:\bar{N}}|f\right) p\left(h\right) p\left(f\right)\\
&\propto p\left(v|h,f,w_{1:N}\right) p\left(w_{1:N}|h,f\right) p\left(\bar{w}_{1:\bar{N}}|f\right) p\left(f\right)\\
&= \text{Binomial}\left(v;\ W, \frac{h}{h+f}\right)\left[\prod_{n=1}^{N}\text{Poisson}\left(w_n;\ \tau\beta(h+f)\right)\right]\\
&\quad\times\left[\prod_{n=1}^{\bar{N}}\text{Poisson}\left(\bar{w}_n;\ \tau\beta f\right)\right]\text{Gamma}\left(f;\ F, \frac{f_{\text{ref}}}{F}\right)\\
&\propto \text{Gamma}\left(f;\ F+W+\bar{W}-v, \frac{1}{\frac{F}{f_{\text{ref}}}+(N+\bar{N})\tau\beta}\right).
\end{aligned}
$$

F.4 Relations in Chapter 6

Derivation of Eq. (*6.8*)

$$\begin{aligned} C(t,t') &= \langle \phi(t)\phi(t')\rangle \\ &= \langle \Delta x(t)\Delta x(t')\rangle - \frac{t}{T}\langle \Delta x(T)\Delta x(t')\rangle - \frac{t'}{T}\langle \Delta x(t)\Delta x(T)\rangle + \frac{tt'}{T^2}\langle \Delta x^2(T)\rangle \\ &= 2D\left(\min(t,t') - \frac{tt'}{T}\right). \end{aligned}$$

Derivation of Eq. (*6.10*)

We start from

$$\begin{aligned} p(f^\star|y) &= \int df^\sharp\, p(f^\star, f^\sharp|y) \\ &= \int df^\sharp\, p(f^\star|f^\sharp, y)p(f^\sharp|y) \\ &= \int df^\sharp\, p(f^\star|f^\sharp)p(f^\sharp|y), \end{aligned}$$

where the last identity follows from a Markov blanket argument, *i.e.*, when conditioned on $f^\sharp$, $f^\star$ is independent of y.

Derivation of Eq. (*6.11*)

As the model reads

$$\begin{aligned} (f^\star, f^\sharp) &\sim \text{Normal}_{N^\star+N^\sharp}\left((\mu^\star,\mu^\sharp), \begin{pmatrix} C^{\star\star} & C^{\star\sharp} \\ C^{\sharp\star} & C^{\sharp\sharp}\end{pmatrix}\right), \\ y_n|f^\sharp &\sim \text{Normal}_{N^\sharp}(f^\sharp, v\mathbb{1}), \end{aligned}$$

then these equations imply

$$\begin{aligned} f^\sharp|y &\sim \text{Normal}_{N^\sharp}\left(\Sigma^{\sharp\sharp}\left[\left(C^{\sharp\sharp}\right)^{-1}\mu^\sharp + (v\mathbb{1})^{-1}y\right], \Sigma^{\sharp\sharp}\right), \\ f^\star|f^\sharp &\sim \text{Normal}_{N^\star}\left(\mu^\star + C^{\star\sharp}\left(C^{\sharp\sharp}\right)^{-1}\left(f^\sharp-\mu^\sharp\right), C^{\star\star} - C^{\star\sharp}\left(C^{\sharp\sharp}\right)^{-1}C^{\sharp\star}\right), \end{aligned}$$

where $\Sigma^{\sharp\sharp} = \left[\left(C^{\sharp\sharp}\right)^{-1} + (v\mathbb{1})^{-1}\right]^{-1}$. In particular, in deriving $p(f^\sharp|y)$ from $p(y|f^\sharp)$, we recognize that $p(f^\sharp|y) \propto p(y|f^\sharp)$ and recomplete the matrix squares in $p(y|f^\sharp)$ as derived in Exercise 1.3. We follow similar logic in deriving $p(f^\star|f^\sharp)$ from $p(f^\star, f^\sharp)$. By now inserting both $p(f^\star|f^\sharp)$ and $p(f^\sharp|y)$ into $\int df^\sharp\, p(f^\star|f^\sharp)p(f^\sharp|y)$ and performing the resulting integral, we arrive at

$$f^\star|y \sim \text{Normal}_{N^\star}\left(\tilde{\mu}, \tilde{C}\right),$$

where

$$\tilde{\mu} = \mu^\star + C^{\star\#}\left(C^{\#\#}\right)^{-1}\left[\Sigma^{\#\#}\left[\left(C^{\#\#}\right)^{-1}\mu^\# + (v\mathbb{1})^{-1}y\right] - \mu^\#\right]$$

and

$$\tilde{C} = C^{\star\#}\left(C^{\#\#}\right)^{-1}\Sigma^{\#\#}\left(C^{\#\#}\right)^{-1}C^{\#\star} + C^{\star\star} - C^{\star\#}\left(C^{\#\#}\right)^{-1}C^{\#\star}.$$

The above mean, $\tilde{\mu}$, and covariance, $\tilde{C}$, can be further simplified.

We start with the mean and note that

$$\begin{aligned}
\tilde{\mu} &= \mu^\star + C^{\star\#}\left(C^{\#\#}\right)^{-1}\left[\Sigma^{\#\#}\left[\left(C^{\#\#}\right)^{-1}\mu^\# + (v\mathbb{1})^{-1}y\right] - \mu^\#\right]\\
&= \mu^\star + C^{\star\#}\left(C^{\#\#}\right)^{-1}\left[\Sigma^{\#\#}\left(C^{\#\#}\right)^{-1}\mu^\# + \Sigma^{\#\#}(v\mathbb{1})^{-1}y - \mu^\#\right]\\
&= \mu^\star + C^{\star\#}\left(C^{\#\#}\right)^{-1}\left[\left[\left(C^{\#\#}\right)^{-1} + (v\mathbb{1})^{-1}\right]^{-1}\left(C^{\#\#}\right)^{-1}\mu^\# + \Sigma^{\#\#}(v\mathbb{1})^{-1}y - \mu^\#\right].
\end{aligned}$$

We now invoke a push-through matrix identity, derived in Exercise 1.3, and obtain

$$\begin{aligned}
&= \mu^\star + C^{\star\#}\left(C^{\#\#}\right)^{-1}\left[\left[C^{\#\#}\left(C^{\#\#}\right)^{-1} + C^{\#\#}(v\mathbb{1})^{-1}\right]^{-1}\mu^\# + \Sigma^{\#\#}(v\mathbb{1})^{-1}y - \mu^\#\right]\\
&= \mu^\star + C^{\star\#}\left(C^{\#\#}\right)^{-1}\left[\left[\mathbb{1} + C^{\#\#}(v\mathbb{1})^{-1}\right]^{-1}\mu^\# + \Sigma^{\#\#}(v\mathbb{1})^{-1}y - \mu^\#\right].
\end{aligned}$$

Applying a Woodbury matrix identity, derived in Exercise 1.3, to the above yields

$$\begin{aligned}
&= \mu^\star + C^{\star\#}\left(C^{\#\#}\right)^{-1}\left[\left[\mathbb{1} - C^{\#\#}\left[\mathbb{1} + (v\mathbb{1})^{-1}C^{\#\#}\right]^{-1}(v\mathbb{1})^{-1}\right]\mu^\# + \Sigma^{\#\#}(v\mathbb{1})^{-1}y - \mu^\#\right]\\
&= \mu^\star + C^{\star\#}\left(C^{\#\#}\right)^{-1}\left[-C^{\#\#}\left[\mathbb{1} + (v\mathbb{1})^{-1}C^{\#\#}\right]^{-1}(v\mathbb{1})^{-1}\mu^\# + \Sigma^{\#\#}(v\mathbb{1})^{-1}y\right].
\end{aligned}$$

Invoking another push-through matrix identity yields

$$\begin{aligned}
&= \mu^\star + C^{\star\#}\left(C^{\#\#}\right)^{-1}\left[-C^{\#\#}\left[v\mathbb{1} + v\mathbb{1}(v\mathbb{1})^{-1}C^{\#\#}\right]^{-1}\mu^\# + \Sigma^{\#\#}(v\mathbb{1})^{-1}y\right]\\
&= \mu^\star + C^{\star\#}\left(C^{\#\#}\right)^{-1}\left[-C^{\#\#}\left[v\mathbb{1} + C^{\#\#}\right]^{-1}\mu^\# + \Sigma^{\#\#}(v\mathbb{1})^{-1}y\right]\\
&= \mu^\star + C^{\star\#}\left[-\left(C^{\#\#}\right)^{-1}C^{\#\#}\left[v\mathbb{1} + C^{\#\#}\right]^{-1}\mu^\# + \left(C^{\#\#}\right)^{-1}\Sigma^{\#\#}(v\mathbb{1})^{-1}y\right]\\
&= \mu^\star + C^{\star\#}\left[-\left[v\mathbb{1} + C^{\#\#}\right]^{-1}\mu^\# + \left(C^{\#\#}\right)^{-1}\Sigma^{\#\#}(v\mathbb{1})^{-1}y\right]\\
&= \mu^\star + C^{\star\#}\left[-\left[v\mathbb{1} + C^{\#\#}\right]^{-1}\mu^\# + \left(C^{\#\#}\right)^{-1}\left[\left(C^{\#\#}\right)^{-1} + (v\mathbb{1})^{-1}\right]^{-1}(v\mathbb{1})^{-1}y\right].
\end{aligned}$$

One more push-through matrix identity returns

$$\begin{aligned}
&= \mu^\star + C^{\star\#}\left[-\left[v\mathbb{1} + C^{\#\#}\right]^{-1}\mu^\# + \left[v\mathbb{1}\left(C^{\#\#}\right)^{-1}C^{\#\#} + v\mathbb{1}(v\mathbb{1})^{-1}C^{\#\#}\right]^{-1}y\right]\\
&= \mu^\star + C^{\star\#}\left[-\left[v\mathbb{1} + C^{\#\#}\right]^{-1}\mu^\# + \left[v\mathbb{1} + C^{\#\#}\right]^{-1}y\right]\\
&= \mu^\star + C^{\star\#}\left[v\mathbb{1} + C^{\#\#}\right]^{-1}\left(y - \mu^\#\right).
\end{aligned}$$

Next, the covariance, $\tilde{C}$, simplifies as follows,

$$\begin{aligned}\tilde{C} &= C^{\star\sharp}\left(C^{\sharp\sharp}\right)^{-1}\Sigma^{\sharp\sharp}\left(C^{\sharp\sharp}\right)^{-1}C^{\sharp\star} + C^{\star\star} - C^{\star\sharp}\left(C^{\sharp\sharp}\right)^{-1}C^{\sharp\star}\\ &= C^{\star\star} + C^{\star\sharp}\left[\left(C^{\sharp\sharp}\right)^{-1}\Sigma^{\sharp\sharp}\left(C^{\sharp\sharp}\right)^{-1} - \left(C^{\sharp\sharp}\right)^{-1}\right]C^{\sharp\star}\\ &= C^{\star\star} + C^{\star\sharp}\left[\left(C^{\sharp\sharp}\right)^{-1}\left[\left(C^{\sharp\sharp}\right)^{-1} + (v\mathbb{1})^{-1}\right]^{-1}\left(C^{\sharp\sharp}\right)^{-1} - \left(C^{\sharp\sharp}\right)^{-1}\right]C^{\sharp\star}.\end{aligned}$$

Invoking a push-through matrix identity on the above yields

$$\begin{aligned}&= C^{\star\star} + C^{\star\sharp}\left[\left[C^{\sharp\sharp}\left(C^{\sharp\sharp}\right)^{-1}C^{\sharp\sharp} + C^{\sharp\sharp}(v\mathbb{1})^{-1}C^{\sharp\sharp}\right]^{-1} - \left(C^{\sharp\sharp}\right)^{-1}\right]C^{\sharp\star}\\ &= C^{\star\star} + C^{\star\sharp}\left[\left[C^{\sharp\sharp} + C^{\sharp\sharp}(v\mathbb{1})^{-1}C^{\sharp\sharp}\right]^{-1} - \left(C^{\sharp\sharp}\right)^{-1}\right]C^{\sharp\star}.\end{aligned}$$

Invoking a Woodbury matrix identity on this returns

$$\begin{aligned}&= C^{\star\star} + C^{\star\sharp}\left[\left(C^{\sharp\sharp}\right)^{-1}\right.\\ &\quad\left. - \left(C^{\sharp\sharp}\right)^{-1}C^{\sharp\sharp}\left[v\mathbb{1} + C^{\sharp\sharp}\left(C^{\sharp\sharp}\right)^{-1}C^{\sharp\sharp}\right]^{-1}C^{\sharp\sharp}\left(C^{\sharp\sharp}\right)^{-1} - \left(C^{\sharp\sharp}\right)^{-1}\right]C^{\sharp\star}\\ &= C^{\star\star} - C^{\star\sharp}\left[v\mathbb{1} + C^{\sharp\sharp}\right]^{-1}C^{\sharp\star}.\end{aligned}$$

Derivation of Eq. (*6.15*)

We start from Eq. (6.13) and Eq. (6.14) in order to write the product of the likelihood and prior as $p(y|f^u)p(f^u)$,

$$\begin{aligned}p(f^u|y) &\propto p(y|f^u)p(f^u)\\ &= \text{Normal}_{N^u}\left(f^u; M^{-1}b, M^{-1}\right)\text{Normal}_{N^u}\left(f^u; 0, C^{uu}\right),\end{aligned}$$

where we have completed the squares of the likelihood and where

$$b = v^{-1}(C^{uu})^{-1}C^{u\sharp}y, \qquad M = v^{-1}(C^{uu})^{-1}C^{u\sharp}C^{\sharp u}(C^{uu})^{-1}.$$

A multiplication of the likelihood and prior now return the result in Eq. (*6.15*) upon completing the squares.

Derivation of Eq. (*6.17*)

$$\begin{aligned}p(\pi^{\text{new}}_{\sigma_m}|b^{\text{old}}_{\sigma_m}) &\propto p(b^{\text{old}}_{\sigma_m}|\pi^{\text{new}}_{\sigma_m})p(\pi^{\text{new}}_{\sigma_m})\\ &= \text{Bernoulli}(b^{\text{old}}_{\sigma_m}; \pi^{\text{new}}_{\sigma_m})\text{Beta}\left(\pi^{\text{new}}_{\sigma_m}; \alpha\frac{1}{M}, \beta\frac{M-1}{M}\right)\\ &= \left(\pi^{\text{new}}_{\sigma_m}\right)^{(b^{\text{old}}_{\sigma_m})}\left(1-\pi^{\text{new}}_{\sigma_m}\right)^{(1-b^{\text{old}}_{\sigma_m})}\frac{\Gamma\left(\alpha\frac{1}{M}+\beta\frac{M-1}{M}\right)}{\Gamma\left(\alpha\frac{1}{M}\right)\Gamma\left(\beta\frac{M-1}{M}\right)}\\ &\quad\left(\pi^{\text{new}}_{\sigma_m}\right)^{\alpha\frac{1}{M}-1}\left(1-\pi^{\text{new}}_{\sigma_m}\right)^{\beta\frac{M-1}{M}-1}\end{aligned}$$

$$= \frac{\Gamma\left(\alpha\frac{1}{M} + \beta\frac{M-1}{M}\right)}{\Gamma\left(\alpha\frac{1}{M}\right)\Gamma\left(\beta\frac{M-1}{M} - 1\right)} \left(\pi^{\text{new}}_{\sigma_m}\right)^{\alpha\frac{1}{M}+b^{\text{old}}_{\sigma_m}-1} \left(1 - \pi^{\text{new}}_{\sigma_m}\right)^{\beta\frac{M-1}{M}-b^{\text{old}}_{\sigma_m}+1}$$
$$\propto \text{Beta}\left(\pi^{\text{new}}_{\sigma_m}; \alpha', \beta'\right),$$

where $\alpha' = \alpha\frac{1}{M} + b^{\text{old}}_{\sigma_m}$ and $\beta' = \beta\frac{M-1}{M} - b^{\text{old}}_{\sigma_m} + 1$.

Derivation of Eq. (*6.18*)

$$\begin{aligned}
p\left(b_{\sigma_m}\right) &= \int_0^1 p\left(b_{\sigma_m}|\pi_{\sigma_m}\right) p\left(\pi_{\sigma_m}\right) d\pi_{\sigma_m} \\
&= \int_0^1 \frac{\Gamma\left(\alpha+\beta\right)}{\Gamma\left(\alpha\right)\Gamma\left(\beta\right)} \left(\pi_k\right)^{\alpha+b_{\sigma_m}-1} \left(1-\pi_k\right)^{\beta+b_{\sigma_m}-1} d\pi_{\sigma_m} \\
&= \frac{\Gamma\left(\alpha+\beta\right)}{\Gamma\left(\alpha\right)\Gamma\left(\beta\right)} \frac{\Gamma\left(\alpha+b_{\sigma_m}\right)\Gamma\left(\beta+b_{\sigma_m}\right)}{\Gamma\left(\alpha+\beta+1\right)} \underbrace{\int_0^1 \text{Beta}\left(\alpha+b_{\sigma_m}, \beta+b_{\sigma_m}\right) d\pi_{\sigma_m}}_{1} \\
&= \frac{\Gamma\left(\alpha+b_{\sigma_m}\right)\Gamma\left(\beta+b_{\sigma_m}\right)}{\left(\alpha+\beta\right)\Gamma\left(\alpha\right)\Gamma\left(\beta\right)} \\
&= \begin{cases} \frac{\alpha}{\alpha+\beta}, & b_{\sigma_m} = 1, \\ \frac{\beta}{\alpha+\beta}, & b_{\sigma_m} = 0, \end{cases}
\end{aligned}$$

which is the Bernoulli density

$$p\left(b_{\sigma_m}\right) = \text{Bernoulli}\left(b_{\sigma_m}; \frac{\alpha}{\alpha+\beta}\right).$$

F.5 Relations in Chapter 7

Derivation of Eq. (*7.2*)

$$\begin{aligned}
p(w_{1:N}, s_{1:N}|\boldsymbol{\pi}, \boldsymbol{\phi}) &= p(w_{1:N}|s_{1:N}, \boldsymbol{\phi})p(s_{1:N}|\boldsymbol{\pi}) \\
&= \prod_{n=1}^{N} p(w_n|s_n, \boldsymbol{\phi})p(s_n|\boldsymbol{\pi}) \\
&= \prod_{n=1}^{N} G_{\phi_{s_n}}(w_n)\text{Categorical}_{\sigma_{1:M}}\left(s_n; \pi_{\sigma_{1:M}}\right) \\
&= \prod_{n=1}^{N} G_{\phi_{s_n}}(w_n)\pi_{s_n}.
\end{aligned}$$

Derivation of Eq. (*7.3*)

$$
\begin{aligned}
p(w_{1:N}|\pi, \phi) &= \sum_{s_{1:N}} p(w_{1:N}, s_{1:N}|\pi, \phi) \\
&= \sum_{s_{1:N}} p(w_{1:N}|s_{1:N}, \pi, \phi)p(s_{1:N}|\pi, \phi) \\
&= \sum_{s_{1:N}} p(w_{1:N}|s_{1:N}, \phi)p(s_{1:N}|\pi) \\
&= \sum_{s_{1:N}} \prod_{n=1}^{N} p(w_n|s_n, \phi)p(s_n|\pi) \\
&= \prod_{n=1}^{N} \sum_{s_n} p(w_n|s_n, \phi)p(s_n|\pi) \\
&= \prod_{n=1}^{N} \sum_{s_n} G_{\phi_{s_n}}(w_n)\pi_{s_n} \\
&= \prod_{n=1}^{N} \sum_{m=1}^{M} G_{\phi_{\sigma_m}}(w_n)\pi_{\sigma_m}.
\end{aligned}
$$

Derivation of Eq. (*7.4*)

The proportionality constant in Eq. (*7.4*), treated as normalization parameter c, follows from

$$
\begin{aligned}
1 &= \sum_{s_1} \cdots \sum_{s_{N-1}} \sum_{s_N} p(s_{1:N}|w_{1:N}) \\
&= \sum_{s_1} \cdots \sum_{s_{N-1}} \sum_{s_N} \left[c \prod_{n=1}^{N} f_{s_n}(w_n)\pi_{s_n} \right] \\
&= c \sum_{s_1} \cdots \sum_{s_{N-1}} \sum_{s_N} \left[\left(\prod_{n=1}^{N-1} f_{s_n}(w_n)\pi_{s_n} \right) f_{s_N}(w_N)\pi_{s_N} \right] \\
&= c \sum_{s_1} \cdots \sum_{s_{N-1}} \left[\left(\prod_{n=1}^{N-1} f_{s_n}(w_n)\pi_{s_n} \right) \sum_{s_N} f_{s_N}(w_N)\pi_{s_N} \right] \\
&= c \left(\sum_{s_1} f_{s_1}(w_1)\pi_{s_1} \right) \cdots \left(\sum_{s_N} f_{s_N}(w_N)\pi_{s_N} \right) \\
&= c \prod_{n=1}^{N} \sum_{s_n} f_{s_n}(w_n)\pi_{s_n}.
\end{aligned}
$$

Thus,

$$
c = \frac{1}{\prod_{n=1}^{N} \sum_{s_n} f_{s_n}(w_n)\pi_{s_n}} = \frac{1}{\prod_{n=1}^{N} \sum_{m=1}^{M} f_{\sigma_m}(w_n)\pi_{\sigma_m}}.
$$

As our model's posterior is

$$p(s_{1:N}|w_{1:N}) = c\prod_{n=1}^{N} f_{s_n}(w_n)\pi_{s_n},$$

after normalization the posterior becomes

$$p(s_{1:N}|w_{1:N}) = \frac{\prod_{n=1}^{N} f_{s_n}(w_n)\pi_{s_n}}{\prod_{n=1}^{N}\sum_{m=1}^{M} f_{\sigma_m}(w_n)\pi_{\sigma_m}} = \prod_{n=1}^{N}\frac{f_{s_n}(w_n)\pi_{s_n}}{\sum_{m=1}^{M} f_{\sigma_m}(w_n)\pi_{\sigma_m}}.$$

Derivation of Eq. (*7.9*)

$$\begin{aligned}
p(s_n|s_{-n},\boldsymbol{\pi},w_{1:N}) &\propto p(w_{1:N}|s_n,s_{-n},\boldsymbol{\pi})p(s_n|s_{-n},\boldsymbol{\pi})\\
&= p(w_{1:N}|s_n,s_{-n})p(s_n|s_{-n},\boldsymbol{\pi})\\
&= p(w_{1:N}|s_n,s_{-n})p(s_n|\boldsymbol{\pi})\\
&= \Big(p(w_1|s_n,s_{-n})\cdots p(w_n|s_n,s_{-n})\cdots p(w_N|s_n,s_{-n})\Big)p(s_n|\boldsymbol{\pi})\\
&= \Big(p(w_1|s_1)\cdots p(w_n|s_n)\cdots p(w_N|s_N)\Big)p(s_n|\boldsymbol{\pi})\\
&\propto p(w_n|s_n)p(s_n|\boldsymbol{\pi})\\
&= \text{Normal}\Big(w_n;\mu_{s_n},v\Big)\,\text{Categorical}_{\sigma_{1:4}}(s_n;\boldsymbol{\pi})\\
&= \frac{e^{-\frac{(w_n-\mu_{s_n})^2}{2v}}}{\sqrt{2\pi v}}\pi_{s_n}.
\end{aligned}$$

Derivation of Eq. (*7.10*)

$$\begin{aligned}
p(\boldsymbol{\pi}|s_{1:N},w_{1:N}) &= p(\boldsymbol{\pi}|s_{1:N}) \propto p(s_{1:N}|\boldsymbol{\pi})p(\boldsymbol{\pi})\\
&= \left(\prod_{n=1}^{N} p(s_n|\boldsymbol{\pi})\right)p(\boldsymbol{\pi}) = \left(\prod_{n=1}^{N}\pi_{s_n}\right)p(\boldsymbol{\pi})\\
&= \pi_{\sigma_1}^{c_{\sigma_1}}\pi_{\sigma_2}^{c_{\sigma_2}}\pi_{\sigma_3}^{c_{\sigma_3}}\pi_{\sigma_4}^{c_{\sigma_4}}p(\boldsymbol{\pi}) = \left(\prod_{m=1}^{4}\pi_{\sigma_m}^{c_{\sigma_m}}\right)p(\boldsymbol{\pi})\\
&= \left(\prod_{m=1}^{4}\pi_{\sigma_m}^{c_{\sigma_m}}\right)\text{Dirichlet}_{\sigma_{1:4}}\Big(\boldsymbol{\pi};\alpha\beta_{\sigma_{1:4}}\Big)\\
&= \left(\prod_{m=1}^{4}\pi_{\sigma_m}^{c_{\sigma_m}}\right)\frac{\Gamma(\alpha)}{\prod_{m=1}^{4}\Gamma(\alpha\beta_{\sigma_m})}\prod_{m=1}^{4}\pi_{\sigma_m}^{\alpha\beta_{\sigma_m}-1} \propto \prod_{m=1}^{4}\pi_{\sigma_m}^{\alpha\beta_{\sigma_m}+c_{\sigma_m}-1},
\end{aligned}$$

where c_{σ_m} is the total count in the state m, *i.e.*, the number of $s_n = \sigma_m$ over all $s_{1:N}$.

F.6 Relations in Chapter 8

Derivation of Eq. (*8.9*)

$$
\begin{aligned}
\mathcal{A}_n(s_n) &= p(w_{1:n}, s_n|\boldsymbol{\rho}, \boldsymbol{\Pi}, \boldsymbol{\phi}) \\
&= \sum_{s_{n-1}} p(w_{1:n}, s_{n-1}, s_n|\boldsymbol{\rho}, \boldsymbol{\Pi}, \boldsymbol{\phi}) \\
&= \sum_{s_{n-1}} p(w_n|w_{1:n-1}, s_{n-1}, s_n, \boldsymbol{\rho}, \boldsymbol{\Pi}, \boldsymbol{\phi}) \\
&\quad \times p(s_n|w_{1:n-1}, s_{n-1}, \boldsymbol{\rho}, \boldsymbol{\Pi}, \boldsymbol{\phi}) p(w_{1:n-1}, s_{n-1}|\boldsymbol{\rho}, \boldsymbol{\Pi}, \boldsymbol{\phi}) \\
&= \sum_{s_{n-1}} p(w_n|s_n, \boldsymbol{\phi}) p(s_n|s_{n-1}, \boldsymbol{\Pi}) p(w_{1:n-1}, s_{n-1}|\boldsymbol{\rho}, \boldsymbol{\Pi}, \boldsymbol{\phi}) \\
&= p(w_n|s_n, \boldsymbol{\phi}) \sum_{s_{n-1}} p(s_n|s_{n-1}, \boldsymbol{\Pi}) p(w_{1:n-1}, s_{n-1}|\boldsymbol{\rho}, \boldsymbol{\Pi}, \boldsymbol{\phi}) \\
&= G_{\phi_{s_n}}(w_n) \sum_{s_{n-1}} \pi_{s_{n-1}\to s_n} \mathcal{A}_{n-1}(s_{n-1}).
\end{aligned}
$$

Derivation of Eq. (*8.10*)

$$
\begin{aligned}
\mathcal{A}_1(s_1) &= p(w_1, s_1|\boldsymbol{\rho}, \boldsymbol{\Pi}, \boldsymbol{\phi}) \\
&= p(w_1|s_1, \boldsymbol{\rho}, \boldsymbol{\Pi}, \boldsymbol{\phi}) p(s_1|\boldsymbol{\rho}, \boldsymbol{\Pi}, \boldsymbol{\phi}) \\
&= p(w_1|s_1, \boldsymbol{\phi}) p(s_1|\boldsymbol{\rho}) \\
&= G_{\phi_{s_1}}(w_1)\, \rho_{s_1}.
\end{aligned}
$$

Derivation of Eq. (*8.11*)

$$
\begin{aligned}
p(s_N|w_{1:N}, \boldsymbol{\rho}, \boldsymbol{\Pi}, \boldsymbol{\phi}) &= \frac{p(w_{1:N}, s_N|\boldsymbol{\rho}, \boldsymbol{\Pi}, \boldsymbol{\phi})}{p(w_{1:N}|\boldsymbol{\rho}, \boldsymbol{\Pi}, \boldsymbol{\phi})} \\
&= \frac{\mathcal{A}_N(s_N)}{p(w_{1:N}|\boldsymbol{\rho}, \boldsymbol{\Pi}, \boldsymbol{\phi})} \\
&\propto \mathcal{A}_N(s_N).
\end{aligned}
$$

Derivation of Eq. (*8.12*)

$$
\begin{aligned}
p\left(s_n|w_{1:N}, \boldsymbol{\rho}, \boldsymbol{\Pi}, \boldsymbol{\phi}\right) &= \frac{p\left(w_{1:n}, s_n, w_{n+1:N}|\boldsymbol{\rho}, \boldsymbol{\Pi}, \boldsymbol{\phi}\right)}{p\left(w_{1:N}|\boldsymbol{\rho}, \boldsymbol{\Pi}, \boldsymbol{\phi}\right)} \\
&= \frac{p\left(w_{1:n}, s_n|\boldsymbol{\rho}, \boldsymbol{\Pi}, \boldsymbol{\phi}\right) p\left(w_{n+1:N}|w_{1:n}, s_n, \boldsymbol{\rho}, \boldsymbol{\Pi}, \boldsymbol{\phi}\right)}{p\left(w_{1:N}|\boldsymbol{\rho}, \boldsymbol{\Pi}, \boldsymbol{\phi}\right)}
\end{aligned}
$$

$$
\begin{aligned}
&= \frac{p\left(w_{1:n}, s_n|\boldsymbol{\rho}, \boldsymbol{\Pi}, \boldsymbol{\phi}\right) p\left(w_{n+1:N}|s_n, \boldsymbol{\Pi}, \boldsymbol{\phi}\right)}{p\left(w_{1:N}|\boldsymbol{\rho}, \boldsymbol{\Pi}, \boldsymbol{\phi}\right)} \\
&= \frac{\mathcal{A}_n(s_n)\mathcal{B}_n(s_n)}{p\left(w_{1:N}|\boldsymbol{\rho}, \boldsymbol{\Pi}, \boldsymbol{\phi}\right)} \\
&\propto \mathcal{A}_n(s_n)\mathcal{B}_n(s_n).
\end{aligned}
$$

Derivation of Eq. (*8.14*)

$$
\begin{aligned}
\mathcal{B}_n(s_n) &= p\left(w_{n+1:N}|s_n, \boldsymbol{\Pi}, \boldsymbol{\phi}\right) \\
&= \sum_{s_{n+1}} p\left(w_{n+1:N}, s_{n+1}|s_n, \boldsymbol{\Pi}, \boldsymbol{\phi}\right) \\
&= \sum_{s_{n+1}} p\left(w_{n+2:N}|w_{n+1}, s_{n+1}, s_n, \boldsymbol{\Pi}, \boldsymbol{\phi}\right) \\
&\quad \times p\left(w_{n+1}|s_{n+1}, s_n, \boldsymbol{\Pi}, \boldsymbol{\phi}\right) p\left(s_{n+1}|s_n, \boldsymbol{\Pi}, \boldsymbol{\phi}\right) \\
&= \sum_{s_{n+1}} p\left(w_{n+2:N}|s_{n+1}, \boldsymbol{\Pi}, \boldsymbol{\phi}\right) p\left(w_{n+1}|s_{n+1}, \boldsymbol{\phi}\right) p\left(s_{n+1}|s_n, \boldsymbol{\Pi}\right) \\
&= \sum_{s_{n+1}} \mathcal{B}_{n+1}(s_{n+1}) G_{\phi_{s_{n+1}}}\left(w_{n+1}\right) \pi_{s_n \to s_{n+1}}.
\end{aligned}
$$

Derivation of Eq. (*8.16*)

$$
\begin{aligned}
p(s_n|s_{n+1:N}, w_{1:N}, \boldsymbol{\rho}, \boldsymbol{\Pi}, \boldsymbol{\phi}) &= p(s_n|s_{n+1}, w_{1:n}, \boldsymbol{\rho}, \boldsymbol{\Pi}, \boldsymbol{\phi}) \\
&= \frac{p(s_n, s_{n+1}|w_{1:n}, \boldsymbol{\rho}, \boldsymbol{\Pi}, \boldsymbol{\phi})}{p(s_{n+1}|w_{1:n}, \boldsymbol{\rho}, \boldsymbol{\Pi}, \boldsymbol{\phi})} \\
&= \frac{p(s_{n+1}|s_n, w_{1:n}, \boldsymbol{\rho}, \boldsymbol{\Pi}, \boldsymbol{\phi})}{p(s_{n+1}|w_{1:n}, \boldsymbol{\rho}, \boldsymbol{\Pi}, \boldsymbol{\phi})} p(s_n|w_{1:n}, \boldsymbol{\rho}, \boldsymbol{\Pi}, \boldsymbol{\phi}) \\
&= \frac{p(s_{n+1}|s_n, w_{1:n}, \boldsymbol{\rho}, \boldsymbol{\Pi}, \boldsymbol{\phi})}{p(s_{n+1}|w_{1:n}, \boldsymbol{\rho}, \boldsymbol{\Pi}, \boldsymbol{\phi})} \frac{p(w_{1:n}, s_n|\boldsymbol{\rho}, \boldsymbol{\Pi}, \boldsymbol{\phi})}{p(w_{1:n}|\boldsymbol{\rho}, \boldsymbol{\Pi}, \boldsymbol{\phi})} \\
&= \frac{p(s_{n+1}|s_n, \boldsymbol{\Pi})}{p(s_{n+1}|w_{1:n}, \boldsymbol{\rho}, \boldsymbol{\Pi}, \boldsymbol{\phi})} \frac{p(w_{1:n}, s_n|\boldsymbol{\rho}, \boldsymbol{\Pi}, \boldsymbol{\phi})}{p(w_{1:n}|\boldsymbol{\rho}, \boldsymbol{\Pi}, \boldsymbol{\phi})} \\
&= \frac{\pi_{s_n \to s_{n+1}}}{p(s_{n+1}|w_{1:n}, \boldsymbol{\rho}, \boldsymbol{\Pi}, \boldsymbol{\phi})} \frac{\mathcal{A}_n(s_n)}{p(w_{1:n}|\boldsymbol{\rho}, \boldsymbol{\Pi}, \boldsymbol{\phi})} \\
&\propto \pi_{s_n \to s_{n+1}} \mathcal{A}_n(s_n) \\
&= \mathcal{A}_n(s_n) \pi_{s_n \to s_{n+1}}.
\end{aligned}
$$

Derivation of Eq. (*8.17*)

$$
\begin{aligned}
\log p\left(s_{1:N}, w_{1:N}|\boldsymbol{\rho}, \boldsymbol{\Pi}, \boldsymbol{\phi}\right) &= \log p\left(s_{1:N}|\boldsymbol{\rho}, \boldsymbol{\Pi}, \boldsymbol{\phi}\right) + \log p\left(w_{1:N}|s_{1:N}, \boldsymbol{\rho}, \boldsymbol{\Pi}, \boldsymbol{\phi}\right) \\
&= \log p\left(s_{1:N}|\boldsymbol{\rho}, \boldsymbol{\Pi}\right) + \log p\left(w_{1:N}|s_{1:N}, \boldsymbol{\phi}\right)
\end{aligned}
$$

$$= \log p\left(s_1|\boldsymbol{\rho}\right) + \log \prod_{n=2}^{N} p\left(s_n|s_{n-1}, \boldsymbol{\Pi}\right)$$

$$+ \log \prod_{n=1}^{N} p\left(w_n|s_n, \boldsymbol{\phi}\right)$$

$$= \log p\left(s_1|\boldsymbol{\rho}\right) + \sum_{n=2}^{N} \log p\left(s_n|s_{n-1}, \boldsymbol{\Pi}\right)$$

$$+ \sum_{n=1}^{N} \log p\left(w_n|s_n, \boldsymbol{\phi}\right)$$

$$= \log \rho_{s_1} + \sum_{n=2}^{N} \log \pi_{s_{n-1}\to s_n} + \sum_{n=1}^{N} \log G_{\phi_{s_n}}\left(w_n\right).$$

Derivation of Eq. (*8.18*)

$$Q_{\boldsymbol{\rho}^{\mathrm{old}},\boldsymbol{\Pi}^{\mathrm{old}},\boldsymbol{\phi}^{\mathrm{old}}}\left(\boldsymbol{\rho}, \boldsymbol{\Pi}, \boldsymbol{\phi}\right) = \sum_{s_{1:N}} p\left(s_{1:N}|w_{1:N}, \boldsymbol{\rho}^{\mathrm{old}}, \boldsymbol{\Pi}^{\mathrm{old}}, \boldsymbol{\phi}^{\mathrm{old}}\right)$$

$$\times \log p\left(s_{1:N}, w_{1:N}|\boldsymbol{\rho}, \boldsymbol{\Pi}, \boldsymbol{\phi}\right)$$

$$= \sum_{s_{1:N}} p\left(s_{1:N}|w_{1:N}, \boldsymbol{\rho}^{\mathrm{old}}, \boldsymbol{\Pi}^{\mathrm{old}}, \boldsymbol{\phi}^{\mathrm{old}}\right) \log \rho_{s_1}$$

$$+ \sum_{s_{1:N}} p\left(s_{1:N}|w_{1:N}, \boldsymbol{\rho}^{\mathrm{old}}, \boldsymbol{\Pi}^{\mathrm{old}}, \boldsymbol{\phi}^{\mathrm{old}}\right) \sum_{n=2}^{N} \log \pi_{s_{n-1}\to s_n}$$

$$+ \sum_{s_{1:N}} p\left(s_{1:N}|w_{1:N}, \boldsymbol{\rho}^{\mathrm{old}}, \boldsymbol{\Pi}^{\mathrm{old}}, \boldsymbol{\phi}^{\mathrm{old}}\right) \sum_{n=1}^{N} \log G_{\phi_{s_n}}\left(w_n\right)$$

$$= \sum_{s_1} p\left(s_1|w_{1:N}, \boldsymbol{\rho}^{\mathrm{old}}, \boldsymbol{\Pi}^{\mathrm{old}}, \boldsymbol{\phi}^{\mathrm{old}}\right) \log \rho_{s_1}$$

$$+ \sum_{s_{1:N}} \sum_{n=2}^{N} p\left(s_{1:N}|w_{1:N}, \boldsymbol{\rho}^{\mathrm{old}}, \boldsymbol{\Pi}^{\mathrm{old}}, \boldsymbol{\phi}^{\mathrm{old}}\right) \log \pi_{s_{n-1}\to s_n}$$

$$+ \sum_{s_{1:N}} \sum_{n=1}^{N} p\left(s_{1:N}|w_{1:N}, \boldsymbol{\rho}^{\mathrm{old}}, \boldsymbol{\Pi}^{\mathrm{old}}, \boldsymbol{\phi}^{\mathrm{old}}\right) \log G_{\phi_{s_n}}\left(w_n\right)$$

$$= \sum_{s_1} p\left(s_1|w_{1:N}, \boldsymbol{\rho}^{\mathrm{old}}, \boldsymbol{\Pi}^{\mathrm{old}}, \boldsymbol{\phi}^{\mathrm{old}}\right) \log \rho_{s_1}$$

$$+ \sum_{s_{n-1}} \sum_{n=2}^{N} \sum_{s_n} p\left(s_{n-1}, s_n|w_{1:N}, \boldsymbol{\rho}^{\mathrm{old}}, \boldsymbol{\Pi}^{\mathrm{old}}, \boldsymbol{\phi}^{\mathrm{old}}\right)$$

$$\times \log \pi_{s_{n-1}\to s_n}$$

$$+ \sum_{s_n} \sum_{n=1}^{N} p\left(s_n|w_{1:N}, \boldsymbol{\rho}^{\mathrm{old}}, \boldsymbol{\Pi}^{\mathrm{old}}, \boldsymbol{\phi}^{\mathrm{old}}\right) \log G_{\phi_{s_n}}\left(w_n\right).$$

Derivation of Eq. (*8.19*)

$$
\begin{aligned}
& p\left(s_n | w_{1:N}, \boldsymbol{\rho}^{\text{old}}, \boldsymbol{\Pi}^{\text{old}}, \boldsymbol{\phi}^{\text{old}}\right) \\
& \quad = \frac{p\left(w_{1:n}, s_n, w_{n+1:N} | \boldsymbol{\rho}^{\text{old}}, \boldsymbol{\Pi}^{\text{old}}, \boldsymbol{\phi}^{\text{old}}\right)}{p\left(w_{1:N} | \boldsymbol{\rho}^{\text{old}}, \boldsymbol{\Pi}^{\text{old}}, \boldsymbol{\phi}^{\text{old}}\right)} \\
& \quad = \frac{p\left(w_{1:n}, s_n | \boldsymbol{\rho}^{\text{old}}, \boldsymbol{\Pi}^{\text{old}}, \boldsymbol{\phi}^{\text{old}}\right) p\left(w_{n+1:N} | w_{1:n}, s_n, \boldsymbol{\rho}^{\text{old}}, \boldsymbol{\Pi}^{\text{old}}, \boldsymbol{\phi}^{\text{old}}\right)}{p\left(w_{1:N} | \boldsymbol{\rho}^{\text{old}}, \boldsymbol{\Pi}^{\text{old}}, \boldsymbol{\phi}^{\text{old}}\right)} \\
& \quad = \frac{p\left(w_{1:n}, s_n | \boldsymbol{\rho}^{\text{old}}, \boldsymbol{\Pi}^{\text{old}}, \boldsymbol{\phi}^{\text{old}}\right) p\left(w_{n+1:N} | s_n, \boldsymbol{\Pi}^{\text{old}}, \boldsymbol{\phi}^{\text{old}}\right)}{p\left(w_{1:N} | \boldsymbol{\rho}^{\text{old}}, \boldsymbol{\Pi}^{\text{old}}, \boldsymbol{\phi}^{\text{old}}\right)} \\
& \quad = \frac{\mathcal{A}_n^{\text{old}}(s_n) \mathcal{B}_n^{\text{old}}(s_n)}{p\left(w_{1:N} | \boldsymbol{\rho}^{\text{old}}, \boldsymbol{\Pi}^{\text{old}}, \boldsymbol{\phi}^{\text{old}}\right)}.
\end{aligned}
$$

Derivation of Eq. (*8.20*)

$$
\begin{aligned}
& p\left(s_{n-1}, s_n | w_{1:N}, \boldsymbol{\rho}^{\text{old}}, \boldsymbol{\Pi}^{\text{old}}, \boldsymbol{\phi}^{\text{old}}\right) \\
& \quad = \frac{p\left(w_{1:n}, w_{n+1:N}, s_{n-1}, s_n | \boldsymbol{\rho}^{\text{old}}, \boldsymbol{\Pi}^{\text{old}}, \boldsymbol{\phi}^{\text{old}}\right)}{p\left(w_{1:N} | \boldsymbol{\rho}^{\text{old}}, \boldsymbol{\Pi}^{\text{old}}, \boldsymbol{\phi}^{\text{old}}\right)} \\
& \quad = \frac{p\left(w_{n+1:N} | w_{1:n}, s_{n-1}, s_n, \boldsymbol{\rho}^{\text{old}}, \boldsymbol{\Pi}^{\text{old}}, \boldsymbol{\phi}^{\text{old}}\right) p\left(w_{1:n}, s_{n-1}, s_n | \boldsymbol{\rho}^{\text{old}}, \boldsymbol{\Pi}^{\text{old}}, \boldsymbol{\phi}^{\text{old}}\right)}{p\left(w_{1:N} | \boldsymbol{\rho}^{\text{old}}, \boldsymbol{\Pi}^{\text{old}}, \boldsymbol{\phi}^{\text{old}}\right)} \\
& \quad = \frac{p\left(w_{n+1:N} | s_n, \boldsymbol{\rho}^{\text{old}}, \boldsymbol{\Pi}^{\text{old}}, \boldsymbol{\phi}^{\text{old}}\right) p\left(w_{1:n}, s_{n-1}, s_n | \boldsymbol{\rho}^{\text{old}}, \boldsymbol{\Pi}^{\text{old}}, \boldsymbol{\phi}^{\text{old}}\right)}{p\left(w_{1:N} | \boldsymbol{\rho}^{\text{old}}, \boldsymbol{\Pi}^{\text{old}}, \boldsymbol{\phi}^{\text{old}}\right)} \\
& \quad = \frac{\mathcal{B}_n^{\text{old}}(s_n) p\left(w_{1:n}, s_{n-1}, s_n | \boldsymbol{\rho}^{\text{old}}, \boldsymbol{\Pi}^{\text{old}}, \boldsymbol{\phi}^{\text{old}}\right)}{p\left(w_{1:N} | \boldsymbol{\rho}^{\text{old}}, \boldsymbol{\Pi}^{\text{old}}, \boldsymbol{\phi}^{\text{old}}\right)} \\
& \quad = \frac{\mathcal{B}_n^{\text{old}}(s_n) p\left(w_n | w_{1:n-1}, s_{n-1}, s_n, \boldsymbol{\rho}^{\text{old}}, \boldsymbol{\Pi}^{\text{old}}, \boldsymbol{\phi}^{\text{old}}\right) p\left(w_{1:n-1}, s_{n-1}, s_n | \boldsymbol{\rho}^{\text{old}}, \boldsymbol{\Pi}^{\text{old}}, \boldsymbol{\phi}^{\text{old}}\right)}{p\left(w_{1:N} | \boldsymbol{\rho}^{\text{old}}, \boldsymbol{\Pi}^{\text{old}}, \boldsymbol{\phi}^{\text{old}}\right)} \\
& \quad = \frac{\mathcal{B}_n^{\text{old}}(s_n) p\left(w_n | s_n, \boldsymbol{\phi}^{\text{old}}\right) p\left(w_{1:n-1}, s_{n-1}, s_n | \boldsymbol{\rho}^{\text{old}}, \boldsymbol{\Pi}^{\text{old}}, \boldsymbol{\phi}^{\text{old}}\right)}{p\left(w_{1:N} | \boldsymbol{\rho}^{\text{old}}, \boldsymbol{\Pi}^{\text{old}}, \boldsymbol{\phi}^{\text{old}}\right)} \\
& \quad = \frac{\mathcal{B}_n^{\text{old}}(s_n) G_{\phi_{s_n}^{\text{old}}}(w_n) p\left(w_{1:n-1}, s_{n-1}, s_n | \boldsymbol{\rho}^{\text{old}}, \boldsymbol{\Pi}^{\text{old}}, \boldsymbol{\phi}^{\text{old}}\right)}{p\left(w_{1:N} | \boldsymbol{\rho}^{\text{old}}, \boldsymbol{\Pi}^{\text{old}}, \boldsymbol{\phi}^{\text{old}}\right)} \\
& \quad = \frac{\mathcal{B}_n^{\text{old}}(s_n) G_{\phi_{s_n}^{\text{old}}}(w_n) p\left(s_n | w_{1:n-1}, s_{n-1}, \boldsymbol{\rho}^{\text{old}}, \boldsymbol{\Pi}^{\text{old}}, \boldsymbol{\phi}^{\text{old}}\right) p\left(w_{1:n-1}, s_{n-1} | \boldsymbol{\rho}^{\text{old}}, \boldsymbol{\Pi}^{\text{old}}, \boldsymbol{\phi}^{\text{old}}\right)}{p\left(w_{1:N} | \boldsymbol{\rho}^{\text{old}}, \boldsymbol{\Pi}^{\text{old}}, \boldsymbol{\phi}^{\text{old}}\right)}
\end{aligned}
$$

$$
\begin{aligned}
&= \frac{\mathcal{B}_n^{\text{old}}(s_n) G_{\phi_{s_n}^{\text{old}}}(w_n)\, p\left(s_n | s_{n-1}, \mathbf{\Pi}^{\text{old}}\right) p\left(w_{1:n-1}, s_{n-1} | \boldsymbol{\rho}^{\text{old}}, \mathbf{\Pi}^{\text{old}}, \boldsymbol{\phi}^{\text{old}}\right)}{p\left(w_{1:N} | \boldsymbol{\rho}^{\text{old}}, \mathbf{\Pi}^{\text{old}}, \boldsymbol{\phi}^{\text{old}}\right)} \\
&= \frac{\mathcal{B}_n^{\text{old}}(s_n) G_{\phi_{s_n}^{\text{old}}}(w_n)\, \pi_{s_{n-1}\to s_n}^{\text{old}} \mathcal{A}_{n-1}^{\text{old}}(s_{n-1})}{p\left(w_{1:N} | \boldsymbol{\rho}^{\text{old}}, \mathbf{\Pi}^{\text{old}}, \boldsymbol{\phi}^{\text{old}}\right)} \\
&= \frac{\mathcal{A}_{n-1}^{\text{old}}(s_{n-1}) \mathcal{B}_n^{\text{old}}(s_n) G_{\phi_{s_n}^{\text{old}}}(w_n)\, \pi_{s_{n-1}\to s_n}^{\text{old}}}{p\left(w_{1:N} | \boldsymbol{\rho}^{\text{old}}, \mathbf{\Pi}^{\text{old}}, \boldsymbol{\phi}^{\text{old}}\right)}.
\end{aligned}
$$

Derivation of Eq. (*8.24*)

The gradient of the Lagrangian is equal to

$$
\begin{aligned}
\frac{\partial \mathbb{L}\left(\lambda, \rho_{\sigma_1}, \ldots, \rho_{\sigma_M}\right)}{\partial \lambda} &= 1 - \sum_{m=1}^{M} \rho_{\sigma_m}, \\
\frac{\partial \mathbb{L}\left(\lambda, \rho_{\sigma_1}, \ldots, \rho_{\sigma_M}\right)}{\partial \rho_{\sigma_m}} &= \frac{\zeta_1^{\text{old}}(\sigma_m)}{\rho_{\sigma_m}} - \lambda, \qquad m = 1:M.
\end{aligned}
$$

Solving $\partial \mathbb{L}/\partial \rho_{\sigma_m} = 0$ implies $\rho_{\sigma_m} = \zeta_1^{\text{old}}(\sigma_m)/\lambda$. Using $\partial \mathbb{L}/\partial \lambda = 0$ gives $\lambda = \sum_{m=1}^{M} \zeta_1^{\text{old}}(\sigma_m)$.

Derivation of Eq. (*8.25*)

The gradient of the Lagrangian is equal to

$$
\begin{aligned}
\frac{\partial \mathbb{K}_m\left(\kappa_m, \pi_{\sigma_m\to\sigma_1}, \ldots, \pi_{\sigma_m\to\sigma_M}\right)}{\partial \kappa_m} &= 1 - \sum_{m'=1}^{M} \pi_{\sigma_m\to\sigma_{m'}}, \\
\frac{\partial \mathbb{K}_m\left(\kappa_m, \pi_{\sigma_m\to\sigma_1}, \ldots, \pi_{\sigma_m\to\sigma_M}\right)}{\partial \pi_{\sigma_m\to\sigma_{m'}}} &= \frac{\sum_{n=2}^{N} \eta_n^{\text{old}}(\sigma_m, \sigma_{m'})}{\pi_{\sigma_m\to\sigma_{m'}}} - \kappa_m, \quad m' = 1:M.
\end{aligned}
$$

Solving $\partial \mathbb{K}_m/\partial \pi_{\sigma_m\to\sigma_{m'}} = 0$ implies $\pi_{\sigma_m\to\sigma_{m'}} = \sum_{n=2}^{N} \eta_n^{\text{old}}(\sigma_m, \sigma_{m'})/\kappa_m$. Using $\partial \mathbb{K}/\partial \kappa_m = 0$ gives $\kappa_m = \sum_{m'=1}^{M} \sum_{n=2}^{N} \eta_n^{\text{old}}(\sigma_m, \sigma_{m'})$.

Derivation of Eq. (*8.28*)

$$
\begin{aligned}
\hat{\mathcal{A}}_n(s_n) &= p(s_n | w_{1:n}, \boldsymbol{\rho}, \mathbf{\Pi}, \boldsymbol{\phi}) \\
&= \frac{p(w_n, s_n | w_{1:n-1}, \boldsymbol{\rho}, \mathbf{\Pi}, \boldsymbol{\phi})}{p(w_n | w_{1:n-1}, \boldsymbol{\rho}, \mathbf{\Pi}, \boldsymbol{\phi})} \\
&= \frac{p(w_n | s_n, w_{1:n-1}, \boldsymbol{\rho}, \mathbf{\Pi}, \boldsymbol{\phi})}{p(w_n | w_{1:n-1}, \boldsymbol{\rho}, \mathbf{\Pi}, \boldsymbol{\phi})} p(s_n | w_{1:n-1}, \boldsymbol{\rho}, \mathbf{\Pi}, \boldsymbol{\phi}) \\
&= \frac{p(w_n | s_n, w_{1:n-1}, \boldsymbol{\rho}, \mathbf{\Pi}, \boldsymbol{\phi})}{p(w_n | w_{1:n-1}, \boldsymbol{\rho}, \mathbf{\Pi}, \boldsymbol{\phi})} \sum_{s_{n-1}} p(s_n, s_{n-1} | w_{1:n-1}, \boldsymbol{\rho}, \mathbf{\Pi}, \boldsymbol{\phi})
\end{aligned}
$$

$$
\begin{aligned}
&= \frac{p(w_n|s_n, w_{1:n-1}, \boldsymbol{\rho}, \boldsymbol{\Pi}, \boldsymbol{\phi})}{p(w_n|w_{1:n-1}, \boldsymbol{\rho}, \boldsymbol{\Pi}, \boldsymbol{\phi})} \\
&\quad \times \sum_{s_{n-1}} p(s_n|s_{n-1}, w_{1:n-1}, \boldsymbol{\rho}, \boldsymbol{\Pi}, \boldsymbol{\phi}) p(s_{n-1}|w_{1:n-1}, \boldsymbol{\rho}, \boldsymbol{\Pi}, \boldsymbol{\phi}) \\
&= \frac{p(w_n|s_n, \boldsymbol{\phi})}{p(w_n|w_{1:n-1}, \boldsymbol{\rho}, \boldsymbol{\Pi}, \boldsymbol{\phi})} \sum_{s_{n-1}} p(s_n|s_{n-1}, \boldsymbol{\Pi}) p(s_{n-1}|w_{1:n-1}, \boldsymbol{\rho}, \boldsymbol{\Pi}, \boldsymbol{\phi}) \\
&= \frac{G_{\phi_{s_n}}(w_n)}{p(w_n|w_{1:n-1}, \boldsymbol{\rho}, \boldsymbol{\Pi}, \boldsymbol{\phi})} \sum_{s_{n-1}} \pi_{s_{n-1}\to s_n} \hat{\mathcal{A}}_{n-1}(s_{n-1}) \\
&= \frac{1}{\hat{\mathcal{C}}_n} G_{\phi_{s_n}}(w_n) \sum_{s_{n-1}} \pi_{s_{n-1}\to s_n} \hat{\mathcal{A}}_{n-1}(s_{n-1}).
\end{aligned}
$$

Derivation of Eq. (*8.29*)

$$
\begin{aligned}
\check{\mathcal{B}}_n(s_n) &= \frac{p\left(w_{n+1:N}|s_n, \boldsymbol{\Pi}, \boldsymbol{\phi}\right)}{p\left(w_{n+1:N}|w_{1:n}, \boldsymbol{\rho}, \boldsymbol{\Pi}, \boldsymbol{\phi}\right)} \\
&= \frac{\sum_{s_{n+1}} p\left(w_{n+1:N}, s_{n+1}|s_n, \boldsymbol{\Pi}, \boldsymbol{\phi}\right)}{p\left(w_{n+2:N}|w_{1:n+1}, \boldsymbol{\rho}, \boldsymbol{\Pi}, \boldsymbol{\phi}\right) p\left(w_{n+1}|w_{1:n}, \boldsymbol{\rho}, \boldsymbol{\Pi}, \boldsymbol{\phi}\right)} \\
&= \frac{\sum_{s_{n+1}} p\left(w_{n+1}|w_{n+2:N}, s_{n+1}, s_n, \boldsymbol{\Pi}, \boldsymbol{\phi}\right) p\left(w_{n+2:N}|s_{n+1}, s_n, \boldsymbol{\Pi}, \boldsymbol{\phi}\right) p\left(s_{n+1}|s_n, \boldsymbol{\Pi}, \boldsymbol{\phi}\right)}{p\left(w_{n+2:N}|w_{1:n+1}, \boldsymbol{\rho}, \boldsymbol{\Pi}, \boldsymbol{\phi}\right) p\left(w_{n+1}|w_{1:n}, \boldsymbol{\rho}, \boldsymbol{\Pi}, \boldsymbol{\phi}\right)} \\
&= \frac{1}{p\left(w_{n+1}|w_{1:n}, \boldsymbol{\rho}, \boldsymbol{\Pi}, \boldsymbol{\phi}\right)} \sum_{s_{n+1}} p\left(w_{n+1}|s_{n+1}, \boldsymbol{\phi}\right) \\
&\quad \times \frac{p\left(w_{n+2:N}|s_{n+1}, \boldsymbol{\Pi}, \boldsymbol{\phi}\right)}{p\left(w_{n+2:N}|w_{1:n+1}, \boldsymbol{\rho}, \boldsymbol{\Pi}, \boldsymbol{\phi}\right)} p\left(s_{n+1}|s_n, \boldsymbol{\Pi}\right) \\
&= \frac{1}{\hat{\mathcal{C}}_{n+1}} \sum_{s_{n+1}} \check{\mathcal{B}}_{n+1}(s_{n+1}) G_{\phi_{s_{n+1}}}\left(w_{n+1}\right) \pi_{s_n\to s_{n+1}}.
\end{aligned}
$$

Derivation of Eq. (*8.30*)

$$
p(w_{1:N}|\boldsymbol{\rho}, \boldsymbol{\Pi}, \boldsymbol{\phi}) = p(w_1|\boldsymbol{\rho}, \boldsymbol{\phi}) \prod_{n=2}^{N} p(w_n|w_{1:n-1}, \boldsymbol{\rho}, \boldsymbol{\Pi}, \boldsymbol{\phi}) = \prod_{n=1}^{N} \hat{\mathcal{C}}_n.
$$

Derivation of Eq. (*8.38*)

$$
\begin{aligned}
p\left(s_n|s_{n+1:N}, \boldsymbol{\rho}, \boldsymbol{\Pi}, \boldsymbol{\phi}, w_{1:N}\right) &= p\left(s_n|s_{n+1}, \boldsymbol{\rho}, \boldsymbol{\Pi}, \boldsymbol{\phi}, w_{1:n}\right) \\
&= \frac{p\left(s_n, s_{n+1}|\boldsymbol{\rho}, \boldsymbol{\Pi}, \boldsymbol{\phi}, w_{1:n}\right)}{p\left(s_{n+1}|\boldsymbol{\rho}, \boldsymbol{\Pi}, \boldsymbol{\phi}, w_{1:n}\right)} \\
&= \frac{p\left(s_{n+1}|s_n, \boldsymbol{\rho}, \boldsymbol{\Pi}, \boldsymbol{\phi}, w_{1:n}\right) p\left(s_n|\boldsymbol{\rho}, \boldsymbol{\Pi}, \boldsymbol{\phi}, w_{1:n}\right)}{p\left(s_{n+1}|\boldsymbol{\rho}, \boldsymbol{\Pi}, \boldsymbol{\phi}, w_{1:n}\right)}
\end{aligned}
$$

$$
\begin{aligned}
&= \frac{p\left(s_{n+1} | s_n, \mathbf{\Pi}\right) p\left(s_n | \boldsymbol{\rho}, \mathbf{\Pi}, \boldsymbol{\phi}, w_{1:n}\right)}{p\left(s_{n+1} | \boldsymbol{\rho}, \mathbf{\Pi}, \boldsymbol{\phi}, w_{1:n}\right)} \\
&= \frac{\pi_{s_n \to s_{n+1}} \mathcal{A}_n(s_n)}{p\left(s_{n+1} | \boldsymbol{\rho}, \mathbf{\Pi}, \boldsymbol{\phi}, w_{1:n}\right)} \\
&\propto \pi_{s_n \to s_{n+1}} \mathcal{A}_n(s_n) \\
&= \mathcal{A}_n(s_n) \pi_{s_n \to s_{n+1}}.
\end{aligned}
$$

Although unnecessary, normalization can be recovered by $\sum_{s_n} p(s_n | s_{n+1:N}, \boldsymbol{\rho}, \mathbf{\Pi}, \boldsymbol{\phi}, w_{1:N}) = 1$.

F.7 Relations in Chapter 9

Derivations in this section rely on two identities,

$$
\begin{aligned}
&\int dz \, \text{Normal}_Y \left(y; b + zA, C\right) \text{Normal}_Z \left(z; m, V\right) \\
&\quad = \text{Normal}_Y \left(y; b + mA, C + A^T VA\right), \\
&\text{Normal}_Y \left(y; b + zA, C\right) \text{Normal}_Z \left(z; m, V\right) \\
&\quad \propto \text{Normal}_Z \left(z; m + (y - b - mA)K, V(\mathbb{1} - AK)\right),
\end{aligned}
$$

where $K = (C + A^T VA)^{-1} AV$. In these, Y and Z denote the sizes of y and z.

Derivation of Eq. (*9.7*)

$$
\begin{aligned}
p(w_{1:N}, r_{1:N}) &= p(w_N | w_{1:N-1}, r_{1:N}) p(w_{1:N-1}, r_{1:N}) \\
&= p(w_N | r_N) p(w_{1:N-1}, r_{1:N}) \\
&= p(w_N | r_N) p(r_N | w_{1:N-1}, r_{1:N-1}) p(w_{1:N-1}, r_{1:N-1}) \\
&= p(w_N | r_N) p(r_N | r_{N-1}) p(w_{1:N-1}, r_{1:N-1}) \\
&\;\;\vdots \\
&= p(w_N | r_N) p(r_N | r_{N-1}) \cdots p(w_2 | r_2) p(r_2 | r_1) p(w_1 | r_1) p(r_1) \\
&= \left(\prod_{n=2}^{N} p(w_n | r_n) p(r_n | r_{n-1}) \right) p(w_1 | r_1) p(r_1) \\
&= p(r_1) p(w_1 | r_1) \left(\prod_{n=2}^{N} p(r_n | r_{n-1}) p(w_n | r_n) \right).
\end{aligned}
$$

Derivation of Eq. (*9.11*)

$$\begin{aligned}\hat{\mathcal{A}}_n(r_n) = p\left(r_n|w_{1:n}\right) &= \frac{p\left(w_n|r_n, w_{1:n-1}\right)p\left(r_n|w_{1:n-1}\right)}{p\left(w_n\right)}\\ &= \frac{p\left(w_n|r_n\right)p\left(r_n|w_{1:n-1}\right)}{p\left(w_n\right)}\\ &\propto p\left(w_n|r_n\right)p\left(r_n|w_{1:n-1}\right).\end{aligned}$$

Derivation of Eq. (*9.12*)

$$\begin{aligned}p(w_1|r_1)p(r_1) &= \text{Normal}_L(w_1; b_1 + r_1 B_1, U_1)\text{Normal}_K(r_1; \bar{\mu}_1, \bar{v}_1)\\ &= \text{Normal}_L(w_1 - b_1; r_1 B_1, U_1)\text{Normal}_K(r_1; \bar{\mu}_1, \bar{v}_1)\\ &= \text{Normal}_L(r_1 B_1; w_1 - b_1, U_1)\text{Normal}_K(r_1; \bar{\mu}_1, \bar{v}_1)\\ &\propto \text{Normal}_K\left(r_1; (w_1 - b_1)B_1^{-1}, \left(B_1 U_1^{-1} B_1^T\right)^{-1}\right)\\ &\quad\times \text{Normal}_K(r_1; \bar{\mu}_1, \bar{v}_1)\\ &\propto \text{Normal}_K\left(r_1; \left((w_1 - b_1)^T U_1^{-1} B_1 + \bar{\mu}_1 \bar{v}_1^{-1}\right)\right.\\ &\quad\left.\times\left(B_1 U_1^{-1} B_1^T + \bar{v}_1^{-1}\right)^{-1}, \left(B_1 U_1^{-1} B_1^T + \bar{v}_1^{-1}\right)^{-1}\right)\\ &= \text{Normal}_K\left(r_1; \hat{\mu}_1, \hat{v}_1\right).\end{aligned}$$

The filter's variance is simplified via the Woodbury matrix identity,

$$\begin{aligned}\hat{v}_1 &= \left(B_1 U_1^{-1} B_1^T + \bar{v}_1^{-1}\right)^{-1}\\ &= \bar{v}_1 - \bar{v}_1 B_1\left(U_1 + B_1^T \bar{v}_1 B_1\right)^{-1} B_1^T \bar{v}_1\\ &= \bar{v}_1\left(\mathbb{1} - \bar{v}_1 B_1\left(U_1 + B_1^T \bar{v}_1 B_1\right)^{-1} B_1^T\right)\\ &= \bar{v}_1\left(\mathbb{1} - B_1 G_1\right),\end{aligned}$$

and the Kalman gain is given by $G_1 = \left(U_1 + B_1^T \bar{v}_1 B_1\right)^{-1} B_1 \bar{v}_1$. The filter's mean is also simplified,

$$\begin{aligned}\hat{\mu}_1 &= \left((w_1 - b_1)U_1^{-1} B_1 + \bar{\mu}_1 \bar{v}_1^{-1}\right)\left(B_1 U_1^{-1} B_1^T + \bar{v}_1^{-1}\right)^{-1}\\ &= \left((w_1 - b_1)U_1^{-1} B_1 + \bar{v}_1^{-1}\bar{\mu}_1\right)\left(\mathbb{1} - B_1^T G_1\right)\bar{v}_1\\ &= \left((w_1 - b_1)U_1^{-1} B_1 \bar{v}_1 + \bar{\mu}_1\right)\left(\mathbb{1} - B_1^T G_1\right)\\ &= (w_1 - b_1)U_1^{-1} B_1 \bar{v}_1 + \bar{\mu}_1 - (w_1 - b_1)U_1^{-1} B_1 \bar{v}_1 B_1^T G_1 - \bar{\mu}_1 B_1^T G_1\\ &= \bar{\mu}_1 + (w_1 - b_1)U_1^{-1} B_1 \bar{v}_1\left(\mathbb{1} - B_1^T G_1\right) - \bar{\mu}_1 B_1 G_1\\ &= \bar{\mu}_1 + (w_1 - b_1)G_1 - \bar{\mu}_1 B_1^T G_1\\ &= \bar{\mu}_1 + \left(w_1 - b_1 - \bar{\mu}_1 B_1\right)G_1.\end{aligned}$$

Derivation of Eq. (*9.13*)

$$
\begin{aligned}
p(r_2|w_1) &= \int dr_1\, p(r_2, r_1|w_1) \\
&= \int dr_1\, p(r_2|r_1, w_1)\, p(r_1|w_1) \\
&= \int dr_1\, p(r_2|r_1)\, p(r_1|w_1) \\
&= \int dr_1\, \mathrm{Normal}_K(r_2; a_2 + r_1 A_2, V_2)\, \mathrm{Normal}_K(r_1; \hat{\mu}_1, \hat{\upsilon}_1) \\
&= \mathrm{Normal}_K\left(r_2; a_2 + \hat{\mu}_1 A_2, V_2 + A_2^T \hat{\upsilon}_1 A_2\right) \\
&= \mathrm{Normal}_K\left(r_2; \bar{\mu}_2, \bar{\upsilon}_2\right).
\end{aligned}
$$

Derivation of Eq. (*9.14*)

$$
\begin{aligned}
\check{\mathcal{D}}_N(r_N) = p(r_N|w_{1:N}) = \hat{\mathcal{A}}_N(r_N) &= \mathrm{Normal}_K\left(r_N; \hat{\mu}_N, \hat{\upsilon}_N\right) \\
&= \mathrm{Normal}_K\left(r_N; \check{\mu}_N, \check{\upsilon}_N\right).
\end{aligned}
$$

Derivation of Eq. (*9.15*)

$$
\begin{aligned}
\check{\mathcal{D}}_{N-1}(r_{N-1}) &= p(r_{N-1}|w_{1:N}) \\
&= \int dr_N\, p(r_{N-1}, r_N|w_{1:N}) \\
&= \int dr_N\, p(r_{N-1}|r_N, w_{1:N})p(r_N|w_{1:N}) \\
&= \int dr_N\, p(r_{N-1}|r_N, w_{1:N-1})p(r_N|w_{1:N}) \\
&\propto \int dr_N\, p(r_{N-1}, r_N, w_{1:N-1})p(r_N|w_{1:N}) \\
&= \int dr_N\, p(r_N|r_{N-1}, w_{1:N-1})p(r_{N-1}, w_{1:N-1})p(r_N|w_{1:N}) \\
&= \int dr_N\, p(r_N|r_{N-1})p(r_{N-1}, w_{1:N-1})p(r_N|w_{1:N}) \\
&= \int dr_N\, p(r_N|r_{N-1})p(r_{N-1}|w_{1:N-1})p(w_{1:N-1})p(r_N|w_{1:N}) \\
&\propto \int dr_N\, p(r_N|r_{N-1})p(r_{N-1}|w_{1:N-1})p(r_N|w_{1:N}) \\
&= \int dr_N\, p(r_N|r_{N-1})\hat{\mathcal{A}}_{N-1}(r_{N-1})\check{\mathcal{D}}_N(r_N)
\end{aligned}
$$

$$
\begin{aligned}
&= \int dr_N \, \text{Normal}_K\left(r_N; a_N + r_{N-1}A_N, V_N\right) \\
&\quad \times \text{Normal}_K\left(r_{N-1}; \hat{\mu}_{N-1}, \hat{\upsilon}_{N-1}\right) \text{Normal}_K\left(r_N; \check{\mu}_N, \check{\upsilon}_N\right) \\
&= \int dr_N \, \text{Normal}_K\Big(r_{N-1}; \hat{\mu}_{N-1} \\
&\quad + \left(r_N - a_N - \hat{\mu}_{N-1}A_N\right) J_{N-1}, \hat{\upsilon}_{N-1}(\mathbb{1} - A_N J_{N-1})\Big) \\
&\quad \times \text{Normal}_K\left(r_N; \check{\mu}_N, \check{\upsilon}_N\right) \\
&= \text{Normal}_K\Big(r_{N-1}; \hat{\mu}_{N-1} + \left(\check{\mu}_N - a_N - \hat{\mu}_{N-1}A_N\right) \\
&\quad \times J_{N-1}, (\mathbb{1} - A_N J_{N-1})\hat{\upsilon}_{N-1} + J_{N-1}^T \check{\upsilon}_N J_{N-1}\Big) \\
&= \text{Normal}_K\left(r_{N-1}; \check{\mu}_{N-1}, \check{\upsilon}_{N-1}\right).
\end{aligned}
$$

The Rauch–Tung–Striebel gain is given by

$$
J_{N-1} = \left(V_N + A_N^T \hat{\upsilon}_{N-1} A_N\right)^{-1} A_N \hat{\upsilon}_{N-1} = \bar{\upsilon}_N^{-1} A_N \hat{\upsilon}_{N-1}.
$$

The smoother's mean is given by

$$
\check{\mu}_{N-1} = \hat{\mu}_{N-1} + \left(\check{\mu}_N - a_N - \hat{\mu}_{N-1}A_N\right) J_{N-1} = \hat{\mu}_{N-1} + J_{N-1}\left(\check{\mu}_N - \bar{\mu}_N\right).
$$

The smoother's variance is given by

$$
\check{\upsilon}_{N-1} = \hat{\upsilon}_{N-1}(\mathbb{1} - A_N J_{N-1}) + J_{N-1}^T \check{\upsilon}_N J_{N-1} = \hat{\upsilon}_{N-1} + J_{N-1}^T \left(\check{\upsilon}_N - \bar{\upsilon}_N\right) J_{N-1}.
$$

Derivation of Eq. (*9.16*)

$$
\begin{aligned}
\bar{\bar{\mathcal{E}}}_{N+1}(r_{N+1}) &= p\left(r_{N+1} | w_{1:N}\right) \\
&= \int dr_N \, p\left(r_{N+1}, r_N | w_{1:N}\right) \\
&= \int dr_N \, p\left(r_{N+1} | r_N\right) p\left(r_N | w_{1:N}\right) \\
&= \int dr_N \, p\left(r_{N+1} | r_N\right) \hat{\mathcal{A}}_N\left(r_N\right) \\
&= \int dr_N \text{Normal}_K\left(r_{N+1}; a_{N+1} + r_N A_{N+1}, V_{N+1}\right) \\
&\quad \times \text{Normal}_K\left(r_N; \hat{\mu}_N, \hat{\upsilon}_N\right) \\
&= \text{Normal}_K\left(r_{N+1}; a_{N+1} + \hat{\mu}_N A_{N+1}, V_{N+1} + A_{N+1}^T \hat{\upsilon}_N A_{N+1}\right) \\
&= \text{Normal}_K\left(r_{N+1}; \bar{\bar{\mu}}_{N+1}, \bar{\bar{\upsilon}}_{N+1}\right).
\end{aligned}
$$

Derivation of Eq. (*9.17*)

$$
\begin{aligned}
\overline{\overline{\mathcal{E}}}_{N+2}(r_{N+2}) &= p\left(r_{N+2}|w_{1:N}\right) \\
&- \int dr_{N+1}\, p\left(r_{N+2}, r_{N+1}|w_{1:N}\right) \\
&= \int dr_{N+1}\, p\left(r_{N+2}|r_{N+1}, w_{1:N}\right) p\left(r_{N+1}|w_{1:N}\right) \\
&= \int dr_{N+1}\, p\left(r_{N+2}|r_{N+1}\right) p\left(r_{N+1}|w_{1:N}\right) \\
&= \int dr_{N+1}\, p\left(r_{N+2}|r_{N+1}\right) \overline{\overline{\mathcal{E}}}_{N+1}\left(r_{N+1}\right) \\
&= \int dr_{N+1}\, \mathrm{Normal}_K\left(r_{N+2}; a_{N+2} + r_{N+1}A_{N+2}, V_{N+2}\right) \\
&\times \mathrm{Normal}_K\left(r_{N+1}; \overline{\overline{\mu}}_{N+1}, \overline{\overline{\upsilon}}_{N+1}\right) \\
&= \mathrm{Normal}_K\left(r_{N+2}; a_{N+2} + \overline{\overline{\mu}}_{N+1}A_{N+2}, V_{N+2} + A_{N+2}^T\overline{\overline{\upsilon}}_{N+1}A_{N+2}\right) \\
&= \mathrm{Normal}_K\left(r_{N+2}; \overline{\overline{\mu}}_{N+2}, \overline{\overline{\upsilon}}_{N+2}\right).
\end{aligned}
$$

Derivation of Eq. (*9.18*)

$$
\begin{aligned}
p(r_{1:N}|w_{1:N}) &= p(r_1|r_{2:N}, w_{1:N})p(r_{2:N}|w_{1:N}) \\
&= p(r_1|r_2, w_1)p(r_{2:N}|w_{1:N}) \\
&= p(r_1|r_2, w_1)p(r_2|r_{3:N}, w_{1:N})p(r_{3:N}|w_{1:N}) \\
&= p(r_1|r_2, w_1)p(r_2|r_3, w_{1:2})p(r_{3:N}|w_{1:N}) \\
&= p(r_1|r_2, w_1)p(r_2|r_3, w_{1:2}) \cdots p(r_{N-1}|r_N, w_{1:N-1})p(r_N|w_{1:N}) \\
&= p(r_1|r_2, w_1)p(r_2|r_3, w_{1:2}) \cdots p(r_{N-1}|r_N, w_{1:N-1})\hat{\mathcal{A}}_N(r_N).
\end{aligned}
$$

Derivation of Eq. (*9.19*)

$$
\begin{aligned}
p(r_n|r_{n+1}, w_{1:n}) &= \frac{p(r_n, r_{n+1}|w_{1:n})}{p(r_{n+1}|w_{1:n})} \\
&\propto p(r_n, r_{n+1}|w_{1:n}) \\
&= p(r_{n+1}|r_n, w_{1:n})p(r_n|w_{1:n}) \\
&= p(r_{n+1}|r_n)p(r_n|w_{1:n}) \\
&= p(r_{n+1}|r_n)\hat{\mathcal{A}}_n(r_n) \\
&= \mathrm{Normal}_K\left(r_{n+1}; a_{n+1} + r_nA_{n+1}, V_n\right) \mathrm{Normal}_K\left(r_n; \hat{\mu}_n, \hat{\upsilon}_n\right) \\
&= \mathrm{Normal}_K\left(r_n; \hat{\mu}_n + \left(r_{n+1} - a_{n+1} - \hat{\mu}_nA_{n+1}\right)J_n, \hat{\upsilon}_n(\mathbb{1} - A_{n+1}J_n)\right) \\
&= \mathrm{Normal}_K\left(r_n; \hat{\mu}_n + (r_{n+1} - \bar{\mu}_{n+1})J_n, \hat{\upsilon}_n - J_n^T\bar{\upsilon}_{n+1}J_n\right).
\end{aligned}
$$

Derivation of Eq. (*9.20*)

$$
\begin{aligned}
p\left(\eta|r_{1:N}\right) &\propto p\left(r_{1:N}|\eta\right) p(\eta)\\
&= \left(\prod_{n=2}^{N} p\left(r_n|r_{n-1},\eta\right)\right) p(\eta)\\
&= \left(\prod_{n=2}^{N} \text{Normal}_3\left(r_n; r_{n-1}, 2\frac{t_n - t_{n-1}}{\eta}\mathbb{1}\right)\right)\text{Gamma}(\eta;\phi,\psi)\\
&\propto \left(\prod_{n=2}^{N} \eta^{\frac{3}{2}} e^{-\frac{\eta}{2}(r_n - r_{n-1})(r_n - r_{n-1})^T}\right)\eta^{A-1}e^{-\frac{\eta}{B}}\\
&= \eta^{\phi+\frac{3}{2}(N-1)-1} e^{-\eta\left(\frac{1}{\psi}+\frac{1}{2}\sum_{n=2}^{N}(r_n - r_{n-1})(r_n - r_{n-1})^T\right)}\\
&\propto \text{Gamma}\left(\eta; \phi + \frac{3}{2}(N-1), \frac{1}{\frac{1}{\psi}+\frac{1}{2}\sum_{n=2}^{N}(r_n - r_{n-1})(r_n - r_{n-1})^T}\right).
\end{aligned}
$$

Derivation of Eq. (*9.21*)

$$
\begin{aligned}
p(\tau|r_{1:N}, w_{1:N}) &\propto p(w_{1:N}|r_{1:N},\tau)p(\tau|r_{1:N})\\
&= \left(\prod_{n=1}^{N} p(w_n|r_n,\tau)\right)p(\tau)\\
&= \left(\prod_{n=1}^{N} \text{Normal}_2\left(w_n; Br_n, \frac{1}{\tau}\mathbb{1}\right)\right)\text{Gamma}(\tau;\Phi,\Psi)\\
&\propto \left(\prod_{n=1}^{N} \tau e^{-\frac{\tau}{2}(w_n - Br_n)(w_n - Br_n)^T}\right)\tau^{\Phi-1}e^{-\frac{\tau}{\Psi}}\\
&= \tau^{\Phi+N-1} e^{-\tau\left(\frac{1}{\Psi}+\frac{1}{2}\sum_{n=1}^{N}(w_n - Br_n)(w_n - Br_n)^T\right)}\\
&\propto \text{Gamma}\left(\tau; \Phi + N, \frac{1}{\frac{1}{\Psi}+\frac{1}{2}\sum_{n=1}^{N}(w_n - Br_n)(w_n - Br_n)^T}\right).
\end{aligned}
$$

F.8 Relations in Chapter 10

Derivation of Eq. (*10.2*)

$$
\begin{aligned}
\mathcal{A}_\ell(\bar{s}_\ell) &= p(\bar{w}_{1:\ell}, \bar{s}_\ell|\boldsymbol{\rho}, \boldsymbol{\Lambda}, \boldsymbol{\phi})\\
&= \sum_{\bar{s}_{\ell-1}} p(\bar{w}_{1:\ell}, \bar{s}_{\ell-1}, \bar{s}_\ell|\boldsymbol{\rho}, \boldsymbol{\Lambda}, \boldsymbol{\phi})
\end{aligned}
$$

$$
\begin{aligned}
&= \sum_{\bar{s}_{\ell-1}} p(\bar{w}_\ell|\bar{w}_{1:\ell-1}, \bar{s}_{\ell-1}, \bar{s}_\ell, \boldsymbol{\rho}, \boldsymbol{\Lambda}, \boldsymbol{\phi}) \\
&\quad \times p(\bar{s}_\ell|\bar{w}_{1:\ell-1}, \bar{s}_{\ell-1}, \boldsymbol{\rho}, \boldsymbol{\Lambda}, \boldsymbol{\phi}) p(\bar{w}_{1:\ell-1}, \bar{s}_{k-1}|\boldsymbol{\rho}, \boldsymbol{\Lambda}, \boldsymbol{\phi}) \\
&= \sum_{\bar{s}_{\ell-1}} p(\bar{w}_\ell|\bar{s}_\ell, \boldsymbol{\phi}) p(\bar{s}_\ell|\bar{s}_{\ell-1}, \boldsymbol{\Lambda}) p(\bar{w}_{1:\ell-1}, \bar{s}_{\ell-1}|\boldsymbol{\rho}, \boldsymbol{\Lambda}, \boldsymbol{\phi}) \\
&= p(\bar{w}_\ell|\bar{s}_\ell, \boldsymbol{\phi}) \sum_{\bar{s}_{\ell-1}} p(\bar{s}_\ell|\bar{s}_{\ell-1}, \boldsymbol{\Lambda}) p(\bar{w}_{1:\ell-1}, \bar{s}_{\ell-1}|\boldsymbol{\rho}, \boldsymbol{\Lambda}, \boldsymbol{\phi}) \\
&= p(\bar{w}_\ell|\bar{s}_\ell, \boldsymbol{\phi}) \sum_{\bar{s}_{\ell-1}} p(\bar{s}_\ell|\bar{s}_{\ell-1}, \boldsymbol{\Lambda}) \mathcal{A}_{\ell-1}(\bar{s}_{\ell-1}).
\end{aligned}
$$

Derivation of Eq. (*10.3*)

$$
\begin{aligned}
\mathcal{A}_\ell(\bar{s}_1) &= p(\bar{w}_1, \bar{s}_1|\boldsymbol{\rho}, \boldsymbol{\Pi}, \boldsymbol{\phi}) \\
&= p(\bar{w}_1|\bar{s}_1, \boldsymbol{\rho}, \boldsymbol{\Pi}, \boldsymbol{\phi}) p(\bar{s}_1|\boldsymbol{\rho}, \boldsymbol{\Pi}, \boldsymbol{\phi}) \\
&= p(\bar{w}_1|\bar{s}_1, \boldsymbol{\phi}) p(\bar{s}_1|\boldsymbol{\rho}).
\end{aligned}
$$

Index

χ^2-minimization, 216
Bernoulli
  definition, 357
  density, 16
  distribution, 15
  sampling, 22
beta
  conjugacy, 133
  definition, 350
BetaNegBinomial
  definition, 360
  modeling time scales, 292
beta prime
  definition, 353
BetaBernP
  conjugacy, 230
  definition, 231
  estimating molecule counts, 238
  hyperparameters, 233
  load, 232
  pixelization, 240
  prior, 231
  regression, 230
  single molecule localization, 241
  step finding, 239
beta distribution
  prior, 133
Binomial
  definition, 357
categorical
  conjugacy, 252
  density, 16
  distribution, 14
Cauchy
  definition, 355
  prior, 179
Dirichlet
  aggregate property, 255
  base distribution, 255
  concentration hyperparameter, 255
  hierarchical, 298
  marginal, 254
  Markov jump process prior, 340
  prior, 136, 252
  process prior, 258
  reparametrization, 255
  sampling, 253
exponential
  definition, 9, 351
  density, 9
  memoryless, 365
  sampling, 20
GaussianP
  blocked Gibbs-like sampler, 228
  Brownian bridge, 222
  Brownian prior, 222
  conjugacy, 224
  definition, 17
  elliptical slice sampler, 230
  Gibbs-like sampler, 228
  inducing point, 225, 226
  Langevin dynamics, 237
  learning point spread function, 243
  Metropolis–Hastings, 227
  predictive distribution, 225
  process prior, 219
  regression, 219
  stream bed profile, 236
  structured kernel, 226
gamma
  conjugacy, 145, 166, 176
  definition, 352
  prior, 145
geometric
  definition, 358
Laplace
  definition, 356
NegBinomial
  definition, 359
Normal_M
  covariance, 220
  definition, 10
  locally periodic kernel covariance, 221
  posterior, 218
  prior, 218
  squared exponential kernel covariance, 220
  stationary kernel covariance, 220
normal
  Box–Muller sampler, 25
  center, 7
  conjugacy, 144, 166, 176
  covariance matrix, 218
  definition, 7, 353
  distribution, 7
  likelihood, 144
  mean, 7
  precision, 145

spread, 7
standard deviation, 7
truncated, 172
variance, 7
Poisson
conjugacy, 142
definition, 358
likelihood, 142
StudentT
definition, 354
uniform
definition, 349
sampling, 18

algorithm
ancestral sampling, 27
ancestral sampling for Gaussian state-space models, 321
ancestral sampling for mixture distributions, 248
backward recursion (stable version), 279
backward recursion (unstable version), 272
backward recursion for Rauch–Tung–Striebel smoothers, 326
Baum–Welch, 273
BIC change point, 156
Box–Muller, 25
conjugate gradient, 116
elliptical slice sampler, 230
expectation-maximization for likelihood, 121
expectation-maximization for posterior, 147
finite state projection, 107
forward filtering backward sampling, 285
forward filtering backward sampling for the linear Gaussian state-space model, 328
forward recursion (stable version), 279
forward recursion (unstable version), 269
forward recursion for Kalman filters, 324
fundamental theorem of simulation for continuous variables, 19
fundamental theorem of simulation for discrete variables, 21
Gibbs sampler, 184
Gibbs sampler for continuous state-space models, 328
Gibbs sampling for Bayesian hidden Markov model, 283
Gillespie simulation, 66
gradient descent, 115
greedy, 156
Hamiltonian Monte Carlo, 207
Hungarian, 250, 291
Markov jump process parameter sampling with trajectory marginalization, 341
Metropolis–Hastings sampler, 171
Metropolis–Hastings sampling for Bayesian hidden Markov model, 286
Metropolis–Hastings within Gibbs for the Markov jump process, 340
Nelder–Mead, 112
Newton–Raphson, 114
sampling Dirichlet random variables, 253
sampling Brownian motion, 93
sampling mixture model constitutive states, 251
trajectory sampling for Markov jump processes, 337
uniformization, 336
Viterbi recursion for hidden Markov model, 273
array
definition, 348
lexicographical ordering, 55, 348

balance
detailed, 168
full, 168
Bayesian nonparametric paradigm
BetaBernP, 231
definition, 132
Dirichlet process, 258
GaussianP, 17, 219
hierarchical Dirichlet, 298
model selection, 157
prior effect, 138
regression, 215
Bayesian paradigm, 109
Bayes' factor, 135
Bayes' rule, 27, 375
conjugacy in exponential family, 143
conjugate prior-likelihood pair, 141
continuous state-space model, 328
credible interval, 138
evidence, 133
hidden Markov model, 282
hierarchical formulation, 148
hyperparameter, 132
information criterion, 152
linear Gaussian state-space model, 329
maximum a posteriori estimate, 133
method, 131
mixture model, 250
model selection, 152
nonparametrics, 132, 138, 157
outliers, 160
posterior, 132, 139
prior, 132
biology
central dogma, 102
feedback, 102
genetic toggle switch, 103
stochastic binary decision, 102
stochastic bistability, 103

biology (*cont.*)
  stochastic simulation, 102
  transcription factor binding, 104
boundary condition
  Dirichlet, 106
  open, 97
Brownian motion, 92

Cartesian product, 55, 59, 71
central limit theorem, 145
chemistry
  bimolecular reaction, 63
  chemical master equation, 71
  chemical species, 63
  chemical system, 62
  chemical system modeling, 62
  chemical system simulation, 65
  excited state lifetime, 129
  Förster resonance energy, 315, 316
  fluorescence spectroscopy, 299
  fluorophore, 45, 299
  ground state, 299
  Jablonski diagram, 299
  kinetic scheme, 76
  mass action law, 72
  photo-bleaching, 45, 294
  photo-blinking, 295
  photo-state, 295
  photo-switching, 45
  propensity, 57, 61, 64
  reactions, 52
  singlet state, 299
  stoichiometric array, 62
  time-resolved spectroscopy, 299
  triplet state, 299
  well-stirred, 65
continuous
  generator matrix, 68
  space homogeneous dynamics, 81
  space increment, 86
  space inhomogenous dynamics, 81
  space transition kernel, 81
  state-space, 80, 93, 318
  time, 93
  time propagator, 69
  transition rate matrix, 53
covariance
  kernel, 220
  locally periodic kernel, 221
  matrix, 220
  squared exponential kernel, 220
  stationary kernel, 220

data
  analysis, 7
  classification, 245
  clustering, 245
  definition, 29
density
  definition, 8
  invariant, 168
  joint, 24
  marginal, 24
  reversible, 168
  stationary, 168
  target, 164
diffusion
  anisotropic, 97
  Brownian motion, 5, 17, 92
  coefficient, 91
  drift, 97
  equation, 96
  hop, 104, 105
  isotropic, 96
  open boundary condition, 97
  Ornstein–Uhlenbeck process, 104
  rotational, 106
  stationary, 96
  temperature dependence, 92
Dirac delta, 14, 361, 363
discrete
  regular times, 74
  state-space, 44, 74, 264
  time, 74, 80, 300
distribution
  banana, 211
  base, 255
  conditional, 25
  confidence interval, 138
  emission, 117, 248
  emission level, 248
  exponential family, 141
  full conditional, 183
  joint, 23
  K-parameter exponential family, 143
  marginal, 23
  mixture, 246
  mother, 248
  observation, 117, 118
  predictive, 135, 136
  target, 164, 169
  transition, 86
  Weibull, 36
dynamical system
  birth process, 48
  birth-death process, 51, 72, 126
  death process, 49
  hidden Markov model, 265
  modeling, 27
  switching process, 53
  trajectory, 42
  trajectory marginalization, 66

equation
  continuity, 97
  Fokker–Planck, 96
  generalized Langevin, 89
  Langevin, 87
  Langevin SDE, 95
  master, 67, 68

overdamped Langevin, 91
Smoluchowski, 99
stochastic differential, 94
underdamped Langevin, 89

filtering
backward variables, 271
continuous time, 342
correction, 323
definition, 269
filter, 322
forecaster, 326
forward filtering backward sampling, 285
forward filtering backward sampling for the hidden Markov model, 284
forward filtering backward sampling for the linear Gaussian state-space model, 327
forward recursion for Kalman filters, 324
forward variables, 269
Gaussian state-space model, 322
hidden Markov model, 269
Kalman, 323
prediction, 323
Rauch–Tung–Striebel smoother, 325
frequentist paradigm
confidence interval, 138
definition, 109
hidden Markov model, 268
mixture model, 248
function
continuous, 201
cumulative distribution, 20
definition, 377
differentiable, 201
dissimilarity, 291
domain, 377
error, 9
gamma, 136, 254
generalized, 363
image set, 377
indicator, 361
inverse cumulative distribution, 20
involution, 201
objective, 217
probability density, 20
range, 377

Gibbs
conditionally conjugate, 188
continuous state-space models, 328
full conditional, 183
hidden Markov model, 283
Metropolis–Hastings within Gibbs, 197, 198
sampler, 182
update, 184
graphical representation
definition, 148
node, 148, 149

Hadamard product, 270
hidden Markov model
Baum–Welch algorithm, 273
Bayesian, 282
decoding, 268
estimation, 268
evaluation, 268
factorial, 296
forward filtering backward sampling, 285
frequentist, 268
Gibbs, 283
infinite, 296
joint decoding, 272
left-to-right, 295
likelihood, 268
marginal decoding, 271
Metropolis–Hastings, 286
parameter, 267
stable backward recursion, 279
stable forward recursion, 279
state sequence decoding, 270
sticky, 293
variant, 291
Viterbi recursion, 273

information theory
entropic prior, 141
entropy, 157
information formula, 157
maximum entropy principle, 158, 159
maximum entropy signal deconvolution, 159
mutual information, 159
rate-distortion theory, 159
Shore–Johnson axioms, 158

Kronecker
delta, 122, 361
product, 55, 59
sum, 59, 71

Laplace's method, 153
likelihood
completed, 119
Cramer–Rao lower bound, 126
definition, 29
dependencies, 109
estimator, 30, 109
hidden Markov model, 268
logarithm, 110
marginal, 119
maximum, 110
ratio, 135
state-space labeling invariance, 280
true, 119
usage, 109

list
  definition, 348

Markov
  assumption, 28, 47
  blanket, 151
  chain, 76
  jump process definition, 50
  jump process notation, 53
  process, 47
  renewal process, 49
  transition probability matrix, 48
Markov chain
  convergence, 192
  feasibility, 169
  invariance, 169
  irreducibility, 169
  mixing, 191
  processing and interpretation, 189
  requirements, 169
  reversibility, 169, 174
  trace, 191
Markov jump
  collapsed, 60, 71
  composite, 54
  composite holding period, 58
  definition, 50
  discretization, 77
  elementary reaction, 57
  hidden, 333
  incongruent composite, 58
  memoryless, 50
  Metropolis–Hastings within Gibbs, 340
  reparametrization, 51
  trajectory, 46
  trajectory marginalization, 341
  uniformization, 335, 336
  virtual jump, 335, 338
matrix operation
  Cholesky decomposition, 223
  Cholesky factor, 223
  Cholesky–Banachiewicz algorithm, 223
  inverse of sum, 33
  matrix exponential, 69
  push through identity, 33
  singular value decomposition, 223
  Woodbury identity, 33
mean square displacement, 91
memoryless
  definition, 364
  exponential distribution, 365
  Markov jump, 50
  random variable properties, 365
Metropolis–Hastings
  definition, 171
  GaussianP, 227
  hidden Markov model, 286
  proposal, 171
microscopy
  aberrated point spread function, 241
  Airy probability density, 38
  astigmatism, 242
  camera (EMCCD), 38, 129
  defocus, 241
  estimating molecule counts, 238
  excited state lifetime, 129
  Förster resonance energy transfer photon crossover, 315
  fluorescence, 198, 299
  fluorescence background, 198
  fluorescence lifetime, 300
  fluorophore, 299
  ground state, 299
  imaging plane, 320
  instrumental response function, 262
  integrative detector, 343
  photo-bleaching, 45, 294
  photo-blinking, 295
  photo-state, 295
  photo-switching, 45
  pixelization, 240
  point spread function, 38, 241, 243
  pupil function, 241
  scalar diffraction theory, 241
  single molecule localization, 241
  single molecule tracking, 332
  singlet state, 299
  triplet state, 299
mixture model
  ancestral sampling, 248
  Bayesian, 250
  categories, 247
  definition, 245, 246
  frequentist, 248
  normal, 209
model
  definition, 3
  estimation, 8
  forward, 41, 139
  forward versus generative, 139
  generative, 29, 139
  generative with measurement uncertainty, 118
  nested, 152
  parameter, 6
model selection
  Bayesian, 136, 137, 152
  Bayesian information criterion (BIC), 152, 153
Monte Carlo
  autocorrelation, 193
  burn-in, 192
  definition, 163
  ergodicity, 168
  expectation, 164
  feasibility, 168
  Hamiltonian, 202
  integration, 163

Markov chain, 166, 169, 174, 189, 191, 192
optimization, 163
proposal, 178
sampler, 163, 170
thinning, 193
warm-up, 192

noise
additive, 248
definition, 84
measurement, 5, 117
multiplicative, 248
time-decorrelated white, 89, 220

optimization
conjugate gradient, 116
expectation-maximization for likelihood, 120
expectation-maximization for posterior, 147
gradient, 113
gradient descent, 115
Hessian, 113
Nelder–Mead, 112

parameter
definition, 108
emission, 117
latent, 108, 117
model, 6
nuisance, 108
observation, 117
scale, 349
physics
Brownian motion, 5, 17, 92
fluctuation–dissipation theorem, 90, 104
force spectroscopy, 314
free energy, 111
friction coefficient, 88
Kramers–Moyall expansion, 98
leapfrog integrator, 203
maximum entropy, 158
moment closure relation, 101
Newton's second law, 82, 83
nonradiative transition, 300
optical trap, 104
partition function, 111
phase-space, 82
photo-electric effect, 6
photo-electron counting, 367
point mass, 81
quantum efficiency, 199
scalar diffraction theory, 241
state dependent force, 83
state independent force, 81
temperature, 90
thermal, 99
thermodynamic entropy, 158
point
change, 154, 156
estimate, 109, 131
statistic, 144
prior
conditional conjugacy, 146
conjugate, 141
definition, 132, 139
hierarchical process, 298
hyper-, 143, 148
hyper-hyperparameter, 143
hyperparameter, 136, 143
improper, 140
informative, 141
Jeffreys, 141
pseudocount, 134, 143, 144
pseudoestimate, 144
strength, 143
uninformative, 140
probability
σ-algebra, 369, 379
Borel σ-algebra, 370
chain rule, 27
conditional distribution, 26, 375
Cox's theorem, 372
cumulative function, 18
density, 8, 11, 374
distribution, 8, 380
distribution function, 20
event space, 368
expectation, 164
generating events, 371
image set, 377
independent event, 374
joint density, 24
marginal density, 24
measurable, 379
measurable space, 379
measure, 371
normalization, 11
pre-image set, 377
sample space, 366
space, 376
support, 349
transition, 76
transition matrix, 48
problem
data classification, 245
data clustering, 245
direct, 31
ill-posed, 31
inverse, 31, 139, 163
regression, 215
well-posed, 31
process
BetaBernP, 231, 258
definition, 40
Dirichlet, 258
GaussianP, 219, 258

process (*cont.*)
hierarchical Dirichlet, 298
hierarchical prior, 298
Ornstein–Uhlenbeck, 108, 128
prior, 219
random, 40
renewal, 49
Wiener, 85, 94
prototype experiment, 7

random variable, 5
completion, 24
continuous, 5, 18
coordinate transformation, 13
de-marginalization, 24
definition, 5
discrete, 5, 14, 20
iid, 23
latent, 108, 117
marginalization, 24
rescaling, 12
sampling, 8
stochastic event, 5
stochastic realization, 6
support, 349
transformation, 12
versus deterministic, 5
random walk
additive, 180
circular, 77
definition, 76
kinetic scheme, 76
MCMC scaling, 181
Metropolis, 180, 196
Metropolis–Hastings, 195
multiplicative, 195
rate
memoryless, 365
parameters, 349
transition, 77
regression
BetaBernP, 230
continuous, 219
discrete, 231
GaussianP, 219
input point, 217
least squares estimator, 216
linear, 215
nonlinear, 216
nonparametric, 215
objective function, 217
parametric, 215
regularization parameter, 217
test point, 217
rule
assessment, 44, 265, 319
Bayes', 27, 375
initialization, 44, 265, 319
transition, 44, 265, 319
sampler
ancestral continuous state-space model, 321
ancestral definition, 26
ancestral for mixture models, 248
auxiliary variable, 198
beam, 299
elliptical slice, 229, 230
forward filtering backward sampling for hidden Markov model, 283
forward filtering backward sampling for the linear Gaussian state-space model, 327
Gibbs, 182
Gibbs for continuous state-space models, 328
Gibbs for hidden Markov model, 283
Gibbs for mixture model, 257
Hamiltonian Monte Carlo, 202
Markov jump process parameter sampling with trajectory marginalization, 341
Metropolis, 178
Metropolis–Hastings, 171
Metropolis–Hastings for GaussianP, 227
Metropolis–Hastings for hidden Markov model, 286
Metropolis–Hastings within Gibbs, 197
Metropolis–Hastings within Gibbs for Markov jump process, 340
Monte Carlo, 170
multiplicative random walk, 195, 211
slice, 200
trajectory sampling for Markov jump processes, 337
uniformization, 336
set theory
element, 368
empty set, 369
image set, 377
pre-image set, 377
set, 368
set intersection, 369
set unions, 369
simulation
Box–Muller, 25
definition, 18
fundamental theorem, 18
Gillespie, 65
state
Bayesian state-space model, 328
collapsed state-space, 61
constitutive, 42
continuous state-space, 318, 319
definition, 41
discrete state-space, 45, 264
dwell phase, 79
dwell time, 80
epoch, 45
escape rate, 50

finite state projection, 107, 129
Gaussian state-space model, 320
hidden, 43
holding, 46
holding period, 46
infinite state truncation, 106
joint decoding, 272
jump time, 46
labeling, 45
latent, 43
linear Gaussian state-space model, 322
marginal decoding, 271
passing, 42
persistence, 78
phase, 45
relabeling, 250
sequence decoding, 270
state-space, 42
state-space label invariance, 280
statistics
sufficient, 144
stochastic
definition, 40
operation, 366
system, 41
structured Markov chain
collapsed state, 60
composite system, 54
congruent system, 54
elementary state-space, 54
elements, 54

underflow, 120, 175

vectors, 348

Back Cover

Data analysis courses reaching beyond elementary topics, such as p-values or fitting residuals, are rarely offered to students of the natural sciences. As a result, data analysis, much like programming, remains improvised. Yet, with an explosion of experimental methods generating diverse data, we believe that students and researchers alike would benefit from a clear presentation of probabilistic modeling methods, many of which have only recently been inspired by theoretical and computational advances in statistics, machine learning, and data science.

As such, this book is developed as a one or two semester course for natural scientists interested in probabilistic data modeling. We begin with the basics of probability, and develop models of dynamical systems and likelihoods in order to build a foundation for Bayesian inference, Monte Carlo samplers, and filtering. We subsequently explore modeling paradigms including mixture models and the Dirichlet process, regression models with Gaussian and Beta-Bernoulli processes, hidden Markov models with an exposition of recent variants, state-space models and Kalman filtering, continuous time processes, and uniformization.

In starting from the basics and building toward current topics, we present probabilistic data modeling and, within it, error propagation as cornerstones of the scientific method critical toward deriving insights from experiments: from the intricate nanoscopic realm revealed through the latest advances in microscopy to the scales spanned by galaxy clusters pouring in from space telescopes.